Virus Bioinformatics

Virus Bioinformatics

Special Issue Editors

Manja Marz
Bashar Ibrahim
Franziska Hufsky
David L. Robertson

MDPI • Basel • Beijing • Wuhan • Barcelona • Belgrade

Special Issue Editors

Manja Marz
Friedrich Schiller University Jena
Germany

Bashar Ibrahim
Gulf University for Science and Technology
Kuwait

Franziska Hufsky
Friedrich Schiller University Jena
Germany

David L. Robertson
MRC-University of Glasgow Centre for Virus Research
UK

Editorial Office
MDPI
St. Alban-Anlage 66
4052 Basel, Switzerland

This is a reprint of articles from the Special Issue published online in the open access journal *Viruses* (ISSN 1999-4915) in 2019 (available at: https://www.mdpi.com/journal/viruses/special_issues/virus_bioinformatics).

For citation purposes, cite each article independently as indicated on the article page online and as indicated below:

LastName, A.A.; LastName, B.B.; LastName, C.C. Article Title. *Journal Name* **Year**, *Article Number*, Page Range.

ISBN 978-3-03921-882-0 (Pbk)
ISBN 978-3-03921-883-7 (PDF)

Contents

About the Special Issue Editors

Manja Marz is Professor of RNA Bioinformatics and High-Throughput Analysis at the Friedrich Schiller University Jena. She is Managing Director of the European Virus Bioinformatics Center. Her research focuses on the identification and annotation of noncoding RNAs and the bioinformatic analysis of viruses. Her group tries to approach these fields in an interdisciplinary way by combining state-of-the-art bioinformatic high-throughput methods with wet lab approaches.

Bashar Ibrahim is Professor of Mathematical Systems Biology at the Gulf University of Science and Technology (GUST). He is also an Affiliate Professor of the University of Jena in Germany and the European Virus Bioinformatics Center. He is interested in multiscale and unconventional modeling and analysis of complex biological systems, such as cell cycle transition control mechanisms and virus dynamics. His research interests lie in the interface between mathematics, molecular biology, medicine, and computational modeling in providing a systems approach to scientific discovery. Furthermore, his group focuses on the application of formal sciences in biology, chemistry, physics, and medicine.

Franziska Hufsky completed her Ph.D. in the field of computational mass spectrometry involving the identification of unknown small molecules and is currently working in the RNA Bioinformatics and High-Throughput Analysis group at Friedrich Schiller University Jena. She is Scientific Coordinator of the European Virus Bioinformatics Center.

David L. Robertson is based at the University of Glasgow Centre of Virus Research, where he is Head of the Viral Bioinformatics group. His research interests include viral evolution and diversity, systems virology, and the trade-offs between molecular evolution and disease. Prof. Robertson has made significant contributions to understanding the importance of viral diversity and recombination, for example, in publishing the first report of widespread recombination in HIV-1 and explanations for the nonrandom distribution of recombination breakpoints. Currently, his research focuses on the study of longitudinal microbiome/virome data, virus–host coevolution, and host specificity.

Preface to "Virus Bioinformatics"

Viruses are the cause of a considerable burden to human and animal health. In recent years, we have witnessed both the emergence of new viral diseases and the re-emergence of known diseases in new geographical areas. The power of new genome-sequencing technologies in association with new tools to handle "big data" provide unprecedented opportunities to address fundamental questions in virology. Virus bioinformatics has become central to virology research, and advances in bioinformatics have led to improved approaches to investigate viral infections and outbreaks, being successfully used to detect, control, and treat infections of humans and animals.

The European Virus Bioinformatics Center (EVBC) was established in 2017 to bring together experts in virology and virus bioinformatics in Europe. During our annual meetings, we have observed that virus bioinformatics is evolving and succeeding as a research area in its own right, representing the interface of virology and computer science. As part of the Third Annual Meeting of the European Virus Bioinformatics Center (EVBC), we have published this Special Issue on Virus Bioinformatics.

Manja Marz, Bashar Ibrahim, Franziska Hufsky, David L. Robertson
Special Issue Editors

Article

The Third Annual Meeting of the European Virus Bioinformatics Center

Franziska Hufsky [1,2,*,†], Bashar Ibrahim [1,3,†], Sejal Modha [4,†], Martha R. J. Clokie [5], Stefanie Deinhardt-Emmer [1,6,7,8], Bas E. Dutilh [1,9,10], Samantha Lycett [11], Peter Simmonds [12], Volker Thiel [1,13,14], Aare Abroi [15], Evelien M. Adriaenssens [1,16], Marina Escalera-Zamudio [17], Jenna Nicole Kelly [13,14], Kevin Lamkiewicz [1,2], Lu Lu [18], Julian Susat [19], Thomas Sicheritz [20], David L. Robertson [1,4,†] and Manja Marz [1,2,*,†]

1   European Virus Bioinformatics Center, 07743 Jena, Germany; bashar.ibrahim@uni-jena.de (B.I.); Stefanie.Deinhardt-Emmer@med.uni-jena.de (S.D.-E.); bedutilh@gmail.com (B.E.D.); volker.thiel@vetsuisse.unibe.ch (V.T.); Evelien.Adriaenssens@quadram.ac.uk (E.M.A.); kevin.lamkiewicz@uni-jena.de (K.L.); David.L.Robertson@glasgow.ac.uk (D.L.R.)
2   RNA Bioinformatics and High-Throughput Analysis, Friedrich Schiller University Jena, 07743 Jena, Germany
3   Chair of Bioinformatics, Matthias-Schleiden-Institute, Friedrich Schiller University Jena, 07743 Jena, Germany
4   MRC-University of Glasgow Centre for Virus Research, Glasgow G61 1QH, UK; s.modha.1@research.gla.ac.uk
5   Department of Genetics and Genome Biology, University of Leicester, Leicester LE1 7RH, UK; mrjc1@leicester.ac.uk
6   Institute of Medical Microbiology, Jena University Hospital, Am Klinikum 1, D-07747 Jena, Germany
7   Section for Experimental Virology, Jena University Hospital, Hans-Knöll-Straße 2, D-07745 Jena, Germany
8   Center for Sepsis Control and Care, Jena University Hospital, D-07747 Jena, Germany
9   Theoretical Biology and Bioinformatics, Science4Life, Utrecht University, Padualaan 8, Utrecht 3584 CH, The Netherlands
10 Centre for Molecular and Biomolecular Informatics, Radboud Institute for Molecular Life Sciences, Radboud University Medical Centre, Geert Grooteplein 26, Nijmegen 6525 GA, The Netherlands
11 Infection & Immunity Division, Roslin Institute, University of Edinburgh, Midlothian EH25 9RG, UK; samantha.lycett@ed.ac.uk
12 Nuffield Department of Medicine, University of Oxford, Peter Medawar Building, South Parks Road, Oxford OX1 3SY, UK; peter.simmonds@ndm.ox.ac.uk
13 Institute of Virology and Immunology, 3012 Bern, Switzerland; jenna.kelly@vetsuisse.unibe.ch
14 Department of Infectious Diseases and Pathobiology, Vetsuisse Facility, University of Bern, 3012 Bern, Switzerland
15 University of Tartu, Institute of Technology, 50411 Tartu, Estonia; aare.abroi@ut.ee
16 Quadram Institute Bioscience, Norwich Research Park, Norwich NR4 7UQ, UK
17 Department of Zoology, University of Oxford, Parks Rd, Oxford OX1 3PS, UK; marina.escalerazamudio@zoo.ox.ac.uk
18 Usher Institute of Population Health Sciences & Informatics, Ashworth Laboratories, Kings Buildings, University of Edinburgh, Charlotte Auerbach Road, Edinburgh EH9 3FL, UK; lu.lu@ed.ac.uk
19 Institute of Clinical Molecular Biology, Kiel University, 24118 Kiel, Germany; jsusat@ikmb.uni-kiel.de
20 Natural History Museum of Denmark, University of Copenhagen, DK-1123 Copenhagen, Denmark; thomassp@snm.ku.dk
*   Correspondence: franziska.hufsky@uni-jena.de (F.H.); manja@uni-jena.de (M.M.); Tel.: +49-3641-9-46482 (M.M.)
†   These authors organized the conference.

Received: 28 April 2019; Accepted: 29 April 2019; Published: 5 May 2019

Abstract: The Third Annual Meeting of the European Virus Bioinformatics Center (EVBC) took place in Glasgow, United Kingdom, 28–29 March 2019. Virus bioinformatics has become central to virology research, and advances in bioinformatics have led to improved approaches to investigate viral infections and outbreaks, being successfully used to detect, control, and treat infections of humans and animals. This active field of research has attracted approximately 110 experts in virology and bioinformatics/computational biology from Europe and other parts of the world to attend the two-day meeting in Glasgow to increase scientific exchange between laboratory- and computer-based researchers. The meeting was held at the McIntyre Building of the University of Glasgow; a perfect location, as it was originally built to be a place for "rubbing your brains with those of other people", as Rector Stanley Baldwin described it. The goal of the meeting was to provide a meaningful and interactive scientific environment to promote discussion and collaboration and to inspire and suggest new research directions and questions. The meeting featured eight invited and twelve contributed talks, on the four main topics: (1) systems virology, (2) virus-host interactions and the virome, (3) virus classification and evolution and (4) epidemiology, surveillance and evolution. Further, the meeting featured 34 oral poster presentations, all of which focused on specific areas of virus bioinformatics. This report summarizes the main research findings and highlights presented at the meeting.

Keywords: virology; virus bioinformatics; software; systems virology; metagenomics; virome; viral taxonomy; virus classification; genome evolution; bacteriophage; virosphere

---

1. Introduction

The European Virus Bioinformatics Center (EVBC) was conceived of in 2017 to bring together experts in virology and virus bioinformatics in Europe [1,2]. EVBC's member numbers have increased steadily since then with currently 151 members from 78 research institutions distributed over 26 countries across Europe and internationally. This spring, the Annual Meeting of the EVBC was held for the third time (Table 1). The Third Annual Meeting of the EVBC attracted experts at all career stages to attend the two-day meeting in Glasgow in an inspiring and interactive scientific environment to promote discussion, exchange of ideas and collaboration and to inspire and suggest new research directions and opportunities.

Table 1. History of the Annual Meeting of the European Virus Bioinformatics Center(EVBC).

Date	Location	# of Participants	Key outcomes
6–8 March 2017	Friedrich Schiller University Jena, Germany	~100	Founding of the Center; Discussion of the role of EVBC; Election of the first Board of Directors; Insights into EU policy and funding opportunities.
9–10 April 2018	Utrecht University, Netherlands	~120	Extending of the EVBC network to include America and Asia; Discussion and design of joint projects; Insights on first applied European fund among EVBC members [3].
28–29 March 2019	University of Glasgow, United Kingdom	~110	Inclusion of contributed talks in themed sections in the scientific programme; Establishment of travel, poster and best contributed talk awards for junior scientists; Need for greater coordination and communication within the European virology community.

2. Sessions and Oral Presentations

During the two-day conference, about 110 participants from 20 countries contributed in productive discussion on the four topics: (1) systems virology, (2) virus-host interactions and the virome, (3) virus classification and evolution and (4) epidemiology, surveillance and evolution. A number of high quality presentations were given by leading virologists and junior scientists. In addition to the eight invited speakers, we had twelve talks selected from the contributed submissions (see http://evbc.uni-jena.de/events/3rd-evbc-meeting). It was clear that the distinction between laboratory and computer researchers is often blurred. That collaborating teams of individuals with different skill sets are often a road to success, while individuals working alone can still make massive contributions. Data-driven research is now mainstream, and the scale and complexity of datasets is ever increasing. Discussions highlighted how virology, like all of biology, is now a data science, exploiting methods from dimensionality reduction of large datasets to data visualisation. We took from this that virus bioinformatics is evolving and succeeding as an area of research in its own right at the interface of virology and computer science and that there are many ways to be a successful researcher.

2.1. Systems Virology

This session was chaired by Philippe Le Mercier (University of Geneva Medical School, Switzerland), board member of the EVBC. Two speakers have been invited on this topic. Volker Thiel (University of Bern, Switzerland), board member of the EVBC, presented about host proteins composing the microenvironment of coronavirus replicase complexes. EVBC member Stefanie Deinhardt-Emmer (Jena University Hospital, Germany) presented about co-infection between *Staphylococcus aureus* and influenza virus. From the submitted abstracts, we selected talks by Jenna Nicole Kelly (University of Bern, Switzerland) on single-cell analysis of influenza virus infection, Florian Erhard (University of Würzburg, Germany) on tools revealing core features of CMV-induced regulation in single cells and Daniel Blanco Melo (Icahn School of Medicine at Mount Sinai, New York, USA) on in-depth transcriptomic analysis in influenza A virus infection. Studying virus infections at the molecular level is as complex as studying the host systems they infect.

2.1.1. Determination of Host Proteins Composing the Microenvironment of Coronavirus Replicase Complexes, by Volker Thiel

Coronaviruses are positive-sense RNA viruses that infect a variety of mammalian and avian species and are mainly associated with respiratory and enteric diseases. In humans, there are four coronaviruses known to cause rather mild respiratory symptoms; however, the appearance of zoonotic viruses, such as the Severe Acute Respiratory Syndrome (SARS) and Middle East Respiratory Syndrome (MERS) coronaviruses, exemplified that coronaviruses can also cause severe and lethal diseases in humans. Within their target cells, coronaviruses replicate their RNA genome at host-derived membranes in the host cell cytoplasm. The Replicase Complex (RC) that is synthesizing the viral RNA is encoded on the genomic RNA and comprises a set of 15–16 non-structural proteins (nsps). Besides canonical functions associated with RNA synthesis, such as RNA-dependent RNA polymerase, helicase and methyltransferases, a wealth of additional enzymatic activities, such as endoribonuclease, ADP-ribosylation and de-ubiquitination, are included within the coronaviral RC, suggesting that various virus–host interactions are taking place at the site of viral RNA synthesis. However, our knowledge about host factors at the interface between the RC and the host cell cytoplasm is rudimentary. To identify the composition of the viral RC and adjacent host cell proteins composing the RC-microenvironment, we engineered a biotin ligase into a coronaviral RC. This allowed us to biotinylate, affinity-purify and identify specifically all viral components constituting the coronavirus RC and host cell proteins that are in close proximity (Figure 1). Amongst the >500 host proteins constituting the RC-microenvironment, we identified numerous proteins associated with vesicular trafficking pathways, ubiquitin-dependent and autophagy-related processes and translation initiation. Notably, following the detection of translation initiation factors at the RC, we were able to visualize and demonstrate active translation proximal to the site of viral RNA synthesis of several coronaviruses. Collectively, our work established a spatial link between viral RNA synthesis and diverse host factors of unprecedented breadth. Many of the coronavirus RC-proximal host proteins and pathways have also documented roles in the life cycle of other positive-stranded RNA viruses, suggesting considerable commonalities and conserved virus-host interactions at the RCs of a broad range of RNA viruses.

Our data may thus serve as a paradigm for other RNA viruses and provide a starting point for a comprehensive analysis of critical virus-host interactions that represent targets for therapeutic intervention [4].

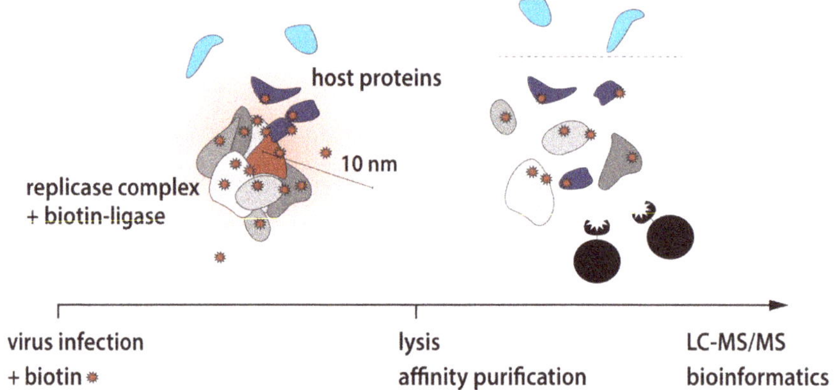

Figure 1. Illustration of the experimental design to determine the microenvironment of coronavirus Replicase Complexes (RCs) (adapted from V'kovski et al. [4]).

2.1.2. Co-Infection between *Staphylococcus aureus* and Influenza Virus Reduces Endothelial Barrier Function, by Stefanie Deinhardt-Emmer

Pneumonia is the most serious inflammatory disease of the respiratory tract and also the most common infectious disease. The classification of pneumonia into Hospital-Acquired Pneumonia (HAP), Community-Acquired Pneumonia (CAP) and Ventilated-Acquired Pneumonia (VAP) indicates the source of disease by a wide variety of microorganisms including bacteria, viruses and fungi. Respiratory tract infections and in particular pneumonia represent the most common cause of sepsis [5]. Long-time associated with bacterial infection, sepsis definition became more in focus as a multifaceted host response to an infecting pathogen, which leads to organ failure [6]. However, Influenza Virus (IV) as a pneumotropic virus can lead to lung failure and systemic host reaction with subsequent multiple organ failure. IV circulates worldwide and causes highly contagious respiratory diseases characterized by mild to severe symptoms. The seasonal IV-associated bronchopneumonia is one of these infectious diseases with the highest population-based mortality rates [7]. Besides virulence factors, the sudden increase of pathogenicity is the most striking problem of influenza accompanied by bacterial co-infection. In a single-centre study conducted at the Jena University Hospital during the winter season 2017/2018, we detected 1197 influenza-virus-positive samples and 89 *S. aureus*-positive respiratory specimens. However, the diagnosis of a co-infection was significantly lower with 17 samples. Interestingly, the mortality rate increased dramatically from single infection (approximately 20%) to co-infection (approximately 80%). Even larger studies indicating similarly dates and also the Spanish flu of 1918 showed that co-infection results in high mortality rates [8]. While the pathogen–host interaction-induced severe dysregulations of the immune response is under investigation in many studies, the regulatory effects between the different pathogens and the subsequent impact on the host are barely understood. In a multifactorial process, a wide range of pathogen factors and pathogen-regulated signalling events are involved in co-pathogenesis. This process is associated with elevated host-response, changed repair-processes, and modifications in the cellular immune response [9]. It is shown that primary IV-infection inhibits the apoptosis mechanism and the following infection with *S. aureus* inhibits IV-induced apoptosis by procaspase-8 activation [10]. Various models are available for studying the mechanisms of the viral–bacterial interference. However, the use of murine models is adversely regarded because of obvious discrepancies between men and mice despite the attempts of humanized murine models to fill the gaps. New methods enable investigations with cost-saving and efficient cell culture models as an excellent supplement to animal experiments. Organ-on-a-chip technology allows species-specific investigations for different cell types and also immune cells. Using this method, viral-bacterial interference can be investigated in a human-specific manner.

2.1.3. Single Cell Analysis of iNfluenza Virus Infection in Its Natural Target Cells Reveals Cell Type-Specific Host Responses and Disparate Viral Burden, by Jenna Nicole Kelly

The human respiratory epithelium is a pseudostratified epithelium that constitutes the first line of defence against invading respiratory pathogens, including influenza viruses. Although several studies have now shown that both viral transcript production and the innate immune response to infection vary widely among single influenza-infected cells, the cause of this extreme heterogeneity remains unclear [11,12]. More specifically, it remains unknown how key innate immune components are distributed among the different cell populations found in the respiratory epithelium and how the latter may influence the host response to infection. To determine the distribution of these innate immune components and to examine how specific cell types respond to influenza infection, we used single-cell RNA sequencing to acquire transcriptomes from primary human Airway Epithelial Cells (hAEC) infected with Influenza A Virus (IAV) (Figure 2) [13]. A low MOI was used to infect hAECs with either Wild-Type (WT) pandemic IAV or an NS1mutated form of the virus (NS1R38A) that impairs its ability to counteract Interferon (IFN) and produces an amplified innate immune response. We then annotated both host and viral transcriptomes of more than 19,000 single cells across the five major hAEC cell types for mock, WT, and NS1R38A conditions. We observed a large heterogeneity in viral burden; however, in contrast to what was found in previous studies, no absence of viral genes was detected. Interestingly, in both WT- and NS1R38A-infected cultures, there was a significant decrease in the fraction of ciliated and goblet cells compared to mock hAECs. We also identified a number of cell-type-specific innate immune responses, including the expression of type I and III IFNs in all major cell types. Collectively, our results represent the first comprehensive report on how individual cells contribute to the antiviral response during IAV infection in the context of the human respiratory epithelium.

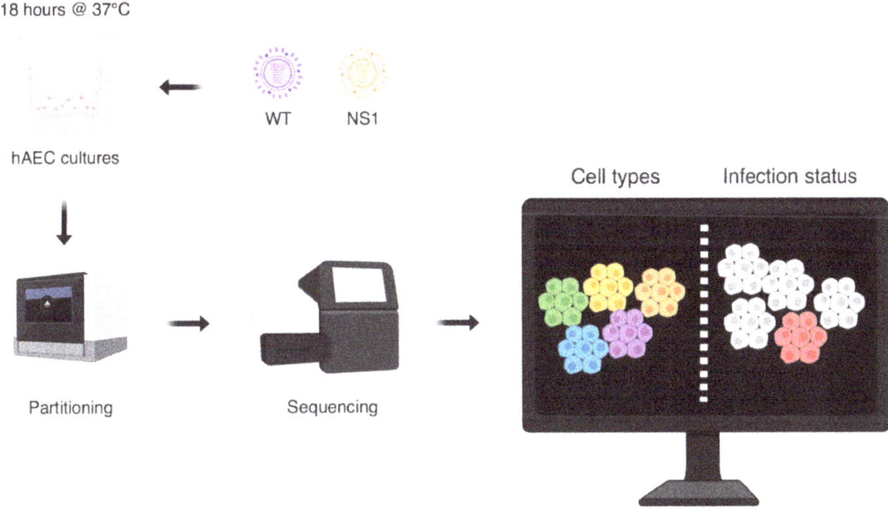

Figure 2. Sequencing and annotation workflow for single influenza-infected cells in the human respiratory epithelium. hAEC, human Airway Epithelial Cells.

2.2. Virus–Host Interactions and the Virome

This session was chaired by David Robertson (MRC-University of Glasgow Centre for Virus Research, United Kingdom), the local organizer of the meeting. Bacteriophages, the viruses of bacteria, are an important and usually neglected component of microbiome studies. Two speakers have been invited on this topic. Martha Clokie (University of Leicester, United Kingdom) presented about the roles of phages in impacting infectious diseases in human microbiomes. EVBC member Bas Dutilh (Utrecht University, Netherlands) presented about global phylogeography and the ancient evolution of the widespread human gut virus crAssphage. From the submitted abstracts, we selected talks by Katherine Brown (University of Cambridge, United Kingdom) on viral transcripts in RNA-seq datasets from bees, mites, and ants, Evelien Adriaenssens (Quadram Institute Bioscience, Norwich, United Kingdom) on genome-resolved metaviromics for the detection of pathogenic viruses in the environment and Josquin Daron (Centre National de la Recherche Scientifique, Montpellier, France) on codon usage preference similarity among human-infecting viruses and their hosts.

2.2.1. Roles of Phages in Impacting Infectious Diseases in Human Microbiomes, by R. J. Martha Clokie

Most of the roles of phages in human health and disease are yet to be unravelled. However, phages in all environments including the human microbiome are increasingly acknowledged to be the puppeteers of their bacterial hosts, shaping their structure and evolution and physiology. Phages associated with bacterial pathogens have multiple, often complex interactions with their bacterial hosts, forcing them to interact differently with other bacterial and human cells. Besides being the ultimate bacterial killers, phages can change bacterial surfaces to prevent recognition by the human immune system. In cystic fibrosis, they can allow their hosts to cope with anaerobic conditions found in mucus-laden lungs, and in many bacteria, they encode potent toxins [14]. There is indeed a plethora of unknown phage-mediated bacterial phenotypes that could be critical for our understanding of disease. Their ability to be developed as targeted removers of pathogenic bacteria is likely to be critical to solving the antimicrobial resistance crisis.

A major limitation for our ability to develop therapeutic phages and also understand fully the ways that phages impact bacteria is that the vast majority of phage gene functions are hypothetical or unknown. In bacterial genomes, there are around 25% unknown genes, or genes that have no known ascribed function, but in phage genomes, only around 25% of the genes are generally known! Thus, when trying to establish how phages specifically interact with their hosts, there is large number of genes of which we need to try and make sense.

To illustrate the diversity within one specific phage set, Martha Clokie presented the work from her lab on phages that infect the gut pathogen *Clostridium difficile* [15–17]. They have identified sets of phages that target clinically-relevant and prevalent strains. Despite the most effective phage set being isolated from one geographical location, they are strikingly variable (Figure 3) with very few identifiable genes in common.

Martha Clokie's group is currently in the process of creating and examining genetic mutants to identify phenotypes and conducting structural work on novel proteins, for example to identify tail fibres. However, this work is time consuming and technically demanding. Choosing which genes to focus on is key, as downstream work is key to unravelling critical phenotypes. Martha Clokie presented data on the efficacy of this phage set to treat disease along with a framework for their ongoing work to use different machine learning approaches to examine the genomes of these phages and their associated bacteria robustly in order to identify hard-to-identify features, for example shared and unique genes of interest. These approaches will direct work to unravel the mechanics of phage efficacy for virulent phages and modes of action for lysogens.

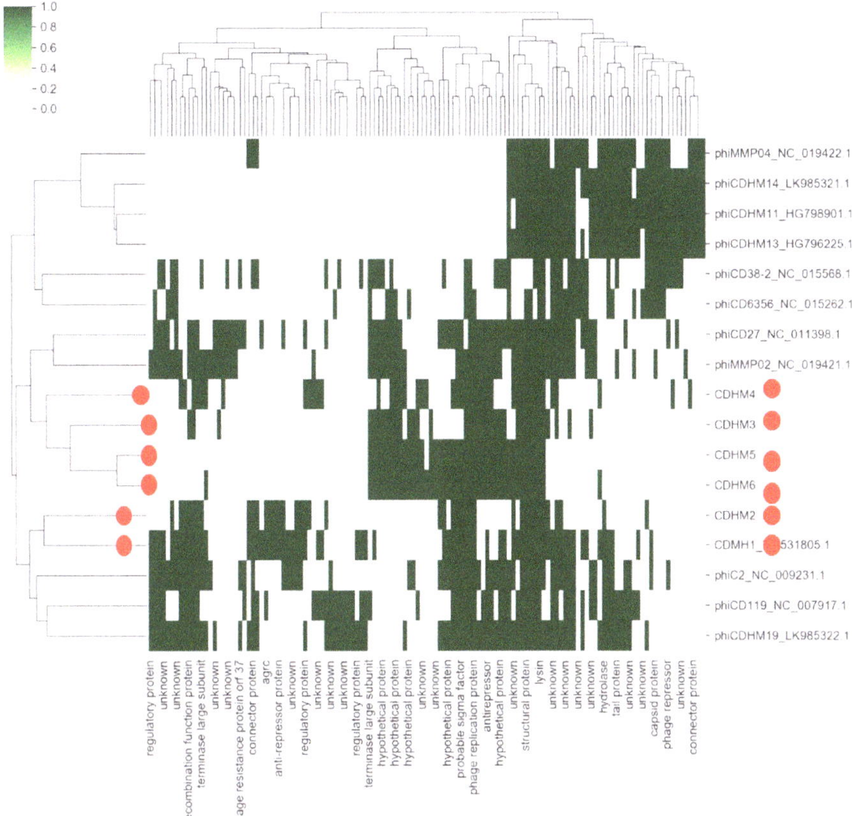

Figure 3. Set of *Clostridium difficile* phages on the vertical axis, which includes six well-characterised myoviruses from Martha Clokies' laboratory (red dots). The genes commonly identified in *C. difficile* phages are shown on the horizontal axis and homologous genes represented by a green line. It is clear that these phages do not share a large common gene set.

2.2.2. Global Phylogeography and Ancient Evolution of the Widespread Human Gut Virus crAssphage, by Bas E. Dutilh

While viruses are vastly abundant and ubiquitous throughout the biosphere, they have remained a relatively unexplored superkingdom of life. Early findings of genomic mosaicism [18] and enhanced mutation rates of especially RNA viruses [19] have led to the conception of viruses as genomically highly variable entities. This was further supported as metagenomics unveiled the extent of genetic diversity of viruses, initially in marine water and human faeces [20], and in many different biomes since. Images of an unparalleled diversity that is dominated by unknown sequences has been the common theme of viral metagenomic explorations. However, while the virosphere is undoubtedly diverse, ubiquitous viruses are increasingly being discovered by metagenomic analysis of globally-distributed, ecologically-stable ecosystems, including once again the global oceans [21,22] and the human gut [23–25].

Moreover, the genome sequence in individual viral lineages may be more conserved than could previously be recognized. Recently, large-scale comparisons of gene order in the genome sequences of dsDNA bacteriophages revealed a surprisingly conserved genomic structure [26,27]. A possible mechanism at play is the genomic encoding of different transcriptional regions with promoters that govern the expression of early, middle and late specific genes, such as known from the well-studied case of the T4 bacteriophage [28]. Together, these findings suggest a highly-optimized genomic encoding of gene expression regulation that is consistent across globally-diverse viral populations.

While the conservation of genomic architecture between distantly-related bacteriophages as outlined above is a striking observation, many open questions remain. For example, it remains unclear to what extent the observations of conserved genomic architecture described above reflect a biased sampling, for example of temperate, dsDNA and/or tailed bacteriophages that have been observed to dominate, e.g., marine systems [29]. Indeed, the modes of genome evolution differ for viruses with different lifestyles [30]. Nevertheless, viruses have vast global population sizes that result in highly-efficient evolutionary selection pressures and optimized genomes. Moreover, viruses and their cellular hosts have been co-evolving for billions of years, allowing ample time for optimization of their genome structures.

Viral mutation rates (including recombination rates) have remained difficult to quantify due to a lack of evolutionary calibration points. For example, on a short time scale of thirty years, a constant recombination rate of five events per year has been observed for Siphoviridae bacteriophages [31], but when longer timespans are assessed, mutation rate estimates may drop dramatically by orders of magnitude [32]. One way of obtaining ancient calibration points in viral evolution in the absence of fossil data is by exploiting their association to hosts. One of the most conserved constituents of the human gut virome is the widespread and abundant bacteriophage crAssphage [23]. Recently, near-complete genome sequences of crAss-like viruses were detected in faecal samples of a range of wild non-human primates living on different continents, including Old-World monkeys, New-World monkeys and apes [33]. Strikingly, these genomes revealed a strong collinearity with human-associated crAss-like viruses, suggesting that the association of crAss-like viruses with the primate gut biome may be millions of years old. Moreover, these findings open the door to investigations into viral mutation rates at long time-scales, once again illustrating how viral metagenomics opens up a treasure trove for virus discovery [34], as well as evolutionary analyses of these smallest and most abundant biological entities on Earth.

2.2.3. Genome-Resolved Metaviromics for the Detection of Pathogenic Viruses in the Environment: Will Eating Shellfish Make You Ill?, by Evelien M. Adriaenssens

Viromics or viral metagenomics has been proposed as an alternative method to qPCR-based approaches for the detection of pathogenic viruses linked to food- and water-borne illness in the aquatic environment [35,36]. The main advantage is that viral communities can be investigated without prior knowledge of the genome sequences or genotypes of the viruses present in the sample. There are, however, several drawbacks associated with viromics, such as laboratory and computational costs, scalability and the issue of viral dark matter in which sequence data are classified as "unknown". In her presentation, Evelien Adriaenssens focused on the latter aspect and showed that reconstruction of Uncultivated Virus Genomes (UViGs) [37] and classification into families reduced the fraction of completely unknown sequences, particularly for RNA viruses. Using read mapping approaches followed by visualisation and analysis with Anvi'o [38], she showed that they can identify pathogenic virus genomes present in the Conwy River catchment area, mainly found in wastewater [39], and showed changing abundance patterns between sample sites and types. Using species-level clustering and differential read mapping, comparative genomics and phylogenetics, she could gradually descend from the bigger picture of viral diversity to strain-level resolution, identifying the genotype of potentially pathogenic viruses. This workflow is ideally suited to find new pathogenic viral species and identify markers for wastewater contamination of the environment.

Evelien M. Adriaenssens was funded by the Biotechnology and Biological Sciences Research Council (BBSRC) under the BBSRC Institute Strategic Programme Gut Microbes and Health BB/R012490/1.

2.3. Virus Classification and Evolution

This session was chaired by Darren Obbard (University of Edinburgh, United Kingdom). Two speakers have been invited on this topic. Peter Simmonds (University of Oxford, United Kingdom) presented about classification of viruses in metagenomic datasets. Unfortunately, Olga Kalinina (Max Planck Institute for Informatics, Saarbrücken, Germany) was unable to make it to the meeting. Instead, Manja Marz (Friedrich Schiller University Jena, Germany), Managing director of the EVBC, presented about machine learning applied to virus data. From the submitted abstracts, we selected talks by Julian Susat (Institute of Clinical Molecular Biology, Kiel, Germany) on the detection of viruses in ancient human remains, Aare Abroi (University of Tartu, Estonia) on the relation between virosphere and biosphere, and Kevin Lamkiewicz (Friedrich Schiller University Jena, Germany) on RNA secondary structures in whole genome alignments of viruses. Based on this submission, Kevin was competitively awarded the PhD travel award.

2.3.1. The Classification of Viruses in Metagenomic Datasets—Where Do You Draw the Line?, by Peter Simmonds

Methodological advances, such as High-Throughput Sequencing (HTS), and new capabilities to recover and assemble genome sequences has unearthed vast numbers of previously-undescribed viruses from environmental, human clinical, veterinary and plant samples. How such viruses can be incorporated into the current virus taxonomy is a major challenge, especially at the family and species levels, which have been historically based largely on descriptive taxon definitions of phenotypic properties that "sequence-only" viruses often lack. These assignments typically encapsulate descriptions of replication strategies, virion structure, and clinical and epidemiological features, such as host range, geographical distribution and disease outcomes. If "sequence-only" viruses are to be formally placed into the classification maintained by the International Committee on the Taxonomy of Viruses (ICTV) as recently proposed [40], then their assignments will have to be based largely or entirely on metrics of genetic relatedness and any other features that might be inferred from their genome sequences. However, there are no published guidelines in the ICTV code on how similar or how divergent viruses must be in order to be considered as new species or new families (https://talk.ictvonline.org/information/w/ictv-information/383/ictv-code).

Peter Simmonds described their investigations of the extent to which the existing virus taxonomy could be reproduced by the recoverable genetic relationships between sequences of viruses currently classified by the ICTV. Comparisons of viruses were based on extraction of protein coding gene signatures and genome organisational features from virus sequences and using these to construct a metric of genetic relatedness through computation of Composite Generalised Jaccard (CGJ) distances between each pair of viruses [41]. For eukaryotic viruses, there was large-scale consistency between such genetic relationships and their current family- and genus-level taxonomic assignments, irrespective of genome configurations and genome sizes. The analysis pipeline, "Genome Relationships Applied to Virus Taxonomy" (GRAViTy), diagrammatically summarised in Figure 4, predicted family membership of eukaryotic viruses with close to 100% accuracy and specificity; this method should therefore enable the vast collection of metagenomic sequences to be classified in a manner consistent with the current ICTV taxonomy. Preliminary analysis of such datasets revealed that over one half (460/921) of (near)-complete genome sequences from recently-generated eukaryotic virus datasets could be assigned to 127 novel family-level groupings, more than double the number of eukaryotic virus families in the ICTV taxonomy.

The taxonomy of the 20 currently-classified prokaryotic virus families differs substantially [42]. Members of three families in particular (*Podoviridae*, *Siphoviridae* and *Myoviridae*) were far more divergent from each other than observed within eukaryotic and archaeal virus families. Applying a CGJ distance threshold of 0.8, prokaryotic viruses form over 100 groupings equivalent to eukaryotic virus families. The use of a common benchmark with which to compare taxonomies of eukaryotic and prokaryotic viruses supports ongoing efforts by the ICTV to revise thoroughly the phage taxonomy so that assignment criteria are consistent across all virus groups. Developing a consistent classification of viruses in which assignments at family and other taxonomic levels extending the current framework, but which will be underpinned both by metrics of genomic relatedness, is essential for future, evidence-based classification of metagenomic viruses.

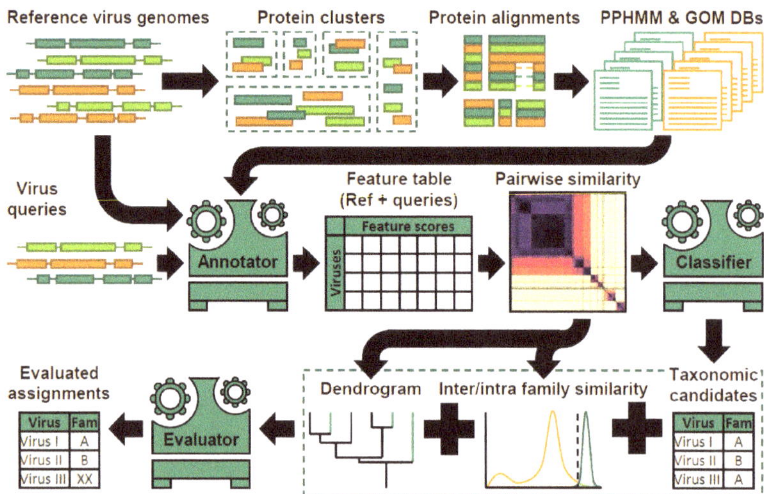

Figure 4. Overview of virus taxonomy prediction by "Genome Relationships Applied to Virus Taxonomy" (GRAViTy). A simplified diagram of the steps used to construct profile tables from sequences of viruses with assigned taxonomic status (reference virus genomes). It further illustrates the steps to classify viruses of of undetermined taxonomic relationships. The method is based on extraction of protein sequences from reference virus genomes and their clustering using pairwise BLASTp bit scores. Sequences in each cluster are then aligned and turned into a Protein Profile Hidden Markov Model (PPHMM). Reference genomes are subsequently scanned against the database of PPHMMs to determine the locations of their genes, and Genomic Organisation Models (GOMs) for each virus family are constructed. These models form the core of the genome annotator (Annotator), which is used to annotate query sequences with information on the presence of genes and the degree of similarity of their genomic organisation to reference virus sequences. From this, genome relationships can be extracted by computation of various genetic distance metrics, including composite generalised Jaccard similarity, which forms the basis for heat maps and dendrograms that depict the relationships of query sequences to the dataset of classified viruses (Classifier) and recommendations for their taxonomic assignments (Evaluator).

2.3.2. Detecting Viruses in Ancient Human Remains, by Julian Susat

The field of ancient DNA covers a wide range of research topics, spanning from human evolution, megafauna to pathogen evolution. Despite the recent advantages in ancient DNA techniques and modern metagenomic screening tools, the identification of authentic viral sequences from ancient material is still challenging. The materials that are mainly used in ancient DNA research, teeth and petrous bones, already limit the number of detectable viruses by their nature. Only viruses that are present in the bloodstream can be detected. The fast evolution of viral pathogens and therefore the comparability to modern variability in viruses makes it even more difficult to identify their ancestors reliably. The highly-fragmented and degraded nature of ancient genetic material and the high risk of modern contamination are causing further problems in the analysis. For the detection of viruses, a wide variety of software utilizing different approaches like HMMs, dedicated marker genes and complete genome references are available to screen these ancient samples for the presence of pathogens. Each of these approaches has its own characteristic strengths and weaknesses. In a competitive alignment approach using all complete virus genomes as a reference, we were able to detect three Hepatitis-B Viruses (HBV) during our regular screening. All three samples originated in Germany and dated to the mediaeval times (1000 BP) and the Neolithic (5000 and 7000 BP). After sequencing and competitive mapping against 16 HBV references, complete HBV genomes could be recovered from all three samples. This resulted in the oldest human pathogenic viral genome that is known up to know. Phylogenetic analysis revealed that the medieval strain was genotype D and surprisingly conserved. The ancient Neolithic strains were closer together than to any other modern and closest to strains from Old-World monkeys. These findings might suggest reciprocal cross-species transmission between human and ape. Furthermore, we could show that the genomic structure of ancient strains closely resembles the structure of modern HBV strains. Since publishing these results, we and others detected more HBV-positive samples, supporting the notion that viruses will become more important for the aDNAcommunity (Figure 5). The new HBV genomes we reconstructed support our earlier findings. A bigger number of HBV cases spanning over longer time frames opens the door for reliable diachronic analysis and maybe even epidemiological analysis. Besides the recent findings of ancient viruses (e.g., Parvovirus), an open question still remains how we could detect and reconstruct extinct or highly-altered virus genomes. Bioinformatic protocols for the detection of unknown viral protein families based on long sequencing reads and high coverage data are published and available, but due to the above-described nature of aDNA, applying these methods is not straightforward, and strong optimization needs to be carried out. Still, these HBV and other findings have opened a new door within the aDNA community and blazed a trail for upcoming viral ancient DNA studies. This work was done by a team composed of Ben Krause-Kyora, Julian Susat, Felix M. Key, Denise Kühnert, Alexander Immel, Alexander Herbig, Almut Nebel and Johannes Krause.

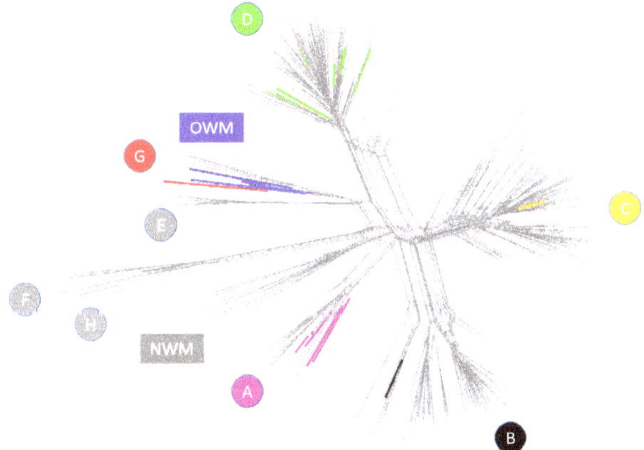

Figure 5. Network of 493 modern genomes, 15 published ancient strains and 12 newly-discovered ancient strains. Single letters indicate HBV genotypes (A–H); coloured strains are of ancient origin; OWM = Old-World Monkey HBV strains, NWM = New-World Monkey HBV strains. D: five new ancient strains, six ancient strains [43–45]; C: one ancient strain [46]; B: one ancient strain [45]; A: two new ancient strains, three ancient strains [45]; G: one new ancient strain; OWM: three new ancient strains, four ancient strains [44,45].

2.3.3. Virosphere and Biosphere—How Related They Are? A Protein (Domain) Based View, by Aare Abroi

Viruses are not always pathogens, and they are also an important and inseparable part of the biosphere and should be studied as such. Unfortunately, the wider functional and evolutionary role of viruses in the biosphere is not yet widely accepted in most disciplines, a good exception being marine biology/ecology, where viruses are already accepted as important players. How the virosphere is related to the rest of the biosphere can be examined in several different ways. One of these ways is a protein domain-based view. We analysed how virosphere protein domain occurrence is related to the occurrence of protein domains in all (sequenced) organisms (we called the last the phylogenomic space of protein domains). This is based on the distribution of protein domains in viruses and in organisms (by superkingdom), i.e., which protein domains are found in viruses (or a specific set of viruses) and to what extent and where these domains are found elsewhere in organisms. In our analysis, we used predefined protein domain databases Pfam, Superfamily and Gene3D. Domains found in the virosphere can be found in a different number of organisms, starting from a few organisms for some viral domains up to all organisms in the others. However, if we specify a narrower set of viruses (Baltimore class, viral family or host range), differences between viral taxons appear. Therefore, the heterogeneity of viruses is also very clearly expressed by where in the phylogenomic space the domains that are found in different viral taxons are located. A few examples are shown in Figure 6. An important conclusion from our analysis is the existence of virosphere-specific protein domains (domains not found in cellular organisms), even at the level of structural homology. Several evolutionary routes that may lead to virosphere specificity (absence in cellular organisms) will be discussed. Considering the new knowledge on virus-to-host gene transfers in eukaryotes during the last ten years, it is clear that the virosphere is a source of functional and structural novelties also for this superkingdom. A possible route for the genesis of novel domains in viruses (as well as in organisms) is double coding or overprinted genes. We have developed a web-tool cRegions (http://bioinfo.ut.ee/cRegions/), which helps to find potential double coding regions (and other embedded functional elements) in coding sequences [47,48]. Of course, there exist many domains that are shared by viruses and organisms. Beside others, virus-to-host gene transfer is one process leading to shared domains. A number of examples for this kind of transfer have been described; however, they are all based on sequence-to-sequence comparison. Taking into account the very fast evolution of viruses, the sequence similarity may fall below the confidential detection limit relatively fast. We applied structure-guided information to detect more ancestral virus-to-host transfers. Our data show that "as a proof of principle", using protein structure-guided HMM models, it is possible to detect V2Htransfers not "visible"; with BLAST analysis.

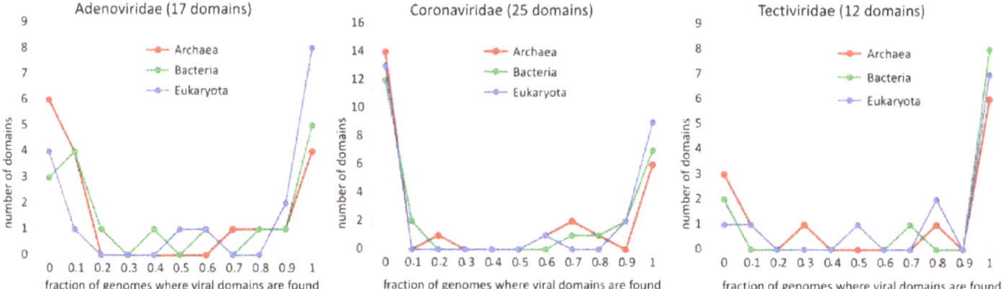

Figure 6. Distribution of the protein domains found in three viral families according to their occurrence in different superkingdoms. Protein domains as they are defined in SCOPat the superfamily level and the occurrence of these domains according to Superfamily assignment (www.supfam.org). For example, Coronaviridae encodes 13 protein domains not found in eukaryotic genomes and nine domains found in more than 90% of eukaryotic genomes.

2.3.4. RNA Secondary Structures in Whole Genome Alignments of Viruses, by Kevin Lamkiewicz

RNA secondary structures are known to play important roles in viruses, and especially in RNA viruses, since they can initiate and facilitate transcription, translation and replication. Several studies indicate that structures are cis-acting regulators for transcription. However, only looking at local structures is not sufficient to capture all RNA–RNA interactions of one molecule. Long-Range Interactions (LRI) are described in a few RNA virus families [49], but are computationally intensive to predict. Further, studies show that a single nucleotide changing can disrupt the replication of a coronavirus completely [50]. Thus, a deep understanding of conserved RNA structures is necessary to develop anti-viral therapies.

In order to increase the confidence of predictions, Multiple Sequence Alignments (MSA) are needed, since they provide conservation information between viruses. Identifying conserved secondary structures in whole genomes of viruses is computationally challenging, as the whole genome has to be considered for possible structures and interactions.

Here, we give an overview of the landscape of RNA secondary structures in viruses and provide a pipeline that generates whole genome alignments with structure annotation for downstream analyses. Our pipeline distinguishes itself from other tools by considering both the sequence and structure of input genomes for the final alignment. Therefore, for the first time, the generation of structure-annotated whole genome alignments for viruses enables sophisticated and comprehensive downstream analysis for RNA structures and RNA functions. This is achieved with an iterative combination of the sequence-based aligner MAFFT [51] and the structure-based aligner LocARNA [52]. For our example case, we were able to predict structures in the genus *Flavivirus* [53] that are consistent with described structures in the literature (Figure 7). Further, we predicted novel structural elements in coding regions of genomes.

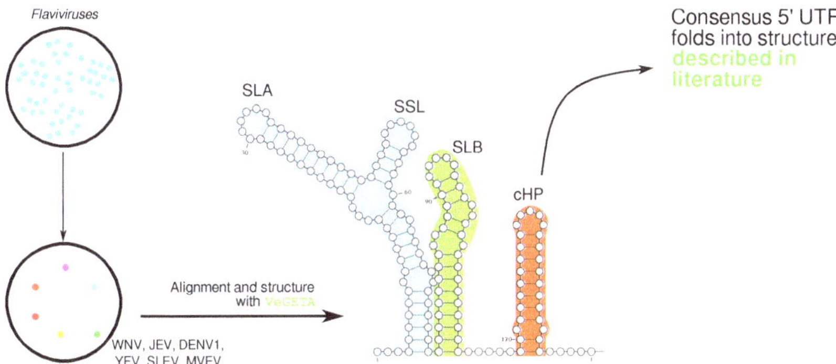

Figure 7. First results of our VeGETApipeline on an example input set consisting of flaviviruses. We were able to identify the West-Nile Virus (WNV), Dengue Virus 1 (DENV1), Japanese Encephalitis Virus (JEV), Yellow Fever Virus (YFW), Saint Louis Encephalitis Virus (SLEV) and Murray Valley Encephalitis Virus (MVEV) as representative viruses from downloaded virus genomes [53]. The resulting alignment calculated by VeGETA has structure annotations for the complete genomes, including 5′ UTR, coding regions and 3′ UTR. Here, we extracted the 5′ UTR from the alignment and visualized the annotated structure elements. These elements agree with the literature [54], as we were able to reconstruct the SLA, SLL, SLBand cHPelements accurately. The first two elements were recognized by the viral replication mechanism (NS5) [55]. The sequence embedded in the SLB structure is known to play a role in the genome circularization of flaviviruses [56], whereas the cHP facilitates the translation of the coding region by pausing the translation machinery and finding the correct starting triplet [57].

2.4. Epidemiology, Surveillance and Evolution

This session was chaired by Edward Hutchinson (MRC-University of Glasgow Centre for Virus Research, United Kingdom). Two speakers have been invited on this topic. Samantha Lycett (University of Edinburgh, United Kingdom) presented about phylodynamics for tracking epidemic, endemic and evolving viral strains. Roman Biek (University of Glasgow, United Kingdom) presented about leveraging pathogen genomics to reveal and control the spread of rabies virus. From the submitted abstracts, we selected talks by Marina Escalera-Zamudio (University of Oxford, United Kingdom) on parallel evolution and the emergence of highly-pathogenic avian influenza A viruses, David Bauer

(University of Oxford, United Kingdom) on the structure of the influenza A virus genome and Lu Lu (University of Edinburgh, United Kingdom) on the evolutionary origins of the epidemic potential among human RNA viruses.

2.4.1. Phylodynamics for Tracking Epidemic, Endemic and Evolving Viral Strains, by Samantha Lycett

Infectious diseases caused by viral pathogens in animal and livestock populations can have important economic and health consequences globally. The ability to foresee where, in which host species and under what conditions outbreaks could occur is key to developing prevention and control strategies. Sequencing pathogens from infected animals has become much more affordable and widespread in recent years, especially during outbreaks and in endemic disease settings with targeted surveillance programmes. Consequently, there are growing collections of animal virus sequences from around the globe. In this talk, the use of viral sequence data together with phylodynamic methodologies for understanding the transmission patterns in animal populations was discussed, using Avian Influenza (AI), Foot-and-Mouth Disease (FMD) and Porcine Reproductive and Respiratory Syndrome (PRRS) as examples [58–60].

Since RNA viruses have fast mutation rates and variable sequences, transmission routes between places and host species can be inferred [59,60]. One approach is to group sequences from individual hosts into discrete locations and/or host species and consider these as discrete traits or subpopulations on time-resolved phylogenetic trees, with the goal to infer which group infected which. Alternatively, locations may be represented as continuous traits (latitude and longitude) in order to estimate spatial diffusion rates and routes.

Using avian influenza as an example of a widespread multi-species disease system, it was shown that wild birds (wild Anseriformes) were responsible for long-range transmissions of highly-pathogenic H5N8, by using a combination of discrete host traits and continuous spatial traits on time-resolved phylogenetic trees [58]. Furthermore the clade to which the H5N8 strains belong is unusual because unlike the highly-pathogenic H5N1 strains, they reassort frequently, picking up different neuraminidase subtypes. By using both host and neuraminidase subtype as discrete traits, it was also shown that reassortment was preferentially occurring in Anseriformes species (ducks, geese, etc.).

To conclude, phylodynamic methods using viral sequence data with time, space and species metadata reveal complex transmission patterns and can be used to understand, track, model and ultimately inform disease control measures.

2.4.2. Parallel Evolution and the Emergence of Highly-Pathogenic Avian Influenza A Viruses, by Marina Escalera-Zamudio

Avian Influenza A Viruses (AIVs) circulate among wild and domestic bird populations worldwide. While some strains only cause mild to asymptomatic infections, known as Low Pathogenicity avian influenza viruses (LP), High Pathogenicity avian influenza viruses (HP) can have an extremely high mortality rate in both domestic and wild bird populations, leading to huge economic loses (Figure 8A) [61]. Thus, surveillance of AIVs is crucial for early detection of outbreaks. Although virulence is a polygenic trait, molecular determinants of virulence have been well characterised for AIVs, such as a polybasic proteolytic cleavage site within the hemagglutinin protein, which enables a systemic viral spread within the host [62]. We hypothesise that the parallel evolution of HP lineages from LP ancestors may have been facilitated by permissive or compensatory secondary mutations occurring anywhere in the viral genome, preceding or following the appearance of a polybasic proteolytic cleavage site. We used a comparative phylogenetic and structural approach to detect shared mutations evolving under positive selection across the whole genome of HP AIVs of the H7NX and H5NX subtypes and developed a model that statistically assesses genotype-phenotype associations. We present cumulative evolutionary and structural evidence that supports the association between parallel mutations and the evolution of the HP phenotype. Parallel mutations occur frequently among HP lineages of the same viral subtype (Figure 8B). Many of the mutations have been previously determined to increase viral fitness in terms of their biological properties, whilst most of these are ranked as stabilising to protein structure, supporting that these are rather permissive/compensatory. The mutational panel provided here may function as an early detection system for transitional virulence stages. Circulating AIVs that do not have a polybasic cleavage site yet, but show all or some of the amino acid changes ranked, should remain under surveillance.

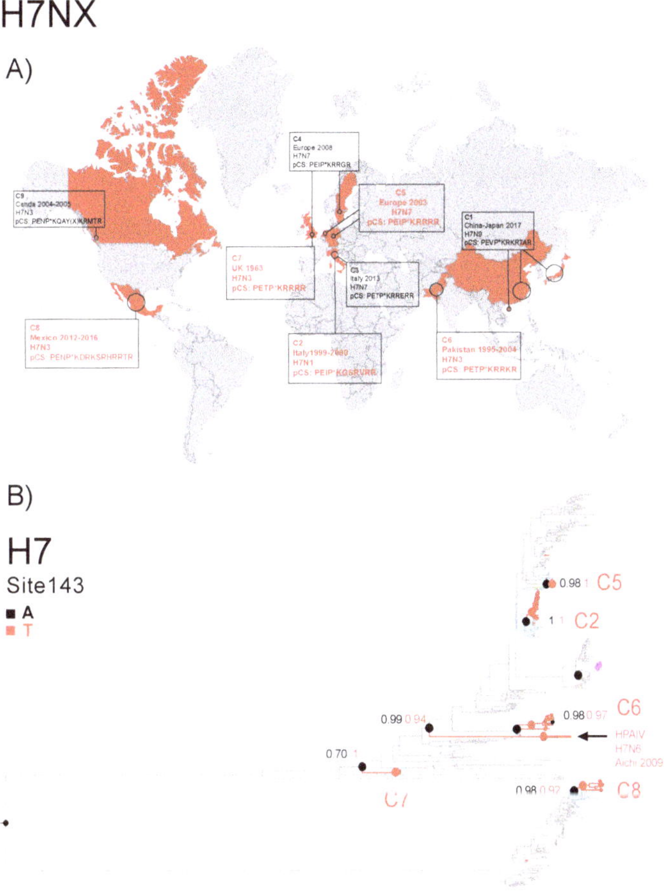

Figure 8. (**A**) Geographical occurrence of historic Highly-Pathogenic (HP) outbreaks for the H7NX viruses. Countries of emergence are highlighted in red. Year of circulation, virus subtype and consensus sequence for the polybasic Cleavage Site (pCS) within the Hemagglutinin (HA) protein are indicated for the selected outbreaks used in this work (C1-C9). Each outbreak corresponds to a distinct genotype, defined as well- supported clusters within all viral genome segment trees (data not shown). (**B**) MCCtree for the HA protein with reconstruction of ancestral states for site 143, as mutation A143T was found to be evolving under parallel evolution and to be associated with the HP phenotype, occurring in 4/9 of the HP clusters analysed. This mutation is a non-conservative amino acid change located within an antigenic pocket site. Branches within the trees are coloured according to the corresponding amino acid states in nodes (tip states not shown). Ancestral nodes preceding the emergence of a mutation associated with the HP lineages are represented with coloured circles. The probabilities of a given amino acid state occurring within ancestral/descending nodes are indicated. The HP clusters of interest are highlighted with blue circles. Mutations strongly associated with an HP phenotype may function as an early detection system for transitional virulence stages.

2.4.3. Evolutionary Origins of Epidemic Potential among Human RNA Viruses, by Lu Lu

For a virus to have epidemic potential in human populations, an infected individual must be capable of transmitting the infection to other individuals. However, for the majority of human RNA virus species, human infections are acquired only from non-human reservoirs. The evolution of human transmissibility is poorly understood. Through parallel analyses of 1755 RNA viruses, we identified at least 90 nodes across 39 genus-level phylogenies associated with transitions involving the gain of human infectivity and/or transmissibility. Human-infective and human-transmissible viruses evolve independently, and at least 73% of human-transmissible RNA virus lineages emerged directly from non-human virus lineages in diverse mammal or bird taxa. Negative sense single-stranded RNA virus lineages generate a higher proportion of strictly zoonotic viruses. Our analysis demonstrates that RNA viruses from mammal/bird lineages not currently known to be infective to humans are a likely source of future epidemics in human populations, a public health threat recently designated "Disease X".

3. Poster Session

Another important facet of this year's annual EVBC meeting was the poster session on Thursday evening. The standard of the research presented was extremely high and, combined with a networking event in the Glasgow University Union, provided plenty of opportunity to meet the presenters. The relaxed atmosphere was instrumental to promoting discussions and developing new interactions between attendees. The list of poster presenters and titles can be found online (http://evbc.uni-jena.de/events/3rd-evbc-meeting).

4. Conclusions

The Third Annual Meeting of the European Virus Bioinformatics Center brought together scientists in the field with expertise in different disciplines for scientific exchange and provided the opportunity for discussing ongoing and new collaborations. The meeting attracted new researchers to virus bioinformatics, which was reflected by several first-time attendees. The presentations strongly underlined the interdisciplinary "virology meets bioinformatics" character of the meeting. We enjoyed lively discussions after the speakers' presentations, in the breaks, during the poster session and at the social events.

We hope that speakers summaries provided in this report will give an interesting insight into the field of virus bioinformatics and will encourage interested researchers to join us at the Fourth Annual Meeting of the EVBC to be held in Switzerland in 2020. For more information, do not hesitate to contact us via evbc@uni-jena.de.

Author Contributions: F.H. drafted the manuscript. The speakers contributed the paragraphs on their own talks. All authors commented on and edited the manuscript.

Funding: This research received no external funding.

Acknowledgments: The conference organisers acknowledge sponsorship from Viruses, an open access journal from MDPI, MSD, Biomex, Theolytics and the MRC-University of Glasgow Centre for Virus Research. Their support paid for the refreshments, room and poster board hire.

Conflicts of Interest: The authors declare no conflict of interest. The sponsors had no role in the decision to publish this report.

References

1. Hufsky, F.; Ibrahim, B.; Beer, M.; Deng, L.; Le Mercier, P.; McMahon, D.P.; Palmarini, M.; Thiel, V.; Marz, M. Virologists—Heroes need weapons. *PLoS Pathog.* **2018**, *14*, e1006771. [CrossRef] [PubMed]
2. Ibrahim, B.; McMahon, D.P.; Hufsky, F.; Beer, M.; Deng, L.; Le Mercier, P.; Palmarini, M.; Thiel, V.; Marz, M. A new era of virus bioinformatics. *Virus Res.* **2018**, *251*, 86–90. [CrossRef]

3. Ibrahim, B.; Arkhipova, K.; Andeweg, A.; Posada-Céspedes, S.; Enault, F.; Gruber, A.; Koonin, E.; Kupczok, A.; Lemey, P.; McHardy, A.; et al. Bioinformatics Meets Virology: The European Virus Bioinformatics Center's Second Annual Meeting. *Viruses* **2018**, *10*, 256. [CrossRef] [PubMed]

4. V'kovski, P.; Gerber, M.; Kelly, J.; Pfaender, S.; Ebert, N.; Lagache, S.B.; Simillion, C.; Portmann, J.; Stalder, H.; Gaschen, V.; et al. Determination of host proteins composing the microenvironment of coronavirus replicase complexes by proximity-labeling. *eLife* **2019**, *8*, e42037. [CrossRef] [PubMed]

5. Mayr, F.B.; Yende, S.; Angus, D.C. Epidemiology of severe sepsis. *Virulence* **2014**, *5*, 4–11. [CrossRef] [PubMed]

6. Singer, M.; Deutschman, C.S.; Seymour, C.W.; Shankar-Hari, M.; Annane, D.; Bauer, M.; Bellomo, R.; Bernard, G.R.; Chiche, J.D.; Coopersmith, C.M.; et al. The Third International Consensus Definitions for Sepsis and Septic Shock (Sepsis-3). *JAMA* **2016**, *315*, 801–810. [CrossRef] [PubMed]

7. Iuliano, A.D.; Roguski, K.M.; Chang, H.H.; Muscatello, D.J.; Palekar, R.; Tempia, S.; Cohen, C.; Gran, J.M.; Schanzer, D.; Cowling, B.J.; et al. Estimates of global seasonal influenza-associated respiratory mortality: A modelling study. *Lancet (London, England)* **2018**, *391*, 1285–1300. [CrossRef]

8. Papanicolaou, G.A. Severe influenza and S. aureus pneumonia: For whom the bell tolls? *Virulence* **2013**, *4*, 666–668. [CrossRef] [PubMed]

9. Klemm, C.; Bruchhagen, C.; van Krüchten, A.; Niemann, S.; Löffler, B.; Peters, G.; Ludwig, S.; Ehrhardt, C. Mitogen-activated protein kinases (MAPKs) regulate IL-6 over-production during concomitant influenza virus and *Staphylococcus aureus* infection. *Sci. Rep.* **2017**, *7*, 42473. [CrossRef]

10. van Krüchten, A.; Wilden, J.J.; Niemann, S.; Peters, G.; Löffler, B.; Ludwig, S.; Ehrhardt, C. *Staphylococcus aureus* triggers a shift from influenza virus-induced apoptosis to necrotic cell death. *FASEB J.* **2018**, *32*, 2779–2793. [CrossRef] [PubMed]

11. Russell, A.B.; Trapnell, C.; Bloom, J.D. Extreme heterogeneity of influenza virus infection in single cells. *eLife* **2018**, *7*, e32303. [CrossRef] [PubMed]

12. Steuerman, Y.; Cohen, M.; Peshes-Yaloz, N.; Valadarsky, L.; Cohn, O.; David, E.; Frishberg, A.; Mayo, L.; Bacharach, E.; Amit, I.; et al. Dissection of Influenza Infection In Vivo by Single-Cell RNA Sequencing. *Cell Syst.* **2018**, *6*, 679–691. [CrossRef] [PubMed]

13. Jonsdottir, H.R.; Dijkman, R. Characterization of Human Coronaviruses on Well-Differentiated Human Airway Epithelial Cell Cultures. In *Coronaviruses*; Springer: New York, NY, USA, 2015; pp. 73–87. [CrossRef]

14. James, C.E.; Davies, E.V.; Fothergill, J.L.; Walshaw, M.J.; Beale, C.M.; Brockhurst, M.A.; Winstanley, C. Lytic activity by temperate phages of *Pseudomonas aeruginosa* in long-term cystic fibrosis chronic lung infections. *ISME J.* **2015**, *9*, 1391–1398. [CrossRef] [PubMed]

15. Shan, J.; Ramachandran, A.; Thanki, A.M.; Vukusic, F.B.I.; Barylski, J.; Clokie, M.R.J. Bacteriophages are more virulent to bacteria with human cells than they are in bacterial culture; insights from HT-29 cells. *Sci. Rep.* **2018**, *8*, 5091. [CrossRef]

16. Nale, J.Y.; Redgwell, T.A.; Millard, A.; Clokie, M.R.J. Efficacy of an Optimised Bacteriophage Cocktail to Clear *Clostridium difficile* in a Batch Fermentation Model. *Antibiotics* **2018**, *7*, 13. [CrossRef]

17. Nale, J.Y.; Spencer, J.; Hargreaves, K.R.; Buckley, A.M.; Trzepiński, P.; Douce, G.R.; Clokie, M.R.J. Bacteriophage Combinations Significantly Reduce *Clostridium difficile* Growth In Vitro and Proliferation In Vivo. *Antimicrob. Agents Chemother.* **2016**, *60*, 968–981. [CrossRef]

18. Hendrix, R.W.; Smith, M.C.; Burns, R.N.; Ford, M.E.; Hatfull, G.F. Evolutionary relationships among diverse bacteriophages and prophages: All the world's a phage. *Proc. Natl. Acad. Sci. USA* **1999**, *96*, 2192–2197. [CrossRef]

19. Sanjuán, R.; Nebot, M.R.; Chirico, N.; Mansky, L.M.; Belshaw, R. Viral mutation rates. *J. Virol.* **2010**, *84*, 9733–9748. [CrossRef]

20. Breitbart, M.; Salamon, P.; Andresen, B.; Mahaffy, J.M.; Segall, A.M.; Mead, D.; Azam, F.; Rohwer, F. Genomic analysis of uncultured marine viral communities. *Proc. Natl. Acad. Sci. USA* **2002**, *99*, 14250–14255. [CrossRef]

21. Roux, S.; Brum, J.R.; Dutilh, B.E.; Sunagawa, S.; Duhaime, M.B.; Loy, A.; Poulos, B.T.; Solonenko, N.; Lara, E.; Poulain, J.; et al. Ecogenomics and potential biogeochemical impacts of globally abundant ocean viruses. *Nature* **2016**, *537*, 689. [CrossRef]

22. Breitbart, M.; Rohwer, F. Here a virus, there a virus, everywhere the same virus? *Trends Microbiol.* **2005**, *13*, 278–284. [CrossRef] [PubMed]

23. Dutilh, B.E.; Cassman, N.; McNair, K.; Sanchez, S.E.; Silva, G.G.Z.; Boling, L.; Barr, J.J.; Speth, D.R.; Seguritan, V.; Aziz, R.K.; et al. A highly abundant bacteriophage discovered in the unknown sequences of human faecal metagenomes. *Nat. Commun.* **2014**, *5*, 4498. [CrossRef] [PubMed]

24. Stern, A.; Mick, E.; Tirosh, I.; Sagy, O.; Sorek, R. CRISPR targeting reveals a reservoir of common phages associated with the human gut microbiome. *Genome Res.* **2012**, *22*, 1985–1994. [CrossRef] [PubMed]

25. Manrique, P.; Bolduc, B.; Walk, S.T.; van der Oost, J.; de Vos, W.M.; Young, M.J. Healthy human gut phageome. *Proc. Natl. Acad. Sci. USA* **2016**, *113*, 10400–10405. [CrossRef] [PubMed]

26. Mahmoudabadi, G.; Phillips, R. A comprehensive and quantitative exploration of thousands of viral genomes. *eLife* **2018**, *7*, e31955. [CrossRef] [PubMed]

27. Kang, H.S.; McNair, K.; Cuevas, D.; Bailey, B.; Segall, A.; Edwards, R.A. Prophage genomics reveals patterns in phage genome organization and replication. *bioRxiv* **2017**, 114819. [CrossRef]

28. Miller, E.S.; Kutter, E.; Mosig, G.; Arisaka, F.; Kunisawa, T.; Ruger, W. Bacteriophage T4 Genome. *Microbiol. Mol. Biol. Rev.* **2003**, *67*, 86–156. [CrossRef]

29. Brum, J.R.; Schenck, R.O.; Sullivan, M.B. Global morphological analysis of marine viruses shows minimal regional variation and dominance of non-tailed viruses. *ISME J.* **2013**, *7*, 1738–1751. [CrossRef] [PubMed]

30. Mavrich, T.N.; Hatfull, G.F. Bacteriophage evolution differs by host, lifestyle and genome. *Nat. Microbiol.* **2017**, *2*. [CrossRef]

31. Kupczok, A.; Neve, H.; Huang, K.D.; Hoeppner, M.P.; Heller, K.J.; Franz, C.M.A.P.; Dagan, T. Rates of Mutation and Recombination in Siphoviridae Phage Genome Evolution over Three Decades. *Mol. Biol. Evol.* **2018**, *35*, 1147–1159. [CrossRef]

32. Simmonds, P.; Aiewsakun, P.; Katzourakis, A. Prisoners of war — host adaptation and its constraints on virus evolution. *Nat. Rev. Microbiol.* **2018**, *17*, 321–328. [CrossRef] [PubMed]

33. Edwards, R.; Vega, A.; Norman, H.; Ohaeri, M.C.; Levi, K.; Dinsdale, E.; Cinek, O.; Aziz, R.; McNair, K.; Barr, J.; et al. Global phylogeography and ancient evolution of the widespread human gut virus crAssphage. *bioRxiv* **2019**, 527796. [CrossRef]

34. Mokili, J.L.; Rohwer, F.; Dutilh, B.E. Metagenomics and future perspectives in virus discovery. *Curr. Opin. Virol.* **2012**, *2*, 63–77. [CrossRef] [PubMed]

35. Symonds, E.M.; Breitbart, M. Affordable Enteric Virus Detection Techniques Are Needed to Support Changing Paradigms in Water Quality Management. *Clean* **2014**, *43*, 8–12. [CrossRef]

36. Bibby, K. Metagenomic identification of viral pathogens. *Trends Biotechnol.* **2013**, *31*, 275–279. [CrossRef] [PubMed]

37. Roux, S.; Adriaenssens, E.M.; Dutilh, B.E.; Koonin, E.V.; Kropinski, A.M.; Krupovic, M.; Kuhn, J.H.; Lavigne, R.; Brister, J.R.; Varsani, A.; et al. Minimum Information about an Uncultivated Virus Genome (MIUViG). *Nat. Biotechnol.* **2018**. [CrossRef]

38. Eren, A.M.; Esen, Ö.C.; Quince, C.; Vineis, J.H.; Morrison, H.G.; Sogin, M.L.; Delmont, T.O. Anvi'o: An advanced analysis and visualization platform for 'omics data. *PeerJ* **2015**, *3*, e1319. [CrossRef]

39. Adriaenssens, E.; Farkas, K.; Harrison, C.; Jones, D.; Allison, H.E.; McCarthy, A.J. Viromic analysis of wastewater input to a river catchment reveals a diverse assemblage of RNA viruses. *bioRxiv* **2018**, 248203. [CrossRef]

40. Simmonds, P.; Adams, M.J.; Benkő, M.; Breitbart, M.; Brister, J.R.; Carstens, E.B.; Davison, A.J.; Delwart, E.; Gorbalenya, A.E.; Harrach, B.; et al. Consensus statement: Virus taxonomy in the age of metagenomics. *Nat. Rev. Microbiol.* **2017**, *15*, 161–168. [CrossRef]

41. Aiewsakun, P.; Adriaenssens, E.M.; Lavigne, R.; Kropinski, A.M.; Simmonds, P. Evaluation of the genomic diversity of viruses infecting bacteria, archaea and eukaryotes using a common bioinformatic platform: Steps towards a unified taxonomy. *J Gen Virol* **2018**, *99*, 1331–1343. [CrossRef]

42. Aiewsakun, P.; Simmonds, P. The genomic underpinnings of eukaryotic virus taxonomy: Creating a sequence-based framework for family-level virus classification. *Microbiome* **2018**, *6*, 38. [CrossRef] [PubMed]

43. Patterson Ross, Z.; Klunk, J.; Fornaciari, G.; Giuffra, V.; Duchêne, S.; Duggan, A.T.; Poinar, D.; Douglas, M.W.; Eden, J.S.; Holmes, E.C.; et al. The paradox of HBV evolution as revealed from a 16th century mummy. *PLoS Pathog.* **2018**, *14*, e1006750. [CrossRef]

44. Krause-Kyora, B.; Susat, J.; Key, F.M.; Kühnert, D.; Bosse, E.; Immel, A.; Rinne, C.; Kornell, S.C.; Yepes, D.; Franzenburg, S.; et al. Neolithic and medieval virus genomes reveal complex evolution of hepatitis B. *eLife* **2018**, *7*, e36666. [CrossRef] [PubMed]

45. Mühlemann, B.; Jones, T.C.; Damgaard, P.d.B.; Allentoft, M.E.; Shevnina, I.; Logvin, A.; Usmanova, E.; Panyushkina, I.P.; Boldgiv, B.; Bazartseren, T.; et al. Ancient hepatitis B viruses from the Bronze Age to the Medieval period. *Nature* **2018**, *557*, 418–423. [CrossRef] [PubMed]

46. Bar-Gal, G.K.; Kim, M.J.; Klein, A.; Shin, D.H.; Oh, C.S.; Kim, J.W.; Kim, T.H.; Kim, S.B.; Grant, P.R.; Pappo, O.; et al. Tracing hepatitis B virus to the 16th century in a Korean mummy. *Hepatology* **2012**, *56*, 1671–1680. [CrossRef]

47. Puustusmaa, M.; Abroi, A. Conservation of the E8 CDS of the E8^{E2} protein among mammalian papillomaviruses. *J. Gen. Virol.* **2016**, *97*, 2333–2345. [CrossRef] [PubMed]

48. Puustusmaa, M.; Abroi, A. cRegions—a tool for detecting conserved cis-elements in multiple sequence alignment of diverged coding sequences. *PeerJ* **2019**, *6*, e6176. [CrossRef]

49. Nicholson, B.L.; White, K.A. Functional long-range RNA-RNA interactions in positive-strand RNA viruses. *Nat. Rev. Microbiol.* **2014**, *12*, 493–504. [CrossRef] [PubMed]

50. Madhugiri, R.; Karl, N.; Petersen, D.; Lamkiewicz, K.; Fricke, M.; Wend, U.; Scheuer, R.; Marz, M.; Ziebuhr, J. Structural and functional conservation of cis-acting RNA elements in coronavirus 5′-terminal genome regions. *Virology* **2018**, *517*, 44–55. [CrossRef] [PubMed]

51. Kuraku, S.; Zmasek, C.M.; Nishimura, O.; Katoh, K. aLeaves facilitates on-demand exploration of metazoan gene family trees on MAFFT sequence alignment server with enhanced interactivity. *Nucleic Acids Res.* **2013**, *41*, W22–W28. [CrossRef] [PubMed]

52. Will, S.; Reiche, K.; Hofacker, I.L.; Stadler, P.F.; Backofen, R. Inferring noncoding RNA families and classes by means of genome-scale structure-based clustering. *PLoS Comput. Biol.* **2007**, *3*, e65. [CrossRef]

53. Pickett, B.E.; Sadat, E.L.; Zhang, Y.; Noronha, J.M.; Squires, R.B.; Hunt, V.; Liu, M.; Kumar, S.; Zaremba, S.; Gu, Z.; et al. ViPR: An open bioinformatics database and analysis resource for virology research. *Nucleic Acids Res.* **2012**, *40*, D593–D598. [CrossRef] [PubMed]

54. Fernández-Sanlés, A.; Ríos-Marco, P.; Romero-López, C.; Berzal-Herranz, A. Functional Information Stored in the Conserved Structural RNA Domains of Flavivirus Genomes. *Front. Microbiol.* **2017**, *8*, 546. [CrossRef]

55. Filomatori, C.V.; Lodeiro, M.F.; Alvarez, D.E.; Samsa, M.M.; Pietrasanta, L.; Gamarnik, A.V. A 5′ RNA element promotes dengue virus RNA synthesis on a circular genome. *Genes Dev.* **2006**, *20*, 2238–2249. [CrossRef] [PubMed]

56. Clyde, K.; Barrera, J.; Harris, E. The capsid-coding region hairpin element (cHP) is a critical determinant of dengue virus and West Nile virus RNA synthesis. *Virology* **2008**, *379*, 314–323. [CrossRef] [PubMed]

57. Kozak, M. Downstream secondary structure facilitates recognition of initiator codons by eukaryotic ribosomes. *Proc. Natl. Acad. Sci. USA* **1990**, *87*, 8301–8305. [CrossRef]

58. Global Consortium for H5N8 and Related Influenza Viruses. Role for migratory wild birds in the global spread of avian influenza H5N8. *Science* **2016**, *354*, 213–217. [CrossRef]

59. Lycett, S.; Tanya, V.N.; Hall, M.; King, D.; Mazeri, S.; Mioulet, V.; Knowles, N.; Wadsworth, J.; Bachanek-Bankowska, K.; Victor, N.N.; et al. The evolution and phylodynamics of serotype A and SAT2 foot-and-mouth disease viruses in endemic regions of Africa. *bioRxiv* **2019**, 572198. [CrossRef]

60. Duchatel, F.; Bronsvoort, M.; Lycett, S. Phylogeographic analysis and identification of factors impacting the diffusion of Foot-and-Mouth disease virus in Africa. *bioRxiv* **2018**, 358044. [CrossRef]

61. Dhingra, M.S.; Artois, J.; Dellicour, S.; Lemey, P.; Dauphin, G.; Von Dobschuetz, S.; Van Boeckel, T.P.; Castellan, D.M.; Morzaria, S.; Gilbert, M. Geographical and Historical Patterns in the Emergences of Novel Highly Pathogenic Avian Influenza (HPAI) H5 and H7 Viruses in Poultry. *Front. Vet. Sci.* **2018**, *5*, 84. [CrossRef]

62. Abdelwhab, E.M.; Veits, J.; Ulrich, R.; Kasbohm, E.; Teifke, J.P.; Mettenleiter, T.C. Composition of the Hemagglutinin Polybasic Proteolytic Cleavage Motif Mediates Variable Virulence of H7N7 Avian Influenza Viruses. *Sci. Rep.* **2016**, *6*, 3950. [CrossRef] [PubMed]

Article

Base-By-Base Version 3: New Comparative Tools for Large Virus Genomes

Shin-Lin Tu [1], Jeannette P. Staheli [2], Colum McClay [1], Kathleen McLeod [1], Timothy M. Rose [2,3] and Chris Upton [1,*]

[1] Biochemistry and Microbiology, University of Victoria, Victoria, BC V8W 2Y2, Canada; cindytu@uvic.ca (S.-L.T.); colummcclay@hotmail.com (C.M.); kathleen_mcleod14@hotmail.com (K.M.)

[2] Center for Global Infectious Disease Research, Seattle Children's Research Institute, Seattle, WA 98101, USA; jeannette.staheli@seattlechildrens.org (J.P.S.); trose@u.washington.edu (T.M.R.)

[3] Department of Pediatrics, University of Washington, Seattle, WA 98195, USA

* Correspondence: cupton@uvic.ca; Tel.: +1-250-721-6507

Received: 25 October 2018; Accepted: 13 November 2018; Published: 15 November 2018

Abstract: Base-By-Base is a comprehensive tool for the creation and editing of multiple sequence alignments that is coded in Java and runs on multiple platforms. It can be used with gene and protein sequences as well as with large viral genomes, which themselves can contain gene annotations. This report describes new features added to Base-By-Base over the last 7 years. The two most significant additions are: (1) The recoding and inclusion of "consensus-degenerate hybrid oligonucleotide primers" (CODEHOP), a popular tool for the design of degenerate primers from a multiple sequence alignment of proteins; and (2) the ability to perform fuzzy searches within the columns of sequence data in multiple sequence alignments to determine the distribution of sequence variants among the sequences. The intuitive interface focuses on the presentation of results in easily understood visualizations and providing the ability to annotate the sequences in a multiple alignment with analytic and user data.

Keywords: bioinformatics; virus; comparative genomics; software; Base-By-Base; BBB; poxvirus; ASFV; MSA

1. Introduction

Base-By-Base (BBB) [1,2], a multiple sequence alignment (MSA) editor, has been under development for more than 15 years and forms an integral component in the Viral Bioinformatics Resource Center (VBRC) platform (www.4virology.net) that supports comparative genomics of large DNA viruses. Although the viruses supported by VBRC are primarily poxviruses and African swine fever virus (ASFV) because of our research interests, BBB is equally valuable for other viral genomes and nucleic acid and protein sequences, which can be imported into BBB from FASTA or GenBank files. The consistent theme running throughout the development of VBRC's tools has been to provide the virologist/biologist with easy-to-use graphical tools that let the user visualize and interact with the raw sequence data. The simplest example of this is providing the ability to quickly and visually scan a viral genome MSA for alignment errors and make manual corrections or use a second alignment algorithm to realign a section of a larger MSA. Without visually reviewing the output of a bioinformatics analysis, users may unknowingly use algorithms and parameters that are not appropriate for their analysis. For example, different tools may include or ignore gaps when calculating percent identity between sequence pairs; for example, counting a single gap of 100 nucleotides as 100 mismatches in one of a pair of sequences (1 kb alignment) that are otherwise 99% identical would create the illusion of 89% identity. The performance of a visual assessment of MSAs should be a routine "reality check" for researchers, rather than relying solely on the numerical output of alignment tools.

Previously described features of BBB include:

- Java code; uses Java Web Start to launch and automate updating of BBB for users;
- Integration with the VBRC's viral genome database (viral orthologous clusters; VOCs);
- Alignment of sequences or subsequences (MUSCLE, ClustalW, MAFFT);
- MSA editing, with intuitive highlighting of differences between sequences;
- Applicability to gene, protein, and virus genome sequences;
- Ability to edit individual sequences;
- Intuitive graphical user interface (GUI) with ability to view sequence residues or full sequence summaries;
- Display of 6-reading frames; understands and links to gene location (data from GenBank files);
- Multiple methods to annotate sequences and MSAs, treating the BBB file as a "results notebook".

Over the years, as we have needed novel functions in our investigations of viral genomes, we have used BBB as our standard platform for analyses that required manipulation and visual presentation of DNA and protein sequences. Thus, what otherwise might have been a solitary Perl or Python script became a feature within BBB. This enhances the functionality of BBB and provides open user access to new functions as new scripts get added. As a result, BBB has become an integrated platform with multiple features that provides a common user-friendly interface for both input of sequences and output of results.

This communication describes new features that have been incorporated into BBB: New "Advanced/Experimental tools" include j-CODEHOP, **Find Differences**, **SNIP**, and **MAFFT-add**; new "Reports" include **Get Counts**, **Get Unique Positions**, and **Get SNP Counts** (of top 2 sequences); new "Tools" that export alignment after deletion of **Specified Columns** or **Columns Containing Gap(s)**. It should be noted that these single nucleotide differences between viral genomes are not "polymorphisms" in the strictest sense, but we have used the term SNP (single nucleotide polymorphism) as it is a recognizable and understandable term among virologists.

2. Materials and Methods

When the BBB project began more than 15 years ago, Java was chosen as the coding language because (1) web browsers were not as capable as today's JavaScript powered interfaces; (2) Java was the primary language taught to undergraduates at the University of Victoria; and (3) it was relatively platform-independent, promising "code once, deploy everywhere" capability, thereby eliminating the compilation and installation obstacles that hinder users that want to try out new tools.

This updated software application is available from the www.4virology.net website (previously virology.uvic.ca) and code is made available upon request under the GNU General Public License version 3. In addition, the BBB and "consensus-degenerate hybrid oligonucleotide primers" (CODEHOP) source code described below has recently been submitted to the GitHub repository (https://github.com/vbrclab/basebybase; https://github.com/vbrclab/Codehop).

j-CODEHOP can be launched from within BBB (menu: **Advanced**) or from its own web page on the www.4virology.net website. In each case, the initial step is the download of the BBB alignment editor configuration file (*.jnlp), which is started by Java Web Start on the user's own computer by default. Java Web Start requires at least Java 7, but less than Java 11 to run. The Java Runtime Environment (JRE) can be downloaded for free, if needed.

Multiple "help" files for BBB and CODEHOP are available, including a "quick start page", "how to doc" and "help book", which get progressively more detailed. A j-CODEHOP tutorial is also provided. Although users are requested to register their email for use of the VBRC, this is only used to allow the resource to email users occasionally to make them aware of important new features; many users choose to use nonidentifying email addresses.

The cowpox viruses (CPXV) used as examples are: BR (AF482758.2), Norway 1994 MAN (HQ420899.1), Germany 1998 2 (HQ420897.1), Germany 1980 EP4 (HQ420895.1), Germany 2002

MKY (HQ420898.1), EleGri07/1 (KC813507.1), BeaBer04/1 (KC813491.1), RatHei09/1 (KC813504.1), GRI-90 (X94355.2), and HumGra07 (KC813510.1). The core conserved nucleotide alignment (60 kb) was used to generate a maximum-likelihood phylogenetic tree using the GTRGAMMA model in RAxML v.8.2.10 [3].

Currently, BBB has an upper limit of about 500 protein sequences (300 aa each) due to the memory assigned to the tool.

3. Results

3.1. CODEHOP Integration

Even though vast amounts of genomic sequences have been obtained recently, it is unlikely that the complete genome sequence will have been determined for all living species that might provide valuable scientific and medical insights. In order to obtain sequence information for specific genes in unsequenced organisms or pathogens, a primer design strategy for PCR amplification of novel genes using "consensus-degenerate hybrid oligonucleotide primers" (CODEHOPs) was previously developed [4]. CODEHOPs are designed from amino acid sequence motifs that are highly conserved within a gene family, and are used in PCR amplification to identify unknown related family members. Each CODEHOP consists of a pool of primers containing all possible nucleotide sequences within a 3′ degenerate core encoding a stretch of 3–4 highly conserved amino acids (Figure 1). A longer 5′ nondegenerate clamp region in the primers contains the most probable nucleotide predicted for each flanking codon. The degenerate core allows primer binding to all existing target variations in the initial PCR cycles, while the clamp region, once integrated into early PCR products, leads to efficient amplification of the PCR products in later PCR cycles. CODEHOPs designed from two adjacent conserved motifs are used to amplify the gene sequences between these motifs.

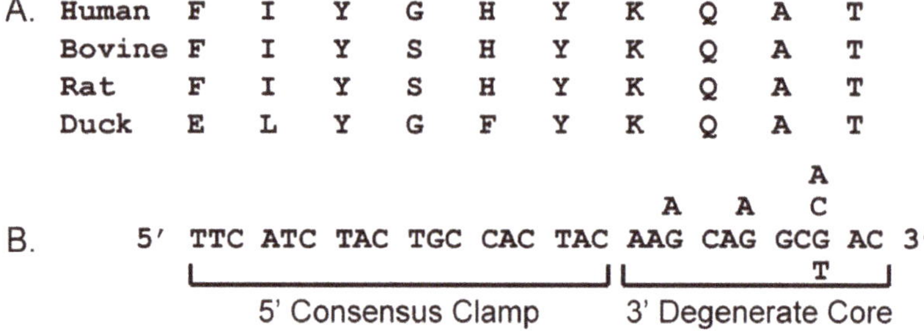

Figure 1. Anatomy of a "consensus-degenerate hybrid oligonucleotide primers" (CODEHOP) PCR primer. A CODEHOP is a pool of related primers containing all possible nucleotide sequences encoding 3 to 4 highly conserved amino acids within a 3′ degenerate core and a 5′ consensus clamp containing the most probable nucleotide at each position for the flanking codons. (**A**) multiple alignment of protein sequences; (**B**) predicted CODEHOP primer pool.

Other methods to identify unknown genes have used degenerate primers, containing most or all of the possible nucleotide sequences encoding amino acid motifs, or a consensus primer containing the most common nucleotide at each codon position within the motifs. However, unlike strictly degenerate or consensus approaches, the CODEHOP PCR approach has proven to be highly successful in amplifying distantly related genes containing significant sequence variations at low copy numbers. The primer design software and the CODEHOP PCR strategy have been utilized for the identification and characterization of new gene orthologs and paralogs in different plant, animal, and bacterial species, as well as for virus typing (e.g., enteroviruses [5]); consequently, the original publication has been cited in more than 800 subsequent publications. In addition, this approach has been successful in identifying new pathogen species and genes, as we have previously published [6–12].

A computer strategy to predict CODEHOP PCR primers from multiply aligned sets of related protein sequences was previously developed, which has been continuously accessible over the internet since 1998 as an integral part of the BLOCKS database developed by Steven Henikoff and hosted by the Fred Hutchinson Cancer Research Center [4]. A description of the CODEHOP program and its uses was published in the 2003 NAR Web services edition [13]. Subsequently, we developed iCODEHOP, an interactive web application independent of the BLOCKS database, to simplify and automate the process of designing CODEHOP PCR primers. The iCODEHOP program added new features, including interactive visualization of predicted CODEHOPs, phylogenetic plots for multiple aligned sequences, and user sessions on the server that allowed data to be stored during the design process [14]. However, due to advances in web browser technology, problems with the stored server sessions, and resource limitations, the iCODEHOP web application could no longer be supported and we have now developed a Java-based iteration called j-CODEHOP for integration into BBB (menu: **Advanced**).

j-CODEHOP guides users through the CODEHOP PCR primer design process, including uploading sequences, creating a multiple alignment, and identifying and visualizing primer pools that match the specified design criteria. A linked tutorial provides a step-by-step guide to demonstrate how to create CODEHOPs, using a sample FASTA file containing related sequences within the uracil DNA glycosylase family. The input to j-CODEHOP can be a set of nonaligned protein sequences or a set of aligned protein sequences. Protein sequence files may be formatted as GenBank (*.gb, *.gbk), EMBL (*.embl), BBB (*.bbb), FASTA (*.fasta, *.fas, *.fa) or CLUSTAL (*.clustal, *.clustalw). The program's output includes a graphic showing predicted CODEHOP primers at their locations along a consensus protein sequence, a graphical representation of the region of the multiple alignment from which they are derived, and a set of metadata about each primer pool (length, degeneracy, and annealing temperature range). j-CODEHOP enables the user to visually scan the entire set of predicted CODEHOP primers to assess their relative positions and orientations within the consensus protein sequence and select individual CODEHOP primers for further analysis.

For the aligned protein sequences and chosen criteria, j-CODEHOP computes all primer possibilities. The initial output shows the consensus amino acid sequence for conserved blocks of the multiple protein alignment (Figure 2A). This sequence is numbered according to the positions in the multiple alignment, with capital letters for amino acids matching the minimum conservation criteria. A second window lists the possible primers to export. The user can view the consensus sequence to visualize the positions of the predicted primers, which are shown as arrows, forward or reverse. The amino acid motif targeted by the $3'$ degenerate core of the primer is aligned with the primer arrow, as are the flanking amino acids specifying the $5'$ non-degenerate clamp. A specific primer can be selected, which will open a third window to show the CODEHOP sequence, the block of aligned sequences used for primer design, and the primer design criteria (Figure 2B). Both forward and reverse CODEHOPs in the correct orientation need to be identified. If an insufficient number of CODEHOPs are predicted, the program can be rerun using more relaxed design criteria or the distance between the group of sequences can be reduced. Detailed methodologies have been previously published that describe the design of CODEHOP PCR primers and their use in identifying novel sequences [4,11,13,14].

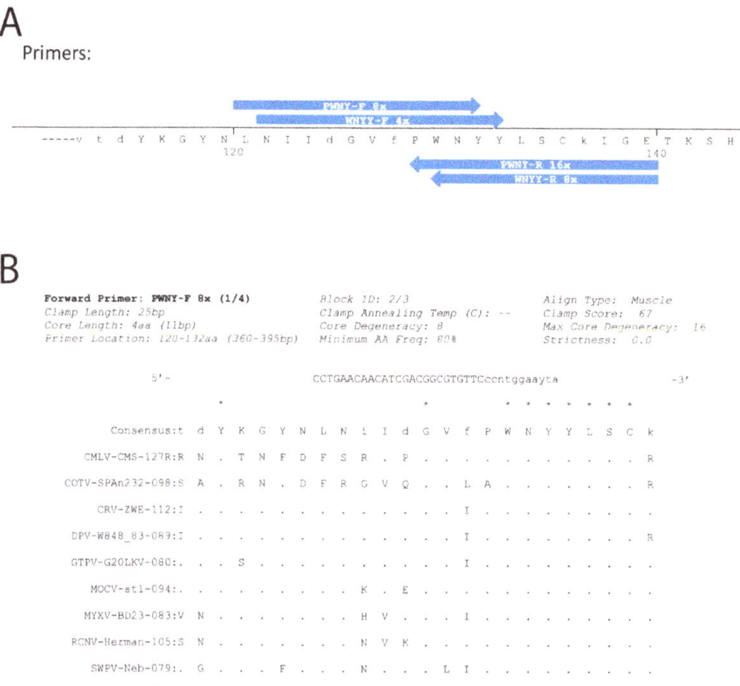

Figure 2. j-CODEHOP primer design output. The uracil DNA glycosylase test data set was used as input for the j-CODEHOP program and the following options were used for primer design: (1) Block making alignment tool—"MUSCLE"; (2) Codon table—"Homo sapiens"; (3) Clamp (nondegenerate 5′ region) length—"25"; (4) Core (degenerate 3′ region)—max degeneracy "16", length of degenerate core in aa "4", strictness (%) "0", min aa conservation (%) "80". Default values were used for the Advanced options: (1) 3′ nucleotide—"Invariant 3′ nt"; (2) Min block length—"5"; Primer concentration (nM)—"50"; (3) Restrict 3′ nucleotide to G or C—"unchecked"; (4) Exclude Leu, Ser, and Arg from 3′ region—"unchecked". (**A**) Initial graphical output showing the consensus amino acid sequence for the ordered blocks of multiply aligned protein sequences. Amino acids showing conservation above the chosen minimum value are capitalized. The positions of predicted CODEHOP PCR primers are indicated showing the extent of the amino acid sequence used for primer design, the direction of the primer (forward (F) or reverse (R)), and the degeneracy of the 3′ degenerate region, i.e., $4\times$, $8\times$ or $16\times$. A CODEHOP with $4\times$ degeneracy is composed of a pool of 4 different primers that provide all possible sequences encoding the 3–4 highly conserved amino acid motif targeted by the CODEHOP primer. The amino acid sequence of the motif is used to name the primer, ex. "PWNY". (**B**) The output obtained by clicking on a primer of interest in the initial graphical output, in this case primer "PWNY-F $8\times$". This output shows the primer sequence (5′ to 3′), with the 5′ nondegenerate consensus region in capital letters and the 3′ degenerate region in small letters, using the international code for ambiguous nucleotides, i.e., "Y" (C,T), "R" (A,G), "N" (A,C,G,T), etc. The codons in the primer sequence are aligned with the block of multiply aligned protein sequences. Amino acid positions showing conservation above the chosen minimum are indicated with an asterisk. Amino acids within the multiple alignment which are identical to the consensus sequence are indicated with a dot. The metadata for the chosen primer design criteria are indicated, as is the primer location in amino acids and base pairs. A third panel (not shown) provides a list of primers predicted from the current amino acid block to export. The primer sequence and metadata can be exported in a "comma-separated values" (CSV) spreadsheet format. The panels shown are high-resolution representations of program output.

3.2. Sequence Characteristics

As the -omics revolution progresses, more and more researchers make use of sequence data from the various databases and, increasingly, the data behind publication claims are not presented. For example, a phylogenetic tree may be published without the MSA that was used to generate it. Given that errors in sequence naming and annotation are common in the databases, it is important that researchers check results if they are going to rely on them. Figure 3 shows a BBB **Visual Summary** of 2 virus genomes (menu: **Reports**), which are almost identical except for four large indels and a block of very poorly matching sequence. This type of visualization is a powerful tool for highlighting inconsistencies in alignments. When we further investigated these sequences (BLAST [15] searches and dotplots [16]), we found that the differences between the two genomes were entirely the result of genome assembly errors.

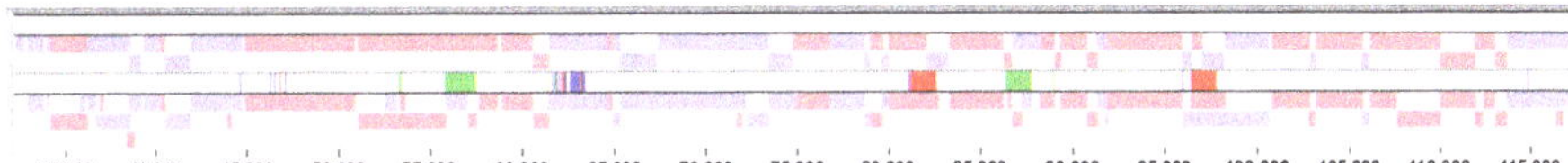

Figure 3. Visual summary from BBB. Pink and pale blue boxes represent genes transcribed to the right and left, respectively, for the genomes of two poxviruses. The centre tract indicates differences between the two sequences: Dark blue lines are SNPs (the abruptly dense SNPs turns out to be falsely assembled out-of-frame sequence from another virus) and green and red blocks show insertions and deletions (erroneously transposed sequences).

Additionally, under the BBB **Reports** menu, the ability to display **Sequence Similarity** and **Sequence Difference** graphs (useful for detecting recombination; not shown) has been supplemented by the plotting of a **Nucleotide Content Graph**. The user has control over which nucleotides are included in the analysis, as well as the size of the sliding window of nucleotides and the number of nucleotides that is used to "step" across the sequence. The tool also allows the user to choose which sequences from an MSA are included in these analyses. Importantly, the option to ignore gapped columns in an MSA has been included.

Additional new **Reports** features that summarize characteristics of an MSA include: (1) **Get Counts**, which counts the number of columns in the MSA with particular features, reporting the number that have a gap, a single nucleotide, two nucleotides (consensus and second type), three nucleotides, and four nucleotides; (2) **Get Unique Positions**, which lists the number of unique positions that are not gaps for each sequence; and (3) **Get SNP Counts**, which examines the top two sequences (sequences can be moved up or down within the MSA to enable sequence selection) and reports the total number of SNPs and the number of each possible substitution.

3.3. Counting Nucleotides Associated with Specific Sequences in MSAs

As noted above, the data supporting a phylogenetic tree are not often provided in manuscripts. Often, it would be useful to know the percent identity between sequences and the numbers of SNPs that distinguish one branch on a tree from another. The ability to generate a nucleotide identity matrix from an MSA is an older feature of BBB. However, now, from within the **Advanced/Experimental Tools** menu, BBB also allows a researcher to query the MSA data that support (or don't support) a phylogenetic tree. The **Find Differences** tool can be used to count the number of SNPs that support a particular branch; e.g., "find nucleotides that are identical in sequences A, B, and C, but different in all other sequences". Figure 4 shows the phylogenetic tree for the central relatively-conserved core (60 kb) of 10 cowpox viruses. For these sequences, the viruses in the DNA sequence identity range from 98.2–99.4%. Counting the number of SNPs unique to each sequence (red numbers in Figure 4) shows that for these cowpox sequences, the branch lengths created may not truly reflect the evolutionary distances. Instead, the lengths were likely compressed due to evidence of recombination shown in Table 1, which artificially reduced distances between distant strains.

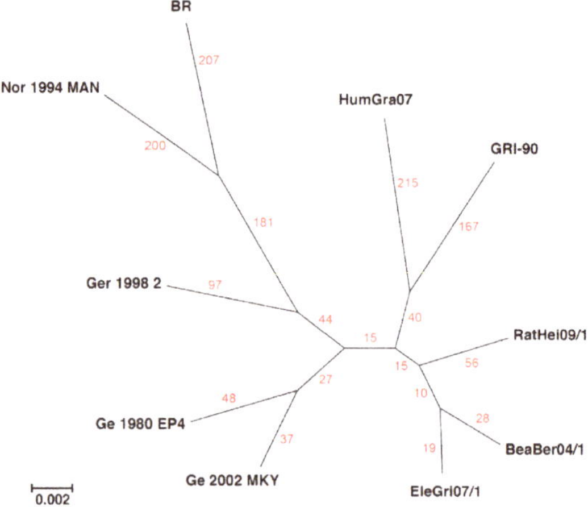

Figure 4. The contrast of unique SNPs found in the genomic core of 10 cowpox viruses (using the BBB **Find Differences** feature) with that of a maximum-likelihood phylogenetic tree. Red numbers denote the number of unique SNPs found for the virus that are not shared with any of the others. The phylogenetic tree branch scale denotes the average number of nucleotide substitution per site.

Table 1. SNPs shared by CPXV-BR, CPXV-Nor1994MAN, and strain noted in the table; all other viruses in Figure 4 have a different nucleotide. SNPs close together are grouped on a single line in the table.

+BeaBer04/1
22,518, 22,519, 22,583
31,870
+RatHei09/1
4677, 4679
4886, 4896, 4899, 4917
9401
16,480
19,731
31,573
35,003
40,781
+Ge 1980 EP4
1204
10,615, 10,618
10,731, 10,747
14,442, 14,457, 14,460, 14,553, 14,574, 14,664, 14,667
15,071, 15,072, 15,076, 15,138, 15,161, 15,163, 15,171
19,409
25,381
30,528, 30,534, 30,547, 30,549, 30,556
32,797, 32,799
35,713, 35,758, 35,812
36,217
41,120
+Ge 2002 MKY,
19,510
47,392

An important feature of the **Find Differences** tool (menu: **Advanced/Experimental Tools**) is that it can allow the matching to be fuzzy. We have termed this feature "tolerance" and it can be viewed as the search "tolerating" one or more (specified by the user) sequences that do not fulfill the query. For example, the query "find nucleotides that are identical in sequences A and B but different in all other sequences, with tolerance = 1" allows any one of the sequences that should be different from A+B to be the same; different sequences are "tolerated" at different positions in the alignment. The software also: (1) Creates a list of all the positions in the alignment that satisfy the query and displays the "tolerated sequence" name if there is one, and (2) displays the distribution of SNPs in the MSA.

These BBB features were created to characterize recombination events among the poxviruses by highlighting the positions of shared SNPs. In any MSA, there will always be coincident SNPs from random events. However, for these cowpox sequences, when "nucleotides that are identical in sequences A, B, and C, but different in all other sequences" are located, some are, as expected, associated with the closest related sequence, but others are from more distant relatives. In addition, many of these coincident SNPs are found to be in nonrandom blocks, suggesting that the arrangements result from recombination among the genomes. Table 1 shows SNPs present only in CPXV-BR and CPXV-Nor1994MAN and one other sequence taken from the tree shown in Figure 4. In several instances, the common SNPs are unexpectedly clustered (Table 1) and likely result from recombination events. The results with CPXV-Ge1980EP4 and CPXV-Ge2002MKY (which are very similar (Figure 4)) as the extra sequence are dramatic; despite their similarity, CPXV-Ge1980EP4 has many more SNPs in common with the other two sequences (Table 1; 33 SNPs) than with CPXV-Ge2002MKY (Figure 4; 3 SNPs).

3.4. Manipulation of Sequences

As previously reported, BBB allows the addition or removal of sequences to an alignment and the removal of columns in an MSA that contains all gap characters which are often generated when removing sequences from an alignment. However, when visually inspecting the relationships between the sequences, it can also be useful to simplify the variation by removing any column that contains a gap character (menu: **Tools/Delete Columns Containing Gap(s) and Export**). Since this action will modify the sequences in use, by deleting residues from some sequences, the program will export the resulting sequence into a new BBB window and prompt the user to enter a new filename. If the sequences in an MSA are very diverged, each will have a relatively large number of unique SNPs. Since these can obscure patterns present among the SNPs shared by subsets of sequences, we also created the **SNIP** feature (menu: **Advanced/Experimental Tools**) that modifies the sequences such that the SNPs that are present only in a single sequence are changed to the consensus nucleotide. Again, because this procedure modifies the actual sequences, users are asked to save the result in a new alignment file.

When using large viral genomes and closely related viruses, SNPs may be relatively infrequent. Therefore, we incorporated a feature into BBB that allows the user to remove any specified column of nucleotides within an MSA. By removing the columns that only contain a single nucleotide, the variation is compressed into a much smaller sequence space and is more easily visualized by the user. First, the **Find Differences** tool (menu: **Advanced/Experimental Tools**) is used to find the columns that are identical (i.e., have no SNPs), then the "Search Log" is used to "List SNP Positions" only. Subsequently, these position values can then be used to delete specified columns (menu: **Tools/Delete Specified Columns and Export**).

3.5. Alignment of Sequences

The options for aligning complete or selected regions of sequences have been updated. Clustal Omega [17] has replaced the option to use ClustalW. Clustal Omega and MUSCLE [18] serve as options to align protein and gene length nucleotide sequences. For the alignment of large viral genomes, MAFFT [19] is the tool of choice. However, the growing number of complete genomes sequenced has translated into a more frequent need to generate larger MSAs, often to update phylogenetic trees. Although MAFFT is available at various web resources and can be easily installed on desktop computers, most users prefer to use MAFFT within BBB (menu: **Tools/Align Selection**). Therefore, we have incorporated the **MAFFT-add** option into the BBB (menu: **Advanced/Experimental Tools**) [20]. This feature allows users to align one or more new sequences to an existing alignment, which significantly reduces the compute time. For example, the alignment of 10 cowpox virus genomes takes approximately 8 min, whereas aligning one new sequence to an alignment of 9 takes a little over 1 min. The **MAFFT-add** function is also useful for scaffolding new contig sequences against a close reference sequence in the process of genome assembly.

4. Discussion

BBB is a foundational program of the Viral Bioinformatics Resource Centre that allows both the viewing and editing of MSAs. It has been developed over many years and serves as a platform for the comparison of large viral genomes, but it is equally useful for small DNA and RNA viral genomes, as well as gene and protein sequence alignments. The data visualization features that it provides are key to its value. These include highlighting differences between sequences in an MSA, plotting graphs of sequence similarity, adding user-comments or primers to sequence regions, and displaying forward and reverse reading frames as well as results of various sequence searches. Importantly, while BBB has provided a common interface to multiple analyses for users, it has also given programmers a standardized data input process and a single visualization canvas. One of the key features of BBB is the ability to read GenBank files so that it becomes aware of the complete set of annotations for the genomes of large viruses, allowing it to display the complete set of annotations for any large virus genome which is extremely useful for comparisons of gene features and single nucleotides.

Here, we have described a significant number of upgrades to BBB that increase the utility of the tool when working with genes, proteins or genomes. The inclusion of j-CODEHOP maintains a functioning version of the CODEHOP algorithm, which otherwise would have been lost to the research community. CODEHOP is a natural fit for BBB, since it is often used to discover novel members of viral families, but also since the generation of the protein MSAs that are used as input by j-CODEHOP to generate CODEHOP primer sequences are already an integral part of BBB. The new plots for similarity are useful for a user looking for recombination events and comparing how different genes are conserved to different degrees. Fuzzy searches for MSA columns that support or fail to support particular phylogenetic relationships bring a new process to the screening of viral genomes for small regions that have been exchanged among viruses. Following the identification of particular MSA nucleotide columns, new editing features in BBB now allow the user to manipulate these MSA columns, thereby simplifying the visualization for the user.

A variety of other tools exist to manipulate and characterize MSAs, including Jalview [21] (primarily for proteins), and IVisTMSA [22] and AliView [23] (primarily for large sequence sets). However, BBB has multiple unique features, including those described here, which make it a valuable multipurpose bioinformatics tool, especially for the comparison and characterization of viral genomes and more.

Author Contributions: Conceptualization, T.M.R. and C.U.; Methodology, C.M., J.P.S., and K.M.; Software, C.M.; Validation, K.M. and S.-L.T.; Writing—Original Draft Preparation, T.M.R., J.P.S., and S.-L.T.; Writing—Review and Editing, S.-L.T. and C.U.; Supervision, C.U.; Funding Acquisition, C.U.

Funding: This research was funded by a Discovery Grant from the Natural Sciences and Engineering Research Council of Canada (RGPIN-2015-04953).

Acknowledgments: The authors would like to thank Richard Boyce for the JavaScript CODEHOP code and the many University of Victoria Co-op students that have contributed to the software available at Viral Bioinformatics Resource Centre.

Conflicts of Interest: The authors declare no conflict of interest.

References

1. Brodie, R.; Smith, A.J.; Roper, R.L.; Tcherepanov, V.; Upton, C. Base-By-Base: Single nucleotide-level analysis of whole viral genome alignments. *BMC Bioinform.* **2004**, *5*, 96–99. [CrossRef] [PubMed]

2. Hillary, W.; Lin, S.-H.; Upton, C. Base-By-Base version 2: Single nucleotide-level analysis of whole viral genome alignments. *Microb. Inform. Exp.* **2011**, *1*, 2. [CrossRef] [PubMed]

3. Stamatakis, A. RAxML version 8: A tool for phylogenetic analysis and post-analysis of large phylogenies. *Bioinformatics* **2014**, *30*, 1312–1313.[CrossRef] [PubMed]

4. Rose, T.M.; Schultz, E.R.; Henikoff, J.G.; Pietrokovski, S.; McCallum, C.M.; Henikoff, S. Consensus-degenerate hybrid oligonucleotide primers for amplification of distantly related sequences. *Nucleic Acids Res.* **1998**, *26*, 1628–1635. [CrossRef] [PubMed]

5. Nix, W.A.; Oberste, M.S.; Pallansch, M.A. Sensitive, Seminested PCR Amplification of VP1 Sequences for Direct Identification of All Enterovirus Serotypes from Original Clinical Specimens. *J. Clin. Microbiol.* **2006**, *44*, 2698–2704. [CrossRef] [PubMed]

6. Van Devanter, D.R.; Warrener, P.; Bennett, L.; Schultz, E.R.; Coulter, S.; Garber, R.L.; Rose, T.M. Detection and analysis of diverse herpesviral species by consensus primer PCR. *J. Clin. Microbiol.* **1996**, *34*, 1666–1671.

7. Børsting, C.; Hummel, R.; Schultz, E.R.; Rose, T.M.; Pedersen, M.B.; Knudsen, J.; Kristiansen, K. *Saccharomyces carlsbergensis* contains two functional genes encoding the acyl-CoA binding protein, one similar to the ACB1 gene from *S. cerevisiae* and one identical to the ACB1 gene from *S. monacensis*. *Yeast* **1997**, *13*, 1409–1421. [CrossRef]

8. Wilson, C.A.; Wong, S.; Muller, J.; Davidson, C.E.; Rose, T.M.; Burd, P. Type C retrovirus released from porcine primary peripheral blood mononuclear cells infects human cells. *J. Virol.* **1998**, *72*, 3082–3087. [PubMed]

9. Rose, T.M. CODEHOP-mediated PCR—A powerful technique for the identification and characterization of viral genomes. *Virol. J.* **2005**, *2*, 20. [CrossRef] [PubMed]

10. Philipp-Staheli, J.; Marquardt, T.; Thouless, M.E.; Bruce, A.G.; Grant, R.F.; Tsai, C.-C.; Rose, T.M. Genetic variability of the envelope gene of Type D simian retrovirus-2 (SRV-2) subtypes associated with SAIDS-related retroperitoneal fibromatosis in different macaque species. *Virol. J.* **2006**, *3*, 11. [CrossRef] [PubMed]

11. Staheli, J.P.; Boyce, R.; Kovarik, D.; Rose, T.M. CODEHOP PCR and CODEHOP PCR primer design. *Methods Mol. Biol.* **2011**, *687*, 57–73. [PubMed]

12. Chouhy, D.; Kocjan, B.J.; Staheli, J.P.; Bolatti, E.M.; Hošnjak, L.; Sagadin, M.; Giri, A.A.; Rose, T.M.; Poljak, M. Detection of novel Betapapillomaviruses and Gammapapillomaviruses in eyebrow hair follicles using a single-tube "hanging droplet" PCR assay with modified pan-PV CODEHOP primers. *J. Gen. Virol.* **2018**, *99*, 109–118. [CrossRef] [PubMed]

13. Rose, T.M.; Henikoff, J.G.; Henikoff, S. CODEHOP (COnsensus-DEgenerate Hybrid Oligonucleotide Primer) PCR primer design. *Nucleic Acids Res.* **2003**, *31*, 3763–3766. [CrossRef] [PubMed]

14. Boyce, R.; Chilana, P.; Rose, T.M. iCODEHOP: A new interactive program for designing COnsensus-DEgenerate Hybrid Oligonucleotide Primers from multiply aligned protein sequences. *Nucleic Acids Res.* **2009**, *37*, W222–W228. [CrossRef] [PubMed]

15. Altschul, S.F.; Gish, W.; Miller, W.; Myers, E.W.; Lipman, D.J. Basic local alignment search tool. *J. Mol. Biol.* **1990**, *215*, 403–410. [CrossRef]

16. Brodie, R.; Roper, R.L.; Upton, C. JDotter: A Java interface to multiple dotplots generated by dotter. *Bioinformatics* **2004**, *20*, 279–281. [CrossRef] [PubMed]

17. Sievers, F.; Higgins, D.G. Clustal Omega, Accurate Alignment of Very Large Numbers of Sequences. In *Multiple Sequence Alignment Methods*; Methods in Molecular Biology; Humana Press: Totowa, NJ, USA, 2013; Volume 1079, pp. 105–116.

18. Edgar, R.C. MUSCLE: Multiple sequence alignment with high accuracy and high throughput. *Nucleic Acids Res.* **2004**, *32*, 1792–1797. [CrossRef] [PubMed]

19. Katoh, K.; Standley, D.M. MAFFT Multiple Sequence Alignment Software Version 7: Improvements in Performance and Usability. *Mol. Biol. Evol.* **2013**, *30*, 772–780. [CrossRef] [PubMed]

20. Kazutaka Katoh, M.C.F. Adding unaligned sequences into an existing alignment using MAFFT and LAST. *Bioinformatics* **2012**, *28*, 3144–3146. [CrossRef] [PubMed]

21. Waterhouse, A.M.; Procter, J.B.; Martin, D.M.A.; Clamp, M.; Barton, G.J. Jalview Version 2—A multiple sequence alignment editor and analysis workbench. *Bioinformatics* **2009**, *25*, 1189–1191. [CrossRef] [PubMed]

22. Pervez, M.T.; Babar, M.E.; Nadeem, A.; Aslam, N.; Naveed, N.; Ahmad, S.; Muhammad, S.; Qadri, S.; Shahid, M.; Hussain, T.; et al. IVisTMSA: Interactive Visual Tools for Multiple Sequence Alignments. *Evol. Bioinform. Online* **2015**, *11*, 35–42. [CrossRef] [PubMed]
23. Larsson, A. AliView: A fast and lightweight alignment viewer and editor for large datasets. *Bioinformatics* **2014**, *30*, 3276–3278. [CrossRef] [PubMed]

Article

Proteogenomics Uncovers Critical Elements of Host Response in Bovine Soft Palate Epithelial Cells Following In Vitro Infection with Foot-And-Mouth Disease Virus

Florian Pfaff [1], Sara Hägglund [2], Martina Zoli [1], Sandra Blaise-Boisseau [3], Eve Laloy [3,4], Susanne Koethe [1], Daniela Zühlke [5], Katharina Riedel [5], Stephan Zientara [3], Labib Bakkali-Kassimi [3], Jean-François Valarcher [2], Dirk Höper [1], Martin Beer [1] and Michael Eschbaumer [1,*]

[1]   Institute of Diagnostic Virology, Friedrich-Loeffler-Institut, Federal Research Institute for Animal Health, 17493 Greifswald, Germany; Florian.Pfaff@fli.de (F.P.); martina.zoli2811@gmail.com (M.Z.); Susanne.Koethe@fli.de (S.K.); Dirk.Hoeper@fli.de (D.H.); Martin.Beer@fli.de (M.B.)
[2]   Swedish University of Agricultural Sciences, Host-pathogen interaction group, Division of Ruminant Medicine, 75007 Uppsala, Sweden; sara.hagglund@slu.se (S.H.); jean-francois.valarcher@slu.se (J.-F.V.)
[3]   Laboratoire de Santé Animale de Maisons-Alfort, UMR 1161 virologie, INRA, Ecole Nationale Vétérinaire d'Alfort, ANSES, Université Paris-Est, 94700 Maisons-Alfort, France; sandra.blaise-boisseau@anses.fr (S.B.-B.); eve.laloy@vet-alfort.fr (E.L.); stephan.zientara@vet-alfort.fr (S.Z.); Labib.BAKKALI-KASSIMI@anses.fr (L.B.-K.)
[4]   Biopôle EnvA, Ecole Nationale Vétérinaire d'Alfort, Université Paris-Est, 94700 Maisons-Alfort, France
[5]   Institute of Microbiology, Department for Microbial Physiology and Molecular Biology, University of Greifswald, 17489 Greifswald, Germany; daniela.zuehlke@uni-greifswald.de (D.Z.); riedela@uni-greifswald.de (K.R.)
*   Correspondence: Michael.Eschbaumer@fli.de; Tel.: +49-38351-7-1211

Received: 20 December 2018; Accepted: 11 January 2019; Published: 12 January 2019

Abstract: Foot-and-mouth disease (FMD) is the most devastating disease of cloven-hoofed livestock, with a crippling economic burden in endemic areas and immense costs associated with outbreaks in free countries. Foot-and-mouth disease virus (FMDV), a picornavirus, will spread rapidly in naïve populations, reaching morbidity rates of up to 100% in cattle. Even after recovery, over 50% of cattle remain subclinically infected and infectious virus can be recovered from the nasopharynx. The pathogen and host factors that contribute to FMDV persistence are currently not understood. Using for the first time primary bovine soft palate multilayers in combination with proteogenomics, we analyzed the transcriptional responses during acute and persistent FMDV infection. During the acute phase viral RNA and protein was detectable in large quantities and in response hundreds of interferon-stimulated genes (ISG) were overexpressed, mediating antiviral activity and apoptosis. Although the number of pro-apoptotic ISGs and the extent of their regulation decreased during persistence, some ISGs with antiviral activity were still highly expressed at that stage. This indicates a long-lasting but ultimately ineffective stimulation of ISGs during FMDV persistence. Furthermore, downregulation of relevant genes suggests an interference with the extracellular matrix that may contribute to the skewed virus-host equilibrium in soft palate epithelial cells.

Keywords: foot-and-mouth disease virus (FMDV); bovine soft palate; nasopharynx; transcriptomics; proteomics; bioinformatics; virus-host interaction; innate immune system; interferon-stimulated genes (ISG)

---

1. Introduction

Foot-and-mouth disease (FMD) is an acute and severe systemic vesicular disease of cloven-hoofed animals (*Artiodactyla*) with tremendous economic impact. The last FMD epizootic in the European Union in the United Kingdom, Ireland, France and the Netherlands in 2001 culminated in the slaughter of more than 6.5 million animals and an economic toll of over €5 billion [1]. The etiological agent is foot-and-mouth disease virus (FMDV), the type species of the genus *Aphthovirus* in the family *Picornaviridae* [2]. FMDV particles comprise a non-enveloped icosahedral capsid that surrounds a single-stranded positive-sense RNA genome with an approximate length of 8.4 kilobases [3]. FMD mainly affects livestock such as cattle, buffalo, pigs, goats, and sheep, but can also be transmitted to deer and wild boar. It is endemic in wild buffalo in Southern Africa [4,5]. Although more than 70 species are known to be susceptible to FMDV, its primary host seem to be buffalo and cattle, in which the disease causes very high morbidity, but only low mortality in adults [6]. During the onset of acute infection, cattle are highly febrile and small vesicles develop on the mucosal membranes of the muzzle, lips, and oral cavity, as well as on the coronary band and interdigital space, and the teats of the udder.

Usually, cattle clinically recover within 2–3 weeks if no secondary infection occurs. At this time, many have completely cleared the virus, however, about 50% of animals may remain subclinically infected for up to three years, depending on the species [7]. The World Organisation for Animal Health (OIE) defines an animal from which infectious FMDV can be recovered by probang sampling later than 28 days post infection (dpi) as persistently infected or a so-called "carrier" [7]. The fear of contagion from carrier animals has severe consequences for trade in live animals and animal products [8]. In vivo studies have shown that in cattle more than 50% of animals will become persistently infected [9], even if they had been vaccinated against FMDV and did not develop clinical disease [10–12]. The exact anatomical sites of persistence are still debated, but different tissues of the upper respiratory tract including the nasopharynx have been suggested. A recent study of the tissue-specific localization of FMDV in persistently infected steers identified the contiguous epithelia of the dorsal soft palate (SP) and the dorsal nasopharynx as the most likely sites of persistence [10]. Evidence for FMDV persistence in lymph nodes and germinal centers was also put forward although no viral replication could be detected [13].

The cellular factors that promote establishment of FMDV persistence and the viral strategies of immune avoidance in the bovine nasopharynx remain currently poorly understood, making it impossible to predict which animals will develop into carriers and which will clear the virus. Previous in vitro work that aimed to decipher cellular responses during persistent FMDV infection hardly reflected persistence in vivo as the used models were either based on immortalized non-bovine cell lines, such as hamster kidney (BHK) cells [14], or bovine cell lines, e.g., bovine kidney (MDBK, EBK) cells [15,16], that are not from the primary site of persistence. O'Donnell et al. [17] used a persistently infected bovine pharynx cell line to examine the gene expression of selected cellular cytokines by RT-qPCR, showing differences in the expression of key antiviral cytokines between acute and persistent infection.

The rationale of this study was therefore to use a novel air–liquid interphase cell culture model based on primary SP cells from cattle [18] that closely resembles the situation in vivo, together with an proteogenomics approach that combined transcriptomics by high-throughput sequencing, RT-qPCR, proteomics and bioinformatics. Our results provide detailed insights into the transcriptional responses of the SP in reaction to acute FMDV infection and reveal long-lasting changes of gene activation throughout persistence. The identified pathways and genes may give rise to further investigations leading to early detection of persistence in cattle and novel vaccines that prevent the carrier state.

2. Materials and Methods

A schematic visualization of the sample processing and analysis workflow can be found in the Supplementary Figure S1.

2.1. Ethics Statement

The tissues used for the study were collected from animals slaughtered for food production. The animals were being processed as part of the normal work of the abattoir, therefore no ethics approval was required.

2.2. Bovine Epithelial Cultures from Soft Palate

Multilayers of bovine dorsal soft palate cells were propagated at the air–liquid interface for 5 weeks before FMDV infection, as described previously [18]. Briefly, bovine dorsal soft palate tissue, collected immediately after slaughter, was dissected and digested at 4 °C overnight in incubation medium supplemented with protease XIV (Sigma-Aldrich, St. Louis, MO, USA). Epithelial cells were thereafter scraped off the underlying tissue, filtered, and incubated in cell culture flasks for 4 h at 37 °C and 5% CO_2. Cells that did not adhere to the plastic were centrifuged at $200\times g$ for 10 min at room temperature, frozen, thawed, and propagated for three to five passages in cell culture flasks before being seeded in 12 mm diameter Corning®Transwell-COL collagen-coated PTFE membrane inserts with 3.0 μm pores (Sigma-Aldrich). The cell culture medium was removed from the upper compartment after five days of culture and changed in the lower compartment every two or three days. The average number of cells that constituted the upper layer of the multilayer was estimated at 750,000.

2.3. Experimental Design and FMDV Infection

After 5 weeks of culture on inserts without passage, cells were infected with a twice-plaque-purified viral clone (FMDV O Clone 2.2, "Cl 2.2") derived from the O/FRA/1/2001 strain that was further propagated on BHK-21 cells (four passages) [16], or negative cell lysate, as described previously [18]. Two experiments (experiment 1 and 2) were performed with SP cells that originated from two different animals (a male and female, respectively) and that were infected at a multiplicity of infection (MOI) of 0.01 (compare Supplementary Table S1). Briefly, the inserts were incubated for one hour with 500 μL of clarified cell lysate from infected or uninfected cell cultures. Following infection and thereafter at a maximum interval of 3 days, the upper compartments were washed with 500 μL cell culture medium containing 10% FCS and, similarly, but only from 2 dpi, the medium in the lower compartments was changed. For each experiment, at days 0, 1 and 28, SP cells from 2 inserts were lysed with 750 μL of TRIzol Reagent (Life Technologies, Carlsbad, CA, USA) and frozen separately at −80 °C for transcriptomic and proteomic analyses.

2.4. RNA and Protein Isolation

For isolation of high-quality total RNA and proteins, 150 μL of trichlormethane (Carl Roth, Karlsruhe, Germany) was added to the lysed cells in 750 μL TRIzol Reagent and the mixture was centrifuged in order to separate the RNA-containing aqueous phase from the DNA- and protein-containing organic phase. The aqueous phase was then mixed with an equal amount of 100% ethanol (Carl Roth) and total RNA was extracted using the RNeasy Mini Kit (Qiagen, Hilden, Germany) with on-column DNase digestion with the RNase-Free DNase Set (Qiagen), following the manufacturer's instructions. The quantity and quality of total RNA was subsequently analyzed using a NanoDrop 1000 spectrophotometer (Peqlab, Erlangen, Germany) and RNA 6000 Pico chips on an Agilent 2100 Bioanalyzer (Agilent Technologies, Böblingen, Germany). All samples were checked for contamination using the 260/280 and 260/230 nm ratios, as well as for RNA degradation using the RNA Integrity Number (RIN).

Polyadenylated mRNA was subsequently isolated from 1–3 μg of high-quality total RNA using the Dynabeads mRNA DIRECT Micro kit (Invitrogen, Carlsbad, CA, USA) following the manufacturer's instructions. Prior to isolation, the ERCC ExFold RNA Spike-In mix 1 (Invitrogen) was supplemented and used as an internal control for all following steps. The quality of the mRNA and the extent of ribosomal RNA contamination was assessed using RNA 6000 Pico chips on the Agilent 2100 Bioanalyzer (Agilent Technologies).

Proteins were extracted from the organic TRIzol phase following the manufacturer's instructions. Briefly, 0.3 mL of 100% ethanol (Carl Roth) were added and DNA was pelleted by centrifugation. To the supernatant, 1.5 mL of isopropanol (Carl Roth) were added and proteins were pelleted by centrifugation. Protein pellets were washed three times with 0.3 M guanidine hydrochloride (Carl Roth) in 95% ethanol (Carl Roth). After a final washing step using 95% ethanol, proteins were air-dried and the pellet was resuspended in freshly prepared 1% SDS (Carl Roth) by ultrasonication. Quality and quantity of the isolated proteins were checked using a 12% polyacrylamide gel (SDS-PAGE) and a colorimetric bicinchoninic acid (BCA) assay.

2.5. Library Preparation and Sequencing

For preparation of whole-transcriptome libraries the Ion Total RNA-Seq Kit v2 (Life Technologies) was used, following the manufacturer's instructions. Briefly, between 1 and 100 ng of the mRNA containing the ERCC spike-in control was treated with RNase III at 37 °C for 10 min. The fragmented mRNA was subsequently purified using the Magnetic Bead Cleanup Module (Life Technologies) and the resulting size distribution was assessed with the Agilent 2100 Bioanalyzer as described above. After hybridization and ligation of appropriate adapters, the fragmented mRNA was reverse transcribed into cDNA using SuperScript III enzyme. The cDNA was purified as described above and amplified for 14 cycles using Platinum PCR SuperMix High Fidelity along with appropriate primers for the generation of barcoded libraries. The resulting libraries were again purified using the method described above and the size distribution was assessed using the Agilent 2100 Bioanalyzer together with the DNA 7500 kit and chip (Agilent Technologies). All libraries were subsequently quantified using the KAPA Library Quantification Kit Ion Torrent (Kapa Biosystems, Wilmington, MA, USA) on a CFX96 Real-Time PCR Detection System (Bio-Rad Laboratories, München, Germany) and pooled at an equimolar ratio. For sequencing an Ion S5XL sequencing system (Life Technologies) along with the Ion 540 OT2 and Chip kit (Life Technologies) for the generation of up to 200 bp reads was used. Each library was sequenced in at least two independent sequencing runs.

2.6. Statistical Analysis of Differential Gene Expression

In order to detect problems and biases during mRNA isolation and library preparation, we used the ERCC_Analysis Plugin (version 5.8.0.1) provided in the Torrent Suite software (version 5.8.0). Only forward strand reads were selected for mapping and the minimum transcript count was set to 50. The raw reads from each sequencing library were quality checked using FastQC (version 0.11.7; Babraham Institute) with a focus on the read length distribution and adapter contamination. In order to quantify the expression of known bovine transcripts in these datasets we used Salmon (version 0.9.1) that uses a lightweight alignment method (quasi-mapping) for rapid transcript abundance estimation [19]. Briefly, we obtained the transcript reference of cattle (GCF000003055.6 Bos Taurus UMD 3.1.1) from NCBI and selected only transcripts that were featured as mRNAs, non-coding RNAs and miscellaneous RNAs (for details see Supplementary Table S2). A Salmon index was created using the option for a perfect hash, rather than a dense hash. Each sequencing library was then used as input for the major Salmon function "quant" using appropriate options for the library type (stranded single-end protocol with reads coming from the forward strand) and the mean read length. The number of bootstraps was set to 100 replicates. Subsequently a "tx2gene" table was prepared using the accessions of each RNA transcript from the aforementioned reference in the first column and the corresponding gene symbol in the second column (see also Supplementary Table S3). Using this table, the transcript abundancy datasets from Salmon "quant" were imported into R workspace (version 3.4.1; [20]) using the "tximport" package (version 1.6.0; [21]) as implemented in the Bioconductor library. For handling and manipulating R scripts, the software RStudio (version 1.0.153) was used. Technical replicates of samples (multiple sequencing of the same library from a single sample) were combined and the datasets were pre-filtered using only genes were more than four samples had raw gene counts greater than or equal to 100. In order to check the dataset for influence of treatment (infection, time and animal) and repeatability (replicates of same treatment) a principal components analysis (PCA) as well as a heatmap clustering was conducted using the regularized log transformed (rlog) read counts. DESeq2 (version 1.18.1; [22]) was then used to identify differentially expressed genes between the treatments based on the negative binomial distribution. The resulting *p*-values were adjusted with the Benjamini–Hochberg procedure and only genes with an adjusted *p*-value below 0.001 and an absolute fold change of >1 were considered significant. The logarithmic fold changes were further shrunken as recommended and described by Love et al. 2014 [22] in order to account for genes with low read counts. Significant differently expressed genes were annotated using the AnnotationDbi package (version 1.40.0) and used for further pathway enrichment analysis. Briefly, gene sets were analyzed using the "enrichPathway" function of the ReactomePA package (version 1.22.0; [23]) and the "enrichKEGG" function of the clusterProfiler package (version 3.6.0; [24]).

2.7. Quantitative Reverse Transcription PCR (RT-qPCR)

In order to confirm the results from the RNA sequencing experiment and to include samples from additional time points, a subset of six target genes (*ANKRD1, CASP7, IDO1, IFIH1, NCAM1, OAS2*) and two reference genes (*ACTB, GAPDH*) was selected for quantitative reverse transcription PCR (RT-qPCR) analysis. For each gene, two intron-spanning primer sets were designed using Primer3 (version 0.4.0; [25]) (for primer sequences see Supplementary Table S4). For RT-qPCR the QuantiTect Probe RT-PCR Kit (Qiagen) was used together with LightCycler 480 ResoLight Dye (Roche, Mannheim, Germany) according to the manufacturers' instructions. The following temperature profile was used on a Bio-Rad CFX96: 50 °C for 30 min, 95 °C for 15 min and 45 cycles of 94 °C for 15 s and 60 °C for 1 min. After each cycle, the fluorescence in the SYBR channel was detected and threshold cycle (Ct) values were deduced after each run. For each newly designed primer set the PCR efficiency was determined using an appropriate dilution series from an independent bovine RNA control. Using the ΔΔCt method, Ct values of target genes were first normalized by subtracting the Ct value of references genes (ΔCt = $Ct_{Target} - Ct_{Reference}$). Mean and standard deviation of ΔCt were calculated from two technical and two biological replicates for each treatment group. Subsequently, the ΔΔCt was calculated by subtracting the ΔCT of the control group from the ΔCt of the treatment group at different time points (ΔΔCt = $ΔCt_{Control} - ΔCt_{Treatment}$) and used for calculation of the fold change of each gene ($2^{-ΔΔCT}$).

2.8. Protein Identification and Quantification, and Statistical Analysis of Differential Protein Expression

Thirty μg of the extracted proteins were separated by SDS-PAGE, stained with Colloidal Coomassie Brilliant Blue G-250 and afterwards lanes were cut into ten equidistant pieces. In-gel digestion using trypsin and purification of tryptic peptides using Ziptips (C18, Millipore) prior to MS analysis were done as described previously [26]. LC-MS/MS were conducted using an EASY-nLC II coupled to a LTQ Orbitrap-Velos mass spectrometer (Thermo Fisher Scientific, Waltham, MA, USA). Peptides were separated at a constant flow rate of 300 nL/min using a binary 76 min gradient from 5% B to 75% B 99.9% ACN, 0.1% acetic acid). Survey scans in the Orbitrap were recorded with a resolution of 60.000 in a m/z range of 300–1700. The 20 most intense peaks per scan cycle were selected for CID fragmentation in the LTQ. Ions with unknown charge state, as well as singly charged ions were excluded from fragmentation. Dynamic exclusion of precursor ions for 30 s was enabled. Internal calibration (lock mass 445.120025) was enabled as well. For protein identification, resulting spectra were searched against a database containing sequences of *Bos taurus* including reverse sequences and common laboratory contaminants (44,376 entries). Database searches using Sorcerer-SEQUEST (version v.27, rev.11; Sage-N Research, Inc., Milpitas, CA, USA) and Scaffold (version v.4.8.4; Proteome Software, Portland, OR, USA) were done as described earlier [27].

In order to identify differentially expressed proteins we imported the raw NSAF values into R workspace and created an ExpressionSet using the Biobase package (version 2.38.0; [28]). We then used the "Power Law Global Error Model" (PLGEM) as implemented in the "plgem" package (version 1.50.0; [29]) and first fitted it to the dataset using default settings and "FMDV infection" (samples from 24 h post infection (hpi) and 28 dpi) as fitting condition. After computation of observed and resampled signal-to-noise ratios, *p*-values for each detected protein were calculated.

2.9. Data Availability

The raw sequencing data along with deduced Salmon read count tables and substantial metadata are available at ArrayExpress (http://www.ebi.ac.uk/arrayexpress) under the accession number E-MTAB-7605. The mass spectrometry proteomics data have been deposited to the ProteomeXchange Consortium (http://www.proteomexchange.org) via the PRIDE [30] partner repository with the dataset identifier PXD012242

3. Results

3.1. A Cell Culture Model for FMDV Persistence

The establishment and full characterization of the primary SP cell culture model are presented by Hägglund et al. [18]. In summary, the primary SP cells formed into multilayers and showed typical features of stratified squamous epithelia, such as tight junctions and impermeability to cell culture media. The cultures were inoculated either with a viral clone derived from the O/FRA/1/2001 strain or negative cell lysate for control. Infection with FMDV at low MOI resulted in limited cytopathic effect, high viral loads and presence of detectable viral antigen. After 28 dpi, the multilayers were still intact, but FMDV antigen and genome remained detectable at very low levels and viable virus could be isolated—giving a clear indication of FMDV persistence.

3.2. RNA Sequencing and Exploratory Data Analysis

The polyadenylated RNA fraction from 21 primary bovine SP cell culture samples was sequenced. From these, 10 and 11 samples originated from a female and male bovine, respectively. Cells for analysis were harvested immediately before inoculation (0 hpi), 24 hpi and 28 dpi (Figure 1A). For each time point (0 hpi, 24 hpi, 28 dpi) and treatment (FMDV, control), a minimum of four biological replicates (two from each animal) was sequenced. The number of reads for each sample ranged from 21.8 to 28.0 million, with an average of 24.7 million. In total, 518.4 million reads were included in the analysis (Supplementary Table S1).

These reads were assigned to a cattle transcript reference and the raw gene count data was transformed with respect to library size and transcript length using an appropriate model (Supplementary Figure S2). A principal components analysis (PCA) based on the normalized gene counts revealed that the samples from non-infected cultures remained closely together independent of the time in culture, while FMDV-infected samples showed a clear time-dependent grouping (Figure 1B). Accordingly, the first principal component (50% of variance) was assumed to represent the transcriptional differences between the samples caused by FMDV infection. Furthermore, transcriptional differences between both animals were clearly visible and represented by principal component 2 (29% of variance). The PCA showed no batch-to-batch effects and biological replicates from the same animal were grouped tightly.

Unsupervised cluster analysis of the 45 most variable gene transcripts clearly separated the FMDV-infected samples from the non-infected controls (Figure 1C). In detail, the infected samples showed a strong positive deviation from the per gene mean count (Figure 1C, green and blue transcript cluster). Interestingly, samples from the acute and persistent phases of infection were clearly separated, because the activation of the aforementioned genes was reduced at 28 dpi (Figure 1C, blue cluster). As in the PCA, the non-infected samples were divided by the differential expression of another set of transcripts (Figure 1C, red transcript cluster) and grouped according to the donor animal they originated from. In summary, the explorative data analysis showed that the FMDV infection has a clear effect on gene expression that is distinct from the animal-dependent effects.

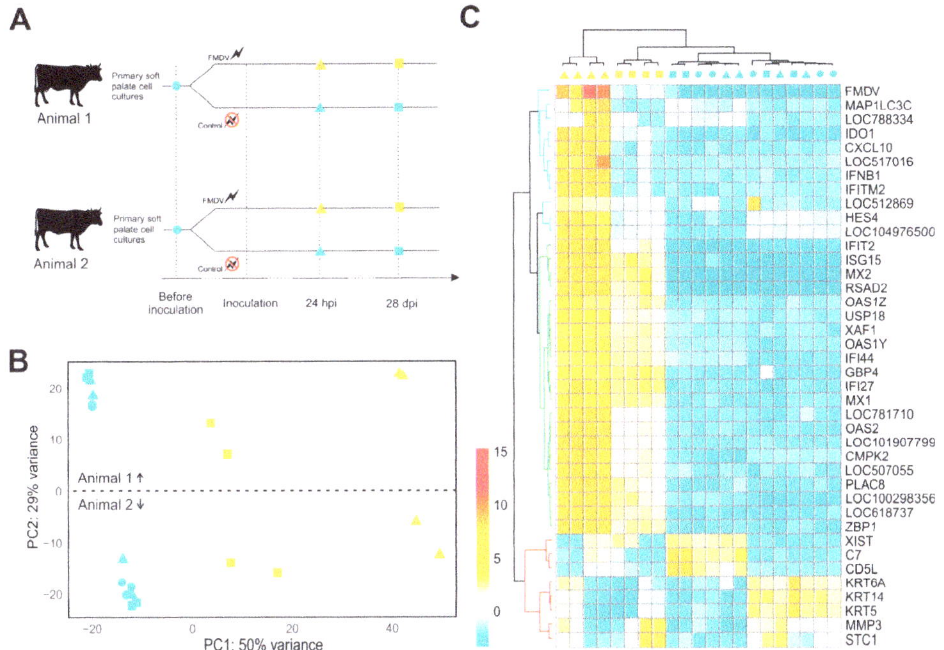

Figure 1. Experimental setup of RNA sequencing and exploratory data analysis. (**A**) Primary soft palate (SP) cell cultures were obtained from two animals. Baseline samples for RNA sequencing were collected immediately before inoculation (circle). Subsequently, the cell cultures were inoculated with foot-and-mouth disease virus (FMDV) (orange symbols) or mock-infected for use as controls (blue symbols). Cells were then harvested for sequencing at 24 h post infection (hpi) (triangles) or 28 days post infection (dpi) (squares), representing acute and persistent infection, respectively. (**B**) Principal components analysis based on normalized gene counts of the 1000 most variable genes. (**C**) The variance of normalized gene counts was calculated for each gene and the 45 genes with the highest variance were selected and visualized in a heat map. The color of the cells indicates the difference from the mean normalized gene count of the corresponding gene. Samples and genes were clustered according to these differences (trees). The transcript cluster highlighted in green is similarly activated by acute and persistent FMDV infection, while the blue cluster highlights transcripts that are only active during the acute phase of infection. The red cluster comprises transcripts whose abundance differs between donor animals used in this experiment.

3.3. Differential Expression during Acute and Persistent FMDV Infection and associated Pathways

In order to confirm the trends from the explorative data analysis and to address the transcriptional changes during acute and persistent infection in more detail, a differential expression analysis was conducted. The transcriptional response of cells from bovine SP tissue to FMDV infection was assessed at 24 hpi and 28 dpi, in contrast to non-infected controls from the same time points (Figure 2). A total of 312 and 73 gene transcripts were differentially expressed at 24 hpi and 28 dpi, respectively. With 305/312 (97.7%) for 24 hpi and 65/73 (89%) for 28 dpi, the majority of differentially expressed transcripts were up-regulated, while only 8/312 (2.5%) and 8/73 (11%) were down-regulated (Figure 2A,B). From a total of 324 differentially expressed transcripts at both time points, 249 and 10 transcripts were solely regulated during either acute or persistent phase, respectively. The expression of 63 transcripts was significantly up-regulated during both stages of infection (Figure 2C). The log2 fold change of these transcripts was much lower at 28 dpi when compared to 24 hpi (Figure 2D), with highest differences observed for *IFIT2*, LOC101907799 and *CMPK2*. In contrast, the log2 fold change for *IFI27*, *PARP9*, *IFI6*, and *OAS1Z* was comparable or nearly equal during both phases of infection. *PLAC8* was slightly stronger regulated at 28 dpi than at 24 hpi. A full list of differentially expressed transcripts can be found in the supplementary material (Supplementary Table S5).

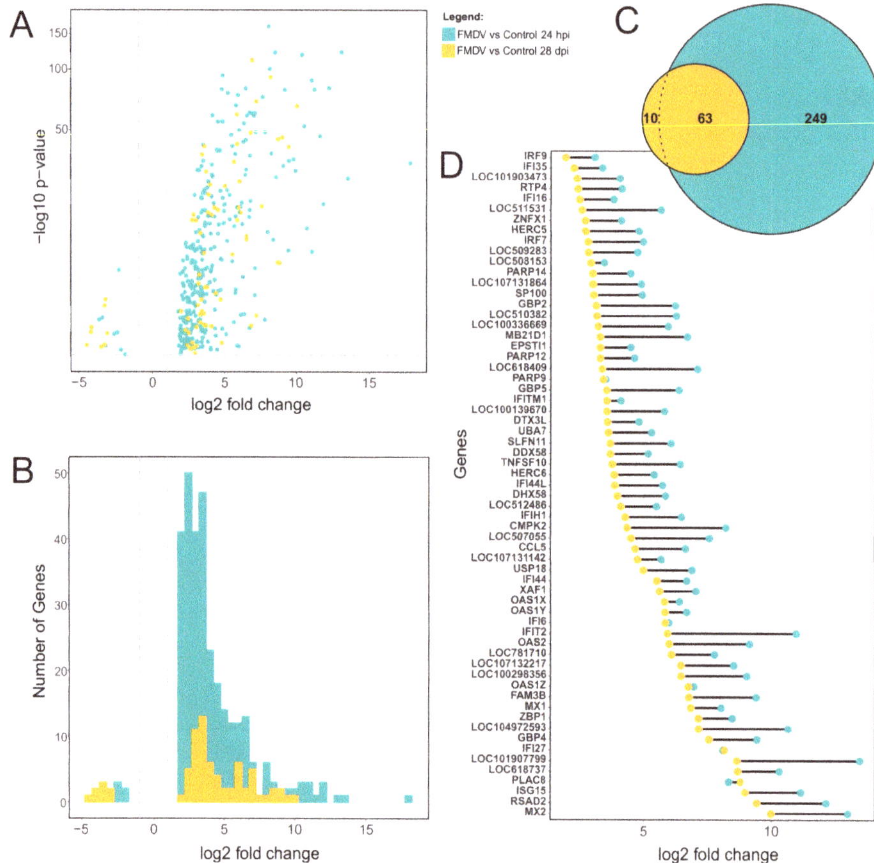

Figure 2. Differential gene expression of SP cells during FMDV infection. (**A**) Volcano plot showing the log2 fold change (*x*-axis) and the adjusted *p*-value (*y*-axis) for all differentially expressed genes during the acute (24 hpi, green) and persistent (28 dpi, orange) phase of FMDV infection. The log2 fold change and adjusted *p*-value are calculated relative to a non-infected control from the same time point. The grey dotted lines indicate the cutoff values: adjusted *p*-value < 0.001 and |log2 fold change| > 1. (**B**) The histogram summarizes the number of genes that have a certain log2 fold change. (**C**) The total number of differentially expressed genes is visualized in a Venn diagram. The overlap indicates the 63 genes that are differentially expressed during both phases of infection. (**D**) The log2 fold change of these 63 genes is compared for acute and persistent infection.

The significantly differently expressed genes during acute and persistent infection were further matched to specific metabolic and signaling pathways using the Reactome database [31] (Figure 3). During both infection stages, most of the regulated genes are associated with interferon signaling, in particular interferon α, β and γ signaling (Figure 3A and Table 1). The induction of these interferons appears to be mediated by the DDX58/IFIH1 pathway. Accordingly, a number of interferon-stimulated genes (ISG), such as the ubiquitin-like family member *ISG15*, that induce strong antiviral mechanisms of the innate immune system are highly activated. During acute infection, transcripts involved in the interleukin-1 (*IL1*) and -10 signaling pathways as well as the programmed cell death (apoptosis) pathway were enriched. However, it is of note that the *IL1* and *IL10* genes were not significantly differently expressed. The MHC I antigen presentation pathway, a part of both the innate and adaptive immune system, was also enriched at 24 hpi. During persistence, most of the genes belonging to the interleukin-1, -10, apoptosis and MHC I antigen presentation pathways were not significantly differentially expressed in comparison to control samples.

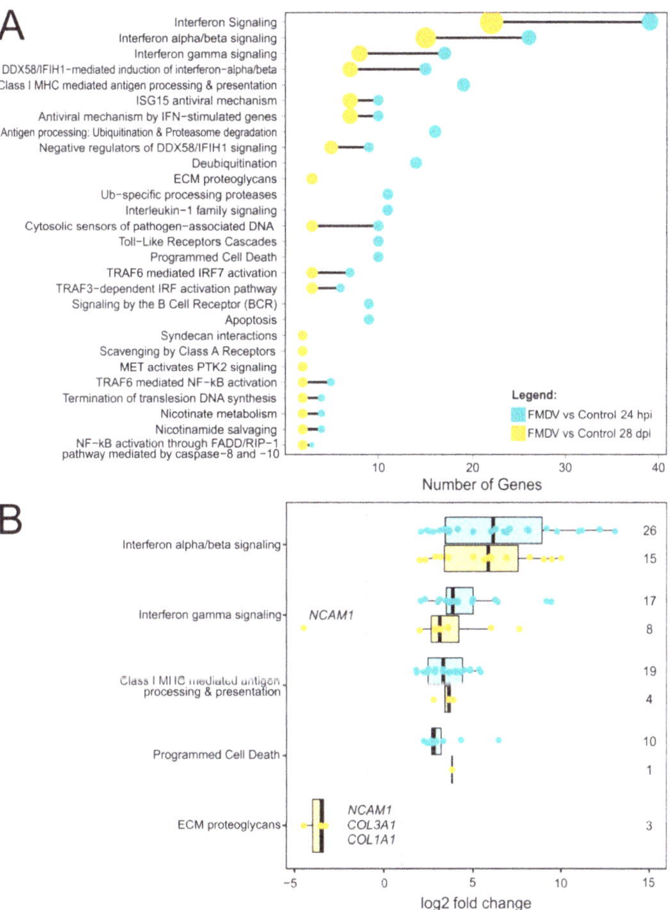

Figure 3. Pathway enrichment analyses of significantly differentially expressed genes during acute and persistent FMDV infection of bovine soft palate cells. (**A**) The number of genes that contribute to a significantly enriched Reactome pathway is shown for 24 hpi (blue dots) and 28 dpi (orange dots). Black lines connect pathways that are enriched in both datasets and the size of the circles represents the ratio of genes found for a certain pathway to the overall gene number. (**B**) The log2 fold change of genes that contribute to selected enriched Reactome pathways are highlighted (24 hpi—blue dots) and 28 dpi—orange dots). The grey dotted lines indicate a |log2 fold change| > 1. The numbers on the right correspond to the number of genes associated with each pathway.

Table 1. Selected enriched pathways and corresponding genes during acute and persistent FMDV infection.

Metabolic Complex	Pathway	24 hpi	28 dpi
Innate immune system	DDX58/IFIH1-mediated induction of interferon-alpha/beta	*DDX58, DHX58, HERC5, IFIH1, IFNB1, IRF1, IRF3, IRF7, ISG15, NFKBIA, NFKBIB, NLRC5, TNFAIP3, TRIM25, UBA7*	*DDX58, DHX58, HERC5, IFIH1, IRF7, ISG15, UBA7*
Cytokine signaling in immune system	Interferon alpha, beta signaling	*ADAR, GBP2, IFI27, IFI35, IFI6, IFIT2, IFIT3, IFITM1, IFITM2, IFNB1, IRF1, IRF3, IRF5, IRF7, IRF9, ISG15, MX1, MX2, OAS2, PSMB8, RNASEL, RSAD2, SOCS1, STAT2, USP18, XAF1*	*GBP2, IFI27, IFI35, IFI6, IFIT2, IFITM1, IRF7, IRF9, ISG15, MX1, MX2, OAS2, RSAD2, USP18, XAF1*
	Interferon gamma signaling	*GBP2, IFI27, IFI35, IFI6, IFIT2, IFITM1, IRF7, IRF9, ISG15, MX1, MX2, OAS2, RSAD2, USP18, XAF1*	*GBP2, GBP4, GBP5, IRF7, IRF9, NCAM1, OAS2, SP100*
	Interleukin-1 family signaling	*IL18, IL18BP, MAP3K8, NFKBIA, NFKBIB, PELI1, PSMB10, PSMB8, PSMB9, PSME2, PSMF1*	
	Interleukin-10 signaling	*CCL2, CCL5, CXCL10, CXCL2, CXCL8, IL18, IL6*	*CCL5*
	Antiviral mechanism by IFN-stimulated genes/ISG15 antiviral mechanism	*DDX58, EIF2AK2, HERC5, IRF3, ISG15, MX1, MX2, TRIM25, UBA7, USP18*	*DDX58, HERC5, ISG15, MX1, MX2, UBA7, USP18*
Adaptive immune system	Class I MHC mediated antigen processing and presentation	*AREL1, CTSS, DTX3L, ERAP2, HECTD2, HERC5, HERC6, PSMB10, PSMB8, PSMB9, PSME2, PSMF1, RBCK1, RNF114, RNF19B, SOCS1, TAP1, TRIM21, UBA7*	*DTX3L, HERC5, HERC6, UBA7*
	Activation of NF-kappaB in B cells	*NFKBIA, NFKBIB, NFKBIE, PSMB10, PSMB8, PSMB9, PSME2, PSMF1*	
Programmed cell death	Programmed cell death	*CASP7, CFLAR, PMAIP1, PSMB10, PSMB8, PSMB9, PSME2, PSMF1, RIPK3, TNFSF10*	*TNFSF10*

All of the genes contributing to the aforementioned pathways were up-regulated during the acute and persistent phases of infection, with a single exception for the nuclear cell adhesion molecule 1 (*NCAM1*) that is related to interferon γ signaling and was down-regulated during persistence (Figure 3B, and Table 2). In addition to *NCAM1*, other proteins that are associated with extracellular membrane (ECM) pathways, such as type I and III collagens (*COL1A1* and *COL3A1*) were also down-regulated during persistence.

Table 2. Genes that are differentially expressed only during the persistent phase of FMDV infection.

Gene	Description	LFC[†]	Adjusted *p*-value	Enriched DAVID Terms
NCAM1	neural cell adhesion molecule 1	−4.48	3.35×10^{-4}	signal peptide, secreted
ANKRD1	ankyrin repeat domain 1	−4.19	6.61×10^{-5}	positive regulation of apoptotic process
SFRP2	secreted frizzled related protein 2	−4.16	3.31×10^{-5}	signal peptide, secreted, positive regulation of apoptotic process
COL1A1	collagen type I alpha 1 chain	−3.50	4.16×10^{-4}	signal peptide, secreted
COL3A1	collagen type III alpha 1 chain	−3.28	3.23×10^{-4}	signal peptide, secreted
MYLK	myosin light chain kinase	−3.26	9.61×10^{-7}	
HTRA3	HtrA serine peptidase 3	−3.21	2.93×10^{-7}	signal peptide, secreted
ALDH1A2	aldehyde dehydrogenase 1 family member A2	−3.09	7.83×10^{-5}	positive regulation of apoptotic process
LY6E	lymphocyte antigen 6 family member E	3.45	2.59×10^{-6}	signal peptide
FBP1	fructose-bisphosphatase 1	4.09	2.94×10^{-5}	

†: log2 fold change.

3.4. RT-qPCR and Quantitative Proteomics

Results from the differential gene expression analysis based on RNA sequencing were confirmed using RT-qPCR. A panel of two reference (*ACTB, GAPDH*) and six target genes (*OAS2, IFIH1, NCAM1, ANKRD1, IDO1, CASP7*) were selected for analysis. These genes were chosen according to their apparent regulation: *OAS2, IFIH1* and *IDO1* were up-regulated during acute and persistent infection, while *NCAM1* and *ANKRD1* were only down-regulated during persistence. *CASP7* was only up-regulated during the acute phase of FMDV infection. The selected genes were also involved in relevant signaling pathways: *IFIH1* is part of the interferon induction cascade, while *OAS2* is an interferon-induced gene with antiviral activity. *CASP7* plays a key role in apoptosis and *IDO1* encodes a metabolically active protein that supports the immune system. *ANKRD1* and *NCAM1* were chosen as interesting target genes that were only regulated during persistence. The RT-qPCR analysis included samples from additional time points (2, 24, and 48 hpi; 7 and 28 dpi) for which log2 fold changes were then calculated using the delta-delta Ct method. For each time point and cell type four biological replicates (two replicates per animal) were used for analysis with the exception for 48 hpi, from which only two replicates derived from a single animal were available. The log2 fold-changes obtained by RT-qPCR of SP cells at the time points 24 hpi and 28 dpi confirm the results of the RNA sequencing (Figure 4A). All of the analyzed genes showed a time-dependent regulation during infection with FMDV (Figure 4B). The genes *IDO1*, *OAS2*, and *IFIH1* were strongly up-regulated from 2 to 24 hpi and were still up-regulated after 28 dpi. In contrast, *CASP7* was only up-regulated during acute infection. *ANKRD1* and *NCAM1* were down regulated as early as 48 hpi and throughout 28 dpi.

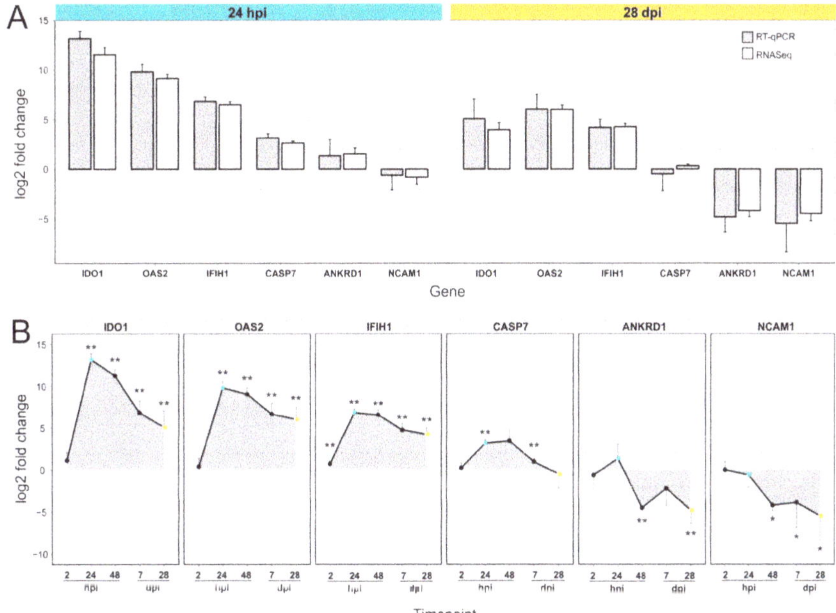

Figure 4. Expression analysis for selected genes by RT-qPCR at different time points. (**A**) Gene expression in FMDV infected soft palate cells compared to uninfected controls, at 24 hpi (blue panel) and 28 dpi (orange panel). The change in gene expression was analyzed for six selected target genes and two reference genes (not shown) using RT-qPCR. Log2 fold changes were calculated in comparison to a non-infected control from the same time points. Grey and white bars indicate the log2 fold change of gene expression using results from RT-qPCR and RNA sequencing (RNASeq), respectively. The grey dotted lines indicate a | log2 fold change | > 1. (**B**) The gene expression of the six selected genes was traced in a time course 2, 24, and 48 hpi; and 7 and 28 dpi. Statistically significant changes in comparison to control samples are highlighted with asterisks (*p*-value < 0.05: *; *p*-value < 0.01: **). Time points 24 hpi and 28 dpi are illustrated in blue and orange, respectively.

Additionally, a GeLC-MS/MS-based comparative proteome analysis was conducted to confirm the results of the RNA sequencing on the protein level, using two biological replicates for each time point and treatment from animal 2. Proteins with significantly different abundance in both experiments were identified and quantified by spectral-counting (Table 3). At the given thresholds, 11 and 8 proteins were differentially expressed when comparing FMDV-infected and control samples at 24 hpi and

28 dpi, respectively. Of these, 6 were differentially expressed at both time points. The Normalized Spectral Abundance Factor (NSAF), measured as the proportion of a single protein in relation to the detected proteins overall revealed an accumulation of the proteins *HERC6*, *IFI44*, *IFI44L*, *ISG15*, *MX1*, *MX2*, *OAS1X*, and *OAS1Y* in cells during persistent infection. In contrast, the proportion of *ATAD1*, *IFIT1*, and *IFIT2* to all other detected proteins was higher during acute than in persistent infection. FMDV polyprotein, *IFIT3* and *RSAD2* were not detected at the protein level at 28 dpi.

Table 3. Proteins with significantly different abundance at 24 hpi or 28 dpi. Only proteins that were also differentially expressed in the RNA sequencing experiment at the given time point are shown.

Protein	Description	24 hpi Mean NSAF	28 dpi Mean NSAF	$\text{NSAF}_{28\,\text{dpi}}/\text{NSAF}_{24\,\text{hpi}}$
ATAD1	ATPase Family, AAA Domain Containing 1	2.50×10^{-4} **	1.56×10^{-4}	0.6
CAG23917.14	FMDV polyprotein	3.16×10^{-4} **	0 [†]	n.a.
HERC6	HECT And RLD Domain Containing E3 Ubiquitin Protein Ligase Family Member 6	2.46×10^{-5}	2.74×10^{-4} **	11.2
IFI44	Interferon Induced Protein 44	1.12×10^{-4} *	4.58×10^{-4} **	4.1
IFI44L	Interferon Induced Protein 44 like	8.32×10^{-5} *	6.98×10^{-4} **	8.4
IFIT1	Interferon-Induced Protein with Tetratricopeptide Repeats 1	5.41×10^{-4} **	7.62×10^{-5} *	0.1
IFIT2	Interferon Induced Protein With Tetratricopeptide Repeats 2	2.31×10^{-4} **	9.34×10^{-5} *	0.4
IFIT3	Interferon Induced Protein With Tetratricopeptide Repeats 3	3.78×10^{-4} **	0 [†]	n.a.
ISG15	ISG15 Ubiquitin-Like Modifier	3.53×10^{-3} **	7.52×10^{-3} **	2.1
MX1	MX Dynamin Like GTPase 1	8.42×10^{-4} **	2.27×10^{-3} **	2.7
MX2	MX Dynamin Like GTPase 2	2.44×10^{-4} **	5.04×10^{-4} **	2.1
OAS1X	2′-5′-Oligoadenylate Synthetase 1 X	2.47×10^{-4} **	4.85×10^{-4} **	2.0
OAS1Y	2′-5′-Oligoadenylate Synthetase 1 Y	2.31×10^{-4} **	3.68×10^{-4} **	1.6
RSAD2	Radical S-Adenosyl Methionine Domain Containing 2	5.69×10^{-4} **	0 [†]	n.a.

*: p-value < 0.01; **: p-value < 0.001; [†]: not significant; n.a. not applicable.

4. Discussion

The failure of the host response to clear virus from the nasopharynx is one of the most important features of FMDV infection, and has been studied previously using monolayers of primary cells and non-primary cell lines from hamsters (BHK-21) [14], swine (SK6) [32], and cattle (EBK, MDBK and pharynx cells) [15–17]. These cells needed regular cell culture passage and were, except for the pharynx cells, not from anatomical locations relevant to natural FMDV infection and persistence, which resulted in a suboptimal reflection of the in vivo situation.

In this study we used multilayers of primary bovine dorsal soft palate (SP) epithelial cells in an air–liquid interface cell culture model [18], as it has been shown that this tissue, along with the dorsal nasopharynx, is the most likely site of primary FMDV infection and persistence in cattle [10,33]. This model mimics several properties of the squamous epithelium that are observed in vivo and allows detailed studies of the target cells of primary FMDV infection. Specialized immune cells, such as B cells, T cells, and natural killer cells, which contribute to host responses in vivo, were depleted by the establishment of the cell culture in air [18], leading to improved standardization and a focus on the primary target cell. In vivo, neutralizing antibodies secreted by B cells are essential for resolving systemic FMDV infection, but have no effect on persistent infection in the nasopharynx [9]. Recent studies of gene expression in nasopharyngeal tissues of cattle have suggested an important role of cytotoxic T cells in the clearance of persistent FMDV infection [34,35]. The transcriptional changes analyzed in the present study, however, are limited to the response of SP epithelial cells themselves to infection with FMDV.

This response was analyzed using RNA sequencing, RT-qPCR and quantitative proteomics. Based on the results from exploratory RNA sequencing alone, infection with FMDV resulted in detectable changes in the transcriptome of the cells and the induced changes were distinct during the acute and persistent phases of infection. Despite that only a small proportion of cells were persistently infected, these influenced the transcriptome significantly. The non-infected controls did not show any comparable, time-dependent changes, indicating stable culture conditions without extensive cell differentiation or degradation. The observed changes induced by FMDV infection were also independent of the donor animals used for the preparation of the primary cell cultures, although transcriptional differences between the animals were detectable, e.g., sex-associated differences in *XIST* and keratin expression [36]. In summary, we observed genes that were highly regulated during both stages of infection (Figure 1C, green cluster), genes exclusively regulated during acute infection (Figure 1C, blue cluster) and animal-specific gene expression (Figure 1C, red cluster). During acute and persistent infection, comparably low numbers of genes were significantly differentially expressed and most of these were up-regulated. Interestingly, 63 genes were differentially expressed during both infection phases, indicating that FMDV infection induces long-lasting changes in the soft palate transcriptome. Furthermore, these genes were generally more up-regulated during acute infection. The transcriptional changes observed by RNA sequencing were supported at the protein level using quantitative proteomics. Although the number of detected differentially expressed proteins was much lower, it was confirmed that overexpression of ISGs leads to detectable levels of these proteins in the cells.

We found a strong activation of the innate immune response at 24 hpi that appears to be triggered by sensing of viral dsRNA over the cytosolic RNA sensors *IFIH1/MDA5* and *RIG-I/DDX58*, as the expression of MDA5 and RIG-I was significantly increased during acute infection. While RIG-I is sensing minus-strand RNA viruses by specific binding to $5'$-triphosphate uncapped RNA genomes, MDA5 specifically senses plus-strand RNA viruses, such as FMDV, by binding to their dsRNA replication intermediates [37,38]. Although MDA5 is thought to be the main cellular detector of FMDV, it has been shown that RIG-I transcription is also elevated during FMDV infection [39]. After specific binding of MDA5 or RIG-I to their ligands they interact with TRIM25 and the mitochondrial protein MAVS that activate the transcription factors IRF3/IRF7 [40] as well as NF-κB [41], which is in accordance with the observed up-regulation of *IRF3*, *IRF7*, *NFKBIA*, *NFKBIB* and *TRIM25* (Table 1). Activated IRF3/IRF7 and NF-κB complexes thereafter induce the expression of type I/III interferons (IFNs) and proinflammatory cytokines, respectively [42]. We found that during acute infection especially IFN-β was expressed at elevated levels with significant overexpression of *IFNB1*, while IFN-α and IFN-λ were only expressed at very low levels (Supplementary Figure S3). However, we observed expression of several ISGs with antiviral activity, such as *IRF1*, *ISG15*, *MX1/2*, *OAS1/2*, and *RSAD2* (see Table 1). This indicates that while IFN-α transcription may be limited, IFN-β alone is able to induce a potent antiviral response in bovine SP cells. It is known that the viral leader proteinase L^{pro} of FMDV can inhibit IFN-β transcription and protein translation, thereby blocking the cells innate immune response [43]. However, at least in this model L^{pro} does not induce a full blocking of IFN-β transcription. A specific blocking of IFN-α transcription has only been shown for swine dendritic cell populations from blood and skin during acute FMDV infection [44]. The induction of IFN-β and associated ISGs was furthermore coincident with high viral genome copy numbers and presence of viral proteins (Table 3) as has been described before [45].

In contrast, no interferon expression was observed during the persistent phase of infection, although the aforementioned virus sensors MDA5/RIG-I and their associated transcription factor *IRF7* were still highly expressed (see *IFIH1* in Figure 3B). A possible factor reducing the transcription of interferons during persistence is the down-regulation of *ANKRD1*, as *ANKRD1* is directly involved in the signal transduction of *IRF3* and *IRF7* by binding IRF7/IRF3 complexes and thereby enhances the expression of type I/III interferons [46]. *ANKRD1* was initially observed in human dermal endothelial cells, where it is induced by inflammatory cytokines [47] and is substantially involved in the fibroblast mediated wound healing [48]. In agreement with our results, the expression of *ANKRD1* has previously been shown to be down-regulated in vivo during persistent FMDV infection of bovine nasopharyngeal tissue [35]. Gene silencing of *ANKRD1* in cells infected with herpes simplex virus resulted in increased viral load and reduced IFN-β (*IFNB1*) and IFN-λ (*IL29*) expression [46]. Therefore, the strong downregulation of *ANKRD1* as early as 48 hpi may be involved in the decrease of IFN-β expression during persistence.

Although no interferon expression was detectable at 28 dpi, the expression of many ISGs with known antiviral activity, such as *MX1/2* and *OAS1/2*, was significantly increased (see Figures 2D and 3B). Furthermore, based on the proteome data, the translation of these genes was stable, as their encoded proteins accumulated in the cells (Table 3). This indicates long-lasting induction of ISGs during persistent infection, with viral protein below the detection limit and low FMDV genome copy numbers. A comparable pattern of highly expressed ISGs as observed during FMDV persistence (*ISG15*, *MX1*, *OAS1/2*, and *USP18*) has been identified in liver samples from chronic infections with hepatitis C virus (HCV) [49,50]. Interestingly, the elevated expression of *ISG15/USP18* during chronic HCV infection, as we also observed during FMDV persistence, correlated with decreased responses to IFN-α treatment. Furthermore, the strong expression of *OAS1/2*, *MX1/2* and *IRF7*, but not of *IRF3* at 28 dpi is in accordance with increased interferon receptor (*IFNR*) signaling [51,52]. Observations from persistent lymphocytic choriomeningitis virus (LCMV) infections revealed that persistence is driven by chronic *IFNR* signaling, characterized by similar ISG activation patterns as observed in the present study [52]. The blockade of *IFNR* by antibodies abolished the expression of *IL-10* and *PD-L1*, two immunosuppressive T-cell exhaustion factors expressed by dendritic cells, and ultimately led to clearance of persistent LCMV infection by activation of IFN-γ expressing CD4+ T cells [52]. In this study, we did not observe upregulation of *IL-10* or *PD-L1*, as no dendritic cells were present in the SP cultures; however, an overexpression of T-cell exhaustion factors was previously found in nasopharyngeal tissues of FMDV carrier animals [35] and therefore may play a role in FMDV persistence in vivo.

The immunosuppressive factors identified in that study also included transforming growth factor β (TGF-β). While an overexpression of TGF-β itself was not evident in our data, lymphocyte antigen 6 family member E (*LY6E*) was one of the two genes that were up-regulated during the persistent phase only. *LY6E*, an ISG, has been implicated in the TGF-β-mediated escape from immune surveillance in many forms of cancer [53]. *LY6E* has been previously shown to promote viral infection [54] and it is essential for clathrin-mediated endocytosis of virus particles [55], a pathway that is also used by FMDV [56].

During acute infection, the interferon-mediated induction of ISGs appears to trigger apoptosis, as indicated by pathway analysis. Interestingly, genes associated with apoptosis, such as *CASP7*, were only expressed during acute infection and their expression waned with increasing time post infection (Figure 3B). The apparent absence of apoptotic processes during persistent infection is in accordance with recent findings that indicated an inhibition of apoptotic pathways in nasopharyngeal tissues from FMDV carrier animals [34,35]. Furthermore, we observed a specific down-regulation of pro-apoptotic genes during persistence, such as *ALDH1A2*, *ANKRD1*, and *SFRP2* (Table 2). Similarly, overexpression of fructose-bisphosphatase 1, the other gene that was only up-regulated during the persistent phase, inhibits many forms of apoptosis by increasing total cellular glutathione [57].

Another interesting observation was the downregulation of *NCAM1* (CD56), a member of the immunoglobulin superfamily, as early as 48 hpi and its further decrease until 28 dpi. Besides its role as a differentiation marker for natural killer cells, *NCAM1* is involved in cell binding and migration [58]. Downregulation of *NCAM1* is associated with decreased cell adhesion capacity, enhanced tumor cell invasiveness and is triggered in other viral infections [59]. The expression of *HTRA3*, a serine protease involved in remodeling of the extracellular matrix (ECM), as well as the expression of the ECM components *COL1A1* and *COL3A1* were significantly down-regulated during persistent infection. The role of these changes during FMDV persistence is currently unknown but lends itself to some speculation. The interaction of cells with the ECM plays a key role in epithelial maturation [60], which in turn is critical for the life cycle of some persistent viruses such as papillomaviruses [61]. Even though FMDV and papillomaviruses are biologically very different, certain features of FMDV persistence in vivo are conspicuously similar to what is observed in papillomavirus infection—particularly the difference between the distribution of viral genome, which is concentrated in the basal stratum germinativum [62], and viral antigen, which is concentrated in the superficial layers of the soft palate epithelium [63]. This segregation helps papillomaviruses evade the immune system as high levels of viral replication and protein synthesis occur only in terminally differentiated cells that are not subject to immune surveillance [64]. Whether a similar mechanism is involved in the maintenance of FMDV persistence remains to be investigated. However, it is difficult to faithfully recreate the complex epithelial structure and ECM interactions in vitro and investigations of the role of the ECM in persistent FMDV infection have to be performed with ex vivo tissue samples.

5. Conclusions

In conclusion, our study independently confirms earlier findings of a polygenic inhibition of apoptosis during persistent FMDV infection, which has been put forward as one of the principal mechanisms for the maintenance of persistence [34,35]. Another proposed mechanism, Th2 polarization and T-cell exhaustion, was not directly represented in our data, because the SP culture model does not include specialized immune cells. Nevertheless, we demonstrated the utility of state-of-the-art proteogenomics for the analysis of transcriptional signatures of acute and persistent FMDV infection in a near-natural in vitro system. We will proceed to apply this technology to tissue samples collected from carrier animals to obtain the first comprehensive picture of the transcriptomic and proteomic alterations associated with FMDV persistence in the natural host. Unraveling the cellular mechanisms of FMDV persistence may ultimately give rise to improved diagnostics and prevention of the FMDV carrier state.

Supplementary Materials: The following are available online at http://www.mdpi.com/1999-4915/11/1/53/s1, Figure S1: Schematic visualization of the proteogenomic workflow, Figure S2: Comparison of different transformation methods for the raw read counts per gene, Figure S3: Comparison of interferon expression, Table S1: Summary of primary soft palate cell culture samples, Table S2: Summary of excluded and included RNA reference transcripts, Table S3: Replaced gene aliases of the *Bos taurus* reference genome, Table S4: Sequences of primers used for the RT-qPCR analysis, Table S5: Differentially expressed genes, File S1: Raw read counts for all significantly differently expressed genes.

Author Contributions: Writing—Original Draft, F.P. and M.E.; Methodology, S.H., F.P. and M.Z.; Investigation, F.P., S.H., M.Z., S.B., E.L., S.K., D.Z. and J.V.; Formal Analysis, Visualization, and Data Curation, F.P.; Supervision, S.Z., L.B., J.V., K.R., D.H., M.B. and M.E.; Conceptualization, S.Z., L.B., J.V., D.H., M.B. and M.E.; Project Administration, S.Z.

Funding: This work was supported by the Swedish Research Council (Formas, Sweden), L'Agence Nationale de la Recherche (ANR, France) and the Federal Office for Agriculture and Food (BLE, Germany) through the ANIHWA ERA-Net project TRANSCRIPTOVAC.

Acknowledgments: We thank Cells for Life Platform, partly funded by the Infrastructure Committee at SLU, Sweden, for providing facilities and equipment. We acknowledge Patrick Zitzow and Jenny Lorke of the Institute of Diagnostic Virology at the Friedrich-Loeffler-Institut for their excellent technical assistance. We are grateful to Dörte Becher and Claudia Hirschfeld from the Institute of Microbiology at the University of Greifswald for helping with the mass spectrometry analyses.

References

1. Knight-Jones, T.J.; Rushton, J. The economic impacts of foot and mouth disease—What are they, how big are they and where do they occur? *Prev. Vet. Med.* **2013**, *112*, 161–173. [CrossRef] [PubMed]
2. Zell, R.; Delwart, E.; Gorbalenya, A.E.; Hovi, T.; King, A.M.Q.; Knowles, N.J.; Lindberg, A.M.; Pallansch, M.A.; Palmenberg, A.C.; Reuter, G.; et al. ICTV Virus Taxonomy Profile: Picornaviridae. *J. Gen. Virol.* **2017**, *98*, 2421–2422. [CrossRef] [PubMed]
3. Forss, S.; Strebel, K.; Beck, E.; Schaller, H. Nucleotide sequence and genome organization of foot-and-mouth disease virus. *Nucleic Acids Res.* **1984**, *12*, 6587–6601. [CrossRef] [PubMed]
4. Alexandersen, S.; Mowat, N. Foot-and-mouth disease: Host range and pathogenesis. *Curr. Top. Microbiol. Immunol.* **2005**, *288*, 9–42. [CrossRef] [PubMed]
5. Thomson, G.R.; Vosloo, W.; Bastos, A.D. Foot and mouth disease in wildlife. *Virus Res.* **2003**, *91*, 145–161. [CrossRef]
6. Grubman, M.J.; Baxt, B. Foot-and-mouth disease. *Clin. Microbiol. Rev.* **2004**, *17*, 465–493. [CrossRef] [PubMed]

7. World Organisation for Animal Health (OIE). Foot and mouth disease (infection with foot and mouth disease virus). In *Manual of Diagnostic Tests and Vaccines for Terrestrial Animals (Terrestrial Manual)*; OIE: Paris, France, 2018; 2.1.8.

8. Arzt, J.; Belsham, G.J.; Lohse, L.; Botner, A.; Stenfeldt, C. Transmission of Foot-and-Mouth Disease from Persistently Infected Carrier Cattle to Naive Cattle via Transfer of Oropharyngeal Fluid. *mSphere* **2018**, *3*. [CrossRef]

9. Alexandersen, S.; Zhang, Z.; Donaldson, A.I. Aspects of the persistence of foot-and-mouth disease virus in animals—The carrier problem. *Microbes Infect.* **2002**, *4*, 1099–1110. [CrossRef]

10. Pacheco, J.M.; Smoliga, G.R.; O'Donnell, V.; Brito, B.P.; Stenfeldt, C.; Rodriguez, L.L.; Arzt, J. Persistent Foot-and-Mouth Disease Virus Infection in the Nasopharynx of Cattle; Tissue-Specific Distribution and Local Cytokine Expression. *PLoS ONE* **2015**, *10*, e0125698. [CrossRef]

11. Cox, S.J.; Voyce, C.; Parida, S.; Reid, S.M.; Hamblin, P.A.; Paton, D.J.; Barnett, P.V. Protection against direct-contact challenge following emergency FMD vaccination of cattle and the effect on virus excretion from the oropharynx. *Vaccine* **2005**, *23*, 1106–1113. [CrossRef]

12. Cox, S.J.; Voyce, C.; Parida, S.; Reid, S.M.; Hamblin, P.A.; Hutchings, G.; Paton, D.J.; Barnett, P.V. Effect of emergency FMD vaccine antigen payload on protection, sub-clinical infection and persistence following direct contact challenge of cattle. *Vaccine* **2006**, *24*, 3184–3190. [CrossRef] [PubMed]

13. Juleff, N.; Windsor, M.; Reid, E.; Seago, J.; Zhang, Z.; Monaghan, P.; Morrison, I.W.; Charleston, B. Foot-and-mouth disease virus persists in the light zone of germinal centres. *PLoS ONE* **2008**, *3*, e3434. [CrossRef] [PubMed]

14. Han, L.; Xin, X.; Wang, H.; Li, J.; Hao, Y.; Wang, M.; Zheng, C.; Shen, C. Cellular response to persistent foot-and-mouth disease virus infection is linked to specific types of alterations in the host cell transcriptome. *Sci. Rep.* **2018**, *8*, 5074. [CrossRef] [PubMed]

15. Zhu, J.; Weiss, M.; Grubman, M.J.; de los Santos, T. Differential gene expression in bovine cells infected with wild type and leaderless foot-and-mouth disease virus. *Virology* **2010**, *404*, 32–40. [CrossRef]

16. Kopliku, L.; Relmy, A.; Romey, A.; Gorna, K.; Zientara, S.; Bakkali-Kassimi, L.; Blaise-Boisseau, S. Establishment of persistent foot-and-mouth disease virus (FMDV) infection in MDBK cells. *Arch. Virol.* **2015**, *160*, 2503–2516. [CrossRef] [PubMed]

17. O'Donnell, V.; Pacheco, J.M.; Larocco, M.; Gladue, D.P.; Pauszek, S.J.; Smoliga, G.; Krug, P.W.; Baxt, B.; Borca, M.V.; Rodriguez, L. Virus-host interactions in persistently FMDV-infected cells derived from bovine pharynx. *Virology* **2014**, *468–470*, 185–196. [CrossRef] [PubMed]

18. Hägglund, S.; Laloy, E.; Näslund, K.; Pfaff, F.; Eschbaumer, M.; Romey, A.; Relmy, A.; Rikberg, A.; Svensson, A.; Huet, H.; et al. Model of persistent foot-and–mouth disease virus infection in multilayered cells derived from bovine dorsal soft palate. *Transbound. Emerg. Dis.* **2018**, under review.

19. Patro, R.; Duggal, G.; Love, M.I.; Irizarry, R.A.; Kingsford, C. Salmon provides fast and bias-aware quantification of transcript expression. *Nat. Methods* **2017**, *14*, 417–419. [CrossRef]

20. R Core Team. *R: A Language and Environment for Statistical Computing*; R Foundation for Statistical Computing: Vienna, Austria, 2017.

21. Soneson, C.; Love, M.I.; Robinson, M.D. Differential analyses for RNA-seq: Transcript-level estimates improve gene-level inferences. *F1000Research* **2015**, *4*, 1521. [CrossRef]

22. Love, M.I.; Huber, W.; Anders, S. Moderated estimation of fold change and dispersion for RNA-seq data with DESeq2. *Genome Biol.* **2014**, *15*, 550. [CrossRef]

23. Yu, G.; He, Q.Y. ReactomePA: An R/Bioconductor package for reactome pathway analysis and visualization. *Mol. Biosyst.* **2016**, *12*, 477–479. [CrossRef] [PubMed]

24. Yu, G.; Wang, L.G.; Han, Y.; He, Q.Y. clusterProfiler: An R package for comparing biological themes among gene clusters. *Omics* **2012**, *16*, 284–287. [CrossRef] [PubMed]

25. Untergasser, A.; Cutcutache, I.; Koressaar, T.; Ye, J.; Faircloth, B.C.; Remm, M.; Rozen, S.G. Primer3—New capabilities and interfaces. *Nucleic Acids Res.* **2012**, *40*, e115. [CrossRef] [PubMed]

26. Eymann, C.; Lassek, C.; Wegner, U.; Bernhardt, J.; Fritsch, O.A.; Fuchs, S.; Otto, A.; Albrecht, D.; Schiefelbein, U.; Cernava, T.; et al. Symbiotic Interplay of Fungi, Algae, and Bacteria within the Lung Lichen *Lobaria pulmonaria* L. Hoffm. as Assessed by State-of-the-Art Metaproteomics. *J. Proteome Res.* **2017**, *16*, 2160–2173. [CrossRef] [PubMed]

27. López-Mondéjar, R.; Zühlke, D.; Větrovský, T.; Becher, D.; Riedel, K.; Baldrian, P. Decoding the complete arsenal for cellulose and hemicellulose deconstruction in the highly efficient cellulose decomposer Paenibacillus O199. *Biotechnol. Biofuels* **2016**, *9*, 104. [CrossRef] [PubMed]

28. Huber, W.; Carey, V.J.; Gentleman, R.; Anders, S.; Carlson, M.; Carvalho, B.S.; Bravo, H.C.; Davis, S.; Gatto, L.; Girke, T.; et al. Orchestrating high-throughput genomic analysis with Bioconductor. *Nat. Methods* **2015**, *12*, 115–121. [CrossRef] [PubMed]

29. Pavelka, N.; Pelizzola, M.; Vizzardelli, C.; Capozzoli, M.; Splendiani, A.; Granucci, F.; Ricciardi-Castagnoli, P. A power law global error model for the identification of differentially expressed genes in microarray data. *BMC Bioinform.* **2004**, *5*, 203. [CrossRef]

30. Vizcaino, J.A.; Csordas, A.; del-Toro, N.; Dianes, J.A.; Griss, J.; Lavidas, I.; Mayer, G.; Perez-Riverol, Y.; Reisinger, F.; Ternent, T.; et al. 2016 update of the PRIDE database and its related tools. *Nucleic Acids Res.* **2016**, *44*, D447–D456. [CrossRef]

31. Fabregat, A.; Jupe, S.; Matthews, L.; Sidiropoulos, K.; Gillespie, M.; Garapati, P.; Haw, R.; Jassal, B.; Korninger, F.; May, B.; et al. The Reactome Pathway Knowledgebase. *Nucleic Acids Res.* **2018**, *46*, D649–D655. [CrossRef]

32. Ni, Z.; Yang, F.; Cao, W.; Zhang, X.; Jin, Y.; Mao, R.; Du, X.; Li, W.; Guo, J.; Liu, X.; et al. Differential gene expression in porcine SK6 cells infected with wild-type and SAP domain-mutant foot-and-mouth disease virus. *Virol. Sin.* **2016**, *31*, 249–257. [CrossRef]

33. Arzt, J.; Juleff, N.; Zhang, Z.; Rodriguez, L.L. The pathogenesis of foot-and-mouth disease I: Viral pathways in cattle. *Transbound. Emerg. Dis.* **2011**, *58*, 291–304. [CrossRef] [PubMed]

34. Stenfeldt, C.; Eschbaumer, M.; Smoliga, G.R.; Rodriguez, L.L.; Zhu, J.; Arzt, J. Clearance of a persistent picornavirus infection is associated with enhanced pro-apoptotic and cellular immune responses. *Sci. Rep.* **2017**, *7*, 17800. [CrossRef] [PubMed]

35. Eschbaumer, M.; Stenfeldt, C.; Smoliga, G.R.; Pacheco, J.M.; Rodriguez, L.L.; Li, R.W.; Zhu, J.; Arzt, J. Transcriptomic Analysis of Persistent Infection with Foot-and-Mouth Disease Virus in Cattle Suggests Impairment of Apoptosis and Cell-Mediated Immunity in the Nasopharynx. *PLoS ONE* **2016**, *11*, e0162750. [CrossRef] [PubMed]

36. Brown, C.J.; Hendrich, B.D.; Rupert, J.L.; Lafreniere, R.G.; Xing, Y.; Lawrence, J.; Willard, H.F. The human XIST gene: Analysis of a 17 kb inactive X-specific RNA that contains conserved repeats and is highly localized within the nucleus. *Cell* **1992**, *71*, 527–542. [CrossRef]

37. Feng, Q.; Hato, S.V.; Langereis, M.A.; Zoll, J.; Virgen-Slane, R.; Peisley, A.; Hur, S.; Semler, B.L.; van Rij, R.P.; van Kuppeveld, F.J. MDA5 detects the double-stranded RNA replicative form in picornavirus-infected cells. *Cell Rep.* **2012**, *2*, 1187–1196. [CrossRef]

38. Kato, H.; Takeuchi, O.; Sato, S.; Yoneyama, M.; Yamamoto, M.; Matsui, K.; Uematsu, S.; Jung, A.; Kawai, T.; Ishii, K.J.; et al. Differential roles of MDA5 and RIG-I helicases in the recognition of RNA viruses. *Nature* **2006**, *441*, 101–105. [CrossRef]

39. Zhu, Z.; Wang, G.; Yang, F.; Cao, W.; Mao, R.; Du, X.; Zhang, X.; Li, C.; Li, D.; Zhang, K.; et al. Foot-and-Mouth Disease Virus Viroporin 2B Antagonizes RIG-I-Mediated Antiviral Effects by Inhibition of Its Protein Expression. *J. Virol.* **2016**, *90*, 11106–11121. [CrossRef]

40. Lin, R.; Heylbroeck, C.; Pitha, P.M.; Hiscott, J. Virus-dependent phosphorylation of the IRF-3 transcription factor regulates nuclear translocation, transactivation potential, and proteasome-mediated degradation. *Mol. Cell. Biol.* **1998**, *18*, 2986–2996. [CrossRef]

41. Liu, S.; Chen, J.; Cai, X.; Wu, J.; Chen, X.; Wu, Y.T.; Sun, L.; Chen, Z.J. MAVS recruits multiple ubiquitin E3 ligases to activate antiviral signaling cascades. *Elife* **2013**, *2*, e00785. [CrossRef]

42. Belgnaoui, S.M.; Paz, S.; Hiscott, J. Orchestrating the interferon antiviral response through the mitochondrial antiviral signaling (MAVS) adapter. *Curr. Opin. Immunol.* **2011**, *23*, 564–572. [CrossRef]

43. De Los Santos, T.; de Avila Botton, S.; Weiblen, R.; Grubman, M.J. The leader proteinase of foot-and-mouth disease virus inhibits the induction of beta interferon mRNA and blocks the host innate immune response. *J. Virol.* **2006**, *80*, 1906–1914. [CrossRef] [PubMed]

44. Nfon, C.K.; Ferman, G.S.; Toka, F.N.; Gregg, D.A.; Golde, W.T. Interferon-alpha production by swine dendritic cells is inhibited during acute infection with foot-and-mouth disease virus. *Viral. Immunol.* **2008**, *21*, 68–77. [CrossRef] [PubMed]

45. Eschbaumer, M.; Stenfeldt, C.; Rekant, S.I.; Pacheco, J.M.; Hartwig, E.J.; Smoliga, G.R.; Kenney, M.A.; Golde, W.T.; Rodriguez, L.L.; Arzt, J. Systemic immune response and virus persistence after foot-and-mouth disease virus infection of naive cattle and cattle vaccinated with a homologous adenovirus-vectored vaccine. *BMC Vet. Res.* **2016**, *12*, 205. [CrossRef] [PubMed]

46. Bin, L.; Li, X.; Richers, B.; Streib, J.E.; Hu, J.W.; Taylor, P.; Leung, D.Y.M. Ankyrin repeat domain 1 regulates innate immune responses against herpes simplex virus 1: A potential role in eczema herpeticum. *J. Allergy Clin. Immunol.* **2018**. [CrossRef] [PubMed]

47. Chu, W.; Burns, D.K.; Swerlick, R.A.; Presky, D.H. Identification and characterization of a novel cytokine-inducible nuclear protein from human endothelial cells. *J. Biol. Chem.* **1995**, *270*, 10236–10245. [CrossRef]

48. Samaras, S.E.; Almodovar-Garcia, K.; Wu, N.; Yu, F.; Davidson, J.M. Global deletion of Ankrd1 results in a wound-healing phenotype associated with dermal fibroblast dysfunction. *Am. J. Pathol.* **2015**, *185*, 96–109. [CrossRef] [PubMed]

49. Chen, L.; Borozan, I.; Feld, J.; Sun, J.; Tannis, L.L.; Coltescu, C.; Heathcote, J.; Edwards, A.M.; McGilvray, I.D. Hepatic gene expression discriminates responders and nonresponders in treatment of chronic hepatitis C viral infection. *Gastroenterology* **2005**, *128*, 1437–1444. [CrossRef] [PubMed]

50. MacQuillan, G.C.; Mamotte, C.; Reed, W.D.; Jeffrey, G.P.; Allan, J.E. Upregulation of endogenous intrahepatic interferon stimulated genes during chronic hepatitis C virus infection. *J. Med. Virol.* **2003**, *70*, 219–227. [CrossRef]

51. Au, W.C.; Moore, P.A.; Lowther, W.; Juang, Y.T.; Pitha, P.M. Identification of a member of the interferon regulatory factor family that binds to the interferon-stimulated response element and activates expression of interferon-induced genes. *Proc. Natl. Acad. Sci. USA* **1995**, *92*, 11657–11661. [CrossRef]

52. Wilson, E.B.; Yamada, D.H.; Elsaesser, H.; Herskovitz, J.; Deng, J.; Cheng, G.; Aronow, B.J.; Karp, C.L.; Brooks, D.G. Blockade of chronic type I interferon signaling to control persistent LCMV infection. *Science* **2013**, *340*, 202–207. [CrossRef]

53. AlHossiny, M.; Luo, L.; Frazier, W.R.; Steiner, N.; Gusev, Y.; Kallakury, B.; Glasgow, E.; Creswell, K.; Madhavan, S.; Kumar, R.; et al. Ly6E/K Signaling to TGFbeta Promotes Breast Cancer Progression, Immune Escape, and Drug Resistance. *Cancer Res.* **2016**, *76*, 3376–3386. [CrossRef] [PubMed]

54. Schoggins, J.W.; Wilson, S.J.; Panis, M.; Murphy, M.Y.; Jones, C.T.; Bieniasz, P.; Rice, C.M. A diverse range of gene products are effectors of the type I interferon antiviral response. *Nature* **2011**, *472*, 481–485. [CrossRef] [PubMed]

55. Hackett, B.A.; Cherry, S. Flavivirus internalization is regulated by a size-dependent endocytic pathway. *Proc. Natl. Acad. Sci. USA* **2018**, *115*, 4246–4251. [CrossRef] [PubMed]

56. Han, S.C.; Guo, H.C.; Sun, S.Q.; Jin, Y.; Wei, Y.Q.; Feng, X.; Yao, X.P.; Cao, S.Z.; Liu, D.X.; Liu, X.T. Productive Entry of Foot-and-Mouth Disease Virus via Macropinocytosis Independent of Phosphatidylinositol 3-Kinase. *Sci. Rep.* **2016**, *6*, 19294. [CrossRef] [PubMed]

57. Voehringer, D.W.; Hirschberg, D.L.; Xiao, J.; Lu, Q.; Roederer, M.; Lock, C.B.; Herzenberg, L.A.; Steinman, L.; Herzenberg, L.A. Gene microarray identification of redox and mitochondrial elements that control resistance or sensitivity to apoptosis. *Proc. Natl. Acad. Sci. USA* **2000**, *97*, 2680–2685. [CrossRef] [PubMed]

58. Moos, M.; Tacke, R.; Scherer, H.; Teplow, D.; Früh, K.; Schachner, M. Neural adhesion molecule L1 as a member of the immunoglobulin superfamily with binding domains similar to fibronectin. *Nature* **1988**, *334*, 701–703. [CrossRef] [PubMed]

59. Blaheta, R.A.; Beecken, W.D.; Engl, T.; Jonas, D.; Oppermann, E.; Hundemer, M.; Doerr, H.W.; Scholz, M.; Cinatl, J. Human cytomegalovirus infection of tumor cells downregulates NCAM (CD56): A novel mechanism for virus-induced tumor invasiveness. *Neoplasia* **2004**, *6*, 323–331. [CrossRef]

60. Bonnans, C.; Chou, J.; Werb, Z. Remodelling the extracellular matrix in development and disease. *Nat. Rev. Mol. Cell Biol.* **2014**, *15*, 786–801. [CrossRef]

61. Feller, L.; Khammissa, R.A.; Wood, N.H.; Lemmer, J. Epithelial maturation and molecular biology of oral HPV. *Infect. Agent Cancer* **2009**, *4*, 16. [CrossRef]

62. Prato Murphy, M.L.; Forsyth, M.A.; Belsham, G.J.; Salt, J.S. Localization of foot-and-mouth disease virus RNA by in situ hybridization within bovine tissues. *Virus Res.* **1999**, *62*, 67–76. [CrossRef]

63. Stenfeldt, C.; Eschbaumer, M.; Rekant, S.I.; Pacheco, J.M.; Smoliga, G.R.; Hartwig, E.J.; Rodriguez, L.L.; Arzt, J. The Foot-and-Mouth Disease Carrier State Divergence in Cattle. *J. Virol.* **2016**, *90*, 6344–6364. [CrossRef] [PubMed]
64. McBride, A.A. Mechanisms and strategies of papillomavirus replication. *Biol. Chem.* **2017**, *398*, 919–927. [CrossRef] [PubMed]

Article

Exploring the Papillomaviral Proteome to Identify Potential Candidates for a Chimeric Vaccine against Cervix Papilloma Using Immunomics and Computational Structural Vaccinology

Satyavani Kaliamurthi [1,2], Gurudeeban Selvaraj [1,2], Sathishkumar Chinnasamy [3], Qiankun Wang [3], Asma Sindhoo Nangraj [3], William CS Cho [4], Keren Gu [1,2] and Dong-Qing Wei [1,3,*]

[1] Center of Interdisciplinary Science-Computational Life Sciences, College of Food Science and Engineering, Henan University of Technology, Zhengzhou 450001, China; satyavani.mkk@haut.edu.cn (S.K.); gurudeeb99@haut.edu.cn (G.S.); gkr@haut.edu.cn (K.G.)
[2] College of Chemistry, Chemical Engineering and Environment, Henan University of Technology, Zhengzhou 450001, China
[3] The State Key Laboratory of Microbial Metabolism, College of Life Sciences and Biotechnology, Shanghai Jiao Tong University, Shanghai 200240, China; sathishimb@gmail.com or sathishimb@sjtu.edu.cn (S.C.); wangqiankun@sjtu.edu.cn (Q.W.); sindhoo_sind@yahoo.com (A.S.N.)
[4] Department of Clinical Oncology, Queen Elizabeth Hospital, Kowloon, Hong Kong; williamcscho@gmail.com or chocs@ha.org.hk
* Correspondence: dqwei@sjtu.edu.cn; Tel.: +86-021-3420-4717; Fax: +86-021-3420-5709

Received: 29 November 2018; Accepted: 10 January 2019; Published: 15 January 2019

Abstract: The human papillomavirus (HPV) 58 is considered to be the second most predominant genotype in cervical cancer incidents in China. HPV type-restriction, non-targeted delivery, and the highcost of existing vaccines necessitate continuing research on the HPV vaccine. We aimed to explore the papillomaviral proteome in order to identify potential candidates for a chimeric vaccine against cervix papilloma using computational immunology and structural vaccinology approaches. Two overlapped epitope segments (23–36) and (29–42) from the N-terminal region of the HPV58 minor capsid protein L2 are selected as capable of inducing both cellular and humoral immunity. In total, 318 amino acid lengths of the vaccine construct SGD58 contain adjuvants (Flagellin and RS09), two Th epitopes, and linkers. SGD58 is a stable protein that is soluble, antigenic, and non-allergenic. Homology modeling and the structural refinement of the best models of SGD58 and TLR5 found 96.8% and 93.9% favored regions in Rampage, respectively. The docking results demonstrated a HADDOCK score of -62.5 ± 7.6, the binding energy (-30 kcal/mol) and 44 interacting amino acid residues between SGD58-TLR5 complex. The docked complex are stable in 100 ns of simulation. The coding sequences of SGD58 also show elevated gene expression in *Escherichia coli* with 1.0 codon adaptation index and 59.92% glycine-cysteine content. We conclude that SGD58 may prompt the creation a vaccine against cervix papilloma.

Keywords: cellular immunity; codon frequency distribution; HPV58; minor capsid protein; TLR agonist; prophylaxis

---

1. Introduction

Currently, viral infection contributes to about 20% of the global burden of human cancer. Among other virus types, the human papillomavirus (HPV) is reported in about 5% of all human cancers, specifically infection associated with the cervix with 250,000 mortalities every year [1]. HPV with its double-stranded DNA contains a non-enveloped small virus, which infects a region of the cutaneous epithelial membrane (skin or integumentary system), or the mucous membrane (i.e., coated as an internal line in hollow spaced organs like the mouth, reproductive organ, urinary tract, or rectum) in the host system [2]. The nomenclature of HPV is distinguished by the International Committee on Taxonomy of Viruses (ICTV), and is based on the suggestion obtained from the study group of papillomavirus [3]. ICTV follows the practice of naming species after a specific virus, such as HPV16, while the related types, namely, the "type species," are designated as strains within the species [3,4]. For example, the frequently used term "HPV species alpha-9" is a synonym for the ICTV term "HPV16 species"; it contains the HPV types 16, 31, 33, -35, 52, 58, and 67 strains, respectively. According to ICTV, the species HPV16 belongs to the family of Papillomaviridae and the genus of Alphapapillomavirus [3]. The genomic relationship between different cancer types has demonstrated that more than 99% of cervical cancer patients are infected with 15 different types of α-clade HPV, defined as "high-risk" or "oncogenic" genital HPVs. The α-clade HPV (6 and 11) causes genital warts while the remaining strains of HPV are related to the risk of cervical cancer. HPV infection is attributed to more than 50% of oropharyngeal and anogenital cancers [5]. Generally, the human immune system can clear the pathogenic infection caused by HPV within 2 years, but this also depends on the efficiency of an individual's immune system and the invading type of HPV. However, in the case of a very weak immune system, it fails to remove the invading high-risk HPVs (hrHPV) that may lead to the development of cervical cancer [6,7]. hrHPV infections are responsible for causing more than 99% of precancerous cervical intraepithelial neoplasias (CIN) and invasive cervical cancers (ICC) [8–10]. In China, HPV-mediated cervical cancer is a substantial public health issue, with 1 million new cervical cancer incidences and 30,000 moralities registered every year [11,12]. In 2018, clinical, epidemiological, and clinicopathological studies reported HPV58 to be the second or third most predominant genotype in precancerous CIN I, II, III, and ICC lesions, a higher grade of squamous intraepithelial or cell carcinoma, and adenocarcinoma of HPV positive patients in different geographical regions of China [11]. Seven provinces of China have reported hrHPV-mediated cervical cancer incidences, namely, Guangdong, Liaocheng, Shanghai, Wenzhou, Wuhan, Southwestern China, and Western China [13–19]. Zhang et al. [20] reported that the HPV16 (6.4%) and HPV58 (5.3%) genotypes were predominantly found in males who had recently been involved in sex, in Shanghai.

Cervarix®, Gardasil®, and Gardasil 9® are the three non-infectious prophylactic Food and Drug Administration (FDA)-approved HPV licensed subunit vaccines in active usage. These vaccines were developed from the major capsid L1 virus-like particles (VLPs) using recombinant DNA technology. Cervarix is a bivalent vaccine based on Baculovirus fermentation and it provides ~70% protection against HPV (16 and 18)-mediated cervical cancer but not against genital warts [21]. Gardasil is a quadrivalent HPV (6, 11, 16, and 18) vaccine based on yeast fermentation technology. It is efficiently used for the prevention of genital warts and gives ~70% protection for cervical cancer [22]. In 2009, the FDA approved a nine-valent Gardasil 9® vaccine that provides protection to HPV types 6, 11, 16, 18, 31, 33, 45, 52, and 58. It has been used for both males and females in the age groups of 9–15 and 9–26 [23]. The new nine-valent vaccine exhibited a positive outcome in high-grade lesions in the absence of HPV (18 and 16) infections [24]. In October 2018, the FDA extended the use of Gardasil 9 to the age group of 27–45 among both the sexes. In addition, the L1 VLP (absence of viral genomic materials)-mediated vaccine production in the eukaryotic (ex. Baculovirus) host system is a complex and tedious process [25,26]. The main limitations of currently available prophylactic vaccines are as follows: they are strain specific, not therapeutic for patients already infected with HPV, they require multiple dosages, and are expensive [27,28]. In addition, the effective straightforward delivery of HPV vaccines can enhance the immunogenic potential against HPVs.

The implementation of the L2 minor capsid protein is a potential alternative in the HPV prophylactic vaccine production. Since the N-terminal region of the L2 protein is highly conserved in low-risk HPV (6 and 11) and 13 different hrHPVs, it is contrasted with the type-specific protection of

L1 prophylactic VLPs [29]. The single copy of the L2 protein ~473 amino acids (AA) is present in each L1 capsomere, resulting in 72 copies per virion [30]. Incidentally, the L2 protein plays a vital role in L1 assembly into the VLPs and enhances the encapsidation of the double-stranded ~8kb circular viral genome [31]. Moreover, the full-length or polypeptides (1–8 or 11–200 AA in length) of the L2 protein enhance the production of neutralizing antibodies in vaccinated experimental models including mice, cattle, and rabbit [32–34]. To date, no L2 VLP-derived prophylactic vaccines have been approved in clinical trials [26] due to their limitation of weak immunogenicity, which imitates the incapability of multimerizing into VLPs.

With this information, in this study, we aimed to design the novel chimeric vaccine from the N-terminal region of the L2 sequence of the HPV58 targets to hrHPVs. The immunomic tools namely, the immune epitope database (IEDB) and NetMHCv4.0, Tepitool, CTLPred, PAComplex, IFNepitope, ABCPred, AllerTOP, AllergenFPv 1.0, ANTIGENpro, program of protein information resource (PIR), and epitopes conservancy were implemented to discover the overlapped epitope segment that is required to induce B-cell and T-cell immunity. Then, the chimeric vaccine (SGD58) was constructed using the overlapped epitope segments, Toll-like receptor (TLR) adjuvants, Th epitopes, and amino acid linkers. The physiochemical and immunological properties of the chimeric vaccine were validated using Protparam, SolPro, VaxiJen, and ANTIGENpro tools. In addition, homology modeling using iterative threading ASSEmbly refinement (I-TASSER), the Robetta beta full chain protein structure prediction server, structural refinement (GalaxyRefine and 3DRefine), and structural validation (protein structure analysis [ProSA], Ramachandran plot, and ERRAT) were performed to obtain the best three-dimensional (3D) model of the chimeric vaccine and the target TLR5 receptor. Then, the interaction of the chimeric vaccine with TLR5 and stability of this complex were determined through protein–protein (PP) docking and molecular dynamic (MD) simulation. Moreover, the virtual cloning and gene expression of the chimeric vaccine in *Escherichia coli (E. coli)* were analyzed to obtain a low-cost HPV vaccine.

2. Materials and Methods

2.1. Protein Sequences

The L2 protein of HPV58 (Accession No.: P26538), the Flagellin protein of *Salmonella enterica* serovar *Dublin* (Accession No.: Q06971), and human TLR5 (Accession No.: O60602) sequences were obtained from the Swiss-Prot reviewed universal protein knowledgebase (UniProt) database [35]. The designed chimeric vaccine was named SGD58, using the name of the first and principal authors along with the strain number.

2.2. Immunomics Analysis

2.2.1. MHC-I Binding Epitope Segments Prediction

Two servers, IEDB and NetMHCv4.0, have been exploited for the identification of major histocompatibility complex class I (MHC-I) binding epitopes from the N-terminal region of the L2 sequence. Specific human MHC-I alleles such as the human leukocyte antigen (HLA)-A* (01:01, 02:01, 02:07, 11:01, 24:01), HLA-B* (46:01, 58:01) and HLA-C* (07:02, 12:03) were abundantly diagnosed in different regions of China, including Guizhou, Henan, Taihu River Basin, Tibetan, Yunnan, Wenzhou, and Wuhan. These alleles were selected for epitope prediction [36–42]. IEDB [43] is a freely available analysis resource with specified algorithms for the identification and determination of immunogenic epitopes. A consensus method was implemented to predict the MHC-I binding epitopes and its production pathway [44]. In this consensus method, three algorithms namely, the neural network (artificial), matrix method (stabilized), and peptide libraries (combinatorial) were combined to predict the promising CTL epitope segments. The epitopes involve proteasomal cleavage (pCle), a transporter associated with antigen processing (TAP), and the MHC-I binding pathway. The lowest percentile rank

(<10%) indicated the good binding efficiency of epitopes with the restricted alleles. NetMHCV4.0 [45] is another potential tool implemented to find MHC-I binding peptides with the best Pearson's correlation coefficient (PCC) of 0.895, based on the combined neural network. The strong and weak binding peptides were predicted based on the thresholds of <0.5 and <2, respectively.

2.2.2. CTL Epitope and TCR -Peptide/Peptide -MHC Interfaces Prediction

The CTLPred tool is a direct method for the prediction of CTL epitope segments instead of MHC binders. The prominent combined approaches were implemented to find the epitopes, based on both the artificial neural networks (ANN) trained by Stuttgart neural network simulator (SNNS) and support vector machine (SVM) methods. The combined methods demonstrate a higher level of accuracy (75.8 %) compared with other individual methods of prediction such as ANN (72.2%) and SVM (75.4%). The default cutoff scores of 0.51 of ANN and 0.36 of SVM were used to find the epitopes or non-epitopes at which the sensitivity and specificity of the predictions are almost similar [46]. A web server PAComplex provides access to examine and visualize the TCR-peptide and peptide–MHC interface (pMHC), respectively. For a given viral protein query sequence, the joint Z-value is obtained with a threshold 2.5. Moreover, it allows the selection of only limited allotypes of MHC class I such as HLA-A*(02:01), HLA-B*(08:01, 35:01, 35:08, 44:05), and HLA-E, respectively. The Z-value was calculated using the following formula:

$$Jz = Z_{MHC} \times Z_{TCR} \tag{1}$$

where Z_{MHC} and Z_{TCR} are the score of a TCR-pMHC complex, calculated by $(E\text{-}\mu)/\sigma$. E denotes the interaction score, μ denotes the mean, and σ denotes the standard deviation from 10,000 random interfaces [47].

2.2.3. MHC Class-II Binding Epitopes Prediction

MHC-II alleles include DQB1*(03:01, 03:03, 06:01), DRB1*(07, 09, 14:01:01, 15:01, 15:07:01), and DPB1*(05:01,05:02:01), specific to the Henan, Taihu River Basin, Tibetan, Yunnan, Wenzhou, and Wuhan provinces of China, which have been selected for epitope prediction [36–42]. The IEDB consensus approaches were used to predict MHC-II binding epitope segments using the neural network-based alignment, stabilized matrix methods-based alignment, and combinatorial library-based algorithms [48]. The peptides with the lowest percentile rank were considered to be of a higher binding affinity. Tepitool [49] is a tool from IEDB analysis resources, which provides accession to the prediction of both class I and II binders. The peptides which show the lowest percentile rank ($IC_{50}<$ or = 500nM) are potentially considered as higher affinity binding peptides.

2.2.4. Interferon-Gamma (INF-γ) Inducing Epitope Prediction

IFNepitope (http://webs.iiitd.edu.in/raghava/ifnepitope/) that is a potential server useful for the prediction and design of INF-γ inducing epitopes. INF-γ inducing epitopes were identified based on motif-based SVM or hybrid algorithms. The hybrid method using residue or dipeptide composition shows 81.39% accuracy [50].

2.2.5. Linear B-Cell Epitope Prediction

ABCPred is used to predict linear B-cell epitopes. It provides 65.93% of accuracy with the involvement of the recurrent neural network (RNN) algorithm. It consists of 700 B-cell and non-B-cell epitope segment datasets each with a length of 20 amino acids [51].

2.2.6. Allergenicity Prediction

AllerTOP is the first proper alignment-free allergenicity server. In this, five machine learning methods such as partial least squares, logistic regression, decision tree, naive Bayes, or *k* nearest neighbors (*k*NN = 1) were implemented to find the allergen. It shows 88.7%, 90.7%, and 86.7% accuracy, specificity, and sensitivity, respectively [52]. AllergenFPv 1.0 is another essential tool for allergenicity prediction based on novel descriptor fingerprint approaches. Twenty naturally existing amino acids in the protein sequences were classified into five descriptors (E) such as E1 (hydrophobicity), E2 (size), E3 (helix-forming propensity), E4 (relative abundance of amino acids), and E5 (β-strand forming propensity). Based on this, the strings were transformed into normal vectors by auto cross covariance (ACC) transformation to find the allergen protein. It exhibits accuracy (87%), specificity (89%), and sensitivity (86%) [53].

2.2.7. Antigenicity

ANTIGENpro is the potential alignment-free and sequence-based antigenicity prediction server with 79% accuracy and an area under curve (AUC) of 0.89. It shows results based on amino acid composition and the random-forest algorithm. The datasets were trained using 5-fold cross-validation. It consists of both protective antigen (193) and non-antigen (193) sequences. It predicts whether the given query epitope segments are antigenic or non-antigenic with their respective probability [54].

2.2.8. Cross-Reactivity Analysis with Human Proteomes

The presence or absence of similarity in predicted epitope segments with the human proteome was analyzed using the peptide-matching program of PIR [55].

2.2.9. Epitopes Conservancy Analysis

Epitopes conservancy (EC) and molecular evolutionary genetic analysis (MEGA) v7.0 tools were used to perform conservancy analysis. The EC tool [56] was employed to find the degree of conservancy of the epitope segments within the set of given hrHPV L2 protein sequences. The selected epitope segments of HPV58 with 15 hrHPV (16, 18, 31, 33, 35, 39, 45, 51, 52, 56, 59, 68, 69, 73, and 82) strains performed EC analysis. ClustalW-based multiple sequence alignment [57] was used to determine the sequence conservation of the overlapped epitope segments with 15 other hrHPV (16, 18, 31, 33, 35, 39, 45, 51, 52, 56, 59, 68, 69, 73, and 82) strains.

2.3. Chimeric Vaccine Design (SGD58) and Validation

2.3.1. Assessment of the Physicochemical Properties of SGD58

The complete chimeric vaccine was designed by joining the optimized epitope segments (02), TLR adjuvants (02), and Th epitopes (02) with suitable amino acid linkers. Moreover, it is required to find the solubility of the designed chimeric vaccine on overexpression in *E. coli*. SOLpro is a useful tool to find the solubility of protein based on the two-stage SVM algorithm. It achieves an overall accuracy of 74%, which develops on standard evaluation metrics with 10-fold cross-validation. It predicts the query protein to be soluble or insoluble at $P \geq 0.5$ [58]. A range of physiochemical characteristics of the designed chimeric vaccine was also determined through ProtParam [59].

2.3.2. Determination of Antigenicity

VaxiJen is the primary server used for the prediction of antigenicity of the input sequence against different targets such as virus, bacteria, fungi, parasites, and tumors. Antigenicity was calculated based on the physicochemical properties of the protein sequences. Every target organism dataset contained 100 antigens and non-antigens. Moreover, the model organisms were validated using leave-one-out cross-validation (LOO-CV); it provides 89% accuracy and an AUC of 0.964 respectively, at the threshold of 0.4 [60].

2.3.3. Analysis of the Tertiary Structure

2.3.3.1. Homology Modeling

For homology modeling, I-TASSER and Robetta were the servers used to design the 3D structure of SGD58 and TLR5. I-TASSER is a potential server that depends on the secondary-structure-mediated program of "Profile-Profile threading alignment (PPA) and iterative implementation of the TASSER." It has predicted a number of protein structures on request basis from 35 countries worldwide. For the query inputs, the user obtains the confidence score, TM score (topology similarity assessments of the two various protein structures), root-mean-square deviation (RMSD), and cluster density values. Nevertheless, the higher C-score (ranging from -5 to $+2$) determines the best model with a higher confidence level. Moreover, the 3D structure of the modeled protein was visualized using UCSF Chimera [61]. The Robetta beta server was used to predict the full chain protein structure. This server (http://robetta.bakerlab.org) gives automated tools for the analysis and prediction of the protein structure. Robetta provides both the *ab initio* and comparative models of protein domains. The comparative models are built from structures detected and aligned by HHSEARCH, SPARKS, and Raptor. The loop regions are assembled from fragments and optimized to fit the aligned template structures. The *de novo* models are built using the Robetta *de novo* protocol. For structure prediction, the submitted query sequences were analyzed minutely into putative domains. For domain prediction, a hierarchical screening method called "Ginzu" was used [62]. Besides, due to the unavailability of the crystal structure of TLR5, we have chosen TLR5 (PDB ID: 3J0A) as a template model to perform the homology modeling.

2.3.3.2. 3D Modeled Structure Refinement

The high C-score model of the designed vaccine from the I-TASSER and model 3 from Robetta beta was further refined using the GalaxyRefine and 3DRefine tools. The GalaxyRefine is a tool that is accessible in the GalaxyWeb server: it is useful to refine the structure of a protein from the given query sequences based on template-based modeling, and undergoes loop and terminus portion refinement through the *ab initio* modeling method. The ninth critical assessment of techniques for protein structure prediction (CASP9) optimizes refinement and produces consistent core structures [63]. Another tool is 3Drefine, which prompts iteration analysis for ~300 amino acid residues efficiently in less than 5 min. It performs post-refinement model analysis with both or single MolProbity and random walk (RW) plus methods. The results are visualized using JSmol [64]. The top five models of each tool were used for further validation.

2.3.3.3. 3D Refined Structures Validation

The refined 3D models from the above steps were validated using three interactive services namely, ProSA, Ramachandran plot analysis, and ERRAT. ProSA-web is a potential tool for the refinement, validation, prediction, and modeling of protein structures. It indicates the difference in the protein structures through the respective score and energy spot. It also facilitates the validation process of the protein structure that is acquired from X-ray scanning, nuclear magnetic resonance (NMR) spectroscopy analysis, and theoretical calculations. As an output, the Z-score corresponds to the overall feature of the validated model [65]. RAMPAGE tool is used to validate the percentage (%) of favored, allowed, and outlier region in the given query chimeric vaccine [66]. The statistics of noncovalent interactions between carbon, nitrogen, and oxygen atoms in the input sequence with best-resolution crystallographic structures were compared using the ERRAT tool. It implements an empirical atom-based approach for verification of the protein structure and is more sensitive to errors (1.5A) [67]. Similar steps were followed to validate the TLR5 model.

2.4. Conformational B-Cell Epitopes Prediction

DiscoTope 2.0 is a potential tool used to analyze the conformational (discontinuous) B-cell epitopes from the input sequence. It showed a highly significant prediction performance with an AUC of 0.824. The default -3.7 threshold limit provides significant specificity (0.75) and sensitivity (0.47). It was selected for the present analysis, and the final score was evaluated by the combined calculation of the propensity score (PS) and contact numbers [68].

2.5. Investigation of the Interaction between SGD58 and TLR5

2.5.1. Protein–Protein (PP) Interaction of the SGD58 with TLR5

The PP interactions are the midpoint for all the biochemical pathways that are involved in the biological functions. HADDOCK v2.2 is a server used for the docking of PP complexes [69]. The scoring function was executed based on the weighted sum of the various energy terms (Van der Waals energy, electrostatic energy, desolvation energy, restraints energy and buried surface area). In addition, intermolecular contacts such as hydrogen bonds (HB) and those that are non-bonded were determined by using the program for automatically plotting protein–protein interactions (LIGPLOT-DIMPLOT) v 4.5.3 [70]. The validated best model of the vaccine construct (Robetta model 3) and TLR5 (Robetta model 5) was chosen as a ligand and receptor, respectively.

2.5.2. MD Simulation

MD simulation determines the strength of the docked complex and the designed vaccine (SGD58). The GROMACS 5.1.2 package, with the CHARMM force field was used to perform MD simulation. The transferable intermolecular potential with three points (TIP3P) and simple point charge (SPC) in the cubic cell of the water model was resolved with protein, and with the addition of appropriate counter ions to satisfy electroneutrality. For the MD simulation, the system was implemented with volume-based canonical (NVT) and pressure-based isothermal-isobaric (NPT) ensembles [71]. The energy minimization of the system was performed with the steepest descent method, which facilitates 50,000 minimization steps and 1000 kJ mol^{-1} of tolerance. The Ewald method with a cutoff for short-range neighbor distance (1.0 nm), and Coulomb (1.0 nm) was used to calculate van der Waals (vdW) and electrostatic interactions [72]. SPC resolved the system and the final minimizations were calculated for a realistic structure concerning the geometry: and solvent orientations were used in the production of the MD simulation. SETTLE and LINCS algorithms were used to assist the geometry of the water molecules and bond angles [72,73]. The pressure of the system (300 K, 1bar) was embraced using the Parrinello–Rahman method and the temperature was regulated using the V-rescale method. Temperature and pressure equilibrated systems were employed for production run (100 ns) and time step (2 fs). The resulting structural coordinates were saved at every 2ps of an interval.

2.6. Analysis of Virtual Gene Expression and Cloning

EMBOSS Backtranseq v1.0 [74] is a suitable tool to uptake the query protein sequences, reverse translate, and return the optimizing coding sequences. Furthermore, the properties of the obtained coding sequences were analyzed to obtain the increased level of gene expression in the respective host. It is well known that the codon plays a crucial role in the expression of the recombinant proteins in various organisms (e.g., *E. coli*, *Homo sapiens*, *Saccharomyces cerevisiae*, *Mus musculus*, and *Rattus norvegicus*). The GenScript rare codon analysis [75] is a prominent tool for codon usage and its distribution analysis (codon adaptation index-CAI, glycine-cystine content-GC, and codon frequency distribution-CFD) in the individual expression host organism is based on the optimum Gene^{TM} algorithm. The endonuclease NdeI (N-terminal) and BamHI (C-terminal) restriction enzyme sites were added to the respective cloning sequences in the host (*E. coli*).

3. Results

3.1. Analysis of Selected Sequences

The retrieved L2 of HPV58, the Flagellin protein of *Salmonella enterica* serovar Dublin, and humanTLR5 contain 472, 505, and 858 amino acids residues, respectively. However, for the potential epitope prediction, an N-terminal region of 12-114 AA residues (highly conserved and virus surface exposed region) were selected from L2 protein sequences. In the case of the Flagellin protein, the N terminal (5–143 amino acids) and C-terminal (419–504 amino acids) regions were selected for the vaccine design.

3.2. Immunomics Analysis

3.2.1. MHC-I Binding Epitopes

The region-specific MHC-I alleles in the Chinese population, namely, the HLA-A* (01:01, 02:01, 02:07, 11:01, 24:01), HLA-B* (46:01, 58:01), and HLA-C* (07:02, 12:03) restricted epitopes, were obtained using the IEDB consensus and NetMHC v.4.0 tools. The epitope segments from the N-terminal region of HPV 58 were overlapped using two different tools as shown in Table S1.

3.2.2. CTL Epitopes and TCR-Peptide/Peptide-MHC Interfaces

The overlapped epitope segments obtained from the above MHC-I prediction were compared with the results of both CTLPred [46] and PAComplex [47] servers. Furthermore, the shared epitope segments obtained from the CTLPred and PAComplex were used for epitope selection, and vaccine design as shown in Table S1.

3.2.3. MHC-II Binding and IFN-γ Producing Epitopes

The lowest percentile rank with strong binding affinity epitope segments with human MHC-II alleles, namely, the DQB1*(03:01, 03:03, 06:01), DRB1*(07, 09, 14:01:01, 15:01, 15:07:01), and DPB1*(05:01, 05:02:01) restricted epitopes were obtained using IEDB consensus and Tepitool servers. The overlapping promiscuous epitope segments from the above prediction (Table S2) were selected and evaluated for their INF-γ production ability. The overlapped INF-γ producing CD4+ (MHC-II) epitope segments are as given in Table S2. Therefore, the shared MHC-II epitope segments could produce IFN-γ against viral infection. Interestingly, the above-obtained overlapped CD4+ epitopes shared the CD8+ epitope segments.

3.2.4. Continuous B-Cell Epitopes

ABCPred predicted the potential antigenic linear B-cell epitope segments. The overlapped B-cell epitopes and their respective position are given in Table S2.

3.2.5. Selection of the Overlapped Epitope Segments

According to the above results (MHC-1, CTL, MHC-11, INF-γ), only four potential antigenic epitope segments were selected from the HPV58 minor capsid protein. The epitope segments, namely,23–36, 30–43, 10–23, and 29–42 from the N-terminal region of the HPV58 L2 protein have been chosen for further studies (Table 1).

Table 1. Selection of overlapped epitope segments from the N-terminal region of HPV58.

Start	End	Overlapped Epitope Segments	IFN-γ Producing Epitopes (Potentiality/ Value)	AllergenFP	AllerTOP	AntigenPro	Cross-Reactivity with Human Proteomes
23	36	**KVEGTTIADQILRY**	+0.528	Non-allergen	Non-allergen	Antigen	Similarity level zero
30	43	ADQILRYGSLGVFF	+1.000	Non-allergen	-	Antigen	Similarity level zero
10	23	CKASGTCPPDVIPK	+1.000	-	-	Antigen	Similarity level zero
29	42	**IADQILRYGSLGVF**	+1.000	Non-allergen	Non-allergen	Antigen	Similarity level zero

The results of the analysis showed four overlapped epitope segments, among these the two epitope segments (23–36 and 29–42) indicated in bold were selected for the vaccine construction. The IFN-γ production ability of the overlapped epitope segments is presented in positive values.

3.2.6. Antigenicity, Allergenicity, and Cross-Reactivity of Selected Epitope Segments

The start and end positions, epitope segments, pro-inflammatory cytokines (INF-γ) productivity, allergenicity, antigenicity, and cross-reactivity with human proteomes of the selected overlapped HPV58 are given in Table 1. Among the four, only AllergenFP and AllerTOP declared segments (23–36 and 29–42) as non-allergen, respectively. AntigenPro shows the selected epitope segments having antigenic potential. In addition, the cross-reactivity result of the epitope segments with the human proteomes has a zero similarity level. It confirmed that there was no distinctive match of overlapped epitope segments as found in *Homo sapiens*. It indicated that these epitope segments would not cause or induce any autoimmune disease or disorders. Based on the above overall analysis, the 23–36 and 29–42 epitope segments were selected to design the vaccine construct.

3.2.7. Epitopes Conservancy

In epitopes-based vaccine design, the conserved epitope segments would be essential in order to give wider cross-protection against various hrHPV strains. Supplementary Table S3 gives the comprehensive analysis of overlapped EP ($\geq$30%), positions, subsequences identity, and hrHPV. The conservation of selected epitopes has cross-protection to the 15 hrHPV as shown in Supplementary Figure S1a. The overlapped epitope segments KVEGTTIADQILRY$_{23\text{-}36}$ and IADQILRYGSLGVF$_{29\text{-}42}$ with 15-hrHPV strains are illustrated in supplementary Figure S1b,c.

3.3. Vaccine Engineering

3.3.1. Designing of Chimeric Vaccine SGD58

The complete vaccine construct consists of (1) two selected epitope sequences (23–36 and 29–42); and (2) two different TLR adjuvants, Flagellin and RS09. Flagellin is recognized as the TLR5 agonist that involves the activation of innate immunity. The head and tail of the vaccine construct contain the N-terminal (5-143 amino acids) and C-terminal regions (419-504 amino acids) of the Flagellin. In addition, RS09, a synthetic short peptide of the TLR4 ligand, is also used; (3) two different T helper (Th) epitopes, namelyPADRE and TpD, are used in the construct. The pan HLA-DR-binding epitope (PADRE) is frequently employed for synthetic or recombinant vaccine development and another universal epitope, TpD, which has 31 amino acids, is also used; and (4) the seven different parts in the vaccine construct were associated with the linkers GPGPG, AAY, and EAAAK. Finally, the designed chimeric vaccine (SGD58) with 318 amino acid sequences was drawn using illustrator for biological sequences (IBS) v1.0 as illustrated in Figure 1

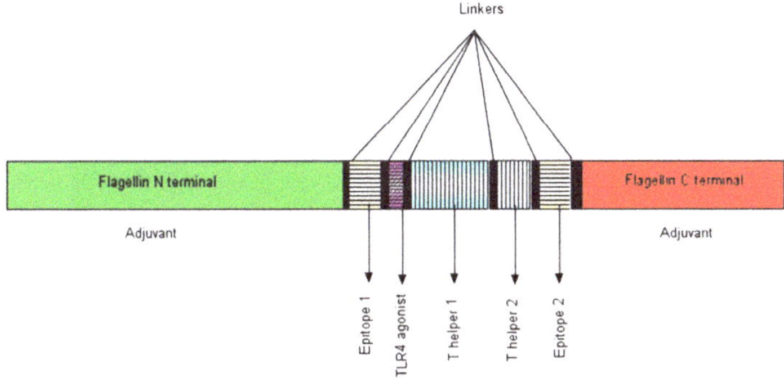

Figure 1. The designed vaccine construct 58 (SGD58). The SGD58 contains seven different segments, namelytwo different adjuvants (N and C-terminal region of Flagellin-TLR5 agonist); two epitope segments (23–36 and 29–42); adjuvant RS09 (TLR4 agonist) and two Th epitopes (PADRE and TpD). All the segments were joined together by using the following linkers (GPGPG, AAY or EAAAK).

3.3.2. Physiochemical and Immunological Properties of SGD58

The various physiochemical properties of the chimeric vaccine (molecular weight, theoretical pI, the total number of negatively and positively charged residues, extinction coefficient, estimated half-life, instability and aliphatic index, GRAVY, and solubility) are demonstrated in Table 2. The designed chimeric vaccine from the L2 protein of HPV58 was highly stable with the computed instability index of 35.88 (<40) and a molecular weight of 33.15 kDa. The highest amino acid composition in the chimeric vaccine is alanine (12.1%), serine (12.1%), leucine (9.3%), asparagine (8.4%), isoleucine (7.6%), glycine (7.0%), glutamine (6.5%), threonine (5.9%), aspartic acid (4.8%), valine (4.5%), glutamic acid (3.7%), proline (3.7%), arginine (3.4%), lysine (3.1%), phenylalanine (2.2%), tyrosine (2.2%), histidine (1.1%), methionine (1.4%), and cysteine (0.3%). The results of the SOLpro analysis indicate that the chimeric vaccine was soluble (0.62) on overexpression in *E. coli*. The antigenic score of the chimeric vaccine was demonstrated as 0.4301 (> threshold of 0.4) and 0.9438 using VaxiJen and ANTIGENpro (Table 2).

Table 2. Evaluation of the various physiochemical and immunological properties of the chimeric vaccine SGD58.

Properties	Results/Values
Number of amino acids	318
Molecular weight	33,394.15
Theoretical pI	8.00
Total number of negatively charged residues (Asp + Glu)	24
Total number of positively charged residues (Arg + Lys)	25
Extinction coefficient M-1 cm-1	12,950
Half-life	20 h (Mammalian reticulocytes, *in vitro*), 30 min (yeast, *in vivo*) and >10h (*E.coli, in vivo*)
Instability index	35.88 (Indicates protein as stable)
Aliphatic index	94.62
Grand average of hydropathicity (GRAVY)	−0.190
Solubility (SolPro)	Soluble with probability 0.621085
Antigenicity (VaxiJen)	0.4301(Probable antigen)
Antigenicity (AntigenPro)	0.943820 (Probable antigen)

3.4. Structural Analysis

3.4.1. Homology Modeling

I-TASSER homology modeling indicated that model 3 of the chimeric vaccine (SGD58) has the highest C-score of −0.39. In addition, the TM and RMSD scores of 0.66 and 7.2Å of model 3 represent the standard similarity and accuracy of the modeled structures. Notably, there is no 3D structure available for the TLR5 on the protein data bank. I-TASSER was used to model the 3D conformational structure of TLR5. The C-score, TM, and RMSD scores of the best model 1 of TLR5 are −0.35, 0.82 and 6.8Å, respectively. Therefore, model 1 is suggested as the best model with a higher confidence level. In addition, modeling with Robetta, we have obtained five different models for TLR5 and the chimeric vaccine candidate. Models 1 and 4 are *de novo* models while models 2, 3, and 5 are *ab initio* models. The Ginzu domain prediction confidence score for TLR5 is 0.9375, and for SGD58 is 0.6502. All the models were selected for structural refinement analysis.

3.4.2. Structural Refinement

The acquired best-modeled 3D structures of SGD58 and TLR5 underwent structural refinement using servers GalaxyWEB and 3D refine. The GalaxyRefine program gives five best-refined models for the whole SGD58 and TLR5. In addition, the lowest 3Drefine score represents the good quality of the refined model, based on the 3D refine force field. All the refined models (1–5) of both GalaxyWEB and 3D refine were used for further structural validation.

3.4.3. Structural Validation

The refined 3D structure obtained in the above section underwent quality improvement using three potential tools: ProSA-web, RAMPAGE, and the ERRAT. The Z-score (ProSA), overall quality factor (ERRAT), and the favored, allowed, and outlier region (RAMPAGE) of the validated 3D structure of SGD58 are given in Table S4 and TLR5 in Table S5. Figure 2 illustrates the Z-score, overall quality factor and the favored, allowed, and outlier region of the selected best SGD58 model. From overall comparison of the obtained results, the Robetta model 3 of SGD58 (Figure S2a) and the Robetta model 5 of TLR5 (Figure S2b) using UCSF Chimera were selected for further analysis.

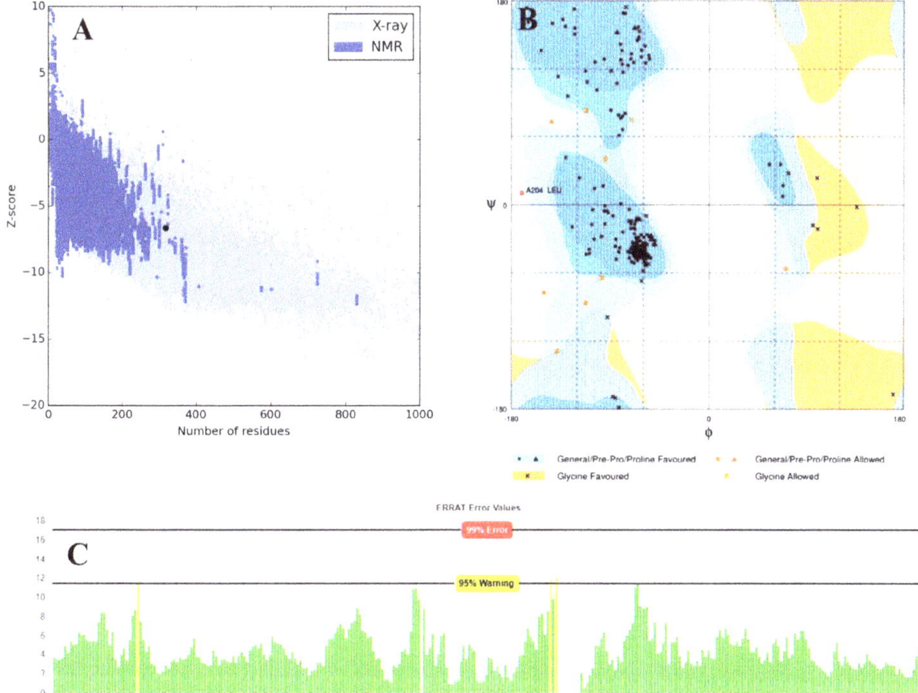

Figure 2. Validation outcome of the refined 3D structure of SGD58. (**A**)The ProSA Z-score of the refined model is -6.65, which is shownby the dark black color spot. The values are presented in the range of native protein conformation. The dark blue and light blue color region represents the Nuclear magnetic resonance and X-ray spectroscopy determination of the experimental protein chains in the protein database (PDB). The X- and Y-axis represent the number of amino acid residues and Z-scores respectively; (**B**) In the Ramachandran plot of the refined model, we illustrated the favored in green circle (96.8%), the allowed in triangle (2.8%) and the outlier in yellow shaded circle (0.3%) regions. (**C**) The ERRAT Plot shows that the overall high quality factor of the refined SGD58 is 99.0033. * On the error axis, two lines are drawn to indicate the confidence with which it is possible to reject regions that exceed that error value. ** Expressed as the percentage of the protein for which the calculated error value falls below the 95% rejection limit. Good high resolution structures generally produce values around 95% or higher. For lower resolutions (2.5 to 3A), the average overall quality factor is around 91%.

3.5. Conformational B-Cell Epitopes

The results of the Discotope 2.0 analysis demonstrated that 18 potential B-cell epitopes were obtained from 318 total residues. Table S6 explains the respective amino acid, residue with contact number, propensity, and Discotope score of the predicted B-cell epitopes.

3.6. Investigation of the Interaction between SGD58 and TLR5

3.6.1. PP Docking Interaction

From the above findings, the best-refined model of SGD58 (Robetta model 3) and TLR5 (Robetta model 5) was used to perform molecular docking using the HADDOCK server. The binding cavity of TLR5 and Flagellin was obtained from a previous report [76,77]. The input TLR5 receptor contains 858 amino acids and SGD58 contains 2923 amino acids. The human TLR5 sequence contains 21 different leucine-rich repeats (LRR) segments with 443 amino acids. Flagellin contains two D1/D0 TLR-binding domains in the N and C terminals of the sequence.

The HADDOCK method directly permits the integration of biophysical information about the protein–protein complex in order to constrain docking. In this study, we docked the target receptor (TLR5) chimeric vaccine candidate (SGD58) to observe the interaction between the complexes. The HADDOCK method clustered 116 structures into 12 clusters, which represents 58.0% of the water-refined models that HADDOCK generated. The TLR5-SGD58 complex shows the highest HADDOCK score of −62.5 ± 7.6, representing the good affinity level between the target and vaccine. The buried surface area (BSA) of cluster 4 of the TLR5-SGD58 complex is 1914.4 ± 124.4, which indicates close proximity and a less water-exposed protein surface. The desolvation energy (43.1 ± 7.9), restraints violation energy (1192.9 ± 96.93), and BSA have a high-quality association with the docking score of the complex. The HADDOCK score, interaction energy, Van der Waals energy, electrostatic energy, desolvation energy, restraints violation energy, and BSA of the top tenclusters are given in Table S7. In all, there are 44 hydrogen bonds between cluster 4 of the TLR5-SGD58 complex. The following amino acids, namely, Lys 148, Asp258, Ser 145, Arg 62, Asn 65, Gln 96, Arg 122, Asn 150, Asn 266, Tyr 120, Lys 177, Gly 76, Glu 80, Gln 72, Thr 73, Ser69, Asn 65, Gln 96, Arg 122, Thr 58, Asp 118, Gly 119, Asp 93, Asn 65, Ser 69, Arg 122, Asn 123, Asp 258, Thr 73, Gln 72, Glu 80, Asn 83, Lys 125, Gln 72, Ser 69, Asn 65, Arg 122, Ser 145, Gln 254, Gly 119, Glu 80, Asn 83, Thr 73, Arg 247, Glu 171, Asp 118, and His 143 act as interacting residues present in the best four clusters of the TLR5-SGD58 complex(Figure 3). Thus, the TLR5-SGD58 complex docking analysis and the intermolecular hydrogen bonding patterns confirm that the interaction of the chimeric vaccine candidate with the target TLR5 can induce both cellular and humoral immunity and inhibit HPV progression.

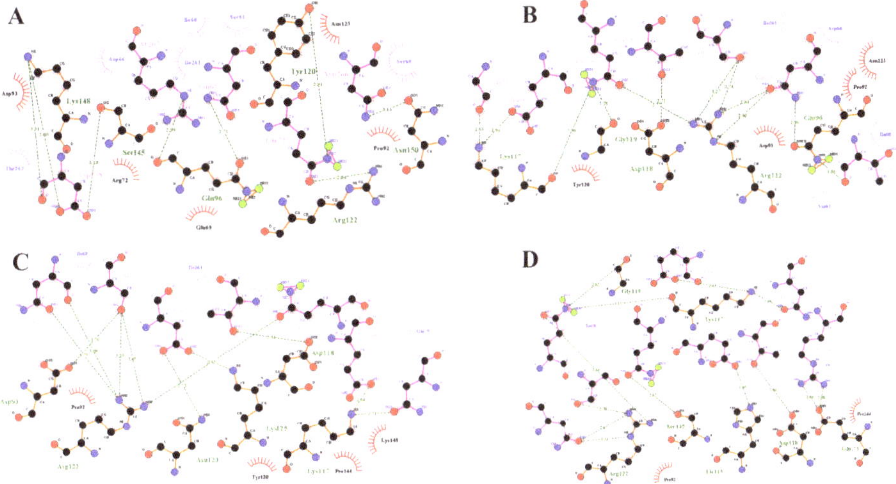

Figure 3. LIGPLOT prepared interacting residues in the TLR5_SGD58 complex. (**A–D**) represent the best structures of the TLR5_SGD58 cluster. The color-coding represents the TLR5 in brown color and the SGD58 in pink color. The dashed lines in green color denote hydrogen-bonding interactions.

3.6.2. MD Simulation

MD simulation demonstrated the stability of the TLR5_SGD58 docked complex in the active site of TLR5. RMSD is the known parameter by which to determine the obtained structure from the MD trajectory. This parameter was evaluated as a preliminary analysis of the backbone atoms of TLR5. The structure and dynamic proprieties of TLR5 were determined using the backbone RMSDs during the simulation period (Figure 4). The RMSD of TLR5 was gradually increased until 20 ns and a nearly 8–100 ns period with 3–4 nm of deviation, and after 30–80 ns, it was stable. The RMSD curve of SGD58 indicates an insignificant variation from 0–20 ns at 5.0 nm, and after 20 ns, it was stable. SGD58 has more a stable RMSD value compared to TLR5 (Figure 4A). The flexibility of each residue in the TLR5-SGD58 complex is calculated based on root mean square fluctuations (RMSF). Figure 4B shows that TLR5 has an insignificant variation of residues, which indicates that this molecule was stable during the MD simulation time of 100 ns. These residues have well known interactions with the vaccine candidate. The ND1b domain (100–200 residues) of the Flagellin fragments of SGD58 shows a low flexibility, which can be attributed to their interaction with the TLR5 protein (Figure 4C). In addition, CD1 and CD0 domains have more fluctuations (200–250 residues). Figure 4D illustrates hydrogen bond interactions throughout the simulation period, to understand the binding efficiency of TLR5 with SGD58. The average number of hydrogen bond interactions was observed in 2.0 nm. Figure S3 illustrates that the potential energy (PE), temperature, total energy (TE), and pressure of SGD58 was stable during the simulation period. The average TE of SGD58 is −7207307343.324 with a standard deviation of 4279.598082. In addition, the average PE of SGD58 is −8985342.697 with a standard deviation of 3370.264894. PE and TE attained equilibrium at a temperature of 300K. The result of the radius of gyration (Rg) analysis is shown in Figure S4. The simultaneous changes in the Rg plots of the complex with TLR5 (Figure S4a) and SGD58 (Figure S4b) indicate that the substantial nature of the complex frequently increases. The Rg plots compression of SGD58 with TLR5 is similar to the RMSD parameter, which indicates the effort of SGD58 to reach internal configuration in TLR5.

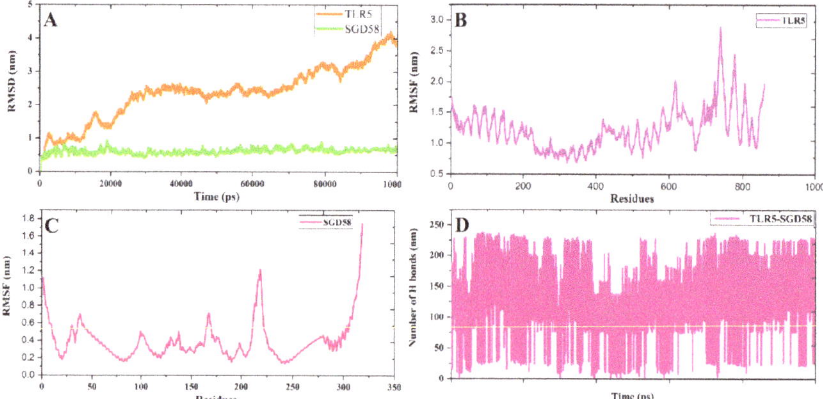

Figure 4. (**A**) The backbone root mean square deviation profile of TLR5 and SGD58 is depicted for the entire 100 ns. (**B**) The root mean square fluctuation of TLR5. (**C**) The root mean square fluctuation of SGD58. (**D**) The total number of intermolecular H bond interactions between SGD58 in complex with TLR5.

3.6.3. Virtual Gene Expression and Cloning

For the given SGD58, the optimized reverse-translated coding sequences were obtained using the EMBOSS Backtranseq tool. The maximal protein expression of this optimized coding sequence in the host (*E. coli*) was analyzed by the GenScript's OptimumGene™ codon optimization tool. Figure S5 illustrates the CAI, GC, and CFD of the gene transcript. The gene (reverse translated coding sequence of the vaccine construct) having an ideal CAI value of 1.00 (>0.8) is more suitable for the above expression (*E. coli*) in the host organism. Moreover, 59.92% of ideal GC content is presented in the gene (between 30% and 70%). However, values outside of these peak ranges would severely inhibit the transcriptional and translational efficiency of the gene products. The CFD value of the gene is 100%, representing their highest codon frequency distribution in the preferred expression organism.

4. Discussion

Immunomics is an integrative area of computer science and experimental immunology and plays a vital role in vaccine development. Immunomics tools and databases are used to forecast the target epitope segments to enhance CTL or B-cell immunity in a cost-effective manner and less experimental time [78–82]. The computational vaccine design involves the engineering of potential non-pathogenic epitope segments with adjuvants to enhance the function of the human immune system against dreadful diseases, including cervical cancer. Unlike conventional vaccines, peptide or epitope vaccines have several advantages; they are synthetic (pathogen-free), have less unwanted side effects, minimize accidental allergenic reactions, and design and predict peptides with self or non-self antigen to elicit and balance the immune responses [83,84]. HPV58 is considered as the most predominant genotype causing cervical cancer incidences in China. HPV has type-restriction, non-targeted delivery, and a high-cost of existing vaccines, which makescontinuing research on HPV vaccine development necessary. Therefore, this study aimed at the design of a chimeric vaccine *via* targeting HPV through immunomics, PP docking, and MD simulation approaches.

Earlier reports suggest that the L2 protein is majorly buried or hidden under the surface of native and matured virions [32,85,86]. The initial interactions between L1/L2 are hydrophobic with coverage of small stretches of amino acid sequences. It exhibits potential effects during *in vitro* assembly [87]. However, the structural relation of L2 minor capsid protein to L1 in the virion particles is not clearly known. In another study, Henio et al. [88] reported that the L1/L2 proteins of HPV have various antigenic epitope segments such as 32–81, 212–231, 272–291, and 347–381 amino acids, and these could be accessible on the surface of L1/L2 virus-like particles. In particular, the N-terminal region of the L2 protein is highly conserved and has diverse functions: it mainly participates in the attachment of the virus particle and its genome assembly in the host system. The N-terminal region 1–12 amino acids are in the DNA binding domain, the 9–12 amino acids are in the furin-cleavage site, and the 22–45 amino acids are in the cyclophilin-B and β-actin-binding domain [89–91]. The L2 protein can act as the prospective target to design the next-generation prophylactic vaccines for HPV [92]. This strategy is prominently supported by the early evidence, such as the production of cross-neutralizing antibodies RG1 [93], K4L2, and K18L2 [94] against the target sites (17–36 AA and 20–38 AA) of L2 in various experimental models. The weak immunogenicity of L2 is a significant obstacles in epitope-based vaccine development; to date, no L2-VLPs-based prophylactic vaccine has been approved for clinical application [30]. Therefore, in this study, we selected the L2 protein to predict the potential epitope candidate and enhance the immunogenicity using adjuvants to design a chimeric vaccine.

Adjuvants have different roles including up-regulation of the immune response, increased action of neutralizing antibodies, processing of cytosolic MHC class-I restricted peptides and target presentation to a specific receptor, acting as an immunogen, and application in the preparation of a single dose [95]. To increase the immunogenicity of L2-derived peptides, the selection of a suitable adjuvant plays a vital role in the vaccine design. Instead of mixing the appropriate adjuvant directly, designing the peptide vaccine candidate using suitable linkers and adjuvants could be highly effective. Alhydrogel® adjuvant 2%, known as "alum," is frequently used as an adjuvant in diphtheria-tetanus-pertussis (DTP), HPV, and hepatitis vaccination [96]. Although the alum adjuvants induced the Th2-mediated immune response, they are ineffective to the pathogen, which is indeed of the Th1 immunity. Later, the emulsion-based incomplete Freund's adjuvant (IFA) was developed, which induces potential Th2, and little Th1-mediated immune responses [97]. However, the application of the emulsion-derived adjuvants is not supported well in the vaccination program due to the induction of autoimmune disorders and an unclear mode of action [98]. To overcome these issues TLR-based ligands were developed and achieved success in the generation of both Th1 and Th2 immune responses in the experimental models. Alphs et al. [99] achieved potential immunogenicity of the synthetic lipo-peptide (HPV16 L2) vaccine through fusion with the Th epitope and TLR ligand. Bacterial Flagellin is a potential TLR5 ligand, which can induce the production of both Th1 and Th2 immune responses.

It is frequently used as an adjuvant in the recombinant vaccine production when fusing with antigenic particles [100,101]. Flagellin, a TLR5 ligand, binds to the particular domain (Toll/interleukin-1 receptor) of the TLR5 receptor in humans. Notably, another newly developed and licensed adjuvant used for human vaccine development is the adjuvant system 04 (AS04). Moreover, AS04 was developed by a combination of 3-O-desacyl-4′-monophosphoryl lipid A (MPL), which is a prominent TLR4 agonist and aluminum salt. In the presence of Cervarix, the AS04 adjuvant induced the function of NF-kappa and cytokine synthesis in cancer cells and animal model. It leads to the appearance of increased antigen-loaded dendritic cells and monocytes followed by CD4$^+$ and B-cells in the injection site [102]. Moreover, as a result of two-dose schedule trial in young girls (9–14 years), the HPV16/18 plus AS04 adjuvant vaccine (Cervarix) was highly immunogenic and has been approved clinically for the prevention of HPV infection, precancerous CIN (I/II/II), and cervical cancer [103]. In 18–25-year-old Chinese women, AS04 adjuvant vaccines were reported as having immunogenicity and an acceptable harmless profile from the randomized-controlled trial [104]. RS09 (short synthetic peptide segments "APPHALS"), is a ligand to TLR4. RS09 does not contain any toxicological effects, and is devoid of skin irritation, serious eye damage, and carcinogenic properties, etc. It successfully enhances the nuclear factor of kappa-light-chain-enhancer of activated B cell (NF-κB) translocation pathways and enhances the pro-inflammatory cytokine and antibody serum concentration in macrophage cells and animal models [105]. TLR adjuvants play an advanced role in commercial vaccines [106]. Two universal Th epitopes (PADRE and TpD) were added to the chimeric vaccine to resolve the deficiency of Th responses. The pan HLA DR-binding epitope is known as "PADRE". It has a binding affinity with more than 15 MHC class-II allotypes and induces proliferative CD4[+] in peripheral blood mononuclear cells (PBMC) from humans [107]. In this manner, it explains the issue raised by HLA polymorphism in the human population [108]. It is extensively studied for synthetic peptide-based vaccine development in C57BL/6 cervical cancer models and Phase I/II clinical trials [109–111]. TpD is another universal memory T-cell helper peptide. The immunization of TpD produces a promising antibody and enhances long-term CD4[+] immune response in mice, *Rhesus macaques*, cynomolgus monkeys, and PBMC in humans [112]. Therefore, we have chosen two different Th epitopes (PADRE and TpD) and two TLR adjuvants (Flagellin and RS04) to enhance the immunogenicity in the chimeric vaccine.

A small flexible linker sequence was employed to join various segments of epitopes, the TLR agonist, and the Th epitopes in the vaccine construct. In this study, GGS linker was used to join the various segments in the vaccine design. The GGS linker facilitates the natural rotation or movement of the epitope segments and adjuvants and ameliorates their free identification by the surface receptor molecules [113,114]. GGS linkers contain nonpolar glycine (Gly) and polar serine (Ser) amino acid residues, which prohibit unnecessary complex formations between the linked partners and retain the function of the chimeric vaccine. A GGS spacer was presented in both natural and artificial linkers, to either increase the stability of the binding domain partners or stabilize the PP complex [115].

The targeted delivery of a vaccine can improve the efficiency and achieve a better outcome. In this study, weselected TLR5 as the target for the chimeric vaccine. Innate immunity-inducing TLR5 are mainly expressed on antigen presenting monocytes and dendritic cells while they encounter the entry of pathogenic microbes [116]. Moreover, TLRs are a well-categorized pattern recognition receptor (PPR) family, which involves the sensing of invading virulent pathogens entry into the host [117]. Horseshoe-shaped LRR are present in each TLRs conserved fold for binding to their respective legends [118]. Numerous studies report that after the binding of TLR5 to the specific ligand, it induces the myeloid differentiation gene 88 (MYD88), which triggers activation of the tumor downstream signaling pathways including NF-κB, mitogen-associated protein kinase (MAPK), and interferon regulatory factors (IRFs) [119]. Once the TLRs recognize their ligand, they become active and induces the production of pro-inflammatory cytokines such as tumor necrosis factor (TNF), interleukins (IL), and INFs [117,118]. In this manner, the host cell increases the capacity to eliminate the invading pathogens. Kim et al. [120] reported that TLR5 is a potential biomarker for the malignant transformation of cervical squamous cells. Therefore, in this study, we selected TLR5 as a target for the chimeric vaccine and its efficiency was determined using PP docking and MD simulation.

TLR5 is an excellent receptor for Flagellin, which is the major component of the bacterial Flagella [116]. Flagellin adjuvant has been extensively used in experimental HPV vaccination. Interestingly, when the host cell responds to Flagellin, TLR5 induces B-cell differentiation into the plasma producing B-cells. Earlier reports demonstrate the significance of Flagellin fused L2-multimer vaccines in experimental rabbits and mice [121,122]. Flagellin is madeup of four important domains: D0, D1, D2, and D3. The D0 and D1 domains are composed of highly conserved N-terminal (1 to 200 amino acids) and C-terminal (405–494 amino acids) regions, which is important for TLR5 agonist action. In addition, the centered hypervariable D2 and D3 regions show the vast differences, by their size and composition, among the various bacterial microorganisms [123–125]. Owing to the higher antigenicity and toxicity caused by the central D2/D3 domain, this antigenic part is removed or replaced by the optimized epitope segments or different adjuvants in the vaccine design. D2/D3 antigen replacement in Flagellin enhances mucosa-immunoglobulin productions in the experimental animal models through intranasal immunization [126]. Therefore, we selected the N and C-terminal regions of Flagellin in the design of the chimeric vaccine. Forstneric et al. [125] reported the appropriate identification of Flagellin by the homology modeled hTLR5 and mTLR5. The crystallographic complex structure of zebra fish TLR5 with the domains (D1, D2 or D3) of *Salmonella* sp. were also studied [77]. However, there was a lack of availability of the crystal structure of TLR5 until now. Therefore, in this study, homology modeling, structural refinement, and validation were performed and found using the Robetta model 5 of TLR5 obtained using the Robetta, 3DGalaxyRefine, ProSA-web, RAMPAGE, and ERRAT. Moreover,

TLR5 is greatly presented in vertebrates [77,125,127]. It facilitates the PP interaction analysis of TLR5 with Flagellin using HADDOCK, as shown in Figure 4b. It shows the interacting residues of this complex observed between the LRR region of TLR5 and the D0/D1 domain of Flagellin. This finding is significantly supported by earlier reports [125,127] concerning TLR5 recognition of the D0 of Flagellin by the inflammasome receptor, for preventing the immune escape of invading pathogenic strains. The MD simulation results depict a constant and stable interaction between SGD58 and TLR5. During the MD simulation study, TLR5 was stable after 15 ns, whereas SGD58 exhibited insignificant variations and was then stable after 10 ns. The structural changes were observed to have gained the optimal sustainability of SGD58 and TLR5. In addition, very slight changes were noticed in the D0/D1 domain regions of SGD58. Finally, the results of the codon optimization and virtual cloning confirms the translated chimeric vaccine sequence in *E. coli* to be capable of regulating the higher level of gene expression. The successful expression of the designed virus-like particles of HPV in *E. coli* is reported in earlier studies [128,129], which can enhance the production of the vaccine at a cheaper cost.

From this report, the new chimeric vaccine candidate was engineered using various immunomics tools, PP docking, and MD simulation, which can reduce the experimental cost and time. The designed chimeric vaccine SGD58, has appropriate structurally refined, physiochemical, and immunological properties that can produce humoral and cellular immune responses against HPV. The designed chimeric vaccine has cross-production with the 15 different hrHPV strains. Further experimental investigation is planned to determine the efficiency of the chimeric vaccine, especially allele-specific for the Chinese population.

Supplementary Materials: The following are available online at http://www.mdpi.com/1999-4915/11/1/63/s1. Figure S1: Conservation of two overlapped epitope segments in fifteen hrHPV strains. Figure S2: Refined 3D structure of the SGD58 and TLR5 by using UCSF Chimera. (a) The 3D structure of the SGD58 was obtained through homology modeling by using Robetta. (b) The 3D structure of the TLR5 was obtained through homology modeling by using Robetta. Figure S3: (a) The total energy, (b) potential energy, (c) temperature and (d) pressure plots of MD simulation for the TLR5-SGD58 complex in simulations of 100 ns. Figure S4: (a) Rg plot of the TLR5 and (b) The Rg plot of SGD58 complex. Figure S5: Codon optimization and *in silico* cloning of the gene. Table S1: The overlapped epitope segments of MHC-I, CTL and TCR from N-terminal region of HPV58 were predicted by using different servers. Table S2: The overlapped epitope segments of MHC-II, INF-gamma producing and B-cell epitopes N-terminal region of HPV58 by using different servers. Table S3: Conservation across-hrHPV strains by the overlapped HPV58 epitope segments. Table S4: Validation of 3D structures of the designed SGD58 obtained by the I-TASSER (I-T) and Robetta and its refinement by the Galaxy Refine (Gal) and 3Drefine (3DR). Table S5: Validation of 3D structures of the TLR5 obtained by the I-TASSER (I-T) and Robetta and its refinement by the GalaxyRefine (Gal) and 3Drefine (3DR). Table S6: Dis-continuous B-cell epitopes identified in the refined 3D structure of designed vaccine constructs of HPV58 by using Discotope 2.0. Table S7: Statistical analysis of the TLR5-SGD58 docking result obtained by HADDOCK.

Author Contributions: G.S., S.K., and D.-Q.W. conceived and designed the experiment. S.K and G.S. performed the immunoinformatics, vaccine design, and molecular docking studies. S.C., Q.W., and A.S.N. performed the molecular dynamics simulation. W.C.C., S.K., G.S., S.C., and D.-Q.W. wrote the main manuscript text. S.K. and A.S.N. formatted the manuscript and figures according to the instructions. G.S., W.C.C., G.K., and D.-Q.W. critically reviewed the manuscript. All the authors approved the final manuscript.

Acknowledgments: The authors duly acknowledge the financial support of the Ministry of Science and Technology of China (Grant No.: 2016YFA0501703), Henan Natural Science (Grant No.: 162300410060), grants from the State Key Lab on Microbial Metabolism, and joint research funds for Medical and Engineering and Scientific Research at Shanghai Jiao Tong University (YG2017ZD14) to DQ.W.; Henan University of Technology (Grant No.: 21450004) to S.K.; Henan University of Technology (Grant No.: 21450003) and China Postdoctoral Science Foundation (Grant No.: 2018M632766) to G.S.; The simulations in this work were supported by the Center for High Performance Computing, Shanghai Jiao Tong University, China.

References

1. Frazer, I.H. Development and implementation of papillomavirus prophylactic vaccines. *J. Immunol.* **2014**, *192*, 4007–4011. [CrossRef] [PubMed]
2. McLaughlin-Drubin, M.E.; Munger, K. Oncogenic activities of human papillomaviruses. *Virus Res.* **2009**, *143*, 195–208. [CrossRef] [PubMed]

3. Bernard, H.U.; Burk, R.D.; Chen, Z.; Van Doorslaer, K.; ZurHausen, H.; de Villiers, E.M. Classification of papillomaviruses (PVs) based on 189 PV types and proposal of taxonomic amendments. *Virology* **2010**, *401*, 70–79. [CrossRef] [PubMed]

4. Fauquet, C.M.; Mayo, M.A.; Maniloff, J.; Desselberger, U.; Ball, LA. Family Papillomaviridae. In *Virus Taxonomy. The Eighth Report of the International Committee on Taxonomy of Viruses*; Elsevier: Amsterdam, The Netherlands, 2005; pp. 239–255.

5. Haedicke, J.; Iftner, T. Human papillomaviruses and cancer. *Radiother. Oncol.* **2013**, *108*, 397–402. [CrossRef]

6. Tao, G.; Yaling, G.; Zhan, G.; Pu, L.; Miao, H. Human papillomavirus genotype distribution among HPV-positive women in Sichuan province, Southwest China. *Arch. Virol.* **2018**, *163*, 65–72. [CrossRef]

7. Dunne, E.F.; Unger, E.R.; Sternberg, M.; McQuillan, G.; Swan, D.C.; Patel, S.S.; Markowitz, L.E. Prevalence of HPV infection among females in the United States. *JAMA* **2007**, *297*, 813–819. [CrossRef]

8. Kenter, G.G.; Welters, M.J.; Valentijn, A.R.; Lowik, M.J.; Berends-van der Meer, D.M.; Vloon, A.P.; Drijfhout, J.W.; Wafelman, A.R.; Oostendorp, J.; Fleuren, G.J.; et al. Phase I immunotherapeutic trial with long peptides spanning the E6 and E7 sequences of high-risk human papillomavirus 16 in end-stage cervical cancer patients shows low toxicity and robust immunogenicity. *Clin. Cancer Res.* **2008**, *14*, 169–177. [CrossRef]

9. de Vos van Steenwijk, P.J.; Ramwadhdoebe, T.H.; Lowik, M.J.; van der Minne, C.E.; Berends-van der Meer, D.M.; Fathers, L.M.; Valentijn, A.R.; Oostendorp, J.; Fleuren, G.J.; Hellebrekers, B.W.; et al. A placebo-controlled randomized HPV16 synthetic long-peptide vaccination study in women with high-grade cervical squamous intraepithelial lesions. *Cancer Immunol. Immunother.* **2012**, *61*, 1485–1492. [CrossRef]

10. Rerucha, C.M.; Caro, R.J.; Wheeler, V.L. Cervical Cancer Screening. *Am. Fam. Physician* **2018**, *97*, 441–448.

11. Hong, Y.; Zhang, C.; Li, X.; Lin, D.; Liu, Y. HPV and cervical cancer related knowledge, awareness and testing behaviors in a community sample of female sex workers in China. *BMC Public Health* **2013**, *13*, 696. [CrossRef]

12. Chen, W.; Zheng, R.; Baade, P.D.; Zhang, S.; Zeng, H.; Bray, F.; Jemal, A.; Yu, X.Q.; He, J. Cancer statistics in China, 2015. *CA Cancer J. Clin.* **2016**, *66*, 115–132. [CrossRef] [PubMed]

13. Zhou, H.L.; Zhang, W.; Zhang, C.J.; Wang, S.M.; Duan, Y.C.; Wang, J.X.; Yang, H.; Wang, X.Y. Prevalence and distribution of human papillomavirus genotypes in Chinese women between 1991 and 2016: A systematic review. *J. Infect.* **2018**, *76*, 522–528. [CrossRef] [PubMed]

14. You, W.; Li, S.; Du, R.; Zheng, J.; Shen, A. Epidemiological study of high-risk human papillomavirus infection in subjects with abnormal cytological findings in cervical cancer screening. *Exp. Ther. Med.* **2018**, *15*, 412–418. [CrossRef] [PubMed]

15. Long, W.; Yang, Z.; Li, X.; Chen, M.; Liu, J.; Zhang, Y.; Sun, X. HPV-16, HPV-58, and HPV-33 are the most carcinogenic HPV genotypes in Southwestern China and their viral loads are associated with severity of premalignant lesions in the cervix. *Virol. J.* **2018**, *15*, 94. [CrossRef] [PubMed]

16. Zhang, C.; Huang, C.; Zheng, X.; Pan, D. Prevalence of human papillomavirus among Wenzhou women diagnosed with cervical intraepithelial neoplasia and cervical cancer. *Infect. Agent. Cancer* **2018**, *13*. [CrossRef] [PubMed]

17. Dai, X.; Chen, L.; Li, J.; Wu, Y.; Hu, Y.; Xiang, F.; Guan, Q. Distribution characteristics of different human papillomavirus genotypes in women in Wuhan, China. *Cancer Med.* **2018**, *32*, e22581. [CrossRef]

18. Zhao, P.; Liu, S.; Zhong, Z.; Hou, J.; Lin, L.; Weng, R.; Su, L.; Lei, N.; Hou, T.; Yang, H. Prevalence and genotype distribution of human papillomavirus infection among women in northeastern Guangdong Province of China. *J. Clin. Lab. Anal.* **2018**, *18*, 204. [CrossRef] [PubMed]

19. Liu, S.; Zhong, Z.; Hou, J.; Lin, L.; Weng, R.; Su, L.; Lei, N.; Hou, T.; Yang, H.; Li, K.; et al. Analysis of HPV distribution in patients with cervical precancerous lesions in Western China. *BMC Infect. Dis.* **2017**, *96*, e7304. [CrossRef]

20. Zhang, C.; Zhang, C.; Huang, J.; Shi, W. The Genotype of Human Papillomavirus and Associated Factors Among High Risk Males in Shanghai, China: A Molecular Epidemiology Study. *Med. Sci. Monit.* **2018**, *24*, 912–918. [CrossRef]

21. FDA licensure of bivalent human papillomavirus vaccine (HPV2, Cervarix) for use in females and updated HPV vaccination recommendations from the Advisory Committee on Immunization Practices (ACIP). *MMWR Morb. Mortal. Wkly. Rep.* **2010**, *59*, 626–629.

22. Recommendations on the use of quadrivalent human papillomavirus vaccine in males–Advisory Committee on Immunization Practices (ACIP), 2011. *MMWR Morb. Mortal. Wkly. Rep.* **2011**, *60*, 1705–1708.

23. Petrosky, E.; Bocchini, J.A., Jr.; Hariri, S.; Chesson, H.; Curtis, C.R.; Saraiya, M.; Unger, E.R.; Markowitz, L.E. Use of 9-valent human papillomavirus (HPV) vaccine: Updated HPV vaccination recommendations of the advisory committee on immunization practices. *MMWR Morb. Mortal. Wkly. Rep.* **2015**, *64*, 300–304. [PubMed]

24. Paz-Zulueta, M.; Alvarez-Paredes, L.; Rodriguez Diaz, J.C.; Paras-Bravo, P.; Andrada Becerra, M.E.; Rodriguez Ingelmo, J.M.; Ruiz Garcia, M.M.; Portilla, J.; Santibanez, M. Prevalence of high-risk HPV genotypes, categorised by their quadrivalent and nine-valent HPV vaccination coverage, and the genotype association with high-grade lesions. *BMC Cancer* **2018**, *18*, 112. [CrossRef] [PubMed]

25. Jiang, R.T.; Schellenbacher, C.; Chackerian, B.; Roden, R.B. Progress and prospects for L2-based human papillomavirus vaccines. *Expert Rev. Vaccines* **2016**, *15*, 853–862. [CrossRef] [PubMed]

26. Chroboczek, J.; Szurgot, I.; Szolajska, E. Virus-like particles as vaccine. *ActaBiochim. Pol.* **2014**, *61*, 531–539.

27. Pandhi, D.; Sonthalia, S. Human papilloma virus vaccines: Current scenario. *Indian J. Sex Transm. Dis. AIDS* **2011**, *32*, 75–85. [CrossRef] [PubMed]

28. Monie, A.; Hung, C.F.; Roden, R.; Wu, T.C. Cervarix: A vaccine for the prevention of HPV 16, 18-associated cervical cancer. *Biologics* **2008**, *2*, 97–105.

29. Angioli, R.; Lopez, S.; Aloisi, A.; Terranova, C.; De Cicco, C.; Scaletta, G.; Capriglione, S.; Miranda, A.; Luvero, D.; Ricciardi, R.; et al. Ten years of HPV vaccines: State of art and controversies. *Crit. Rev. Oncol. Hematol.* **2016**, *102*, 65–72. [CrossRef]

30. Karanam, B.; Jagu, S.; Huh, W.K.; Roden, R.B. Developing vaccines against minor capsid antigen L2 to prevent papillomavirus infection. *Immunol. Cell Biol.* **2009**, *87*, 287–299. [CrossRef]

31. Schiller, J.T.; Muller, M. Next generation prophylactic human papillomavirus vaccines. *Lancet Oncol.* **2015**, *16*, e217–e225. [CrossRef]

32. Wang, J.W.; Roden, R.B. L2, the minor capsid protein of papillomavirus. *Virology* **2013**, *445*, 175–186. [CrossRef] [PubMed]

33. Chandrachud, L.M.; Grindlay, G.J.; McGarvie, G.M.; O'Neil, B.W.; Wagner, E.R.; Jarrett, W.F.; Campo, M.S. Vaccination of cattle with the N-terminus of L2 is necessary and sufficient for preventing infection by bovine papillomavirus-4. *Virology* **1995**, *211*, 204–208. [CrossRef] [PubMed]

34. Gaukroger, J.M.; Chandrachud, L.M.; O'Neil, B.W.; Grindlay, G.J.; Knowles, G.; Campo, M.S. Vaccination of cattle with bovine papillomavirus type 4 L2 elicits the production of virus-neutralizing antibodies. *J. Gen. Virol.* **1996**, *77 (Pt 7)*, 1577–1583. [CrossRef]

35. UniProt: The universal protein knowledgebase. *Nucleic Acids Res.* **2017**, *45*, D158–D169. [CrossRef] [PubMed]

36. Chen, S.; Hong, W.; Shao, H.; Fu, Y.; Liu, X.; Chen, D.; Xu, A. Allelic distribution of HLA class I genes in the Tibetan ethnic population of China. *Int. J. Immunogenet.* **2006**, *33*, 439–445. [CrossRef] [PubMed]

37. Chen, S.; Hu, Q.; Xie, Y.; Zhou, L.; Xiao, C.; Wu, Y.; Xu, A. Origin of Tibeto-Burman speakers: Evidence from HLA allele distribution in Lisu and Nu inhabiting Yunnan of China. *Hum. Immunol.* **2007**, *68*, 550–559. [CrossRef]

38. Chen, S.; Ren, X.; Liu, Y.; Hu, Q.; Hong, W.; Xu, A. Human leukocyte antigen class I polymorphism in Miao, Bouyei, and Shui ethnic minorities of Guizhou, China. *Hum. Immunol.* **2007**, *68*, 928–933. [CrossRef]

39. Wang, X.C.; Sun, L.Q.; Ma, L.; Li, H.X.; Wang, X.L.; Wang, X.; Yun, T.; Meng, N.L.; Lv, D.L. Prevalence and genotype distribution of human papillomavirus among women from Henan, China. *Asian Pac. J. Cancer Prev.* **2014**, *15*, 7333–7336. [CrossRef] [PubMed]

40. Zhang, H.Y.; Fei, M.D.; Jiang, Y.; Fei, Q.Y.; Qian, H.; Xu, L.; Jin, Y.N.; Jiang, C.Q.; Li, H.X.; Tiggelaar, S.M.; et al. The diversity of human papillomavirus infection among human immunodeficiency virus-infected women in Yunnan, China. *Virol. J.* **2014**, *11*, 202. [CrossRef] [PubMed]

41. Lu, J.F.; Shen, G.R.; Li, Q.; Chen, X.; Ma, C.F.; Zhu, T.H. Genotype distribution characteristics of multiple human papillomavirus in women from the Taihu River Basin, on the coast of eastern China. *BMC Infect. Dis.* **2017**, *17*, 226. [CrossRef]

42. Wang, Y.; Xue, J. Distribution and role of high-risk human papillomavirus genotypes in women with cervical intraepithelial neoplasia: A retrospective analysis from Wenzhou, southeast China. *Cancer Med.* **2018**. [CrossRef] [PubMed]

43. Dai, X.; Chen, L.; Li, J.; Wu, Y.; Hu, Y.; Vita, R.; Overton, J.A.; Greenbaum, J.A.; Ponomarenko, J.; Clark, J.D.; et al. The immune epitope database (IEDB) 3.0. *Cancer Med.* **2015**, *43*, D405–D412.

44. Moutaftsi, M.; Peters, B.; Pasquetto, V.; Tscharke, D.C.; Sidney, J.; Bui, H.H.; Grey, H.; Sette, A. A consensus epitope prediction approach identifies the breadth of murine T(CD8+)-cell responses to vaccinia virus. *Nat. Biotechnol.* **2006**, *24*, 817–819. [CrossRef] [PubMed]

45. Nielsen, M.; Lundegaard, C.; Worning, P.; Lauemoller, S.L.; Lamberth, K.; Buus, S.; Brunak, S.; Lund, O. Reliable prediction of T-cell epitopes using neural networks with novel sequence representations. *Protein Sci.* **2003**, *12*, 1007–1017. [CrossRef] [PubMed]

46. Bhasin, M.; Raghava, G.P. Prediction of CTL epitopes using QM, SVM and ANN techniques. *Vaccine* **2004**, *22*, 3195–3204. [CrossRef] [PubMed]

47. Liu, I.H.; Lo, Y.S.; Yang, J.M. PAComplex: A web server to infer peptide antigen families and binding models from TCR-pMHC complexes. *Nucleic Acids Res.* **2011**, *39*, W254–W260. [CrossRef]

48. Wang, P.; Sidney, J.; Dow, C.; Mothe, B.; Sette, A.; Peters, B. A systematic assessment of MHC class II peptide binding predictions and evaluation of a consensus approach. *PLoS Comput. Biol.* **2008**, *4*, e1000048. [CrossRef]

49. Paul, S.; Sidney, J.; Sette, A.; Peters, B. TepiTool: A Pipeline for Computational Prediction of T Cell Epitope Candidates. *Curr. Protoc. Immunol.* **2016**, *114*, 18–19. [CrossRef]

50. Dhanda, S.K.; Vir, P.; Raghava, G.P. Designing of interferon-gamma inducing MHC class-II binders. *Biol. Direct* **2013**, *8*, 30. [CrossRef]

51. Saha, S.; Raghava, G.P. Prediction of continuous B-cell epitopes in an antigen using recurrent neural network. *Proteins* **2006**, *65*, 40–48. [CrossRef]

52. Dimitrov, I.; Bangov, I.; Flower, D.R.; Doytchinova, I. AllerTOP v.2—A server for in silico prediction of allergens. *J. Mol. Model.* **2014**, *20*, 2278. [CrossRef] [PubMed]

53. Dimitrov, I.; Naneva, L.; Doytchinova, I.; Bangov, I. Allergen FP: Allergenicity prediction by descriptor fingerprints. *Bioinformatics* **2014**, *30*, 846–851. [CrossRef] [PubMed]

54. El-Manzalawy, Y.; Dobbs, D.; Honavar, V. Predicting protective bacterial antigens using random forest classifiers. In Proceedings of the ACM Conference on Bioinformatics, Computational Biology and Biomedicine, Orlando, FL, USA, 7–10 October 2012; pp. 426–433.

55. Chen, C.; Li, Z.; Huang, H.; Suzek, B.E.; Wu, C.H. A fast Peptide Match service for UniProt Knowledgebase. *Bioinformatics* **2013**, *29*, 2808–2809. [CrossRef] [PubMed]

56. Bui, H.H.; Sidney, J.; Li, W.; Fusseder, N.; Sette, A. Development of an epitope conservancy analysis tool to facilitate the design of epitope-based diagnostics and vaccines. *BMC Bioinform.* **2007**, *8*, 361. [CrossRef] [PubMed]

57. Kumar, S.; Stecher, G.; Tamura, K. MEGA7: Molecular Evolutionary Genetics Analysis Version 7.0 for Bigger Datasets. *Mol. Biol. Evol.* **2016**, *33*, 1870–1874. [CrossRef] [PubMed]

58. Magnan, C.N.; Randall, A.; Baldi, P. SOLpro: Accurate sequence-based prediction of protein solubility. *Bioinformatics* **2009**, *25*, 2200–2207. [CrossRef]

59. Gasteiger, E.; Hoogland, C.; Gattiker, A.; Duvaud, S.E.; Wilkins, M.R.; Appel, R.D.; Bairoch, A. Protein Identification and Analysis Tools on the ExPASy Server. In *The Proteomics Protocols Handbook*; Walker, J.M., Ed.; Humana Press: Totowa, NJ, USA, 2005; pp. 571–607.

60. Doytchinova, I.A.; Flower, D.R. VaxiJen: A server for prediction of protective antigens, tumour antigens and subunit vaccines. *BMC Bioinform.* **2007**, *8*, 4. [CrossRef]

61. Zhang, Y. I-TASSER server for protein 3D structure prediction. *BMC Bioinform.* **2008**, *9*, 40. [CrossRef] [PubMed]

62. Kim, D.E.; Chivian, D.; Baker, D. Protein structure prediction and analysis using the Robetta server. *Nucleic Acids Res.* **2004**, *32*, W526–W531. [CrossRef]

63. Ko, J.; Park, H.; Heo, L.; Seok, C. GalaxyWEB server for protein structure prediction and refinement. *Nucleic Acids Res.* **2012**, *40*, W294–W297. [CrossRef]

64. Bhattacharya, D.; Cheng, J. i3Drefine software for protein 3D structure refinement and its assessment in CASP10. *PLoS ONE* **2013**, *8*, e69648. [CrossRef] [PubMed]

65. Wiederstein, M.; Sippl, M.J. ProSA-web: Interactive web service for the recognition of errors in three-dimensional structures of proteins. *Nucleic Acids Res.* **2007**, *35*, W407–W410. [CrossRef] [PubMed]

66. Lovell, S.C.; Davis, I.W.; Arendall, W.B., 3rd; de Bakker, P.I.; Word, J.M.; Prisant, M.G.; Richardson, J.S.; Richardson, D.C. Structure validation by Calpha geometry: phi, psi and Cbeta deviation. *Proteins* **2003**, *50*, 437–450. [CrossRef] [PubMed]

67. Colovos, C.; Yeates, T.O. Verification of protein structures: Patterns of nonbonded atomic interactions. *Protein Sci.* **1993**, *2*, 1511–1519. [CrossRef] [PubMed]

68. Kringelum, J.V.; Lundegaard, C.; Lund, O.; Nielsen, M. Reliable B cell epitope predictions: Impacts of method development and improved benchmarking. *PLoS Comput. Biol.* **2012**, *8*, e1002829. [CrossRef] [PubMed]

69. van Zundert, G.C.P.; Rodrigues, J.P.G.L.M.; Trellet, M.; Schmitz, C.; Kastritis, P.L.; Karaca, E.; Melquiond, A.S.J.; van Dijk, M.; de Vries, S.J.; Bonvin, A.M.J.J. The HADDOCK2.2 Web Server: User-Friendly Integrative Modeling of Biomolecular Complexes. *J. Mol. Biol.* **2016**, *428*, 720–725. [CrossRef] [PubMed]

70. Wallace, A.C.; Laskowski, R.A.; Thornton, J.M. LIGPLOT: A program to generate schematic diagrams of protein-ligand interactions. *Protein Eng.* **1996**, *8*, 127–134. [CrossRef]

71. Lei, D.; Zhang, X.; Jiang, S.; Cai, Z.; Rames, M.J.; Zhang, L.; Ren, G.; Zhang, S. Structural features of cholesteryl ester transfer protein: A molecular dynamics simulation study. *Proteins* **2013**, *81*, 415–425. [CrossRef]

72. Schuttelkopf, A.W.; van Aalten, D.M. PRODRG: A tool for high-throughput crystallography of protein-ligand complexes. *ActaCrystallogr. D Biol. Crystallogr.* **2004**, *60*, 1355–1363. [CrossRef]

73. Kirubakaran, P.; Kothandan, G.; Cho, S.J.; Muthusamy, K. Molecular insights on TNKS1/TNKS2 and inhibitor-IWR1 interactions. *Mol. Biosyst.* **2014**, *10*, 281–293. [CrossRef]

74. Rice, P.; Longden, I.; Bleasby, A. EMBOSS: The European Molecular Biology Open Software Suite. *Trends Genet.* **2000**, *16*, 276–277. [CrossRef]

75. The GenScript Rare Codon Analysis. Available online: https://www.genscript.com/tools/rare-codon-analysis (accessed on 1 October 2018).

76. Song, W.S.; Jeon, Y.J.; Namgung, B.; Hong, M.; Yoon, S.I. A conserved TLR5 binding and activation hot spot on flagellin. *Sci. Rep.* **2017**, *7*, 40878. [CrossRef] [PubMed]

77. Yoon, S.I.; Kurnasov, O.; Natarajan, V.; Hong, M.; Gudkov, A.V.; Osterman, A.L.; Wilson, I.A. Structural basis of TLR5-flagellin recognition and signaling. *Science* **2012**, *335*, 859–864. [CrossRef] [PubMed]

78. Negahdaripour, M.; Eslami, M.; Nezafat, N.; Hajighahramani, N.; Ghoshoon, M.B.; Shoolian, E.; Dehshahri, A.; Erfani, N.; Morowvat, M.H.; Ghasemi, Y. A novel HPV prophylactic peptide vaccine, designed by immunoinformatics and structural vaccinology approaches. *Infect. Genet. Evol.* **2017**, *54*, 402–416. [CrossRef] [PubMed]

79. Kaliamurthi, S.; Selvaraj, G.; Kaushik, A.C.; Gu, K.R.; Wei, D.Q. Designing of CD8(+) and CD8(+)-overlapped CD4(+) epitope vaccine by targeting late and early proteins of human papillomavirus. *Biologics* **2018**, *12*, 107–125.

80. Khan, A.; Junaid, M.; Kaushik, A.C.; Ali, A.; Ali, S.S.; Mehmood, A.; Wei, D.Q. Computational identification, characterization and validation of potential antigenic peptide vaccines from hrHPVs E6 proteins using immunoinformatics and computational systems biology approaches. *PLoS ONE* **2018**, *13*, e0196484. [CrossRef]

81. Zhao, M.; Wei, D.Q. Rare Diseases: Drug Discovery and Informatics Resource. *Interdiscip. Sci.* **2018**, *10*, 195–204. [CrossRef]

82. Sabah, S.N.; Gazi, M.A.; Sthity, R.A.; Husain, A.B.; Quyyum, S.A.; Rahman, M.; Islam, M.R. Designing of Epitope-Focused Vaccine by Targeting E6 and E7 Conserved Protein Sequences: An Immuno-Informatics Approach in Human Papillomavirus 58 Isolates. *Interdiscip. Sci.* **2018**, *10*, 251–260. [CrossRef]

83. Kaliamurthi, S.; Selvaraj, G.; Junaid, M.; Khan, A.; Gu, K.; Wei, D.Q. Cancer Immunoinformatics: A Promising Era in the Development of Peptide Vaccines for Human Papillomavirus induced Cervical cancer. *Curr. Pharm. Des.* **2018**. [CrossRef]

84. Purcell, A.W.; McCluskey, J.; Rossjohn, J. More than one reason to rethink the use of peptides in vaccine design. *Nat. Rev. Drug Discov.* **2007**, *6*, 404–414. [CrossRef]

85. Bywaters, S.M.; Brendle, S.A.; Tossi, K.P.; Biryukov, J.; Meyers, C.; Christensen, N.D. Antibody Competition Reveals Surface Location of HPV L2 Minor Capsid Protein Residues 17-36. *Viruses* **2017**, *9*, 336. [CrossRef] [PubMed]

86. Buck, C.B.; Cheng, N.; Thompson, C.D.; Lowy, D.R.; Steven, A.C.; Schiller, J.T.; Trus, B.L. Arrangement of L2 within the papillomavirus capsid. *J. Virol.* **2008**, *82*, 5190–5197. [CrossRef] [PubMed]

87. Finnen, R.L.; Erickson, K.D.; Chen, X.S.; Garcea, R.L. Interactions between papillomavirus L1 and L2 capsid proteins. *J. Virol.* **2003**, *77*, 4818–4826. [CrossRef] [PubMed]

88. Heino, P.; Skyldberg, B.; Lehtinen, M.; Rantala, I.; Hagmar, B.; Kreider, J.W.; Kirnbauer, R.; Dillner, J. Human papillomavirus type 16 capsids expose multiple type-restricted and type-common antigenic epitopes. *J. Gen. Virol.* **1995**, *76*, 1141–1153. [CrossRef]

89. Yang, R.; Yutzy, W.H.T.; Viscidi, R.P.; Roden, R.B. Interaction of L2 with beta-actin directs intracellular transport of papillomavirus and infection. *J. Biol. Chem.* **2003**, *278*, 12546–12553. [CrossRef]

90. Fay, A.; Yutzy, W.H.T.; Roden, R.B.; Moroianu, J. The positively charged termini of L2 minor capsid protein required for bovine papillomavirus infection function separately in nuclear import and DNA binding. *J. Virol.* **2004**, *78*, 13447–13454. [CrossRef]

91. Richards, R.M.; Lowy, D.R.; Schiller, J.T.; Day, P.M. Cleavage of the papillomavirus minor capsid protein, L2, at a furin consensus site is necessary for infection. *Proc. Natl. Acad. Sci. USA* **2006**, *103*, 1522–1527. [CrossRef]

92. Tyler, M.; Tumban, E.; Chackerian, B. Second-generation prophylactic HPV vaccines: Successes and challenges. *Expert Rev. Vaccines* **2014**, *13*, 247–255. [CrossRef]

93. Gambhira, R.; Jagu, S.; Karanam, B.; Gravitt, P.E.; Culp, T.D.; Christensen, N.D.; Roden, R.B. Protection of rabbits against challenge with rabbit papillomaviruses by immunization with the N terminus of human papillomavirus type 16 minor capsid antigen L2. *J. Virol.* **2007**, *81*, 11585–11592. [CrossRef]

94. Rubio, I.; Seitz, H.; Canali, E.; Sehr, P.; Bolchi, A.; Tommasino, M.; Ottonello, S.; Muller, M. The N-terminal region of the human papillomavirus L2 protein contains overlapping binding sites for neutralizing, cross-neutralizing and non-neutralizing antibodies. *Virology* **2011**, *409*, 348–359. [CrossRef]

95. Cox, J.C.; Coulter, A.R. Adjuvants—A classification and review of their modes of action. *Vaccine* **1997**, *15*, 248–256. [CrossRef]

96. Marrack, P.; McKee, A.S.; Munks, M.W. Towards an understanding of the adjuvant action of aluminium. *Nat. Rev. Immunol.* **2009**, *9*, 287–293. [CrossRef] [PubMed]

97. Chan, P.K.; Liu, S.J.; Cheung, T.H.; Yeo, W.; Ngai, S.M.; Cheung, J.L.; Chong, P.; Man, S. T-cell response to human papillomavirus type 58 L1, E6, And E7 peptides in women with cleared infection, cervical intraepithelial neoplasia, or invasive cancer. *Clin. Vaccine Immunol.* **2010**, *17*, 1315–1321. [CrossRef] [PubMed]

98. Ott, G.; Barchfeld, G.L.; Chernoff, D.; Radhakrishnan, R.; van Hoogevest, P.; Van Nest, G. MF59. Design and evaluation of a safe and potent adjuvant for human vaccines. *Pharm. Biotechnol.* **1995**, *6*, 277–296.

99. Alphs, H.H.; Gambhira, R.; Karanam, B.; Roberts, J.N.; Jagu, S.; Schiller, J.T.; Zeng, W.; Jackson, D.C.; Roden, R.B. Protection against heterologous human papillomavirus challenge by a synthetic lipopeptide vaccine containing a broadly cross-neutralizing epitope of L2. *Proc. Natl. Acad. Sci. USA* **2008**, *105*, 5850–5855. [CrossRef] [PubMed]

100. Huleatt, J.W.; Jacobs, A.R.; Tang, J.; Desai, P.; Kopp, E.B.; Huang, Y.; Song, L.; Nakaar, V.; Powell, T.J. Vaccination with recombinant fusion proteins incorporating Toll-like receptor ligands induces rapid cellular and humoral immunity. *Vaccine* **2007**, *25*, 763–775. [CrossRef] [PubMed]

101. Mizel, S.B.; Bates, J.T. Flagellin as an adjuvant: Cellular mechanisms and potential. *J. Immunol.* **2010**, *185*, 5677–5682. [CrossRef] [PubMed]

102. Didierlaurent, A.M.; Morel, S.; Lockman, L.; Giannini, S.L.; Bisteau, M.; Carlsen, H.; Kielland, A.; Vosters, O.; Vanderheyde, N.; Schiavetti, F.; et al. AS04, an aluminum salt- and TLR4 agonist-based adjuvant system, induces a transient localized innate immune response leading to enhanced adaptive immunity. *J. Immunol.* **2009**, *183*, 6186–6197. [CrossRef]

103. Pinto, L.A.; Dillner, J.; Beddows, S.; Unger, E.R. Immunogenicity of HPV prophylactic vaccines: Serology assays and their use in HPV vaccine evaluation and development. *Vaccine* **2018**, *36 Pt A*, 4792–4799. [CrossRef]

104. Zhu, F.C.; Hu, S.Y.; Hong, Y.; Hu, Y.M.; Zhang, X.; Zhang, Y.J.; Pan, Q.J.; Zhang, W.H.; Zhao, F.H.; Zhang, C.F.; et al. Efficacy, immunogenicity, and safety of the HPV-16/18 AS04-adjuvanted vaccine in Chinese women aged 18–25 years: Event-triggered analysis of a randomized controlled trial. *Cancer Med.* **2017**, *6*, 12–25. [CrossRef]

105. Shanmugam, A.; Rajoria, S.; George, A.L.; Mittelman, A.; Suriano, R.; Tiwari, R.K. Synthetic Toll like receptor-4 (TLR-4) agonist peptides as a novel class of adjuvants. *PLoS ONE* **2012**, *7*, e30839. [CrossRef] [PubMed]

106. Reed, S.G.; Hsu, F.C.; Carter, D.; Orr, M.T. The science of vaccine adjuvants: Advances in TLR4 ligand adjuvants. *Curr. Opin. Immunol.* **2016**, *41*, 85–90. [CrossRef] [PubMed]

107. Alexander, J.; Sidney, J.; Southwood, S.; Ruppert, J.; Oseroff, C.; Maewal, A.; Snoke, K.; Serra, H.M.; Kubo, R.T.; Sette, A.; et al. Development of high potency universal DR-restricted helper epitopes by modification of high affinity DR-blocking peptides. *Immunity* **1994**, *1*, 751–761. [CrossRef]

108. Rosa, D.S.; Tzelepis, F.; Cunha, M.G.; Soares, I.S.; Rodrigues, M.M. The pan HLA DR-binding epitope improves adjuvant-assisted immunization with a recombinant protein containing a malaria vaccine candidate. *Immunol. Lett.* **2004**, *92*, 259–268. [CrossRef] [PubMed]

109. Ressing, M.E.; van Driel, W.J.; Brandt, R.M.; Kenter, G.G.; de Jong, J.H.; Bauknecht, T.; Fleuren, G.J.; Hoogerhout, P.; Offringa, R.; Sette, A.; et al. Detection of T helper responses, but not of human papillomavirus-specific cytotoxic T lymphocyte responses, after peptide vaccination of patients with cervical carcinoma. *J. Immunother.* **2000**, *23*, 255–266. [CrossRef] [PubMed]

110. Wu, C.Y.; Monie, A.; Pang, X.; Hung, C.F.; Wu, T.C. Improving therapeutic HPV peptide-based vaccine potency by enhancing CD4+ T help and dendritic cell activation. *J. Biomed. Sci.* **2010**, *17*, 88. [CrossRef] [PubMed]

111. Daftarian, P.; Mansour, M.; Benoit, A.C.; Pohajdak, B.; Hoskin, D.W.; Brown, R.G.; Kast, W.M. Eradication of established HPV 16-expressing tumors by a single administration of a vaccine composed of a liposome-encapsulated CTL-T helper fusion peptide in a water-in-oil emulsion. *Vaccine* **2006**, *24*, 5235–5244. [CrossRef]

112. Fraser, C.C.; Altreuter, D.H.; Ilyinskii, P.; Pittet, L.; LaMothe, R.A.; Keegan, M.; Johnston, L.; Kishimoto, T.K. Generation of a universal CD4 memory T cell recall peptide effective in humans, mice and non-human primates. *Vaccine* **2014**, *32*, 2896–2903. [CrossRef]

113. Li, G.; Huang, Z.; Zhang, C.; Dong, B.J.; Guo, R.H.; Yue, H.W.; Yan, L.T.; Xing, X.H. Construction of a linker library with widely controllable flexibility for fusion protein design. *Appl. Microbiol. Biotechnol.* **2016**, *100*, 215–225. [CrossRef]

114. Yu, K.; Liu, C.; Kim, B.G.; Lee, D.Y. Synthetic fusion protein design and applications. *Biotechnol. Adv.* **2015**, *33*, 155–164. [CrossRef]

115. van Rosmalen, M.; Krom, M.; Merkx, M. Tuning the Flexibility of Glycine-Serine Linkers to Allow Rational Design of Multidomain Proteins. *Biochemistry* **2017**, *56*, 6565–6574. [CrossRef]

116. Hoang, T.X. Identification and Characterization of a Splicing Variant in the 5′ UTR of the Human TLR5 Gene. *BioMed Res. Int.* **2017**, *2017*, 8727434. [CrossRef] [PubMed]

117. Takeuchi, O.; Akira, S. Pattern recognition receptors and inflammation. *Cell* **2010**, *140*, 805–820. [CrossRef]

118. Bell, J.K.; Mullen, G.E.; Leifer, C.A.; Mazzoni, A.; Davies, D.R.; Segal, D.M. Leucine-rich repeats and pathogen recognition in Toll-like receptors. *Trends Immunol.* **2003**, *24*, 528–533. [CrossRef]

119. Kopp, E.; Medzhitov, R. Recognition of microbial infection by Toll-like receptors. *Curr. Opin. Immunol.* **2003**, *15*, 396–401. [CrossRef]

120. Kim, W.Y.; Lee, J.W.; Choi, J.J.; Choi, C.H.; Kim, T.J.; Kim, B.G.; Song, S.Y.; Bae, D.S. Increased expression of Toll-like receptor 5 during progression of cervical neoplasia. *Int. J. Gynecol. Cancer* **2008**, *18*, 300–305. [CrossRef] [PubMed]

121. Cao, L.; Zhang, T.; Zhu, J.; Li, A.; Zheng, K.; Zhang, N.; Su, B.; Xia, W.; Wu, H.; Li, N.; et al. Polymorphism of TLR5 rs5744174 is associated with disease progression in Chinese patients with chronic HBV infection. *Apmis* **2017**, *125*, 708–716. [CrossRef]

122. Kalnin, K.; Tibbitts, T.; Yan, Y.; Stegalkina, S.; Shen, L.; Costa, V.; Sabharwal, R.; Anderson, S.F.; Day, P.M.; Christensen, N.; et al. Low doses of flagellin-L2 multimer vaccines protect against challenge with diverse papillomavirus genotypes. *Vaccine* **2014**, *32*, 3540–3547. [CrossRef]

123. Eaves-Pyles, T.D.; Wong, H.R.; Odoms, K.; Pyles, R.B. Salmonella flagellin-dependent proinflammatory responses are localized to the conserved amino and carboxyl regions of the protein. *J. Immunol.* **2001**, *167*, 7009–7016. [CrossRef]

124. Murthy, K.G.; Deb, A.; Goonesekera, S.; Szabo, C.; Salzman, A.L. Identification of conserved domains in Salmonella muenchenflagellin that are essential for its ability to activate TLR5 and to induce an inflammatory response in vitro. *J. Biol. Chem.* **2004**, *279*, 5667–5675. [CrossRef]

125. Forstneric, V.; Ivicak-Kocjan, K. The role of the C-terminal D0 domain of flagellin in activation of Toll like receptor 5. *PLoS Pathog.* **2017**, *13*, e1006574. [CrossRef] [PubMed]

126. Yang, J.; Zhong, M.; Zhang, Y.; Zhang, E.; Sun, Y.; Cao, Y.; Li, Y.; Zhou, D.; He, B.; Chen, Y.; et al. Antigen replacement of domains D2 and D3 in flagellin promotes mucosal IgA production and attenuates flagellin-induced inflammatory response after intranasal immunization. *Hum. Vaccin. Immunother.* **2013**, *9*, 1084–1092. [CrossRef] [PubMed]

127. Velova, H.; Gutowska-Ding, M.W.; Burt, D.W.; Vinkler, M. Toll-like receptor evolution in birds: Gene duplication, pseudogenisation and diversifying selection. *Mol. Biol. Evol.* **2018**. [CrossRef] [PubMed]

128. Hu, Y.M.; Huang, S.J.; Chu, K.; Wu, T.; Wang, Z.Z.; Yang, C.L.; Cai, J.P.; Jiang, H.M.; Wang, Y.J.; Guo, M.; et al. Safety of an Escherichia coli-expressed bivalent human papillomavirus (types 16 and 18) L1 virus-like particle vaccine: An open-label phase I clinical trial. *Hum. Vaccin. Immunother.* **2014**, *10*, 469–475. [CrossRef] [PubMed]

129. Xie, M.; Li, S.; Shen, W.; Li, Z.; Zhuang, Y.; Mo, X.; Gu, Y.; Wu, T.; Zhang, J.; Xia, N. Expression, purification and immunogenicity analysis of HPV type 18 virus-like particles from *Escherichia coli*. *Sheng Wu Gong Cheng XueBao* **2009**, *25*, 1082–1087.

Technical Note

QuantIF: An ImageJ Macro to Automatically Determine the Percentage of Infected Cells after Immunofluorescence

Lynda Handala [1,†], Tony Fiore [1,†], Yves Rouillé [2,*] and Francois Helle [1,*]

[1] EA4294, Agents Infectieux, Résistance et Chimiothérapie, Centre Universitaire de Recherche en Santé, Centre Hospitalier Universitaire et Université de Picardie Jules Verne, 80054 Amiens, France; l.handala@gmail.com (L.H.); toni.fiore37@gmail.com (T.F.)

[2] University of Lille, CNRS, INSERM, CHU Lille, Pasteur Institute of Lille, U1019-UMR8204-CIIL-Center for Infection and Immunity of Lille, 59019 Lille, France

* Correspondence: yves.rouille@ibl.cnrs.fr (Y.R.); francois.helle@u-picardie.fr (F.H.); Tel.: +33-3-20-87-10-27 (Y.R.); +33-3-22-82-53-51 (F.H.)

† These authors contributed equally to this work.

Received: 21 January 2019; Accepted: 17 February 2019; Published: 19 February 2019

Abstract: Counting labeled cells, after immunofluorescence or expression of a genetically fluorescent reporter protein, is frequently used to quantify viral infection. However, this can be very tedious without a high content screening apparatus. For this reason, we have developed QuantIF, an ImageJ macro that automatically determines the total number of cells and the number of labeled cells from two images of the same field, using DAPI- and specific-stainings, respectively. QuantIF can automatically analyze hundreds of images, taking approximately one second for each field. It is freely available as supplementary data online at MDPI.com and has been developed using ImageJ, a free image processing program that can run on any computer with a Java virtual machine, which is distributed for Windows, Mac, and Linux. It is routinely used in our labs to quantify viral infections in vitro, but can easily be used for other applications that require quantification of labeled cells.

Keywords: virus; infection; fluorescent reporter protein; image quantification; Hepatitis C virus; Yellow Fever Virus; polyomavirus; Coxsackievirus B4

---

1. Introduction

When evaluating viral infections in vitro, fluorescence microscopy is commonly used to monitor the expression of a viral protein following immunostaining. However, this method requires a high content screening apparatus to count large numbers of fluorescent cells. Manual evaluation is feasible when analyzing few images, but it can result in subjective evaluation by the researcher. Furthermore, it is very time-consuming when working with hundreds of images, containing thousands of cells per image.

ImageJ is a free image-processing program that was developed 20 years ago by Wayne S. Rasband at the National Institute of Health, and has become a valuable tool for researchers [1,2]. It is a Java-based software that can run on any computer using a Java virtual machine. It is thus available for Windows, Mac, and Linux. ImageJ can convert images into numerical values that can be exported and further processed with other software for statistical analysis. Furthermore, a major strength of ImageJ is the possibility to record macros that enable the automatization of image analysis.

In this technical note, we present QuantIF, an ImageJ macro for determining the percentage of fluorescent cells following immunofluorescence staining. QuantIF can be used when the specific staining in the cytoplasm and/or nucleus of a cell is diffuse. The macro automatically determines the total number of cells and fluorescently labeled cells for a series of images corresponding to different conditions. For each condition, two pictures of the same field must be taken, the first one corresponding to the specific staining and the second one corresponding to the DAPI staining. In this way, the series of images to be analyzed are placed in the same folder, with images corresponding to the specific staining in odd rank and images of DAPI staining in even rank. When the macro is run, it automatically processes all images in the folder, taking around one second to analyze both images of each field. Ultimately, all results are saved as a ".xls" file that can be processed for statistical analysis.

2. Macro Description

QuantIF was developed using ImageJ version 1.52e and Java version 8. It is freely available as supplementary data online at MDPI.com. In order to use the QuantiIF macro, it is necessary to save the QuantIF.ijm file in the "Plugins" folder of ImageJ. The macro will then appear in the "Plugins" menu. When QuantIF is launched, the folder containing the images for analysis must be selected. Then, parameter values should be entered in a dialog box (Figure 1a), (i) the type/name of the specific staining, (ii) the staining threshold, and (iii) the size limits of nuclei. Once the parameters have been entered, the macro starts analyzing the images. They are first converted to 8-bit images, displaying 256 gray levels. Indeed, we recommend directly exporting images as 8-bit TIFF files, from the microscope software. The background of the images is then removed by running the efficient "Subtract Background" ImageJ command.

QuantIF relies on the "Analyze Particles" tool of ImageJ, which requires binary, black and white, images. For this reason, images are converted to binary masks by implementing the Huang's fuzzy thresholding method [3]. An automatic threshold is set for DAPI staining images since strong and contrasted signals are expected for all these images (Figure 1b,c). In contrast, for immunostaining images, the automatic threshold is generally not applicable since some images may show no signal (in negative controls for instance). For this reason, a manual thresholding is implemented with the staining threshold value entered in the parameter's dialog box (Figure 1a). The threshold value must range between 0 and 255; pixels with values under and above the threshold are converted to white and black, respectively (Figure 1d,e). The "Watershed" command is also applied to the DAPI staining mask in order to separate nearby nuclei [4]. However, it is important to avoid cell overconfluence to obtain interpretable results (see below). Furthermore, the commands "Dilate", "Close", and "Fill Holes" are applied to the mask of the specific staining in order to completely include the area corresponding to the nuclei. To analyze similar particles in the DAPI- and specific-staining masks and avoid counting autofluorescent debris, a new mask corresponding to the nuclei of immunostained cells is created. This is performed by executing the "Image Calculator" command and the "AND" operator using the DAPI- and specific-staining masks (Figure 1f). Finally, the total number of DAPI-stained and immunostained cells' nuclei are counted by implementing the "Analyze Particles" tool to the DAPI staining mask (Figure 1g) and the immunostained cells' nuclei mask (Figure 1h), respectively. The size limits for the nuclei entered in the parameters dialog box correspond to the minimum and maximum pixel area sizes that are taken into account to exclude anything that is not an object of interest. Additionally, to help exclude unwanted objects, roundness values have been set between 0.7 and 1.0.

After processing, the numbers of DAPI-stained nuclei and immunostained cells' nuclei for each condition are saved as a ".xls" file in the folder that has been analyzed (Figure 1i). In addition, "Total Area", "Average Size", and "%Area" values are saved in the file. While "Total Area" values are not useful, the "Average Size" values can help in choosing the size limits for the nuclei that must be entered in the parameters dialog box. Furthermore, "%Area" values of the DAPI staining masks give an idea on cell confluence, which should not typically exceed 30% for optimal results. To help researchers find the best parameters for their analyses, the different masks can be saved in the folder that is being analyzed. To do so, the "//" symbols preceding the "saveAs" line commands must be deleted in the QuantIF.ijm file.

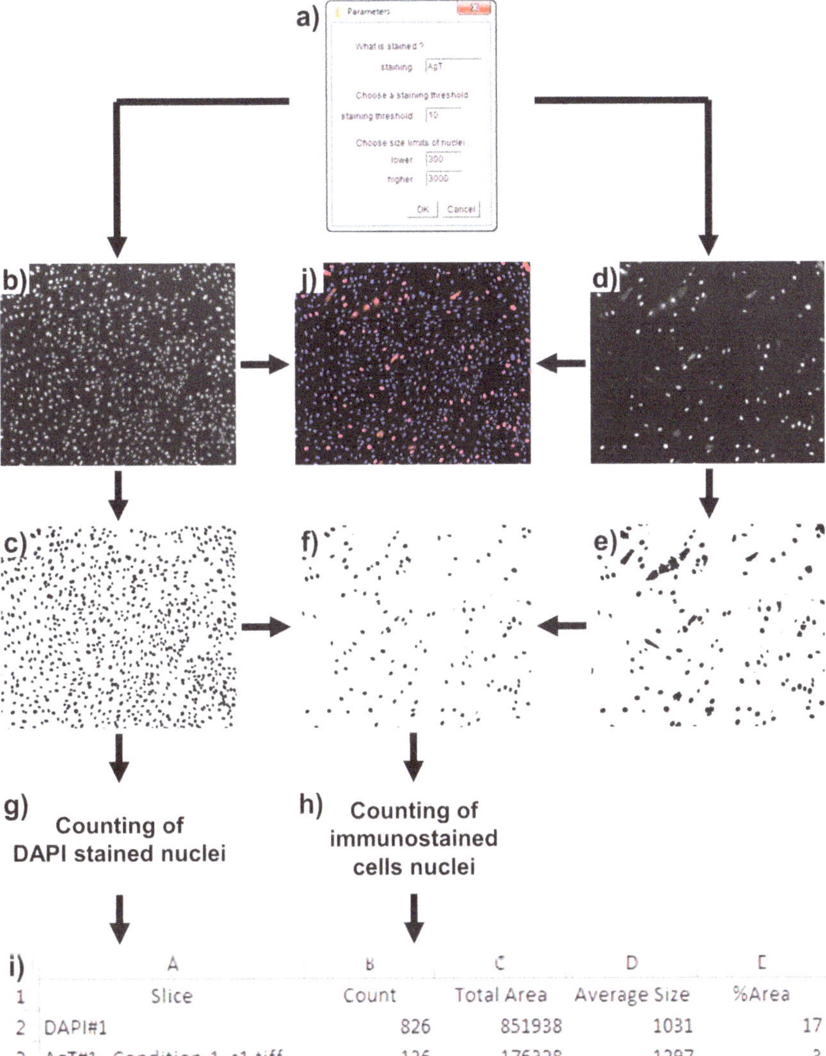

Figure 1. Description of the QuantIF macro. After entering the parameters into the dialog box (**a**), two images of each field are analyzed. The DAPI staining image (**b**) is converted to a DAPI staining mask (**c**), and the specific staining image (**d**) is converted to a specific staining mask (**e**), by implementing the Huang's fuzzy thresholding method. A third mask corresponding to the nuclei of the immunostained cells is created using the "Image Calculator" command and the "AND" operator (**f**). Finally, DAPI stained nuclei and immunostained cell nuclei are counted using the "Analyze Particles" tool (**g,h**). After processing, the numbers of DAPI-stained nuclei and immunostained cell nuclei for each condition are saved as a ".xls" file in the folder that has been analyzed (**i**). A merge of the DAPI and specific staining images is shown for informational purposes (**j**).

3. Discussion and Conclusions

QuantIF is a free, simple, and robust automated tool to estimate the proportion of virally infected cells after immunofluorescence. It is routinely used in our labs to quantify Hepatitis C Viral infections following detection of the E1 envelope glycoprotein that localizes predominantly to the endoplasmic reticulum in HCV-infected cells [5,6]. Similarly, we use it to evaluate Yellow Fever Virus infections using anti-E staining [5]. QuantIF is also used to quantify BKPyV and SV40 infections after detection of the VP1 or AgT proteins that show cytoplasmic and/or nuclear staining patterns (Figure 1d), as well as Coxsackievirus B4 infections using anti-VP1 staining [5]. QuantIF can also be used to quantify infection when using recombinant viruses expressing fluorescent reporter proteins [5]. Furthermore, it can serve many researchers for other applications that require counting labeled cells.

Supplementary Materials: The following is available online at http://www.mdpi.com/1999-4915/11/2/165/s1, File S1: QuantIF.ijm.

Author Contributions: Conceptualization, Y.R. and F.H.; validation, L.H., T.F., Y.R., and F.H.; writing, Y.R. and F.H.

Funding: This work was supported by the Université de Picardie Jules Verne, the Université de Lille, Inserm and CNRS.

Acknowledgments: We thank Zuzana Saidak and Thomas Belcher for proofreading the manuscript.

Conflicts of Interest: The authors declare no conflict of interest.

References

1. Schindelin, J.; Rueden, C.T.; Hiner, M.C.; Eliceiri, K.W. The ImageJ ecosystem: An open platform for biomedical image analysis. *Mol. Reprod. Dev.* **2015**, *82*, 518–529. [CrossRef] [PubMed]
2. Schneider, C.A.; Rasband, W.S.; Eliceiri, K.W. NIH Image to ImageJ: 25 years of image analysis. *Nat. Methods* **2012**, *9*, 671–675. [CrossRef] [PubMed]
3. Huang, L.-K.; Wang, M.-J.J. Image thresholding by minimizing the measures of fuzziness. *Pattern Recognit.* **1995**, *28*, 41–51. [CrossRef]
4. Vincent, L.; Soille, P. Watersheds in digital spaces: An efficient algorithm based on immersion simulations. *IEEE Trans. Pattern Anal. Mach. Intell.* **1991**, *13*, 583–598. [CrossRef]
5. Ferlin, J.; Farhat, R.; Belouzard, S.; Cocquerel, L.; Bertin, A.; Hober, D.; Dubuisson, J.; Rouille, Y. Investigation of the role of GBF1 in the replication of positive-sense single-stranded RNA viruses. *J. Gen. Virol.* **2018**, *99*, 1086–1096. [CrossRef] [PubMed]
6. Rouille, Y.; Helle, F.; Delgrange, D.; Roingeard, P.; Voisset, C.; Blanchard, E.; Belouzard, S.; McKeating, J.; Patel, A.H.; Maertens, G.; et al. Subcellular localization of hepatitis C virus structural proteins in a cell culture system that efficiently replicates the virus. *J. Virol.* **2006**, *80*, 2832–2841. [CrossRef] [PubMed]

Article

A Needle in A Haystack: Tracing Bivalve-Associated Viruses in High-Throughput Transcriptomic Data

Umberto Rosani [1], Maxwell Shapiro [2], Paola Venier [1] and Bassem Allam [3,*]

[1] Department of Biology, University of Padua, 35121 Padua, Italy; umberto.rosani@unipd.it (U.R.); paola.venier@unipd.it (P.V.)
[2] Department of Applied Mathematics and Statistics, Stony Brook University, Stony Brook, NY 11794-5000, USA; maxwell.shapiro@stonybrook.edu
[3] School of Marine and Atmospheric Sciences, Stony Brook University, Stony Brook, NY 11794-5000, USA
* Correspondence: bassem.allam@stonybrook.edu; Tel.: +1-(631)-632-8745

Received: 1 February 2019; Accepted: 25 February 2019; Published: 1 March 2019

Abstract: Bivalve mollusks thrive in environments rich in microorganisms, such as estuarine and coastal waters, and they tend to accumulate various particles, including viruses. However, the current knowledge on mollusk viruses is mainly centered on few pathogenic viruses, whereas a general view of bivalve-associated viromes is lacking. This study was designed to explore the viral abundance and diversity in bivalve mollusks using transcriptomic datasets. From analyzing RNA-seq data of 58 bivalve species, we have reconstructed 26 nearly complete and over 413 partial RNA virus genomes. Although 96.4% of the predicted viral proteins refer to new viruses, some sequences belong to viruses associated with bivalve species or other marine invertebrates. We considered short non-coding RNAs (sncRNA) and post-transcriptional modifications occurring specifically on viral RNAs as tools for virus host-assignment. We could not identify virus-derived small RNAs in sncRNA reads obtained from the oyster sample richest in viral reads. Single Nucleotide Polymorphism (SNP) analysis revealed 938 A-to-G substitutions occurring on the 26 identified RNA viruses, preferentially impacting the AA di-nucleotide motif. Under-representation analysis revealed that the AA motif is under-represented in these bivalve-associated viruses. These findings improve our understanding of bivalve viromes, and set the stage for targeted investigations on the specificity and dynamics of identified viruses.

Keywords: bivalve; virome; RNA-seq; RNA viruses; sncRNA; ADAR; RNAi

---

1. Introduction

Viruses are the most abundant biological entity on the Earth, likely outnumbering bacteria and eukaryotic cells [1], with the oceans being the most likely richest reservoir of virus biodiversity [2]. The only constraint that viruses have is the need for a host for their replication, either to take advantage of the host replication machinery, or to hijack the genome to freely replicate as selfish genetic elements [3]. The evolutionary success of viruses is supported by highly dynamic genomes, which can undergo punctual changes or integration events that enable the circumvention of host immune defenses, the capture of new genes, and even host switching, among other events [4,5]. The frequent exchange of genetic material is evident in the highly variable sizes of viral genomes. While RNA viruses seem to have a ~32 kb size constraint [6], the genomes of DNA viruses can be uncommonly large, with the giant *Mimiviruses* genomes being in the order of megabases and far exceeding the few kilobases of circular single stranded DNA genomes of *cress viruses* [7,8]. The presence of an antiviral system in every living organism further supports the global distribution of viruses [9], although their biological roles go beyond pathogenicity [10]. In fact, viruses are responsible for selective pressures causing evolutionary transitions [11] as they drive the dynamics of host populations and interfere with biological invasions [12]. The gene flux from viruses to eukaryotic organisms is suggested to drive the long-term evolution of host genomes [13]. Conversely, the evolutionary pressure of host antiviral defenses shapes viral genomes in a never-ending arms race [14–17]. According to the sequence data currently available, the viruses identified so far represent a numerically insignificant portion of viral biodiversity, possibly no more than 1% of the extant viruses [1,18]. Thanks to an unprecedented level of sensitivity and accuracy, high-throughput sequencing (HTS) has become the gold standard for viral discovery and for advancements in the characterization of viral metagenomes [19,20], although most of the so-generated viral sequences remain "unclassified" due to uncertainty about authentic virus hosts. As viruses are mostly hidden in the host nucleic acids, and an unusually high sequencing coverage would be necessary to allow their detection, the current representations of the extant virosphere are fragmentary, and they crucially depend on sample preparation strategy, sequencing technology, and sequencing depth [21]. Nevertheless, DNA and RNA sequencing datasets often contain viral sequences, and committed analyses can provide snapshots of the viromes associated with a given organism [5,22].

Although it is steadily expanding, virus discovery and the study of antiviral immunity in invertebrates is biased towards a few model organisms and arthropods of economic and medical importance [23]. In this respect, highly speciose and ecologically important groups like mollusks, and the Lophotrochozoa more widely, remain largely unstudied, leaving huge taxonomic gaps in our knowledge. Since their initial diversification in the early Cambrian (Paleozoic Era), bivalves successfully colonized a variety of aquatic environments, from cold-water seas, to freshwater basins and deep anoxic vents, with some species showing an invasive behavior [24–27]. A number of bivalve species have been investigated for their peculiar adaptation strategies [28], innate immune systems [29], and bio-inspired applications [30] as well as for their use as models for human health [31]. Today, few bivalve genome drafts are available, whereas more than 2,100 transcriptomic datasets have been deposited in public databases (NCBI SRA archive, accessed in November 2018). So far, very few viruses of bivalve mollusks have been described, mainly those that have major negative economic impacts on farmed species. In particular, a herpesvirus (*Ostreid herpesvirus-1*, OsHV-1) associated with temperature-related oyster mortalities [32] has become a pressing issue for the production sector, and today, OsHV-1 and its variants are described in numerous studies [33–38]. Viruses belonging to the *Papovaviridae* and *Iridoviridae* families have been associated with bivalve diseases, whereas a few members of the *Togaviridae*, *Reoviridae*, *Birnaviridae*, and *Picornaviridae* virus families have been reported without evidence of associated disease [39,40]. Until the advent of HTS technologies, the identification of these viruses was mainly based on electron microscopy, and seldom validated by molecular studies [41].

Virome discovery through RNA HTS is challenging when applied to bivalve samples. According to the ability to detect minute quantities of viral nucleic acids, HTS also catches sequences that possibly derive from tissue surface contamination, or from the simple transit of another virus host in bivalve tissues [42,43]. The identification of giant viruses and human viruses in bivalve samples mainly accumulated in the gills and gut by filter-feeding [44–46], and the presence of an algal virus (Heterosigma akashiwo RNA virus-1) in the gills of both *Crassostrea gigas* and *Mytilus galloprovincialis* growing up in association [47], exemplify the importance of developing new approaches for assigning a virus to its authentic host. Ecological role and economic importance, peculiar genome features, differential susceptibility to pathogens, as well as their tendency to accumulate microbes highlights filter-feeding bivalves as fascinating models for virus–host interaction studies. The objective of this work was to explore the diversity and distribution of bivalve RNA viruses through the analysis of available RNA-seq samples. To do so, we performed an extensive analysis of the HTS transcriptome data of bivalves, we recovered bivalve-associated RNA viruses, and we traced their distribution over many bivalve RNA-seq samples. Moreover, we investigated how different RNA selection methods applied during library preparation can affect the performance of viral-oriented HTS analysis.

Finally, we tested two different in silico approaches for assigning bivalve-associated viruses to their genuine hosts.

2. Materials and Methods

The overall analysis pipeline is summarized in Figure 1, and it is detailed through the following paragraphs. The analyses are based either on the available RNA-seq datasets, or on newly produced data that we submitted to the NCBI SRA archive (the corresponding IDs are cited in the text and tables).

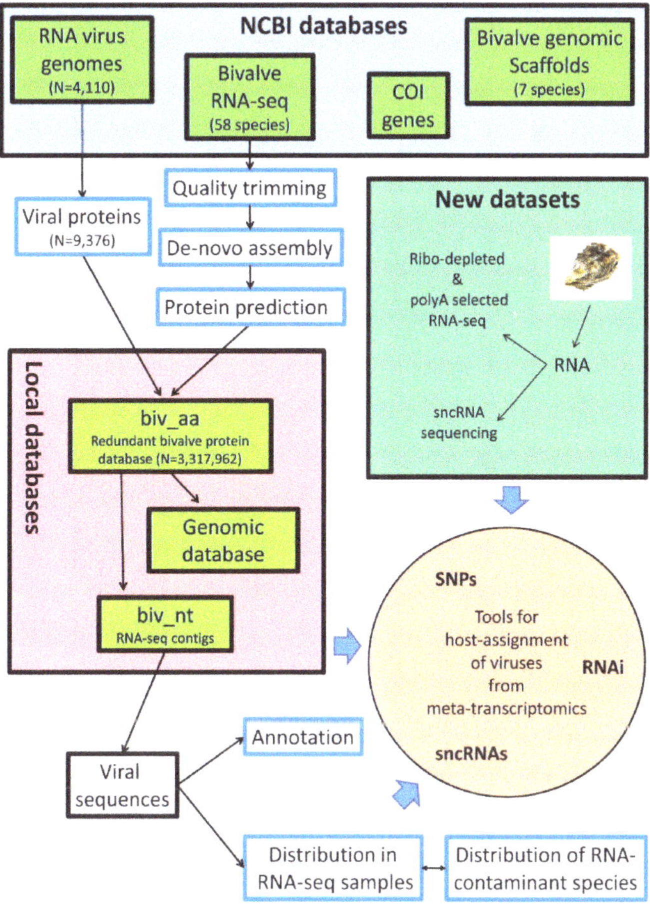

Figure 1. Summary of the analysis pipeline. The graph summarizes all of the steps that were used to extract viral sequences from bivalve RNA-seq datasets. Additional details are reported in corresponding sections of Materials and Methods.

2.1. Data Retrieval

Public sequence datasets were retrieved from NCBI databases in April 2017. A total of 7125 viral genomes, including 3008 RNA viruses, were downloaded from the NCBI Genome database. Additionally 1102 invertebrate-associated RNA virus genomes were downloaded from the NCBI nucleotide database [43,47] for a total of 4110 genomes of RNA viruses. RNA-seq samples referring to 58 bivalve species and four pooled bivalve or gastropod meta-transcriptomic samples were obtained from NCBI SRA archives. Genome drafts of five bivalve species (*Bathymodiolus platifrons*, *C. virginica*, *Mizuhopecten yessoensi*, *Modiolus philippinarum*, and *M. galloprovincialis*) were downloaded from the NCBI WGS database, while *C. gigas* and *Pinctada fucata* genomes were obtained from EnsembleMetazoa release35 and from [48], respectively. Supplementary Table S1 summarizes the sequence datasets used in this work. *Cytochrome C Oxidase Subunit I* (COI) sequences were downloaded from the NCBI nucleotide archive, and their redundancies were reduced using *cd-hit-est* [49], applying a cut-off of 95% similarity. In order to compare two RNA selection methods for their aptitudes of viral read recovery, we retrieved two RNA-seq datasets obtained from a single *C. gigas* sample (described elsewhere, SRR8237210 and SRR7636587 for polyA and ribo-depleted data, respectively).

2.2. Transcriptome de novo Assembly, ORF Prediction, and Protein Domain Mapping

RNA-seq reads were trimmed for the presence of adaptor sequences, and for quality, using TrimGalore! [50], allowing for a maximum of two ambiguous bases and a quality threshold of PHRED20. Trimmed reads were de novo assembled using CLC Genomic Workbench v.10 (Qiagen, Hilden, Germany), setting automatic word and bubble sizes, and a minimal contig length of 200 bp. The resulting contigs were subjected to open reading frame (ORF) prediction, using the transdecoder tool included in the Trinity suite [51], applying a minimal ORF length of 100 codons. HMMer v.3.1 [52] was used to identify the presence of conserved protein domains (Pfam-A models, v.29 [53], applying a cut-off E-value of 10^{-5}.

2.3. Identification of Viral Sequences

A redundant BLAST database [54] was built, using the predicted proteins obtained from all of the transcriptomic assemblies of bivalve species. All the annotated ORFs encoded by 4110 RNA virus genomes were translated into amino acids, and the resulting 9376 protein sequences were used as blast queries against the bivalve database (blastp, cut-off E-value 10^{-50}). Moreover, all the bivalve-derived protein sequences encoding a viral RNA-dependent RNA-polymerase (vRdRp) domain were selected. A vRdRp was identified by using six different PFAM Hidden Markov Models, corresponding to IDs: PF00680, PF00978, PF00998, PF02123, PF07925, and PF04197. Bivalve genome scaffolds were used to compose a genomic database to discriminate between host-encoded and viral sequences. Sequences identified from the redundant protein database, and from the search of the vRdRp domains, and showing no matches against bivalve genomes, were further processed to reduce the redundancy, applying a cut-off of 90% of similarity (*cd-hit*). The resulting protein sequences were used to recover the corresponding nucleotidic contigs from the initial transcriptome assemblies and they were considered as complete or partial genomes of RNA viruses. For the purposes of this paper, a viral genome was considered to be "nearly complete" if it was composed of a unique contig that was longer than 5 kb and encoding at least one complete ORF.

2.4. Distribution of Viruses among RNA-Seq Samples, Expression Analysis, and SNP Calling

The amount of reads mapping to the "nearly complete" viral genomes in selected RNA-seq samples was determined by stringent mapping of the trimmed reads on the viral genome sequences (0.9 both for length and similarity fraction, CLC mapper tool). For a selection of informative RNA-seq samples, the total read counts were used to calculate the percentage of reads mapping to each virus over the total numbers of reads of the sample, thus providing a comparison of the amount of viral RNA between RNA-seq samples that was not biased by different sequencing depths or read layouts. To obtain the expression profiles of selected oyster RNAi-related genes, 183 RNA-seq datasets (Supplementary Table S1) were mapped onto *C. gigas* gene models [55] and used to compute expression values such as transcripts per million (TPM) [56]. RNA-seq data were also used to call single-nucleotide variations (single-nucleotide polymorphisms, SNPs) across viral genomes. Specifically, to detect genuine SNPs, the trimmed reads were mapped onto the "nearly complete" viral genomes, setting 0.5 and 0.8 for the length and similarity fractions, respectively. A SNP was called if it was present in at least 1% of the locally aligned reads after using the following parameters: minimum average quality of the five surrounding bases, PHRED30; minimum required coverage, 50×; minimum required count, 5. The SNPs were annotated according to the neighbor base.

2.5. Estimation of the Contamination Levels of RNA-Seq Samples

To provide an estimation of the fraction of reads that were not related to the declared biological sample (as indicated in the SRA details), we mapped the RNA-seq reads onto a collection of 205,357 non-redundant COI sequences. Reads were mapped applying a similarity fraction of 0.8, over 0.8 of the read length (CLC mapper tool) and the TPM values were computed. Similarly, to estimate the presence of known viruses among the RNA-seq datasets, the amount of reads mapping to the 7125 virus genomes obtained from NCBI was also computed (in this case, by applying 0.9 and 0.9 for the length and similarity fractions, respectively).

2.6. Small RNA Sequencing and Reads Analysis

The fraction of small RNAs (<200 bp) of the *C. gigas* sample that was used to prepare the paired polyA and ribo-depleted libraries were extracted using the Mirvana kit (Thermofisher, Waltham, MA, USA). RNA was quantified by using a Qubit fluorimeter instrument, and the RNA size profile was determined with an Agilent small RNA chip (Agilent, Santa Clara, CA, USA). Library preparation and sequencing (PE150) was outsourced and carried out on an HiSeq Illumina platform (Admera Health, New York, NY, USA), and submitted to the NCBI SRA archive, under the accession ID SRR8587800. The paired reads were trimmed for quality, and for the presence of adaptors, as described for mRNA reads, and the correctly paired reads were joined into fragments. The resulting clean fragments, in a length range of 15–50 nt, were used for the detection of viral-derived small RNAs (vsRNAs) by direct mapping on the identified viral contigs or by using the VirusDetect pipeline [57]. To discriminate between genuine vsRNAs versus RNA degradation products, we correlated the number of mapped sRNA reads with the viral expression levels.

2.7. Analysis of Viral Genome Editing

The genomes of the RNA viruses retrieved from NCBI were analyzed for the distribution of the frequency of di-nucleotides as hallmarks of the virus genome fitness (i.e., adaptive genome changes). To look for these adaptive genome changes, we used the cytidine deaminase under-representation reporter (CDUR) [58]. Given the user-defined motifs and an input coding sequence, CDUR effectively utilizes a permutation test to determine whether the given motif is significantly depleted in the input sequence than one would expect by chance (details below). The two main metrics that are analyzed are: 1. the "below" metric, which determines whether the number of occurrences of a motif is significantly fewer than expected, and 2. the "repTrFrac" metric, which determines the ratio of motifs that would incur a non-synonymous transition mutation, against the total number of those motifs in the sequence, which is significantly fewer than expected. Both of these metrics are determined by shuffling the coding sequence at the third position of each codons, so that the underlying amino acid sequence is unchanged. We chose this method of shuffling as it also preserves the GC content of the input sequence, as changing the GC content has been reported to yield biased results [14]. This shuffling is repeated 1000 times; in each shuffled sequence, we counted the number of user-defined motifs ("below" metric), as well as the ratio of nonsynonymous transition mutations that occur at those motifs, compared to the number of motifs ("repTrFrac" metric). In both cases, we determined the percentage of shuffled sequences with fewer motif counts and repTrFrac counts than that of the input, to yield a statistical *p*-value. A sequence with a *p*-value of <0.05 is said to be under-represented in that motif metric, whereas a *p*-value of >0.95 is said to be over-represented in that motif metric (see Figure 2 in [58]).

A particularly interesting case is when, for a given motif, a sequence is under-represented in the "below" metric, and is over-represented in the "repTrFrac" metric for that motif. This suggests that this sequence has maximally tried to reduce the number of occurrences of that motif, as any further reduction would result in amino acid changes, which may negatively impact that coding sequence. Recent studies have shown that certain gammaherpesvirus may be under such pressures [59]. In this case, the sequence is considered to have attained maximal under-representation. We performed CDUR analysis on 3872 RNA viral genomes with a known host obtained from NCBI (Supplementary Table S1), as well as on the newly recovered "nearly complete viral genomes" presented in this paper.

2.8. Phylogenetic Analysis

Protein sequences referring to vRdRP domains were aligned using MUSCLE [60], and trees were generated with MEGA 6 [61], using neighbor-joining clustering methods with 100 bootstrap replicates. The phylogenetic tree was uploaded to the iTOL server for easier visualization [62].

3. Results

We produced transcriptomic assemblies of 58 bivalve species, and we used all of the predicted proteins to produce a redundant BLAST database, including more than 3 million entries (hereinafter called biv_aa). To identify the putative virus sequences associated with the analyzed RNA-seq samples from different bivalve species, we queried biv_aa with 9376 protein sequences belonging to 4110 known RNA virus genomes (Supplementary Table S1). We extracted additional viral sequences from the same database by searching all six available PFAM domains of viral *RNA-dependent RNA polymerase* (vRdRp). In the absence of a conserved viral gene, we used these domains to identify RNA viruses, since vRdRp is needed for the transcription of the viral genome during productive virus replication [63,64]. For the BLAST searches, we applied a conservative E-value of 10^{-50}, and to further limit false positive results, we discharged the BLAST matches with less than five hits. Moreover, we screened all of the positive hits against a database composed of available genomic scaffolds of bivalve species, to remove genomically encoded sequences (Supplementary Table S1). As a result, 708 biv_aa entries showed a genuine similarity to viral sequences, and the conserved domains included in these proteins further sustained their viral origin, since we found 253 vRdRp, 80 *CRPV capsid protein like*, 73 RNA *helicase*, 78 *Dicistroviridae minor capsid protein*, and 69 *Picornavirus capsid protein* domains (Supplementary Table S2). The removal of similar sequences (>90% of similarity) resulted in 413 unique sequences. Most of the redundant sequences were found in transcriptomes of the same bivalve species, either in RNA-seq samples originating from the same geographical location, as in the case of *M. galloprovincialis* or *C. gigas* samples from Goro (Italy), or obtained from geographically unrelated samples of the same species, although few exceptions are present, and they are discussed below. Despite most sequences being retrieved by BLAST searches using viral sequences as queries, they showed a limited similarity to known viruses (Supplementary Table S2). We could confidently assign only 15 sequences (3.6% of the total) to 11 known viruses (BLASTn with an E-value lower than 10^{-100} and identity >95%), namely, six bivalve-associated RNA viruses from the lagoon of Goro (Italy), two viruses associated with marine invertebrates from China, plus three other RNA viruses, the algal *Heterosigma akashiwo RNA virus*, the plant virus *Zygocactus virus X*, and the *Sacbrood virus* (Supplementary Table S2).

3.1. Effect of RNA-Seq Library Preparation Protocols on the Detection of Viral Sequences

In order to evaluate the effect of the RNA selection method applied during library preparation on the recovery of viral reads, we analyzed two different datasets, each of them derived from a single biological sample by using alternative RNA selection approaches: polyA RNA selection or ribosomal RNA depletion. The first dataset was prepared specifically for viral meta-transcriptomic analysis, starting from two biological samples (named "mix of bivalves", sample ID: SAMN04625952 and "mix of gastropods", sample ID: SAMN04625958 [43]). We analyzed a second dataset obtained from a single *C. gigas* specimen naturally infected with OsHV-1, using the same two RNA selection methods (sample ID: SAMN09760011). The analysis of the four meta-transcriptomic samples showed that, despite a considerable variability in the numbers of raw reads, the assembled contigs, as well as the number of predicted proteins yielded somewhat comparable values, except for SRR3401755, for which only few proteins could be predicted. For these datasets, polyA-selection allowed for the recovery of a higher ratio of viral to total proteins, particularly for the bivalve samples (Table 1). On the contrary, the analysis of the oyster dataset clearly showed the opposite trend in terms of viral read recovery, since we identified 46 contigs encoding a vRdRP domain in the ribo-depleted sample, compared to 10 in the polyA selected one. Although five out of the 10 polyA viral contigs were also found in the ribo-depleted dataset, the longer contig was always generated from the ribo-depleted dataset.

Table 1. Assembly statistics of four viral-metagenomic samples. SRA and sample IDs, RNA selection methods, number of reads in million, and the number of assembled contigs and predicted proteins are reported. The number of viral RNA-dependent-RNA-polymerase (vRdRp) domains, the ratio of viral protein to total proteins, as well as the number of complete viral genomes, are also indicated.

SRA ID	Sample ID	RNA Selection Method	No. of Reads (M)	No. Of Assembled Contig	No. of Predicted Protein	No. of vRdRp	Ratio of Viral Proteins	No. of Complete Genomes *
SRR3401648	SAMN04625952	Ribo-depletion	99.7	96,102	45,343	30	0.00007	2
SRR3401653		polyA-selection	58.3	120,399	38,498	129	0.00034	9
SRR3401753	SAMN04625958	Ribo-depletion	47.9	180,272	48,687	43	0.00009	6
SRR3401755		polyA-selection	60.3	105,611	14,661	54	0.00037	4
SRR7637587	SAMN09760011	Ribo-depletion	54.1	156,166	41,785	46	0.00011	5
SRR8237210		polyA-selection	52.0	93,172	40,301	10	0.00002	0

* additional details on the complete viral genomes are reported in Table 2.

3.2. Identification of "Nearly Complete" Viral Genomes

As mentioned above, we putatively identified 413 viral protein sequences in 364 nucleotidic contigs, indicating that some contigs included more than one viral ORF. Theoretically, each contig can be considered as a viral genome, but if we compare their average length (1.39 kb) with the median lengths of the known RNA virus genomes (4.8 kb), a realistic assumption is that most of these represent incomplete genomes. For the purpose of this paper, we considered "nearly full-length viral genomes" as only being contigs that are longer than 5 kb and encoding at least one complete ORF. Therefore, we identified 26 contigs ranging in length between 5.4 and 9.7 kb as being "nearly complete viral genomes", with 12 contigs encoding two ORFs corresponding to one replicative and one structural protein, while the other contigs (14) encoded a single ORF. These "nearly complete" viruses were named according to the species from which they were assembled (for instance, viruses identified in RNA-seq samples rich in viral sequences and referring to *C. gigas*, *M. galloprovincialis*, *Ruditapes philippinarum*, and *M. edulis*), whereas the unique viral contig found in the *Elliptio complanata* transcriptome (Table 2) was a complete viral genome (Elicom_virus1, 7106 nt). Notably, five nearly complete viral genomes sequences were identified in our ribo-depleted *C. gigas* RNA-seq sample. A total of 10 nearly complete viral genomes sequences could be assigned to a known virus, while other eight other sequences displayed an intermediate/low similarity to known viral sequences, and eight other different sequences referred to completely unknown viruses (the latter being associated with RNA-seq samples of *C. gigas* (4), *E. complanata* (1), *M. galloprovincialis* (1), *Mizuhopecten yessoensis* (1) and *R. philippinarum* (1)). In three viral genomes (*Bivalve RNA virus G1*, *Rudphi virus 4*, and *Heterosigma akashiwo RNA virus*-1) we could identify a polyA tail at the 3′ end of the sequence (Supplementary Figure S1). New sequences or sequences not fully matching the known viral genomes have been deposited in the NCBI database, and the accession IDs are reported in Table 2.

Table 2. Summary of the 26 nearly complete viral genomes. Bivalve species, total number of viral sequences, non-redundant (nr) sequences, and number of "nearly complete viral genomes" are reported. Sequenced tissue and geographical origins of the RNA samples, library type, virus name, and virus distribution in transcriptomes of other bivalves and virus coverage are also reported (as is the total number of viral reads, and as a percentage over the total reads). The BLAST similarities are reported with the *E*-values, NCBI ID, and the description and percentage of identity.

| Species | Viral Sequences | | | Tissue | Geographic Origin | Library Type | Virus Name | Virus Distribution among Bivalve RNA-seq | Total Viral Reads | % of Viral Reads * | NCBI ID | Blastp | | Identity % |
	Total	nr	Nearly Complete									E-Value	Description	
Atrina pectinata	17	15	1	mixed	China	PA	Atrpec_virus1	\	2104	0.00202	MG210792	0	Wenzhou picorna-like virus 26	99.57
Crassostrea gigas	148	109	13	gills	Italy	PA	Cragig_virus3	mytgal	898	0.00135	MG210795	\	\	\
						PA	Bivalve hepelivirus G	mytgal	4058	0.00611	KX158876	0	Bivalve hepelivirus G	99.95
						PA	Bivalve RNA virus G5	\	1382	0.00173	KX158874	0	Bivalve RNA virus G5	100
						PA	Bivalve RNA virus G3	mytgal	1286	0.00161	KX158873	0	Bivalve RNA virus G3	100
						PA	Cragig_virus1	\	890	0.00134	MG210793	\	\	\
						PA	Cragig_virus2	mytgal	468	0.00058	MG210794	0	Wenzhou picorna-like virus 24	93.6
						PA	Bivalve RNA virus G1	mytgal	37902	0.04732	KX158871	0	Bivalve RNA virus G1	100
						PA	AY337486	mytgal	5156	0.00644	AY337486	0	Heterosigma akashiwo RNA virus	100
						RD	Cragig_virus6	\	30047	0.05554	MK561968	1ˉ108	Beihai picorna-like virus 21	66
						RD	Cragig_virus7	\	5493	0.01015	MK561969	3ˉ40	Wenzhou picorna-like virus 41	70
						RD	Cragig_virus8	\	955	0.00177	MK561970	0	Rhizosolenia setigera RNA virus	69
						RD	Cragig_virus9	\	9200	0.01701	MK561971	\	\	\
						RD	Cragig_virus10	\	568	0.00105	MK561972	\	\	\
Elliptio complanata	2	1	1	mixed	USA	PA	Elicom_virus1	\	2268	0.00552	MG210796	\	\	\

Table 2. *Cont.*

| Species | Viral Sequences | | | Tissue | Geographic Origin | Library Type | Virus Name | Virus Distribution among Bivalve RNA-seq | Total Viral Reads | % of Viral Reads * | NCBI ID | Blastp | | |
	Total	nr	Nearly Complete									E-Value	Description	Identity %
Mizuhopecten yessoensis	25	25	1	mixed	China	PA	Mizyes_virus1	\	1974	0.00522	MG210800	\	\	\
Mytilus coruscus			1	mixed	China	PA	Mytcor_virus1	\	4460	0.01028	MG210801	0	Pitaya virus X isolate P37	98.24
Mytilus edulis	37	33	1	mixed	France	PA	Mytedu_virus1	\	1936	0.00694	MG210802	0	Barns Ness breadcrumb sponge aquatic picorna-like virus 2	99
Mytilus galloprovincialis	115	52	3	gills	Italy	PA	Bivalve RNA virus G4	cragig, atrpec	4818	0.00713	KX158875	0	Bivalve RNA virus G4	99.77
						PA	Mytgal_virus1	cragig	\	\	MG210803	\	\	\
						PA	Mytgal_virus2	cragig	1432	0.00212	MG210804	0	Wenzhou picorna-like virus 51	78.7
Ruditapes philippinarum	121	49	5	gills	China	PA	Rudphi_virus1	\	9842	0.02869	MG210805	0	Wenzhou picorna-like virus 38	72.9
				gills	China	PA	Rudphi_virus2	\	1031	0.00301	MG210806	\	\	\
				gills	China	PA	Rudphi_virus3	\	181992	0.29884	MG210807	0	Wenzhou gastropodes virus 2	97.3
				gills	China	PA	Rudphi_virus4	ruddec, ostste, ostlur, cracor, cragig, mytedu	5388365	3.06157	MG210808	0	Wenzhou gastropodes virus 1	92
				larvae	USA	PA	Rudphi_virus5	\	19965	0.02936	MG210809	0	Marine RNA virus BC-4	70

Abbreviations: ruddec, *R. decussatus*; rudphi, *R. philippinaru n*; ostste, *O. stentina*; ostlur, *O. lurida*; cragig, *C. gigas*; atrpec, *A. pectinata*; mytgal, *M. galloprovincialis*; mytedu, *M. edulis*; PA, polyadenylated RNA library; RD, ribo-depleted RNA library; * percentage of reads from the RNA-seq sample with the highest number of mapped reads, mapping to the given virus (counting only correctly paired reads).

Subsequently, we evaluated the distribution of these 26 viruses in 226 RNA-seq samples, referring to their putative host species. Since the initial removal of redundant viral proteins suggested that some of these viruses are distributed over RNA-seq samples of multiple species, or they originated from samples that were possibly contaminated by pathogen-associated (e.g., *Perkinsus spp.*) RNAs, we included additional 24 RNA-seq samples in the distribution analysis, for a total of 250 datasets (Supplementary Table S3).

More than eight million reads were mapped onto the 26 viral genomes, with 862,949 and 7,092,869 reads that matched the *Rudphi_virus3* and *Rudphi_virus4* genomes, respectively. Twenty-two viral genomes were covered by at least 1000 reads, and 82 out of 250 RNA-seq samples included more than 1000 viral reads, and for this reason, they were selected for further consideration (Supplementary Table S3). In these 82 samples, the fraction of viral reads over the total ones per single virus usually did not exceed 1‰, except for *Rudphi_virus4*, which was covered by 30‰ of total reads for a larval *R. philippinarum* RNA-seq sample, and *Rudphi_virus3*, which reached 6‰ in one gastropod meta-transcriptomic sample (Table 2, and Supplementary Table S3). Few viruses showed a distribution over samples of different bivalves, e.g., *Rudphi_virus4* (present in *R. decussatus*, *R. philippinarum*, *C. cortenzinesis*, *C. gigas*, and *M. edulis* samples) and *Bivalve RNA virus G4* (present in *C. gigas*, *M. galloprovincialis*, and *Atrina pectinata*, Figure 2). Moreover, the occurrences of *Rudphi_virus4* and *Rudphi_virus3* go beyond bivalve species, since we traced them both in metagenomic gastropod samples. Since *Rudphi_virus3* originated from a *Perkinsus*-infected sample of *R. philippinarum*, and it was traced in 12 clam datasets, we further investigated the presence of this virus in the publicly available *Perkinsus* transcriptome data (11 RNA-seq samples, Supplementary Table S1). As a result, some reads (3.9‰) of a sample of *Perkinsus olseni* trophozoites exposed to clam plasma (SRR2094558) were mapped to this virus, whereas, only 22 viral reads (<0.00001‰) were detected in the paired control (Supplementary Table S3). In contrast, other viruses were associated to the unique RNA sample, for instance, *Elicom_virus1*, *Rudphi_virus5*, and *Mytedu_virus1* (Figure 2 and Supplementary Table S3).

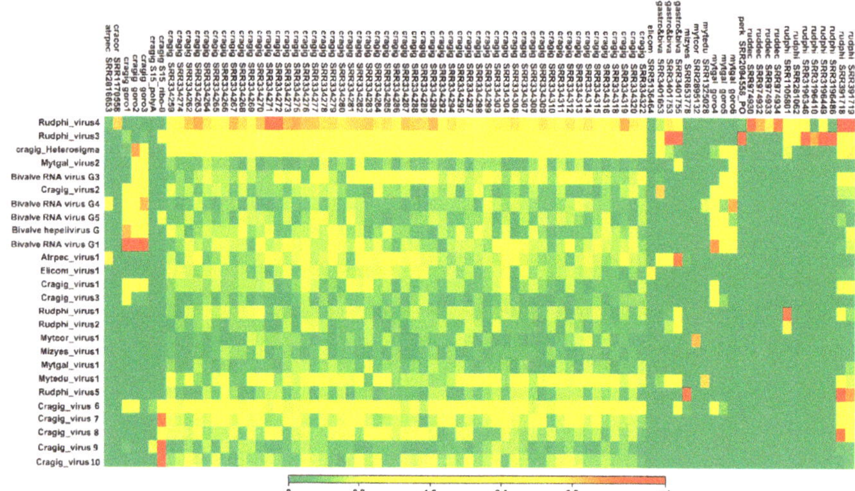

Figure 2. Coverage heat map for the 26 bivalve-associated viruses over 82 RNA-seq samples with at least 1000 viral reads. Data are reported as $\log_{10}$, as depicted in the colored scale. Raw data are reported in Supplementary Table S3.

3.3. Evaluation of Contaminant RNAs in RNA-Seq Samples

We mapped the reads of 16 selected RNA-seq samples to a collection of COI gene sequences, namely, *C. gigas* S15 (Ribo-0 and polyA), a "mix of gastropods" (Ribo-0 and polyA), one *P. olseni*, four *R. decussatus*, and seven *R. philippinarum* samples, since they included reads of multi-host viruses (Figure 2). We calculated the fraction of reads mapping to each COI entry over the total reads that mapped onto the whole COI dataset, and we used it as a tool to evaluate the contribution of biological contaminants in each RNA-seq dataset. As result, 159 COI entries showed at least 0.01% of mapped reads (Supplementary Table S4). Obviously, the first COI entry of each sample corresponded to the sequenced biological sample, thus confirming that the gastropod mix samples were composed of multiple species. For some samples, we observed additional COI entries with lower percentages, as in the two *R. philippinarum* samples with the highest numbers of viral reads (SRR391718-19), where we detected several contaminant species (56 and 57 COI entries with at least 1% of mapped reads, respectively). We noted the presence of a known bivalve-associated tunicate (*Diplosoma listerianum*) in 10 of the tested RNA-seq samples. Although the *D. listerianum* COI value is equal to 100% in the *P. olseni* sample, due to the absence of the Alveolata entries in the COI dataset, our analysis confirmed the absence of clam RNAs in the *Perkinsus* samples, including a high level of *Rudphi_virus3*. Intriguingly, both the Ribo-0 and polyA S15 datasets showed a low contamination of *Lacconectus peguensis* (Coleoptera).

3.4. Tools for the Host-Assignment of Bivalve-Associated Viruses

Our analysis further demonstrated that most of the transcriptome-derived viruses could be only tentatively assigned to a specific host, due to their occurrence in samples of even phylogenetically distant species. Under this context, the application of coverage cut-offs appeared to be an unreliable approach for host-assignment. Therefore, we investigated the feasibility of two alternative approaches for the host-assignment of bivalve-associated viruses obtained from transcriptomic data, as follows.

The first approach investigates the presence of virus-derived RNAi products (vsRNAs), and it is used to reconstruct full-length genomes of viruses infecting arthropods [65–67]. Since the antiviral role of the RNAi system of bivalves has never been demonstrated, we firstly investigated the expression patterns of selected RNAi-related genes (*DICER, DROSHA, ARGONAUTE, PIWI,* and *RNA-dependent RNA polymerase*) in 184 *C. gigas* RNA-seq samples, including some samples that were very rich in viral reads (Supplementary Table S1), to correlate the gene expression values with the presence of actively transcribing RNA and DNA viruses (Supplementary Table S5). We showed that RNAi-related genes are mostly expressed in the early developmental stages of oyster, when two PIWI and one Argonaute transcript showed remarkable expression levels (Supplementary Table S5, panel A), and PIWI1 was preferentially expressed in gonads (Supplementary Table S5, panel C). Apart from these samples, we reported a considerable expression of PIWI1 in three oyster gill samples, and in an additional sample referring to adductor muscles (SRR334286). While the latter result is difficult to explain, the expression of PIWI1 in the oyster gill samples from Goro (Italy) correlated with the presence of RNA viruses (see Figure 2, Supplementary Tables S3 and S5). Although at lower expression levels we reported that one *RNA-dependent RNA polymerase* transcript (EKC38952), belonging to a gene family typically expressed in the digestive gland, showed considerable expression levels in a few other samples, namely two out of three biological replicates of oysters infected with OsHV-1 (12 hours after infection, gills) and a spat sample highly infected by the same virus (Supplementary Table S5, panel B; sample G1). Taken together, these results provide limited evidence for an active role of some components of the RNAi pathway during viral infections in oyster. To further investigate the functionality of RNAi as antiviral system, we sequenced the fraction of small RNAs of the *C. gigas* sample used for library comparison, and found a high number of viral reads belonging both to DNA and RNA viruses (see Figure 2). Small non-coding RNA (sncRNA) sequencing yielded 10.1 million clean fragments in a length range of 15–50 nt. A total of 22,587 sncRNA reads matched the viral contigs identified in this sample, plus the OsHV-1 genome. However, we observed a positive correlation between the expressions of viral genes (using both Ribo-0 and polyA datasets) and the number of sncRNA reads that matched these ORFs (r^2 of 0.994 and 0.996, respectively), suggesting that the reads mostly originated from RNA degradation products, instead of being genuine vsRNAs. We further analyzed the coverage of the sncRNA reads along the five "nearly complete viral genomes" originating from the S15 oyster ribo-depleted data. To do this, we mapped to the viral genomes the sncRNA reads that did not match to the oyster genome (1.436 M reads), and we calculated the size profiles of each of the mapped subsets (Figure 3). Notably, comparing the size profile of the whole sncRNA library with the profile of the sncRNA reads that did not match to the oyster genome, we showed that *C. gigas* sncRNA reads peaked at 21 nt (microRNAs), whereas the unmapped sncRNA reads peaked at 30 nt, indicative of their Piwi-interacting RNA (piRNA) nature. However, only 199 sncRNA reads mapped to one of the five viral genomes reconstructed by using the paired Ribo-depleted RNA-seq reads, and the size profiles showed a low enrichment of 29–30 nt reads with a distribution over the whole viral genome (Figure 3).

Moreover, we subjected the sncRNA reads to VirusDetect, a bioinformatics pipeline that is designed for the identification and reconstruction of viral genomes starting from short reads [57]. Although 169,970 sncRNA reads could be aligned to the viral reference database, and the tool could assemble 40 contigs, the 20–22 nt enrichment fraction was always low, and it did not support their vsRNA nature. According to the presence of numerous OsHV-1 reads in the paired RNA-seq data (the polyA and Ribo-0 datasets), VirusDetect identified several matches to the OsHV-1 genomes, but again, with a low 20–22 nt enrichment fraction.

The second approach that we tested leveraged on the identification of single-nucleotide modifications (SNPs) occurring specifically on viral transcripts produced by the action of host enzymes acting as antiviral defenses. Therefore, we attempted to select and count the subset of total SNPs generated by the host double-stranded RNA (dsRNA) editor enzyme *adenosine deaminase acting on dsRNA* (ADAR), which is assumed to specifically modify viral dsRNAs through A-to-I editing [68]. For each of the 26 viruses, we selected the RNA-seq sample with the higher number of reads, and we called these low-frequency SNPs; among the identified SNPs, we selected the ADAR-compatible ones (A-to-G). We identified 7569 SNPs located on viral coding sequences, and we classified 938 of them as being ADAR-compatible. Considering the 5′ position, we showed that 31% of the selected SNPs had an adenine at the flanking position, while 42% had a thymine (Figure 4a). Also, we searched for the evolutionary footprint of the action of ADAR on viral genomes in parallel. To do this, we used the CDUR tool [58] (see Materials and Methods) to determine under- or over-representation of a motif in a given sequence. Firstly, we used a training set of 3872 genomes of RNA viruses with a known host (Figure 4b and Supplementary Table S1). By analyzing the WA (W = A/T), AA, CA, GA, and TA motifs, the CDUR analysis showed that the TA motif is under-represented in 62.7% of the analyzed ORFs, while the AA, GA, and CA motifs are under-represented in 32.9, 8.1, and 1.5% of ORFs, respectively. Intriguingly, 4% of TA-under-represented ORFs maximized this under-representation, since additional variations will cause non-synonymous SNPs. Although we have to take into consideration that the viral representatives of each of the host classes are variable (Figure 4b), by linking the under-representation values with the viral host, we showed that most (>70%) of the algae, invertebrate, and vertebrate viruses reduced the TA motifs in their coding regions, while we observed moderate percentages (50–60%) for fungal and plant viruses, and lower percentages for bacterial and protozoa viruses (Figure 4c). Accordingly, the sequences with a maximization of the TA reduction were only a small fraction of the ones for fungal, plant, invertebrate, protozoa and vertebrate viruses (Figure 4c). Subsequently, we used the CDUR package to investigate the under-representation of the motif in the ORFs of the nearly complete RNA virus genomes described in this paper (Figure 4d). Consistent with the previous results, only the TA and AA motifs were statistically significantly under-represented. However, we did not observe ORFs with maximized TA reductions, while eight out of 11 ORFs showing AA being under-represented, significantly maximized the AA motif reduction (Figure 4d).

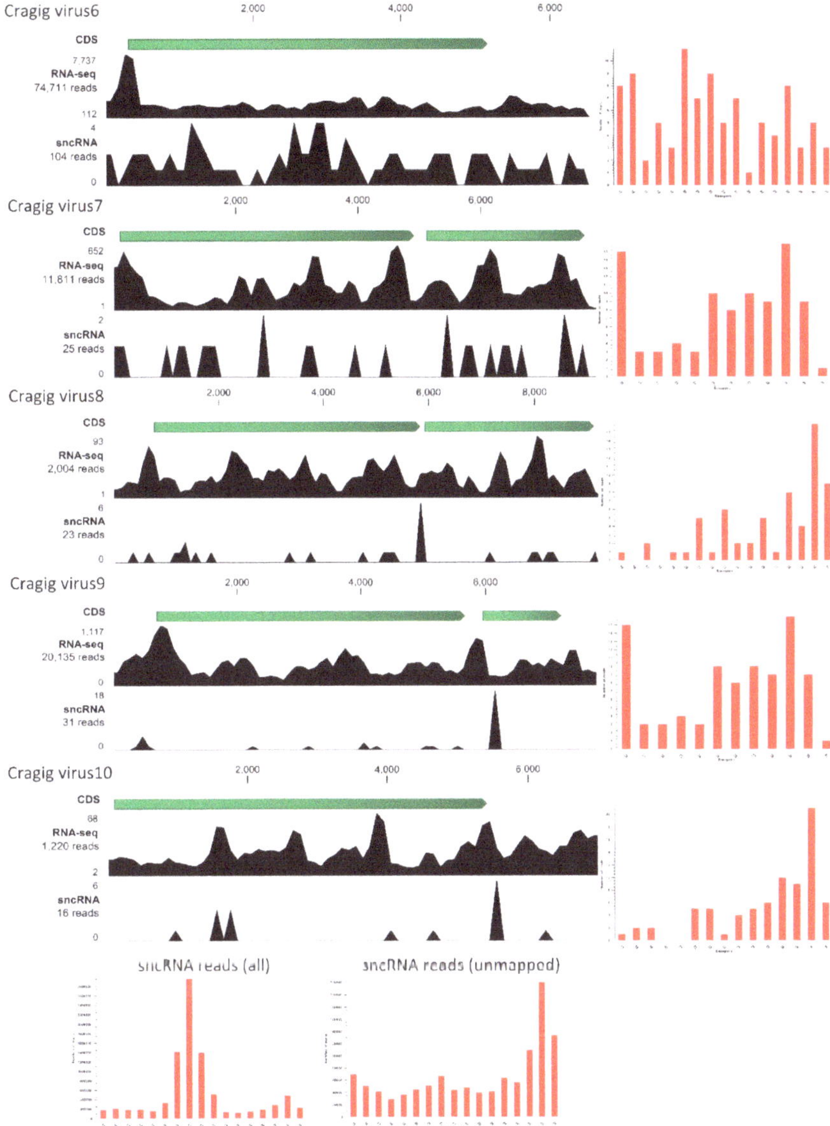

Figure 3. Small non-coding RNA (sncRNA) reads analysis. The RNA-seq and sncRNA read distribution for each of the five "nearly complete viral genomes" reconstructed from the ribo-depleted oyster RNA-seq data are shown. The open reading frames (ORFs) for each virus are shown in green, while the number of mapped reads are reported on the left. The histograms on the right represent the size distributions of the mapped sncRNA reads (in the range of 15–31 nt). The bottom histograms show the size distribution for the whole library (left), and for the reads that did not match the oyster genome (right).

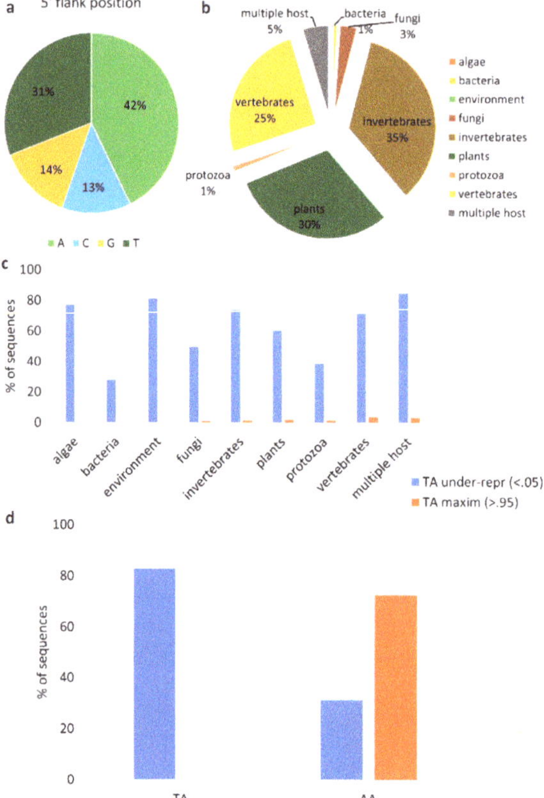

Figure 4. Analysis of virus editing. (**a**). Distribution of the flanking base of ADAR-compatible SNPs. (**b**). Host distribution of the 3872 viral genomes used as a training set. (**c**). Percentage of viral sequences (N = 3872) showing statistically significant TA under-representation (<0.05) and maximization (>0.95), divided per host class. (**d**). Percentage of TA and AA under-representation (<0.05) and maximization (>0.95), measured on the coding sequences of the nearly complete viral genomes reported in this paper.

To better contextualize our results, and to assign the 26 "nearly complete viral genomes" to a viral group, we attempted a phylogenetic analysis based on the regions corresponding to the vRdRP domains (the phylogenetic tree can be visualized at [69], or as Supplementary Data S1). The phylogenetic tree obtained by the comparison of 2019 sequences of viral origins showed poor bootstrap support for most of the nodes, due to the high heterogenicity of the vRdDP sequences. Several of the sequences of the 26 viruses reported herein clustered with picoRNA-like viruses obtained from meta-transcriptomic surveys of mollusk species [43]. These viruses included *Mytedu virus1*, *Myzyes virus1*, *Cragig virus1*, *Cragig virus2*, *Cragig virus3*, *Cragig virus6*, *Rudphi virus4*, *Rudphi virus5*, and *Bivalve RNA virus G3*. *Cragig virus 10* showed similarities with *Bivalve hepelivirus G* (herpes-like viral family, as defined by [43]). Although *Cragig virus7* also clustered in a group of picoRNA-like viruses, it appeared to be separated from the other marine picoRNA-like sequences. Similarly, *Cragig virus8* and *Cragig virus9* formed a cluster including picoRNA-like viruses and one diatom virus (*Chaetoceros socialis f. radians RNA virus1*). None of our viruses grouped in clusters were characterized by the presence of abundant vertebrate viruses, while *Mytcor virus1* was grouped with plant viruses, supporting its BLASY similarity to *Pitaya virus X*.

4. Discussion

Viruses can infect almost every living organism, and viral nucleic acids, either DNA or RNA, are often found when the host sequences are analyzed, making host RNA-seq samples suitable targets for viral discovery [21,70]. In this study, we analyzed bivalve RNA-seq data, and recovered both partial and complete RNA virus genomes, exploiting them to investigate the limits of meta-transcriptomics approaches oriented to viral discovery in bivalves. We identified 413 unique sequences of viral origin, most of them showing a limited similarity with known viruses, demonstrating that bivalve RNA sequencing allows for the identification of viral sequences. These included 26 nearly complete viral genomes. Although the factors that mostly influence the number of viral reads in an RNA-seq sample seem to be the strategies used for sample collection (e.g., the inclusion of water present in the shell cavity), we showed that the RNA selection method used during library preparation also contributed to the recovery of viral reads. For the samples prepared specifically for viral meta-transcriptomics, it was demonstrated that it is possible to recover multiple RNA virus genomes from single samples [43], while analyzing RNA-seq data (polyA-selected) originally designed for host expression analysis, we could identify at most, one complete viral genome per sample, over multiple (partial) viral genomes. Analyzing available meta-transcriptomic data, we found that polyA-enrichment is somewhat more effective than ribosomal-depletion in term of viral read recovery. However, we demonstrated that ribo-depletion is capable of higher performance, since we could reconstruct five nearly complete viral genomes from a RNA sample prepared for oyster expression survey. Arguably, polyA-selection would bias the virus sequence identification in the case of polyadenylated viral genomes (e.g., *Picornavirales*), although we could find evidence of the presence of polyA-tails only in three out of 26 viruses. Overall, our result strongly enforced the use of ribo-depletion for the preparation of RNA-seq libraries targeting viral discovery.

In agreement with a recent study reporting a wide host distribution of invertebrate viruses [43], we traced six out of 26 viruses in RNA-seq samples of different bivalve species, and we reported three bivalve-associated viruses that were very similar to viruses identified from gastropod or sponge meta-transcriptomics data [70,71]. The presence of identical viruses in different bivalve species, or even in phylogenetically distant invertebrates has two possible explanations: either these viruses infect a broad-range of animals, or the species hosting these viruses is shared by different (marine) animals. In support of the first hypothesis, even if invertebrates (arthropods in particular) are rich in viruses [72], strong evidence for host–virus co-evolution was rarely reported [17], and host jumping seems to be common for invertebrate viruses [5]. These attributes are in agreement with the new concepts of RNA virus phylogenesis that are inferred by viral metagenomics, suggesting extensive horizontal virus transfer events and a broad host range for protostome viruses [73]. The second hypothesis, i.e., that these viruses are hosted by an organism that is common in the marine environment, may be the easiest explanation for the presence of identical viruses in samples of different species, and can be further supported by the filter-feeding activity of bivalves. In fact, given the functions that are exerted by the gut and gills (the latter tissue is commonly used for RNA-seq experiments), contamination by RNA originating from waterborne bacteria, fungi, microalgae, or even microeukaryotes, is common in bivalve RNA-seq samples. This situation is well-depicted by one of the complete genomes that we recovered, the algal virus *Heterosigma akashiwo RNA virus*, which we traced in RNA-seq samples of co-cultured *C. gigas* and *M. galloprovincialis*, and even in an unrelated *R. philippinarum* sample. In this study, we exploited the COI reads to identify possible co-occurring organisms of the RNA-seq samples rich in viral reads. Although COI is not a universal gene marker, such an analysis can provide an immediate view of the purity of the samples [71]. In our study, the COI analysis did not identify a contaminant organism that completely matched the distributions of multi-species viruses. The contamination with *D. listerianum* RNA present in several bivalve RNA-seq samples confirmed the wide distribution of this fouling tunicate, but it could represent only a partial explanation for the multi-species distribution of *Rudphi_virus3*. The *R. philippinarum* sample, including 3% of *Rudphi_virus4* reads was shown to be heavily contaminated by this tunicate, and by other non-bivalve species, but none of these species correlated with the distribution of this virus.

The determination of the host of meta-transcriptomics-derived viruses is likely one of the main challenges of viromics based on high-throughput data [23]. At the tissue level, both Transmission Electron Microscope (TEM) imaging of viral particles and in situ hybridization techniques are suitable to confirm host assignment only if infection intensity is sufficiently high, but they are unfeasible in the case of a very large number of samples, or for already-sequenced RNA-seq samples, such as the ones that we analyzed. Therefore, we investigated the feasibility of two alternative approaches that, exploiting the same RNA-seq samples used for viral discovery, can provide evidence that is useful for host-assigning bivalve-associated viruses. RNAi is used as an antiviral defense in plants, insects and nematodes, where efficient RNAi processing of viral genomes into virus-derived small RNAs (vsRNAs) perfectly matching to the original genome activates the RNA-induced silencing complex (RISC), which in turn catalytically marks viral sequences for degradation [74]. Differently from immuno-recognition mechanisms based on antibodies, RNAi is not impacted by viral mutations, and in *C. elegans*, it functions at nano-molar concentrations thanks to the amplification of vsRNA signals using RNA-dependent RNA polymerase [75]. Recently, an RdRP-independent mode of RNAi amplification has been reported in *Drosophila*, although the underpinning genetic mechanism is still unknown [76]. At least in worms, RNAi-based antiviral immunity can be generationally transmitted, and provide a kind of epigenetic immune-memory [77] that recalls the prokaryotic CRISPR-Cas-based adaptive immunity [78]. Recently, the identification of vsRNAs in a molluscan gastropod (*Nucella lapillus*) opens intriguing questions about the phylogenetic distribution of the antiviral defense system, and about the mechanism itself [71]. In insects, the analysis of the fraction of vsRNAs among sncRNA datasets allowed for an unbiased reconstruction of pathogenic viruses [65–67] but, although bivalve antiviral immunity partially resembles that of arthropods [79], the antiviral role of RNAi has never been directly demonstrated [80]. Our analysis suggested that RNAi exerted limited importance in antiviral defense in bivalves, since even if we showed that *C. gigas* PIWI1 is induced in RNA-seq samples containing RNA viruses and a more limited induction of one oyster RDR gene was consistent with the active transcription of OsHV-1, we were not able to clearly identify vsRNAs among sncRNA reads obtained from the same oyster sample, including abundant reads of RNA viruses and of OsHV-1 (dsDNA virus). Considering all of the viral contigs obtained from this RNA-seq dataset, our analyses strongly suggest that the sncRNA reads that mapped on these viral contigs are due to RNA degradation. Differently, looking only at the five complete viral genomes, we could not exclude that weak RNAi activity generated few vsRNAs. Improving the power of our analysis by increasing the coverage or by using chemical treatments to specifically enrich the sncRNA fraction [71], it would be possible to detect genuine vsRNAs, even in bivalves. In particular, the size profiles that we reported for the putative vsRNAs seemed to be biased through a non-Dicer production mechanism [65,71].

In vertebrates, the antiviral role of RNAi is superseded by the interferon pathway, which through the activation of interferon-stimulated genes (ISGs), promotes the recognition of viral-derived products and inhibits viral propagation. Among ISGs, powerful sequence editors like ADAR and *apolipoprotein B mRNA editing enzyme, catalytic polypeptide-like* (APOBEC), enzymatically mutate viral transcripts and genomes [81,82] by acting on target sites that are highly conserved throughout the metazoan evolution [14,83–85]. To counteract these host-mediated editing mechanisms, some viral genomes have evolved to reduce the frequency of sites that are more vulnerable to targeting by the host immune system [58,86,87]. Surprisingly, we demonstrated a diffuse under-representation of the "TA" motif in most of the known RNA viruses, although only few of them maximized this reduction, and we could not link this result to a specific class of hosts. A similar trend of TA under-representation was also present in the 26 viruses described herein, but for these viruses, we showed that there was a tendency to maximize the reduction of the AA motif. Similar to the 26 viruses reported herein, a similar trend characterizes, among others, the *Antarctic picorna-like virus 1* and *3*, *Acute bee paralysis virus* and *Aphid lethal paralysis virus*, which represented *Picornavirales* with a phylogenetic vicinity with some bivalve-derived viruses [47]. SNP analysis did not highlight a predominant fraction of ADAR-compatible variations over the total SNPs in the 26 viruses. Arguably, CDUR analysis only determined under-/over-representation, and the possible role of ADAR as source of these shifts should be confirmed by dedicated experiments.

Overall, This study underlines the heterogeneity and variability of RNA viruses that are associated with marine mollusks, and the limited data that is available on environmental RNA viruses. While the simultaneous analysis of viral products, antiviral host defense processes, and products in the RNA-seq samples could support host assignment, this alone is not enough when dealing with suspension-feeders that are able to accumulate environmental microbes, and their viral symbionts. Given the growing body of knowledge on the role of viruses in host fitness, targeted investigations aimed at unraveling the diversity of "genuine" bivalve viruses are needed for a better understanding of factors affecting the health and well-being of these ecologically- and economically-important species.

Data availability: Short RNA-sequencing reads have been deposited in the NCBI SRA archive with accession ID SRR8587800, as part of the SRA project PRJNA484109.

Supplementary Materials: The following are available online at http://www.mdpi.com/1999-4915/11/3/205/s1, Supplementary Table S1. Summary of the sequence data used for this work. 1. Genome data of bivalves. Species name, order, genome status and source are reported. 2. Transcriptome data. Class, Organism abbreviation, project ID (SRA archive), species name and order, sample origin, tissue, read layout, and number of predicted CDS sequences are reported. 3. *C. gigas* transcriptome data used for RNA-seq expression analysis. Project and SRR IDs, description and millions of reads are reported. 4. Viral genomes. NCBI ID, genome size, virus name, and taxonomical classification are reported. Supplementary Table S2. Annotation of the 413 viral sequences. The E-value, accession, description, and identity of the first BLAST match are reported, as well as the PFAM domain identified on these sequences (region, E-value, PFAM accession, and name). Supplementary Table S3. Virus distribution over 250 RNA-seq samples. The number of reads mapping to the 26 viral genomes are reported for 250 RNA-seq samples. The 82 samples, including more than 1000 viral reads are highlighted in red. Supplementary Table S4. COI mapping results. Mapped reads (reported as per millions) for 16 RNA-seq samples on COI entries (only the entries with at least 0.01 % of mapped reads are reported). Supplementary Table S5. Expression data of 10 RNAi-associated genes in oyster RNA-seq samples. Expression data are reported as TPM. Gene ID, description, and RNA-seq sample IDs, grouped per experimental treatment, are reported. (Panel A-C) Amount of OsHV-1 reads or RNA virus reads are depicted as blue or red dots, respectively, and referred to a secondary $\log_{10}$ bar scale (on the right of each graph). Expression patterns of RNAi-associated genes in oyster RNA-seq samples. Supplementary Figure S1. Genome construction of the 26 "nearly complete viral genomes" described in this work. ORFs are depicted as green arrows, and polyadenylation tails as black boxes. Supplementary Data S1. Phylogenetic tree of vRdRP regions in nexus format.

Author Contributions: Conceptualization, P.V. and B.A.; Data curation, U.R.; Formal analysis, U.R. and M.S.; Funding acquisition, P.V.; Methodology, U.R. and M.S.; Resources, P.V.; Writing—original draft, U.R. and B.A.; Writing—review & editing, U.R., P.V. and B.A.

Funding: This research was supported by grants from the EU Horizon 2020 research and innovation project VIVALDI (GA No. 678589) to P.V., and from the National Science Foundation (IOS-1656753) and New York Sea Grant (R/XG-24) to B.A. U.R. was supported by a biennial grant UNIPD-BIRD 168432.

Conflicts of Interest: The authors declare no conflict of interest.

References

1. Simmonds, P.; Adams, M.J.; Benkő, M.; Breitbart, M.; Brister, J.R.; Carstens, E.B.; Davison, A.J.; Delwart, E.; Gorbalenya, A.E.; Harrach, B.; et al. Consensus statement: Virus taxonomy in the age of metagenomics. *Nat. Rev. Microbiol.* **2017**, *15*, 161–168. [CrossRef] [PubMed]

2. Suttle, C.A. Marine viruses — major players in the global ecosystem. *Nat. Rev. Microbiol.* **2007**, *5*, 801–812. [CrossRef] [PubMed]

3. Koonin, E.V.; Dolja, V.V. Virus world as an evolutionary network of viruses and capsidless selfish elements. *Microbiol. Mol. Biol. Rev. MMBR* **2014**, *78*, 278–303. [CrossRef] [PubMed]

4. Schulz, F.; Yutin, N.; Ivanova, N.N.; Ortega, D.R.; Lee, T.K.; Vierheilig, J.; Daims, H.; Horn, M.; Wagner, M.; Jensen, G.J.; et al. Giant viruses with an expanded complement of translation system components. *Science* **2017**, *356*, 82–85. [CrossRef] [PubMed]

5. Zhang, Y.-Z.; Shi, M.; Holmes, E.C. Using Metagenomics to Characterize an Expanding Virosphere. *Cell* **2018**, *172*, 1168–1172. [CrossRef] [PubMed]

6. Mahy, B.W.J. The Evolution and Emergence of RNA Viruses. *Emerg. Infect. Dis.* **2010**, *16*, 899. [CrossRef]

7. Colson, P.; La Scola, B.; Levasseur, A.; Caetano-Anollés, G.; Raoult, D. Mimivirus: Leading the way in the discovery of giant viruses of amoebae. *Nat. Rev. Microbiol.* **2017**, *15*, 243–254. [CrossRef] [PubMed]

8. Rosario, K.; Duffy, S.; Breitbart, M. A field guide to eukaryotic circular single-stranded DNA viruses: Insights gained from metagenomics. *Arch. Virol.* **2012**, *157*, 1851–1871. [CrossRef] [PubMed]

9. Iranzo, J.; Puigbò, P.; Lobkovsky, A.E.; Wolf, Y.I.; Koonin, E.V. Inevitability of Genetic Parasites. *Genome Biol. Evol.* **2016**, *8*, 2856–2869. [CrossRef] [PubMed]

10. TenOever, B.R. The Evolution of Antiviral Defense Systems. *Cell Host Microbe* **2016**, *19*, 142–149. [CrossRef] [PubMed]

11. Koonin, E.V. Viruses and mobile elements as drivers of evolutionary transitions. *Philos. Trans. R. Soc. Lond. B. Biol. Sci.* **2016**, *371*. [CrossRef] [PubMed]

12. Faillace, C.A.; Lorusso, N.S.; Duffy, S. Overlooking the smallest matter: Viruses impact biological invasions. *Ecol. Lett.* **2017**, *20*, 524–538. [CrossRef] [PubMed]

13. Forterre, P.; Prangishvili, D. The major role of viruses in cellular evolution: Facts and hypotheses. *Curr. Opin. Virol.* **2013**, *3*, 558–565. [CrossRef] [PubMed]

14. Chen, J.; MacCarthy, T. The preferred nucleotide contexts of the AID/APOBEC cytidine deaminases have differential effects when mutating retrotransposon and virus sequences compared to host genes. *PLoS Comput. Biol.* **2017**, *13*. [CrossRef] [PubMed]

15. Coffin, J.M. Virions at the gates: Receptors and the host-virus arms race. *PLoS Biol.* **2013**, *11*, e1001574. [CrossRef] [PubMed]

16. Nakano, Y.; Aso, H.; Soper, A.; Yamada, E.; Moriwaki, M.; Juarez-Fernandez, G.; Koyanagi, Y.; Sato, K. A conflict of interest: The evolutionary arms race between mammalian APOBEC3 and lentiviral Vif. *Retrovirology* **2017**, *14*, 31. [CrossRef] [PubMed]

17. Obbard, D.J.; Dudas, G. The genetics of host–virus coevolution in invertebrates. *Curr. Opin. Virol.* **2014**, *8*, 73–78. [CrossRef] [PubMed]

18. Mokili, J.L.; Rohwer, F.; Dutilh, B.E. Metagenomics and future perspectives in virus discovery. *Curr. Opin. Virol.* **2012**, *2*, 63–77. [CrossRef] [PubMed]

19. Van Aerle, R.; Santos, E.M. Advances in the application of high-throughput sequencing in invertebrate virology. *J. Invertebr. Pathol.* **2017**. [CrossRef] [PubMed]

20. Penaud-Budloo, M.; Lecomte, E.; Guy-Duché, A.; Saleun, S.; Roulet, A.; Lopez-Roques, C.; Tournaire, B.; Cogné, B.; Léger, A.; Blouin, V.; et al. Accurate identification and quantification of DNA species by next-generation sequencing in adeno-associated viral vectors produced in insect cells. *Hum. Gene Ther. Methods* **2017**. [CrossRef] [PubMed]

21. Greninger, A.L. A decade of RNA virus metagenomics is (not) enough. *Virus Res.* **2018**, *244*, 218–229. [CrossRef] [PubMed]

22. Moniruzzaman, M.; Wurch, L.L.; Alexander, H.; Dyhrman, S.T.; Gobler, C.J.; Wilhelm, S.W. Virus-host relationships of marine single-celled eukaryotes resolved from metatranscriptomics. *Nat. Commun.* **2017**, *8*, 16054. [CrossRef] [PubMed]

23. Obbard, D.J. Expansion of the metazoan virosphere: Progress, pitfalls, and prospects. *Curr. Opin. Virol.* **2018**, *31*, 17–23. [CrossRef] [PubMed]

24. Bielen, A.; Bošnjak, I.; Sepčić, K.; Jaklič, M.; Cvitanić, M.; Lušić, J.; Lajtner, J.; Simčič, T.; Hudina, S. Differences in tolerance to anthropogenic stress between invasive and native bivalves. *Sci. Total Environ.* **2016**, *543*, 449–459. [CrossRef] [PubMed]

25. Darrigran, G.; Damborenea, C. Ecosystem engineering impact of Limnoperna fortunei in South America. *Zoolog. Sci.* **2011**, *28*, 1–7. [CrossRef] [PubMed]

26. Karatayev, A.Y.; Burlakova, L.E.; Mastitsky, S.E.; Padilla, D.K. Predicting the spread of aquatic invaders: Insight from 200 years of invasion by zebra mussels. *Ecol. Appl. Publ. Ecol. Soc. Am.* **2015**, *25*, 430–440. [CrossRef]

27. Plazzi, F.; Ceregato, A.; Taviani, M.; Passamonti, M. A molecular phylogeny of bivalve mollusks: Ancient radiations and divergences as revealed by mitochondrial genes. *PLoS ONE* **2011**, *6*, e27147. [CrossRef] [PubMed]

28. Faure, B.; Schaeffer, S.W.; Fisher, C.R. Species distribution and population connectivity of deep-sea mussels at hydrocarbon seeps in the Gulf of Mexico. *PloS ONE* **2015**, *10*, e0118460. [CrossRef] [PubMed]

29. Buchmann, K. Evolution of Innate Immunity: Clues from Invertebrates via Fish to Mammals. *Front. Immunol.* **2014**, *5*. [CrossRef] [PubMed]

30. Petrone, L.; Kumar, A.; Sutanto, C.N.; Patil, N.J.; Kannan, S.; Palaniappan, A.; Amini, S.; Zappone, B.; Verma, C.; Miserez, A. Mussel adhesion is dictated by time-regulated secretion and molecular conformation of mussel adhesive proteins. *Nat. Commun.* **2015**, *6*, 8737. [CrossRef] [PubMed]

31. Fernández Robledo, J.A.; Yadavalli, R.; Allam, B.; Pales Espinosa, E.; Gerdol, M.; Greco, S.; Stevick, R.J.; Gómez-Chiarri, M.; Zhang, Y.; Heil, C.A.; et al. From the raw bar to the bench: Bivalves as models for human health. *Dev. Comp. Immunol.* **2019**, *92*, 260–282. [CrossRef] [PubMed]

32. Farley, C.A.; Banfield, W.G.; Kasnic, G.; Foster, W.S. Oyster herpes-type virus. *Science* **1972**, *178*, 759–760. [CrossRef] [PubMed]

33. Abbadi, M.; Zamperin, G.; Gastaldelli, M.; Pascoli, F.; Rosani, U.; Milani, A.; Schivo, A.; Rossetti, E.; Turolla, E.; Gennari, L.; et al. Identification of a newly described OsHV-1 μvar from the North Adriatic Sea (Italy). *J. Gen. Virol.* **2018**. [CrossRef] [PubMed]

34. Bai, C.; Wang, C.; Xia, J.; Sun, H.; Zhang, S.; Huang, J. Emerging and endemic types of Ostreid herpesvirus 1 were detected in bivalves in China. *J. Invertebr. Pathol.* **2015**, *124*, 98–106. [CrossRef] [PubMed]

35. Burioli, E. a. V.; Prearo, M.; Riina, M.V.; Bona, M.C.; Fioravanti, M.L.; Arcangeli, G.; Houssin, M. Ostreid herpesvirus type 1 genomic diversity in wild populations of Pacific oyster Crassostrea gigas from Italian coasts. *J. Invertebr. Pathol.* **2016**, *137*, 71–83. [CrossRef] [PubMed]

36. Chang, P.H.; Kuo, S.T.; Lai, S.H.; Yang, H.S.; Ting, Y.Y.; Hsu, C.L.; Chen, H.C. Herpes-like virus infection causing mortality of cultured abalone Haliotis diversicolor supertexta in Taiwan. *Dis. Aquat. Organ.* **2005**, *65*, 23–27. [CrossRef] [PubMed]

37. Davison, A.J.; Trus, B.L.; Cheng, N.; Steven, A.C.; Watson, M.S.; Cunningham, C.; Le Deuff, R.-M.; Renault, T. A novel class of herpesvirus with bivalve hosts. *J. Gen. Virol.* **2005**, *86*, 41–53. [CrossRef] [PubMed]

38. Martenot, C.; Lethuillier, O.; Fourour, S.; Oden, E.; Trancart, S.; Travaillé, E.; Houssin, M. Detection of undescribed ostreid herpesvirus 1 (OsHV-1) specimens from Pacific oyster, Crassostrea gigas. *J. Invertebr. Pathol.* **2015**, *132*, 182–189. [CrossRef] [PubMed]

39. Arzul, I.; Corbeil, S.; Morga, B.; Renault, T. Viruses infecting marine molluscs. *J. Invertebr. Pathol.* **2017**. [CrossRef] [PubMed]

40. Zannella, C.; Mosca, F.; Mariani, F.; Franci, G.; Folliero, V.; Galdiero, M.; Tiscar, P.G.; Galdiero, M. Microbial Diseases of Bivalve Mollusks: Infections, Immunology and Antimicrobial Defense. *Mar. Drugs* **2017**, *15*, 182. [CrossRef] [PubMed]

41. Renault, T.; Novoa, B. Viruses infecting bivalve molluscs. *Aquat. Living Resour.* **2004**, *17*, 397–409. [CrossRef]

42. Brum, J.R.; Sullivan, M.B. Rising to the challenge: Accelerated pace of discovery transforms marine virology. *Nat. Rev. Microbiol.* **2015**, *13*, 147–159. [CrossRef] [PubMed]

43. Shi, M.; Lin, X.-D.; Tian, J.-H.; Chen, L.-J.; Chen, X.; Li, C.-X.; Qin, X.-C.; Li, J.; Cao, J.-P.; Eden, J.-S.; et al. Redefining the invertebrate RNA virosphere. *Nature* **2016**, *540*, 539–543. [CrossRef] [PubMed]

44. Andrade, K.R.; Boratto, P.P.V.M.; Rodrigues, F.P.; Silva, L.C.F.; Dornas, F.P.; Pilotto, M.R.; La Scola, B.; Almeida, G.M.F.; Kroon, E.G.; Abrahão, J.S. Oysters as hot spots for mimivirus isolation. *Arch. Virol.* **2015**, *160*, 477–482. [CrossRef] [PubMed]

45. Iaconelli, M.; Purpari, G.; Della Libera, S.; Petricca, S.; Guercio, A.; Ciccaglione, A.R.; Bruni, R.; Taffon, S.; Equestre, M.; Fratini, M.; et al. Hepatitis A and E Viruses in Wastewaters, in River Waters, and in Bivalve Molluscs in Italy. *Food Environ. Virol.* **2015**, *7*, 316–324. [CrossRef] [PubMed]

46. Suffredini, E.; Proroga, Y.T.R.; Di Pasquale, S.; di Maro, O.; Losardo, M.; Cozzi, L.; Capuano, F.; de Medici, D. Occurrence and Trend of Hepatitis A Virus in Bivalve Molluscs Production Areas Following a Contamination Event. *Food Environ. Virol.* **2017**, *9*, 423–433. [CrossRef] [PubMed]

47. Rosani, U.; Gerdol, M. A bioinformatics approach reveals seven nearly-complete RNA-virus genomes in bivalve RNA-seq data. *Virus Res.* **2016**. [CrossRef] [PubMed]

48. Okinawa Institute of Science and Technology (OIST) Marine Genomics Unit Genome Projects. Available online: http://marinegenomics.oist.jp/ (accessed on 1 January 2019).

49. Fu, L.; Niu, B.; Zhu, Z.; Wu, S.; Li, W. CD-HIT: Accelerated for clustering the next-generation sequencing data. *Bioinforma. Oxf. Engl.* **2012**, *28*, 3150–3152. [CrossRef] [PubMed]

50. TrimGalore! Available online: https://www.bioinformatics.babraham.ac.uk/projects/trim_galore/ (accessed on 1 January 2019).

51. Haas, B.J.; Papanicolaou, A.; Yassour, M.; Grabherr, M.; Blood, P.D.; Bowden, J.; Couger, M.B.; Eccles, D.; Li, B.; Lieber, M.; et al. De novo transcript sequence reconstruction from RNA-seq using the Trinity platform for reference generation and analysis. *Nat. Protoc.* **2013**, *8*, 1494–1512. [CrossRef] [PubMed]

52. Eddy, S.R. Accelerated Profile HMM Searches. *PLoS Comput. Biol.* **2011**, *7*, e1002195. [CrossRef] [PubMed]

53. Finn, R.D.; Bateman, A.; Clements, J.; Coggill, P.; Eberhardt, R.Y.; Eddy, S.R.; Heger, A.; Hetherington, K.; Holm, L.; Mistry, J.; et al. Pfam: The protein families database. *Nucleic Acids Res.* **2014**, *42*, D222–D230. [CrossRef] [PubMed]

54. Camacho, C.; Coulouris, G.; Avagyan, V.; Ma, N.; Papadopoulos, J.; Bealer, K.; Madden, T.L. BLAST+: Architecture and applications. *BMC Bioinf.* **2009**, *10*, 421. [CrossRef] [PubMed]

55. Zhang, G.; Fang, X.; Guo, X.; Li, L.; Luo, R.; Xu, F.; Yang, P.; Zhang, L.; Wang, X.; Qi, H.; et al. The oyster genome reveals stress adaptation and complexity of shell formation. *Nature* **2012**, *490*, 49–54. [CrossRef] [PubMed]

56. Wagner, G.P.; Kin, K.; Lynch, V.J. A model based criterion for gene expression calls using RNA-seq data. *Theory Biosci. Theor. Den Biowiss.* **2013**, *132*, 159–164. [CrossRef] [PubMed]

57. Zheng, Y.; Gao, S.; Padmanabhan, C.; Li, R.; Galvez, M.; Gutierrez, D.; Fuentes, S.; Ling, K.-S.; Kreuze, J.; Fei, Z. VirusDetect: An automated pipeline for efficient virus discovery using deep sequencing of small RNAs. *Virology* **2017**, *500*, 130–138. [CrossRef] [PubMed]

58. Shapiro, M.; Meier, S.; MacCarthy, T. The cytidine deaminase under-representation reporter (CDUR) as a tool to study evolution of sequences under deaminase mutational pressure. *BMC Bioinf.* **2018**, *19*, 163.

59. Martinez, T.; Shapiro, M.; Bhaduri-McIntosh, S.; MacCarthy, T. Evolutionary effects of the AID/APOBEC family of mutagenic enzymes on human gamma-herpesviruses. *Virus Evol.* **2019**, *5*. [CrossRef] [PubMed]

60. Edgar, R.C. MUSCLE: A multiple sequence alignment method with reduced time and space complexity. *BMC Bioinf.* **2004**, *5*, 113. [CrossRef] [PubMed]

61. Tamura, K.; Stecher, G.; Peterson, D.; Filipski, A.; Kumar, S. MEGA6: Molecular Evolutionary Genetics Analysis version 6.0. *Mol. Biol. Evol.* **2013**, *30*, 2725–2729. [CrossRef] [PubMed]

62. Interactive Tree Of Life (iTOL). Available online: https://itol.embl.de/ (accessed on 28 February 2019).

63. Venkataraman, S.; Prasad, B.V.L.S.; Selvarajan, R. RNA Dependent RNA Polymerases: Insights from Structure, Function and Evolution. *Viruses* **2018**, *10*, 76. [CrossRef] [PubMed]

64. Dolan, P.T.; Whitfield, Z.J.; Andino, R. Mechanisms and Concepts in RNA Virus Population Dynamics and Evolution. *Annu. Rev. Virol.* **2018**, *5*, 69–92. [CrossRef] [PubMed]

65. Miesen, P.; Joosten, J.; van Rij, R.P. PIWIs Go Viral: Arbovirus-Derived piRNAs in Vector Mosquitoes. *PLoS Pathog.* **2016**, *12*, e1006017. [CrossRef] [PubMed]

66. Vijayendran, D.; Airs, P.M.; Dolezal, K.; Bonning, B.C. Arthropod viruses and small RNAs. *J. Invertebr. Pathol.* **2013**, *114*, 186–195. [CrossRef] [PubMed]

67. Vodovar, N.; Goic, B.; Blanc, H.; Saleh, M.-C. In silico reconstruction of viral genomes from small RNAs improves virus-derived small interfering RNA profiling. *J. Virol.* **2011**, *85*, 11016–11021. [CrossRef] [PubMed]

68. Samuel, C.E. Adenosine deaminases acting on RNA (ADARs) are both antiviral and proviral. *Virology* **2011**, *411*, 180–193. [CrossRef] [PubMed]

69. Interactive Tree Of Life (iTOL). Available online: https://itol.embl.de/shared/umbertorosani (accessed on 28 February 2019).

70. Shi, M.; Zhang, Y.-Z.; Holmes, E.C. Meta-transcriptomics and the evolutionary biology of RNA viruses. *Virus Res.* **2018**, *243*, 83–90. [CrossRef] [PubMed]

71. Waldron, F.M.; Stone, G.N.; Obbard, D.J. Metagenomic sequencing suggests a diversity of RNA interference-like responses to viruses across multicellular eukaryotes. *PLoS Genet.* **2018**, *14*, e1007533. [CrossRef] [PubMed]

72. Li, C.-X.; Shi, M.; Tian, J.-H.; Lin, X.-D.; Kang, Y.-J.; Chen, L.-J.; Qin, X.-C.; Xu, J.; Holmes, E.C.; Zhang, Y.-Z. Unprecedented genomic diversity of RNA viruses in arthropods reveals the ancestry of negative-sense RNA viruses. *eLife* **2015**, *4*, 5378. [CrossRef] [PubMed]

73. Dolja, V.V.; Koonin, E.V. Metagenomics reshapes the concepts of RNA virus evolution by revealing extensive horizontal virus transfer. *Virus Res.* **2017**. [CrossRef] [PubMed]

74. Chin, W.-X.; Ang, S.K.; Chu, J.J.H. Recent advances in therapeutic recruitment of mammalian RNAi and bacterial CRISPR-Cas DNA interference pathways as emerging antiviral strategies. *Drug Discov. Today* **2017**, *22*, 17–30. [CrossRef] [PubMed]

75. Tsai, H.-Y.; Chen, C.-C.G.; Conte, D.; Moresco, J.J.; Chaves, D.A.; Mitani, S.; Yates, J.R.; Tsai, M.-D.; Mello, C.C. A ribonuclease coordinates siRNA amplification and mRNA cleavage during RNAi. *Cell* **2015**, *160*, 407–419. [CrossRef] [PubMed]
76. Petit, M.; Mongelli, V.; Frangeul, L.; Blanc, H.; Jiggins, F.; Saleh, M.-C. piRNA pathway is not required for antiviral defense in Drosophila melanogaster. *Proc. Natl. Acad. Sci. USA* **2016**, *113*, E4218–E4227. [CrossRef] [PubMed]
77. Gammon, D.B.; Ishidate, T.; Li, L.; Gu, W.; Silverman, N.; Mello, C.C. The Antiviral RNA Interference Response Provides Resistance to Lethal Arbovirus Infection and Vertical Transmission in Caenorhabditis elegans. *Curr. Biol. CB* **2017**, *27*, 795–806. [CrossRef] [PubMed]
78. Koonin, E.V. Evolution of RNA- and DNA-guided antivirus defense systems in prokaryotes and eukaryotes: Common ancestry vs convergence. *Biol. Direct* **2017**, *12*, 5. [CrossRef] [PubMed]
79. Green, T.J.; Rolland, J.-L.; Vergnes, A.; Raftos, D.; Montagnani, C. OsHV-1 countermeasures to the Pacific oyster's anti-viral response. *Fish Shellfish Immunol.* **2015**, *47*, 435–443. [CrossRef] [PubMed]
80. Green, T.J.; Speck, P. Antiviral Defense and Innate Immune Memory in the Oyster. *Viruses* **2018**, *10*, 133. [CrossRef] [PubMed]
81. Bass, B.L. RNA editing by adenosine deaminases that act on RNA. *Annu. Rev. Biochem.* **2002**, *71*, 817–846. [CrossRef] [PubMed]
82. Liu, M.-C.; Liao, W.-Y.; Buckley, K.M.; Yang, S.Y.; Rast, J.P.; Fugmann, S.D. AID/APOBEC-like cytidine deaminases are ancient innate immune mediators in invertebrates. *Nat. Commun.* **2018**, *9*, 1948. [CrossRef] [PubMed]
83. Carpenter, J.A.; Keegan, L.P.; Wilfert, L.; O'Connell, M.A.; Jiggins, F.M. Evidence for ADAR-induced hypermutation of the Drosophila sigma virus (Rhabdoviridae). *BMC Genet.* **2009**, *10*, 75. [CrossRef] [PubMed]
84. Porath, H.T.; Carmi, S.; Levanon, E.Y. A genome-wide map of hyper-edited RNA reveals numerous new sites. *Nat. Commun.* **2014**, *5*, 4726. [CrossRef] [PubMed]
85. Porath, H.T.; Schaffer, A.A.; Kaniewska, P.; Alon, S.; Eisenberg, E.; Rosenthal, J.; Levanon, E.Y.; Levy, O. A-to-I RNA Editing in the Earliest-Diverging Eumetazoan Phyla. *Mol. Biol. Evol.* **2017**, *34*, 1890–1901. [CrossRef] [PubMed]
86. Piontkivska, H.; Matos, L.F.; Paul, S.; Scharfenberg, B.; Farmerie, W.G.; Miyamoto, M.M.; Wayne, M.L. Role of Host-Driven Mutagenesis in Determining Genome Evolution of Sigma Virus (DMelSV.; Rhabdoviridae) in Drosophila melanogaster. *Genome Biol. Evol.* **2016**, *8*, 2952–2963. [CrossRef] [PubMed]
87. Piontkivska, H.; Frederick, M.; Miyamoto, M.M.; Wayne, M.L. RNA editing by the host ADAR system affects the molecular evolution of the Zika virus. *Ecol. Evol.* **2017**, *7*, 4475–4485. [CrossRef] [PubMed]

Article

The Transcriptional Landscape of Marek's Disease Virus in Primary Chicken B Cells Reveals Novel Splice Variants and Genes

Luca D. Bertzbach [1,†], Florian Pfaff [2,*,†], Viktoria I. Pauker [3], Ahmed M. Kheimar [1,4], Dirk Höper [2], Sonja Härtle [5], Axel Karger [3,*] and Benedikt B. Kaufer [1,*]

[1]   Institute of Virology, Freie Universität Berlin, Robert von Ostertag-Straße 7-13, 14163 Berlin, Germany; luca.bertzbach@fu-berlin.de (L.D.B.); ahmed.kheimar@fu-berlin.de (A.M.K.)
[2]   Institute of Diagnostic Virology, Friedrich-Loeffler-Institut, Federal Research Institute for Animal Health, Südufer 10, 17493 Greifswald-Insel Riems, Germany; dirk.hoeper@fli.de
[3]   Institute of Molecular Virology and Cell Biology, Friedrich-Loeffler-Institut, Federal Research Institute for Animal Health, Südufer 10, 17493 Greifswald-Insel Riems, Germany; viktoria.pauker@uni-greifswald.de
[4]   Department of Poultry Diseases, Faculty of Veterinary Medicine, Sohag University, Sohag 82424, Egypt
[5]   Department of Veterinary Sciences, Institute for Animal Physiology, Ludwig-Maximilians-Universität München, 80539 Munich, Germany; sonja.haertle@tiph.vetmed.uni-muenchen.de
*   Correspondence: florian.pfaff@fli.de (F.P.); axel.karger@fli.de (A.K.); benedikt.kaufer@fu-berlin.de (B.B.K.)
†   These authors contributed equally to this work.

Received: 18 February 2019; Accepted: 13 March 2019; Published: 16 March 2019

Abstract: Marek's disease virus (MDV) is an oncogenic alphaherpesvirus that infects chickens and poses a serious threat to poultry health. In infected animals, MDV efficiently replicates in B cells in various lymphoid organs. Despite many years of research, the viral transcriptome in primary target cells of MDV remained unknown. In this study, we uncovered the transcriptional landscape of the very virulent RB1B strain and the attenuated CVI988/Rispens vaccine strain in primary chicken B cells using high-throughput RNA-sequencing. Our data confirmed the expression of known genes, but also identified a novel spliced MDV gene in the unique short region of the genome. Furthermore, *de novo* transcriptome assembly revealed extensive splicing of viral genes resulting in coding and non-coding RNA transcripts. A novel splicing isoform of MDV UL15 could also be confirmed by mass spectrometry and RT-PCR. In addition, we could demonstrate that the associated transcriptional motifs are highly conserved and closely resembled those of the host transcriptional machinery. Taken together, our data allow a comprehensive re-annotation of the MDV genome with novel genes and splice variants that could be targeted in further research on MDV replication and tumorigenesis.

Keywords: Marek's disease virus (MDV); RNA-seq; transcriptome; splicing; polycistronic viral transcripts; primary B cells; RB1B; CVI988/Rispens; ICP0

---

1. Introduction

Marek's disease virus (MDV), also known as Gallid alphaherpesvirus 2, causes a deadly lymphoproliferative disease in chickens. Typical clinical symptoms include immunosuppression, paralysis and polyneuritis, acute brain edema, and lymphoma that develop as early as 3 weeks post infection [1,2]. MDV has a big economic impact on the poultry industry worldwide due to animal losses, reduced growth, decreased egg production, and cost of vaccination [3]. Vaccines are crucial for the protection against MDV, as very virulent strains can cause mortalities of up to 100% in susceptible unvaccinated chickens [4]. Live attenuated MDV vaccines such as the gold standard Rispens strain (CVI988) are highly effective in preventing tumor formation [3,5], but do not provide sterilizing immunity.

MDV infection is initiated by the inhalation of the virus from a contaminated environment. Macrophages and other phagocytic cells are thought to take up the virus and transfer it to lymphatic tissues, where B and T cells are infected [6]. B cells are efficiently infected during the initial lytic virus replication, whereas T cell subsets play key roles in MDV pathogenesis [7–11]. MDV establishes latency predominantly in CD4+ T cells, which can also transport the virus to the feather follicle epithelium. These cells efficiently produce the infectious virus and shed it into the environment [3]. Latently infected T cells can also be transformed, resulting in deadly lymphomas.

MDV has a 180 kilo base pairs class E genome consisting of a unique long (U_L) and a unique short (U_S) sequence that are flanked by terminal (TR_L and TR_S) and internal (IR_L and IR_S) inverted repeat regions [12,13]. MDV encodes about 100 proteins that orchestrate the virus life cycle and/or contribute to pathogenesis [12,14]. Until now, analyses of the viral transcriptome has been limited to chicken fibroblasts that are not infected in chickens, ex vivo samples [15,16] and tumor cells [17]; however, the mRNA expression in the primary target cells of lytic replication in vivo remained elusive. This is mainly due to the short lifespan of B and T cells in culture and the low quantity of infected cells in lymphoid organs of chickens [7,18]. To overcome this obstacle, we recently developed an in vitro infection system for primary B and T cells that allows for a prolonged survival and efficient infection of these cells [19]. We used this system to analyze the MDV transcriptome in the most frequent lytically infected cell type in vivo, the B cells.

In this study, we performed next generation RNA-sequencing and protein profiling in primary B cells infected with the very virulent RB1B strain or the vaccine strain CVI988. Our data reveal that the coding capacity of the MDV genome is larger than expected. We identified novel MDV genes and splice variants, and confirmed them either on the protein level or by RT-PCR. This comprehensive approach provides novel insights into the transcriptome of MDV in the natural target cells and supply a basis for future research on MDV pathogenesis.

2. Materials and Methods

2.1. Ethics Statement

Valo specific-pathogen-free (SPF) chickens (VALO BioMedia GmbH, Osterholz-Scharmbeck, Germany) were housed for 6 to 11 weeks of age and humanely euthanized prior to the isolation of the bursa of Fabricius. The animal work was approved by the governmental agency, the Landesamt für Gesundheit und Soziales (LAGeSo) in Berlin, Germany (approval number T0245/14, approval date 23 October 2014).

2.2. Cells

Embryonated SPF Valo chicken eggs (VALO BioMedia GmbH,) were used for the preparation of chicken embryo cells (CEC). CEC were maintained in minimal essential medium (MEM, PAN Biotech; Aidenbach, Germany) supplemented with 1-10% fetal bovine serum (FBS) and penicillin/streptomycin as previously described [20]. B cells were obtained from the bursa of Fabricius by dissociation of the organ and subsequent isolation of the cells by density gradient centrifugation as previously described [21]. Briefly, the bursa of Fabricius was homogenized through a 40 μm cell filter to obtain a uniform single cell suspension. Suspension cells were carefully applied on a Biocoll separating solution (Biochrom; Berlin, Germany), centrifuged for 12 min at 650 × g with slow acceleration, and deactivated deceleration. Lymphocytes at the interphase were carefully transferred to a new tube, washed with PBS, and maintained in RPMI 1640 (PAN Biotech) supplemented with 10% FBS and penicillin [100 U/mL]/streptomycin [100 μg/mL] at 41 °C under a 5% CO_2 atmosphere. B cells were activated using recombinant soluble chicken CD40 ligand (chCD40L) [22], which was expressed in HEK293 cells and purified using a Vivacell 250 ultrafiltration concentrator (Sartorius; Göttingen, Germany).

2.3. Viruses

All viruses were reconstituted by calcium phosphate transfection of CEC with purified bacterial artificial chromosome (BAC) DNA as previously described [23]. The very virulent RB1B strain and the vaccine strain CVI988 both express a green fluorescent protein (GFP) under the control of the early thymidine kinase promotor. The viruses were propagated on CEC for up to six passages, and infected cells were stored in liquid nitrogen until further use. All virus stocks were titrated on fresh CEC.

2.4. Infection of Primary Chicken B Cells

Primary chicken B cells were infected by co-cultivation with infected CEC due to the strictly cell-associated nature of MDV. One million CEC were infected with 30,000 plaque-forming units (PFU) of CVI988, RB1B, or mock infected. After 4 days, one million B cells were seeded on the infected 6-well-plates in the presence of CD40L for 16 h at 41 °C. All B cells were then carefully removed from the CEC monolayer, washed with phosphate buffered saline (PBS), and prepared for fluorescence-activated cell sorting.

2.5. Flow Cytometry

Viable bursal B cells were detected using the eFluor780 fixable viability dye at a dilution of 1:1000 (Affymetrix eBioscience; San Diego, CA, USA) as previously described [19]. Cells were sorted using a FACS Aria III sorter and the FACSDiva software (Becton Dickinson; Franklin Lakes, NJ, USA). For each sample, approximately 10^5 to 10^6 infected B cells were sorted at 4 °C and stored at -80 °C until further analysis. The purity of GFP+ sorted fractions was determined by FACS reanalysis and yielded 99.73% (±0.46 SD) for mock-infected cells, 95.33% (±1.29 SD) for RB1B infected cells and 97.47% (±1.33 SD) for CVI988 infected cells.

2.6. High-Throughput RNA-Sequencing

RNA was isolated from three independent experiments of CVI988 or RB1B infected chicken B cell cultures and sequenced as described [24]. Briefly, total RNA was extracted using TRIzol reagent (Life Technologies; Carlsbad, CA, USA) in combination with the RNeasy Mini Kit (Qiagen; Hilden, Germany) following the manufacturer's instructions. Additionally, RNA was treated with DNase using the RNase-Free DNase Set (Qiagen). Subsequently, ERCC ExFold RNA Spike-In mix 1 (Invitrogen; Carlsbad, CA, USA) was added to the total RNA as an internal control and the polyadenylated (poly(A)) RNA fraction was extracted using the Dynabeads mRNA DIRECT Micro kit (Invitrogen). Whole transcriptome libraries were prepared using the Ion Total RNA-Seq Kit v2 (Life Technologies) following the manufacturer's instructions. Quality and quantity of the nucleic acids was controlled at each step using the NanoDrop 1000 spectrophotometer (Peqlab) or Agilent 2100 Bioanalyzer (Agilent Technologies; Böblingen, Germany) in combination with appropriate chips, respectively. The resulting libraries were finally quantified using the KAPA Library Quantification Kit for Ion Torrent (Kapa Biosystems; Wilmington, MA, USA) on a CFX96 Real-Time PCR Detection System (BioRad Laboratories) and sequenced on an Ion S5XL system (Life Technologies) using the Ion 540 OT2 and Chip kit (Life Technologies).

2.7. Sequence and Bioinformatic Analyses

Reads from separate sequencing runs (technical replicates) of the same library (biological replicates) were combined and quality-trimmed using the 454 Sequencing System Software (v. 3.0; Roche; Mannheim, Germany) along with appropriate Ion Torrent specific adapter sequences. Each quality-trimmed data set was then mapped to a non-redundant version of the MDV reference NC_002229.3 [12] (only segments U_L-IR_L-IR_S-U_S) using STAR (version 2.6.1a; [25]), running in basic two-pass mode. In this manuscript, the MDV genes were designated according to the current gene nomenclature used for the prototype alphaherpesvirus herpes simplex virus 1 (HSV-1) [26,27].

As the mRNA libraries were amplified for several rounds, the sequence duplicates were removed prior to *de-novo* assembly and coverage analysis. Therefore, the duplicated reads were marked and removed in each mapped dataset using Picard (version 2.18.20; http://broadinstitute.github.io/picard). Subsequently, the unique aligned reads were directionally sorted using samtools (version 1.9; [28]) and the sequence depth was deduced from each dataset using bedtools (version 2.15.0; [29]) and samtools. The directionally sorted reads were then used for *de-novo* assembly using the 454 Sequencing System Software (v. 3.0; Roche) running in "-cdna" mode. Assembly was done for each biological replicate separately and with the combined read data from all replicates. Deduplicated and directionally sorted reads were also used as basis for the coverage plots. In order to receive high quality splice junctions, all assembled "isotigs" (transcript variants) were then re-mapped to the non-redundant version of MDV reference NC_002229.3 [12] using STARlong (version 2.6.1a; [25]) in basic two-pass mode. Options were set to allow a single mapping isotig to yield a splice junction with a maximum intron length of 10,000 bp. Only positions corresponding to the high-quality splice junctions were then selected from the splice junctions of the initial mapping approach for further analysis. The frequency of spliced reads was calculated by dividing the number of reads with splice junction by the total number of reads at the respective donor site. The overall splice frequency at a single donor site for CVI988 and RB1B was then averaged from the individual splice frequencies in the replicates. Based on the deduced splice junctions, the positions up- and downstream of these were extracted and visualized using the R (version 3.4.1; [30]) package "ggseqlogo" (version 0.1; [31]) in combination with RStudio (version 1.0.153) in order to receive information on donor and acceptor motif sequences. Polyadenylation cleavage clusters were determined with ContextMap (version 2.7.9; [32]) in combination with bowtie2 (version 2.2.9; [33]) using the complete trimmed dataset as input and the "–polyA" parameter. The resulting poly(A) cleavage sites were then combined into clusters, as the exact position of mRNA cleavage downstream of a cleavage signal can be heterogeneous [34]. Starting with the first poly(A) cleavage site, all other sites on the same strand within a window of the next 30 nt were combined into a single cluster. The window was then moved to the next cleavage site that was not within the last cluster. This was repeated for all cleavage sites. To scan for enriched regulatory motifs within the three prime untranslated region (3′-UTR) of MDV transcripts, the sequences 50 nt up- and 20 nt downstream of the start position of each identified poly(A) cleavage cluster were extracted and analyzed using DREME as part of the MEME suite (version 4.9.0; [35]) using default settings. Subsequently, the identified enriched motif (AWTAAA) was searched in the non-redundant version of MDV reference NC_002229.3 [12] using FIMO (version 4.9.0; [36]) with default settings. Poly(A) cleavage clusters and regulatory motifs were then grouped into relevant pairs based on the FIMO *p*-value and the their distance, allowing a maximum distance of 50 nt. Differential gene expression between RB1B and the vaccine strain CVI988 was conducted using Salmon (version 0.12.0; [37]) in combination with DESeq2 (version 1.18.1; [38]) as described earlier [24]. All relevant MDV CDS sequences were used as transcript reference and genes with an adjusted *p*-value > 0.01 were considered significant. Potential phosphorylation sites in novel protein SORF6 were predicted using the NetPhos 3.1 Server [39] and its DNA-protein binding probability using the DNABIND server [40].

2.8. LC-MALDI TOF/TOF Mass Spectrometry

Infected FACS sorted primary chicken B cells and mock infected primary chicken B cells were lysed in batches of 1.5×10^6 cells using 150 µl of a lysis buffer containing 0.1 M DL-Dithothreitol (DTT) and 2% SDS in 0.1 M Tris-HCl (pH 8.0) at 99 °C for 5 min. Protein contents were determined by densitometry of Coomassie stained SDS gels [41,42]. After cell lysis, 20 µg aliquots were digested using the FASP protocol as described [43]. Samples were differentially labeled by dimethylation [44] using unlabeled and ^{13}C-labeled formaldehyde, respectively, and subjected to nano-LC MALDI-TOF/TOF mass spectrometry as described previously [45]. Briefly, peptides were separated by nano reversed-phase liquid chromatography (EASY-nLC II, Bruker; Bremen, Germany), spotted to a MALDI target (Proteineer fcII, Bruker), and analyzed with an UltrafleXtreme MALDI-TOF/TOF mass spectrometer

(Bruker) as described previously [46]. Peptide spectra were acquired in the m/z range 700 to 3.500 Da with a minimum signal-to-noise (S/N) ratio of 7. Proteins were identified with a Mascot server (version 2.4.1; Matrix Science Ltd; London, UK) and analyzed using ProteinScape software (version 3; Bruker). Oxidation of methionine, acetylation of protein N-termini, and dimethylation of lysine and peptide N-termini with either isotopomer were set as variable modifications, whereas the carbamydomethylation of cysteine residues was set as a fixed modification. Two independent experiments were performed with inverted labeling. As database for the protein identification with the MASCOT search engine (Matrix Science Ltd), the *Gallus gallus* proteome was downloaded from the ENSEMBL website [47] and the viral sequences were added to the FASTA file. Viral protein content was calculated in mol% using the exponentially modified protein abundance index (emPAI) [48]. To identify peptides covering the potential new splicing sites discovered by RNA-sequencing, a database with sequence fragments covering a 67 amino acid region centered on the splicing site was constructed and used for the database search with MASCOT.

2.9. Reverse Transcription and PCR over Splice Junctions

RNA was isolated as described above. cDNA was synthesized after DNAse treatment (Promega; Fitchburg, WI, USA) with the Applied Biosystems High-Capacity cDNA Reverse Transcription Kit (Thermo Fischer; Waltham, MA, USA). Conventional Taq-PCR was performed with primers specific to the respective viral gene (Table S2). Amplification of BAC DNA was used as a positive control. Mock-infected cells and samples without reverse transcriptase to exclude a contamination with genomic DNA were included as negative controls.

2.10. Data Availability

The RNA-seq raw data were deposited in the ArrayExpress database at EMBL-EBI (www.ebi.ac. uk/arrayexpress) under accession number E-MTAB-7772. A supplementary GFF file for the reference sequence NC_002229.3 containing annotations for all detected introns, poly(A) cleavage sites and associated motifs, as well as the novel CDS for SORF6 can be found in the Supplementary File S1.

3. Results and Discussion

3.1. The MDV Transcriptional Landscape

To assess the transcriptional landscape of MDV in the primary target cells of lytic replication in vivo, we used a previously established in vitro infection system for primary chicken B cells [19]. B cells were infected for 16 h with the very virulent MDV strain RB1B and the vaccine strain CVI988 and analyzed by high-throughput RNA sequencing.

The overall RNA-seq dataset consisted of 82.6 million reads from three biological replicates of CVI988 (48.2 million reads) and two biological replicates of RB1B (34.4 million reads). A third replicate of RB1B did not yield sufficient amount of reads and was therefore excluded from our analysis. An average of 10.5% and 11.2% of the CVI and RB1B datasets respectively could be mapped to the MDV reference sequence (see Table S3).

The position and direction of mapped reads fitted very well to the previously annotated MDV genes (Figure 1A). Highly abundant genes like the immediate early gene SORF1 (ICP4) or the UL49 tegument protein (VP22) correlate well with previously published data [49]. Surprisingly, only minor differences were detected between the transcriptome of CVI988 and RB1B. Comparing the 94 detected MDV genes, two variants of MDV075 encoding the 14-kDa polypeptides (pp14), were significantly higher expressed in RB1B infected primary B cells (Figure 1B). These phosphorylated cytoplasmic proteins arise from splice variants of the same gene and are thought to be involved in transcriptional regulation and increased neurovirulence [50–52]. Furthermore, the hypothetical gene MDV082 [53] that is located on the same transcript as the ICP4 gene, was significantly higher expressed in CVI988 infected B cells. However, the transcriptome of RB1B and CVI988 only shows subtle differences in primary chicken B cells, suggesting that the differences in their pathogenesis might be due to sequence changes on the protein level and/or functional differences of virulence factors. Similarly, we only detected marginal quantitative differences between the expression levels of viral proteins in a proteome analysis using LC-MALDI TOF/TOF mass spectrometry (Figure S1 and Table S1). Taken together, our data indicate that there are only minor differences in the mRNA and protein expression levels after B cell infections with the very virulent RB1B or the CVI988 vaccine strain.

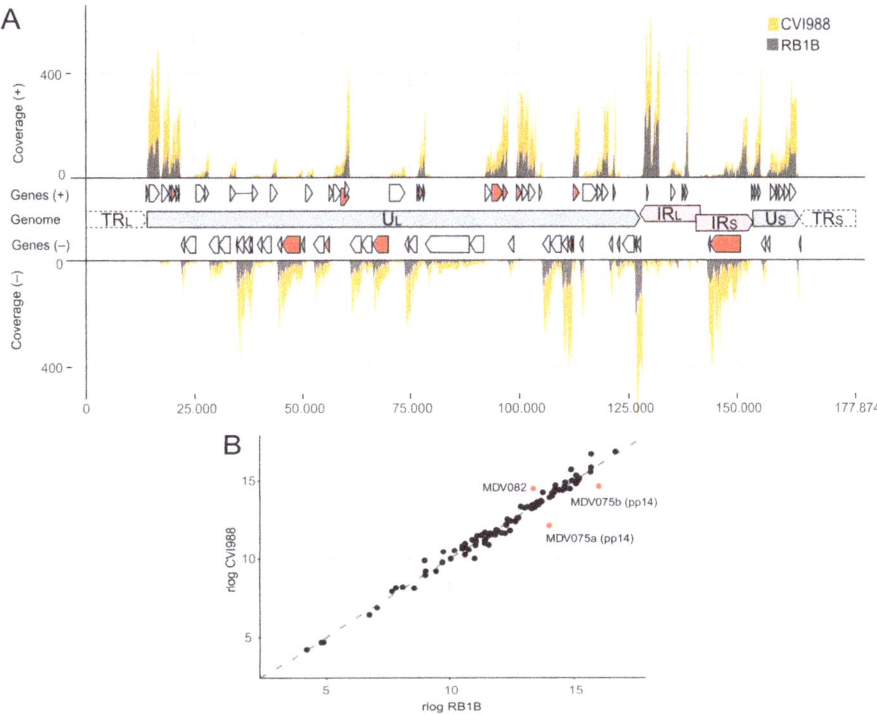

Figure 1. The Marek's disease virus (MDV) transcriptome and proteome in in vitro infected primary chicken B cells. (**A**) Visualization of the deduplicated and strand-specific RNA-seq read coverage of plus (+) and minus (-) strand-encoded genes across the MDV genome. Orange curves indicate CVI988 reads and grey curves indicate reads for RB1B (with the respective annotated genes as grey arrow bars). Red bars depict proteins identified by MS. The two unique regions, unique long (U_L) and short (U_S) are flanked by terminal (TR_L and TR_S) and internal (IR_L and IR_S) inverted repeat regions. Nucleotide position numbers are derived from [12]. (**B**) Gene expression scatterplot comparing normalized expression levels (rlog) in RB1B and CVI988 infected primary chicken B cells. Red dots indicate significantly differentially expressed genes.

3.2. Splicing of Polycistronic MDV Transcripts

In addition to the transcriptional profile, we could readily identify 71 introns that were represented by at least one *de novo* transcript (Figure 2A). Some of the introns and associated spliced genes have been previously described such as the viral lipase (vLIP) [54], LORF2 (MDV012) [55], UL15 [12], UL44 (glycoprotein C) [56], vIL8 [57] and pp14 [51]. However, analysis of the MDV transcriptome revealed a number of novel splice forms (Table S4). These results are in line with previous RNA-seq analysis for other alphaherpesviruses [58] that also revealed a plethora of novel splice products. The detected splice variants could contribute to viral proteomic diversity and could prevent viral mRNA degradation through the virion host shutoff UL41 endoribonuclease [59]. For HSV-1 it has been shown that UL41 not only targets many cellular but also viral mRNAs. Spliced mRNAs are protected from UL41-mediated degradation by bound exon junction complexes (EJCs) [60].

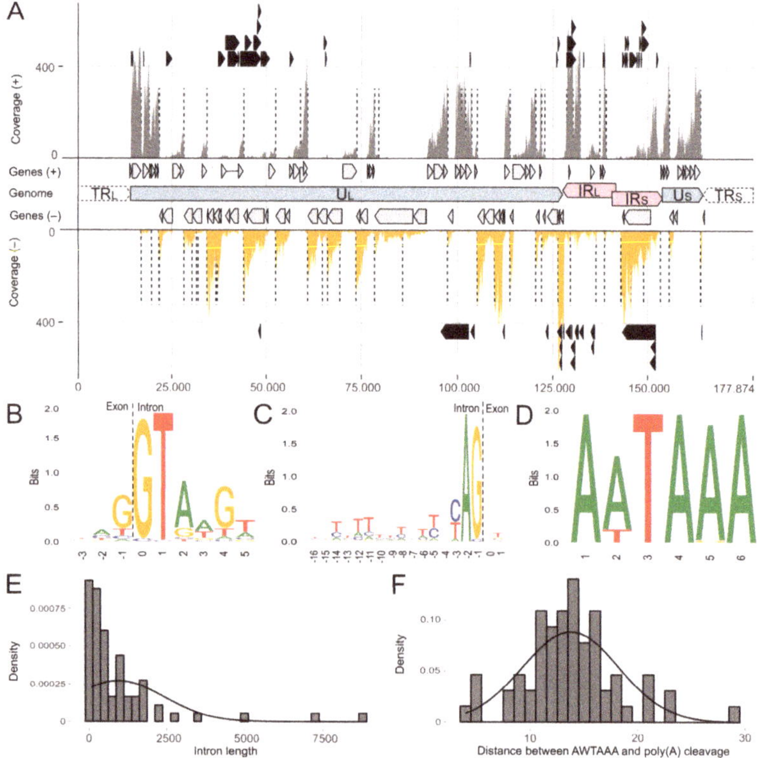

Figure 2. Overview of Marek's disease virus (MDV) mRNA splicing and poly(A) cleavage. (**A**) Visualization of the cumulative CVI988 and RB1B RNA-seq coverage of plus (+) and minus (-) strand-encoded genes across the MDV genome. Black arrow bars indicate introns and dashed lines indicate poly(A) cleavage sites. More detailed information is shown in Table S4 and Table S5. (**B**) Nucleotide frequency maps (sequence logo) of splice donor sites in MDV-encoded mRNAs. The relative heights of letters correspond to frequencies of bases at each position. (**C**) Sequence logo of splice acceptor sites in MDV-encoded mRNAs. (**D**) Sequence logo of polyadenylation signals in MDV-encoded three prime untranslated regions (3′-UTR). (**E**) Histogram depicting MDV intron length distributions. (**F**) Histogram depicting the distance from AWTAAA-like motifs to the poly(A) cleavage site.

The identified splice site sequences mostly represent canonical splicing motifs, containing the GT at the donor and AG at the acceptor sites (Figure 2B,C). The intron length varied between 70 and 8651 nt (Figure 2E). Intriguingly, the intron frequencies differed between RB1B and CVI988 in infected primary chicken B cells (Figures 3–5). By matching the intron positions with our MS data, we identified a peptide that spans the exon-exon junction of UL15 (Figure S2).

The analysis of poly(A) cleavage signals within the RNA-seq data revealed abundant bicistronic and polycistronic MDV transcripts (Table S5). These transcripts encode for two or more proteins and were characterized as regions of high coverage that were not separated by a poly(A) cleavage site. Here we found that the canonical AATAAA motif is the most frequent and functional polyadenylation signal in MDV, followed by ATTAAA (Figure 2D). Interestingly, we also found evidence for alternative non-canonical polyadenylation signals in MDV mRNA 3′ UTRs (Figure 2 and Table S5). The distances between the detected AWTAAA polyadenylation signal motif and the actual poly(A) cleavage site was 13.8 nt (±4.4 SD) and confirmed that not only the polyadenylation signal sequence, but also its distance from the poly(A) is highly conserved (Figure 2D,F) [61,62].

3.3. The Transcriptional Makeup of the MDV Unique Long Region (U_L).

The MDV unique regions mainly harbor genes that are conserved among alphaherpesviruses and are involved in DNA replication and production of progeny virus [26]. The U_L spans over roughly 113,000 base pairs and harbors the majority of the MDV-encoded genes [26]. Within the U_L, we could detect high transcription rates of nearly all annotated genes. Splicing was identified in multiple genes including RLORF14 (pp24), vLIP [63], and LORF2 (MDV012) [55] and in a transcript antisense to UL5 (MDV017) (Figure 3). Only minor differences were observed in the intron frequencies between RB1B and CVI988. To confirm the splice events and frequencies detected by RNA-seq, we performed RT-PCR analyses on several randomly selected genes (Figure 3C, Figure 4B, Figure 5C, Figures S2 and S3). All analyzed genes showed a comparable splice pattern in both RNA-seq and RT-PCR. In addition, we confirmed a novel splice site of UL15 by MS and RT-PCR (Figure S2). UL15 encodes the tripartite terminase subunit that is involved in DNA packaging into the viral capsid. Splicing of UL15 mRNA has already been observed in herpes simplex virus type 1 (HSV-1) [64] and duck enteritis virus (DEV) [65]. However, the observed UL15 isoforms in MDV are to our knowledge unknown and expand the number of potential proteins encoded by UL15 to at least five.

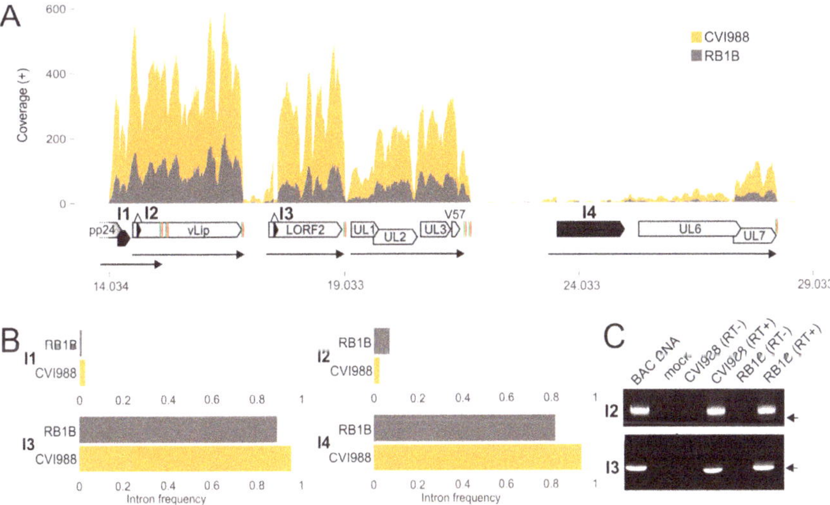

Figure 3. The Marek's disease virus (MDV) unique long (U_L) region. (**A**) Visualization of RNA-Seq coverage across parts of the MDV U_L region with respective introns in black. Green and red arrows indicate the polyadenylation signal and the poly(A) cleavage site respectively. Underlying black arrows suggest unspliced (mono-, bi-, or polycistronic) mRNAs. (**B**) Comparison of intron frequencies in RB1B and CVI988 infected primary chicken B cells. I1: pp24, I2: vLIP, I3: LORF2 (MDV012), I4: transcript antisense to UL5 (MDV017). (**C**) RT-PCR was performed to validate the splicing event. PCR products were derived using forward/reverse primers to amplify the respective intron-flanking regions. The representative gel images illustrate the results of RT-PCR analysis. The black arrows indicate the spliced form of the respective gene.

We could also confirm known splice sites in UL44 (gC) in our analysis (Figure 4) [56]. These splice variants lead to a gC protein that lacks the transmembrane domain and is secreted into the supernatant [56]. Beyond that, we also confirmed novel splice sites like in RLORF14a (pp38) (Figure S3). Several capsid and tegument protein-encoding genes encoded in the U_L and U_S regions, like UL18 (triplex capsid protein 2), UL19 (VP5), UL49 (VP22), UL49.5 (gN), US7 (gI) and US8 (gE) also undergo different splicing events (Table S4), contradicting the long-standing paradigm that splicing is a rare phenomenon throughout the alphaherpesvirus family [13].

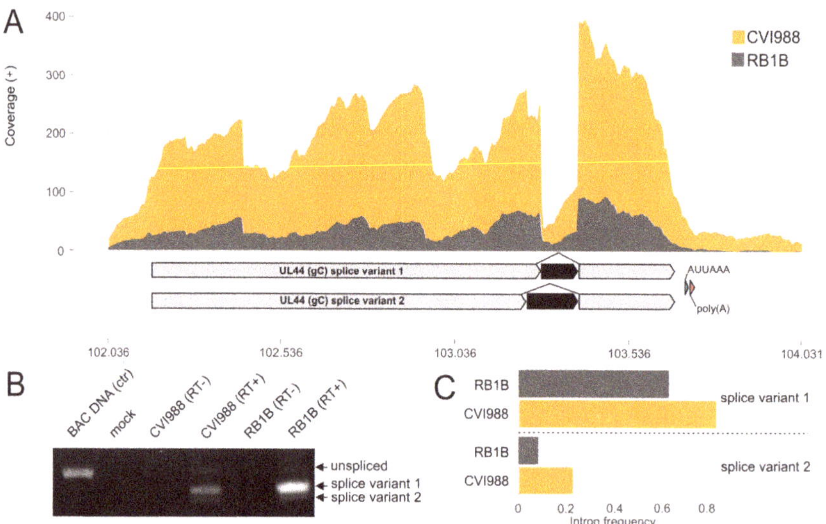

Figure 4. Glycoprotein C (gC) splicing. (**A**) Visualization of RNA-seq coverage across the Marek's disease virus (MDV) gC gene UL44 with respective introns in black. The green and red arrow indicate the canonical ATTAAA polyadenylation signal and the poly(A) cleavage site respectively. (**B**) RT-PCR was performed to validate gC splicing. PCR products were derived using forward/reverse primers to amplify the respective intron-flanking regions. (**C**) Comparison of the gC intron frequencies in RB1B and CVI988 infected primary chicken B cells.

Some identified splice variants would result in proteins with an altered membrane topology (TMHMM Server, v. 2.0). This is for example the case for splice variants of pp24 and of pp38, which results in changes of the previously assessed hydrophobic anchor domains of both proteins [66]. While pp38 seems to exist as splice variants with and without a membrane anchor, splicing of pp28 could retain its membrane association while altering its function (Figure S5).

3.4. The Transcriptional Makeup of the MDV Unique Short Region (U_S).

The MDV U_S region contains many genes that play important roles in the viral life cycle. Intriguingly, we detected splice variants of several envelope glycoproteins as described above. In addition, we identified a hitherto uncharacterized spliced transcript of a gene located downstream of SORF2A, termed SORF6 (Figure 5). This novel gene possesses an upstream TATA box in the transcriptional regulatory region, an intron and exon with respective donor and acceptor site and a downstream polyadenylation signal with the poly(A) cleavage cluster (as described in Figure 2). Furthermore, the resulting protein is predicted to be 85 amino acids in size, shows several predicted phosphorylation sites (Figure S6) and may act as a DNA-binding protein (DNABIND server [40]). The region containing this novel gene (Figure 5) has previously been associated with the virulence of the virus [67]. However, more work needs to be done to understand the contribution of this novel gene and the region in MDV virulence.

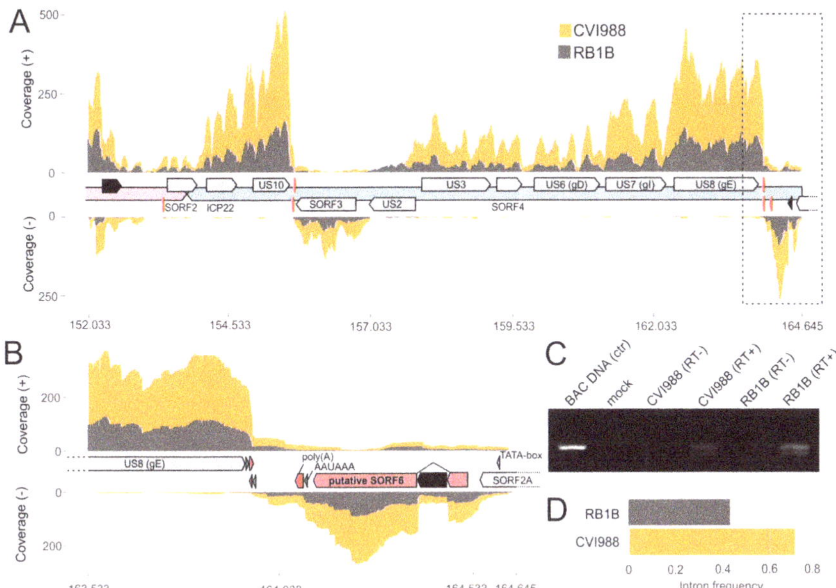

Figure 5. The Marek's disease virus (MDV) unique short (U$_S$) region. (**A**) Visualization of RNA-seq coverage across the MDV U$_S$ region with respective introns in black. (**B**) Zoom into the far-right region of the MDV US with depiction of the novel U$_S$ gene SORF6. Green and red arrows indicate the polyadenylation signals and the poly(A) cleavage sites, respectively. (**C**) RT-PCR was performed to validate the splicing event in the novel gene SORF6. PCR products were derived using forward/reverse primers to amplify the respective intron-flanking regions as full-length (upper band) and spliced (lower band). (**D**) Comparison of the novel U$_S$ gene (SORF6) intron frequencies in RB1B and CVI988 infected primary chicken B cells.

3.5. The Transcriptional Makeup of the MDV Repeat Regions

The repeat regions mostly contain MDV-specific genes encoding for proteins or RNA that play a role in the cell tropism, MDV pathogenesis, latency, and transformation [14]. Here, we observed excessive splicing antisense to ICP4. These transcripts are part of the latency associated transcript (LAT) region, have a complex splice pattern, and their functions remain largely unknown [68,69]. Some of these RNAs function as MDV-encoded micro RNAs and are described elsewhere [70,71].

Only moderate RLORF7 (Meq) and vIL8 splicing was detected in infected primary B cells 16 hpi although more extensive splicing activity has been observed in this region of the MDV genome in vitro and in vivo [72–74]. These splice variants are likely higher expressed in latently infected and transformed cells. The splice variants of the neurovirulence factor pp14 encoded by MDV075 were efficiently detected as published previously [15,44].

In the RNA-seq data, we did not detect any reads complementary to vTR. However, this region is annotated as a hypothetical MDV gene termed RLORF1 (an arginine-rich protein/ICP0-like protein) [75]. RLORF1 is discussed as a potential positional orthologue of alphaherpesviral ICP0 proteins; however, it does not contain typical ICP0 features such as a C3HC4 zinc RING finger at the N-terminus or a nuclear localization signal (NLS). To assess if ICP0 protein is expressed and if it plays a role in replication, we generated recombinant MDV mutants harboring an HA-tagged ICP0 (RB1B_ICP0-HA) or an ICP0 knockout (RB1B_ΔMetICP0). The knockout did not affect MDV replication and cell to cell spread in vitro (Method S1 and Figure S4A) and no ICP0 was detected by western blotting (Method S1 and Figure S4B), suggesting that ICP0 is not expressed and therefore does not play a role in the virus life cycle.

Furthermore, we detected several poly(A) cleavage sites in combination with appropriate motifs within the MDV repeat regions, that are to our knowledge undescribed. The presence of these transcriptional signals in combination with sufficient read coverage suggest the existence of hitherto hypothetical protein coding regions, including RLORF11 and MDV082.

3.6. MDV Noncoding RNAs

Although we enriched for poly(A) mRNA, some newly identified introns do not lie in annotated MDV coding sequences and some splice donor and acceptor sites do not give rise to conclusive protein-encoding mRNAs. Such sequences could be easily regarded as 'nonsense' transcripts that are rapidly degraded. However, the importance of noncoding RNAs (ncRNAs) in MDV infections is expanding. Several viral ncRNAs were found to be expressed by MDV [75–78] but the multitude of functions played by viral ncRNAs, and especially by long ncRNAs (lncRNAs) and stable intronic sequence RNAs (sisRNAs), have not been thoroughly investigated yet and the unexpected transcriptomic complexity may have been overlooked in MDV research so far. Of note is, that antisense transcription was also observed in related alphaherpesviruses like HSV-1 or pseudorabies virus (PRV), and in human herpesvirus 6 (HHV-6) RNA-seq data [79–81].

Taken together, our MDV RNA-seq data provide novel insights into the transcriptional profile of the RB1B and CVI988 strains. Despite stark differences in their pathogenicity, the two viruses show a similar transcriptomic profile in primary chicken B cells.

4. Conclusions

B cells are a major target for lytic MDV replication in vivo [8,10]; however, it remained impossible to assess the MDV transcriptome in primary B cells, due to the short-lived nature of these cell. The aim of this study was to evaluate the gene expression profiles of the very virulent RB1B strain and the commercial MDV vaccine CVI988 in primary chicken B cells by RNA-seq using our recently established in vitro infection system [19]. We developed a bioinformatics pipeline that can be easily transferred to other herpesviruses or large DNA viruses to identify unknown transcript isoforms and associated motifs.

The RNA-seq revealed the expression of 94 MDV transcripts and the presence of 71 introns that lead to mostly novel splice forms and antisense transcripts. In addition, we could identify a novel gene in the U_S region of the MDV genome that we will characterize in future studies. While some of the detected splice sites were previously published, we identified several novel splice variants and confirmed some of them by RT-PCR and/or MS. However, more work is certainly required to dissect their relevance in the MDV life cycle.

We found that MDV produces bicistronic and polycistronic transcripts as a mechanism to maximize its coding capacities. Poly(A) cleavage after the upstream AATAAA motif seems to be the most frequent and functional polyadenylation signal in MDV. The identification of possible alternative transcript termination (ATT) needs further experimental evidence (Figure 2 and Table S5). ATT is a strong regulatory factor in eukaryotes [82], but there is only limited data for ATT in herpesvirus transcription.

The comparison of the transcriptome between the very virulent RB1B strain and the CVI988 vaccine revealed differences in only a few transcripts (Figure 1). However, more work needs to be done to unravel significant differences that could possibly point towards a mechanism of attenuation or provide valuable information for the development of diagnostic tools.

In summary, our data demonstrate that the MDV genome is more complex than previously assumed. It provides a source of reference for MDV transcripts expressed in primary chicken B cells and lays the foundation for future research on MDV-encoded gene products and splice variants.

Supplementary Materials: The following are available online at http://www.mdpi.com/1999-4915/11/3/264/s1: Method S1: Generation and in vitro characterization of RB1BΔMetICP0, Figure S1: Protein concentration (mol%) scatterplot comparing levels of detected viral proteins in RB1B and CVI988 infected primary chicken B cells,

Figure S2: Detected splice variants of UL15, Figure S3: RT-PCR confirmation of splicing events in pp38, Figure S4: Plaques size assay and western blot analysis for the ICP0 mutant RB1BΔMetICP0, Figure S5: Transmembrane helix prediction for spliced and unspliced pp24 and pp38, Figure S6: Prediction of serine, threonine or tyrosine phosphorylation sites in the hypothetical MDV protein SORF6 encoded on the U$_S$ segment, Table S1: Marek's disease virus (MDV) proteins detected by mass spectrometry, Table S2: Primers used in this study, Table S3: Summary of RNA-seq read mapping, Table S4: Introns identified from RNA-seq data, Table S5: Poly(A) cleavage sites, polyadenylation signals and polycistronic transcripts identified in MDV transcriptomes, Supplementary File S1: Annotations for NC_002229.3 in GFF format.

Author Contributions: Conceptualization: L.D.B., F.P., A.M.K., A.K. and B.B.K.; methodology: L.D.B., F.P., V.I.P., A.K. and B.B.K.; formal analysis: L.D.B., F.P., A.K. and B.B.K..; investigation: L.D.B., F.P., V.I.P. and A.M.K.; resources: S.H.; writing—original draft preparation: L.D.B. and F.P.; writing—review and editing: L.D.B., F.P., A.K. and B.B.K.; visualization: L.D.B., F.P., A.K. and B.B.K.; supervision: D.H., A.K. and B.B.K.; project administration: A.K. and B.B.K.; funding acquisition: A.K. and B.B.K.; all authors read and approved the final manuscript.

Funding: This research was funded by the *Deutsche Forschungsgemeinschaft* (DFG), grant numbers KA 3492/3-1 and KA 2900/3-1 awarded to B.B.K. and A.K. respectively.

Acknowledgments: We thank Ann Reum and Yu You (Institut of Virology, Freie Universität Berlin), Marina Kohn (Institute for Animal Physiology, Ludwig-Maximilians-Universität München) and Patrick Zitzow (Institute of Diagnostic Virology, Friedrich-Loeffler-Institut) for outstanding technical assistance.

Conflicts of Interest: The authors declare no conflict of interest.

References

1. Nair, V. Evolution of Marek's disease—A paradigm for incessant race between the pathogen and the host. *Vet. J.* **2005**, *170*, 175–183. [CrossRef]

2. Davison, T.F.; Nair, V. *Marek's Disease: An Evolving Problem*; Elsevier: Amsterdam, The Netherlands, 2004.

3. Davison, F.; Nair, V. Use of Marek's disease vaccines: Could they be driving the virus to increasing virulence? *Expert Rev. Vaccines* **2005**, *4*, 77–88. [CrossRef]

4. Witter, R.L. Increased virulence of Marek's disease virus field isolates. *Avian Dis.* **1997**, *41*, 149–163. [CrossRef]

5. Baigent, S.J.; Nair, V.K.; Le Galludec, H. Real-time PCR for differential quantification of CVI988 vaccine virus and virulent strains of Marek's disease virus. *J. Virol. Methods* **2016**, *233*, 23–36. [CrossRef]

6. Jarosinski, K.W.; Tischer, B.K.; Trapp, S.; Osterrieder, N. Marek's disease virus: Lytic replication, oncogenesis and control. *Expert Rev. Vaccines* **2006**, *5*, 761–772. [CrossRef]

7. Baigent, S.J.; Ross, L.J.; Davison, T.F. Differential susceptibility to Marek's disease is associated with differences in number, but not phenotype or location, of pp38+ lymphocytes. *J. Gen. Virol.* **1998**, *79 Pt 11*, 2795–2802. [CrossRef]

8. Baaten, B.J.; Staines, K.A.; Smith, L.P.; Skinner, H.; Davison, T.F.; Butter, C. Early replication in pulmonary B cells after infection with Marek's disease herpesvirus by the respiratory route. *Viral Immunol.* **2009**, *22*, 431–444. [CrossRef]

9. Shek, W.; Calnek, B.; Schat, K.; Chen, C. Characterization of Marek's disease virus-infected lymphocytes: Discrimination between cytolytically and latently infected cells. *J. Natl. Cancer Inst.* **1983**, *70*, 485–491.

10. Bertzbach, L.D.; Laparidou, M.; Hartle, S.; Etches, R.J.; Kaspers, B.; Schusser, B.; Kaufer, B.B. Unraveling the role of B cells in the pathogenesis of an oncogenic avian herpesvirus. *Proc. Natl. Acad. Sci. USA* **2018**, *115*, 11603–11607. [CrossRef]

11. Calnek, B.W.; Schat, K.A.; Ross, L.J.; Shek, W.R.; Chen, C.L. Further characterization of Marek's disease virus-infected lymphocytes. I. In vivo infection. *Int. J. Cancer* **1984**, *33*, 389–398. [CrossRef]

12. Tulman, E.R.; Afonso, C.L.; Lu, Z.; Zsak, L.; Rock, D.L.; Kutish, G.F. The genome of a very virulent Marek's disease virus. *J. Virol.* **2000**, *74*, 7980–7988. [CrossRef]

13. Davison, A.J. Comparative analysis of the genomes. In *Human Herpesviruses: Biology, Therapy, and Immunoprophylaxis*; Arvin, A., Campadelli-Fiume, G., Mocarski, E., Moore, P.S., Roizman, B., Whitley, R., Yamanishi, K., Eds.; Cambridge University Press: Cambridge, UK, 2007.

14. Bertzbach, L.D.; Kheimar, A.; Ali, F.A.Z.; Kaufer, B.B. Viral Factors Involved in Marek's Disease Virus (MDV) Pathogenesis. *Curr. Clin. Microbiol. Rep.* **2018**, *5*, 238–244. [CrossRef]

15. Heidari, M.; Huebner, M.; Kireev, D.; Silva, R.F. Transcriptional profiling of Marek's disease virus genes during cytolytic and latent infection. *Virus Genes* **2008**, *36*, 383–392. [CrossRef]

16. Neerukonda, S.N.; Tavlarides-Hontz, P.; McCarthy, F.; Pendarvis, K.; Parcells, M.S. Comparison of the Transcriptomes and Proteomes of Serum Exosomes from Marek's Disease Virus-Vaccinated and Protected and Lymphoma-Bearing Chickens. *Genes* **2019**, *10*, 116. [CrossRef]

17. Mwangi, W.N.; Vasoya, D.; Kgosana, L.B.; Watson, M.; Nair, V. Differentially expressed genes during spontaneous lytic switch of Marek's disease virus in lymphoblastoid cell lines determined by global gene expression profiling. *J. Gen. Virol.* **2017**, *98*, 779–790. [CrossRef]

18. Baigent, S.J.; Ross, L.J.; Davison, T.F. A flow cytometric method for identifying Marek's disease virus pp38 expression in lymphocyte subpopulations. *Avian Pathol.* **1996**, *25*, 255–267. [CrossRef]

19. Schermuly, J.; Greco, A.; Hartle, S.; Osterrieder, N.; Kaufer, B.B.; Kaspers, B. In vitro model for lytic replication, latency, and transformation of an oncogenic alphaherpesvirus. *Proc. Natl. Acad. Sci. USA* **2015**, *112*, 7279–7284. [CrossRef]

20. Schat, K.; Purchase, H. *Cell-Culture Methods. A Laboratory Manual for the Isolation and Identification of Avian Pathogens*; American Association of Avian Pathologists: Kennett Square, PA, USA, 1998.

21. Martin, A.; Lillehoj, H.S.; Kaspers, B.; Bacon, L.D. Antigen-specific T cell proliferation following coccidia infection. *Poult Sci.* **1993**, *72*, 2084–2094. [CrossRef]

22. Kothlow, S.; Morgenroth, I.; Tregaskes, C.A.; Kaspers, B.; Young, J.R. CD40 ligand supports the long-term maintenance and differentiation of chicken B cells in culture. *Dev. Comp. Immunol.* **2008**, *32*, 1015–1026. [CrossRef]

23. Engel, A.T.; Selvaraj, R.K.; Kamil, J.P.; Osterrieder, N.; Kaufer, B.B. Marek's disease viral interleukin-8 promotes lymphoma formation through targeted recruitment of B cells and CD4+ CD25+ T cells. *J. Virol.* **2012**, *86*, 8536–8545. [CrossRef]

24. Pfaff, F.; Hägglund, S.; Zoli, M.; Blaise-Boisseau, S.; Laloy, E.; Koethe, S.; Zühlke, D.; Riedel, K.; Zientara, S.; Bakkali-Kassimi, L.; et al. Proteogenomics Uncovers Critical Elements of Host Response in Bovine Soft Palate Epithelial Cells Following In Vitro Infection with Foot-And-Mouth Disease Virus. *Viruses* **2019**, *11*, 53. [CrossRef]

25. Dobin, A.; Davis, C.A.; Schlesinger, F.; Drenkow, J.; Zaleski, C.; Jha, S.; Batut, P.; Chaisson, M.; Gingeras, T.R. STAR: Ultrafast universal RNA-seq aligner. *Bioinformatics* **2013**, *29*, 15–21. [CrossRef]

26. Lee, L.F.; Wu, P.; Sui, D.; Ren, D.; Kamil, J.; Kung, H.J.; Witter, R.L. The complete unique long sequence and the overall genomic organization of the GA strain of Marek's disease virus. *Proc. Natl. Acad. Sci. USA* **2000**, *97*, 6091–6096. [CrossRef]

27. Davison, A.J. Herpesvirus systematics. *Vet. Microbiol.* **2010**, *143*, 52–69. [CrossRef]

28. Li, H.; Handsaker, B.; Wysoker, A.; Fennell, T.; Ruan, J.; Homer, N.; Marth, G.; Abecasis, G.; Durbin, R. 1000 Genome Project Data Processing Subgroup. The Sequence Alignment/Map format and SAMtools. *Bioinformatics* **2009**, *25*, 2078–2079. [CrossRef]

29. Quinlan, A.R.; Hall, I.M. BEDTools: A flexible suite of utilities for comparing genomic features. *Bioinformatics* **2010**, *26*, 841–842. [CrossRef]

30. R Core Team. *R: A Language and Environment for Statistical Computing*; R Core Team: Vienna, Austria, 2017.

31. Wagih, O. ggseqlogo: A versatile R package for drawing sequence logos. *Bioinformatics* **2017**, *33*, 3645–3647. [CrossRef]

32. Bonfert, T.; Kirner, E.; Csaba, G.; Zimmer, R.; Friedel, C.C. ContextMap 2: Fast and accurate context-based RNA-seq mapping. *BMC Bioinform.* **2015**, *16*, 122. [CrossRef]

33. Langmead, B.; Salzberg, S.L. Fast gapped-read alignment with Bowtie 2. *Nat. Methods* **2012**, *9*, 357–359. [CrossRef]

34. Pauws, E.; van Kampen, A.H.; van de Graaf, S.A.; de Vijlder, J.J.; Ris-Stalpers, C. Heterogeneity in polyadenylation cleavage sites in mammalian mRNA sequences: Implications for SAGE analysis. *Nucl. Acids Res.* **2001**, *29*, 1690–1694. [CrossRef]

35. Bailey, T.L.; Boden, M.; Buske, F.A.; Frith, M.; Grant, C.E.; Clementi, L.; Ren, J.; Li, W.W.; Noble, W.S. MEME SUITE: Tools for motif discovery and searching. *Nucl. Acids Res.* **2009**, *37*, W202–W208. [CrossRef]

36. Grant, C.E.; Bailey, T.L.; Noble, W.S. FIMO: Scanning for occurrences of a given motif. *Bioinformatics* **2011**, *27*, 1017–1018. [CrossRef]

37. Patro, R.; Duggal, G.; Love, M.I.; Irizarry, R.A.; Kingsford, C. Salmon provides fast and bias-aware quantification of transcript expression. *Nat. Methods* **2017**, *14*, 417–419. [CrossRef]

38. Love, M.I.; Huber, W.; Anders, S. Moderated estimation of fold change and dispersion for RNA-seq data with DESeq2. *Genome Biol.* **2014**, *15*, 550. [CrossRef]

39. Blom, N.; Gammeltoft, S.; Brunak, S. Sequence and structure-based prediction of eukaryotic protein phosphorylation sites. *J. Mol. Biol.* **1999**, *294*, 1351–1362. [CrossRef]

40. Szilagyi, A.; Skolnick, J. Efficient prediction of nucleic acid binding function from low-resolution protein structures. *J. Mol. Biol.* **2006**, *358*, 922–933. [CrossRef]

41. Neuhoff, V.; Arold, N.; Taube, D.; Ehrhardt, W. Improved staining of proteins in polyacrylamide gels including isoelectric focusing gels with clear background at nanogram sensitivity using Coomassie Brilliant Blue G-250 and R-250. *Electrophoresis* **1988**, *9*, 255–262. [CrossRef]

42. Laemmli, U.K. Cleavage of structural proteins during the assembly of the head of bacteriophage T4. *Nature* **1970**, *227*, 680–685. [CrossRef]

43. Wisniewski, J.R.; Zougman, A.; Nagaraj, N.; Mann, M. Universal sample preparation method for proteome analysis. *Nat. Methods* **2009**, *6*, 359–362. [CrossRef]

44. Boersema, P.J.; Aye, T.T.; van Veen, T.A.; Heck, A.J.; Mohammed, S. Triplex protein quantification based on stable isotope labeling by peptide dimethylation applied to cell and tissue lysates. *Proteomics* **2008**, *8*, 4624–4632. [CrossRef]

45. Pauker, V.I.; Bertzbach, L.D.; Hohmann, A.; Kheimar, A.; Teifke, J.P.; Mettenleiter, T.C.; Karger, A.; Kaufer, B.B. Imaging Mass Spectrometry and Proteome Analysis of Marek's Disease Virus-Induced Tumors. *mSphere* **2019**, *4*. [CrossRef]

46. Henning, A.K.; Groschup, M.H.; Mettenleiter, T.C.; Karger, A. Analysis of the bovine plasma proteome by matrix-assisted laser desorption/ionisation time-of-flight tandem mass spectrometry. *Vet. J.* **2014**, *199*, 175–180. [CrossRef]

47. Aken, B.L.; Achuthan, P.; Akanni, W.; Amode, M.R.; Bernsdorff, F.; Bhai, J.; Billis, K.; Carvalho-Silva, D.; Cummins, C.; Clapham, P.; et al. Ensembl 2017. *Nucl. Acids Res.* **2017**, *45*, D635–D642. [CrossRef]

48. Ishihama, Y.; Oda, Y.; Tabata, T.; Sato, T.; Nagasu, T.; Rappsilber, J.; Mann, M. Exponentially modified protein abundance index (emPAI) for estimation of absolute protein amount in proteomics by the number of sequenced peptides per protein. *Mol. Cell. Proteom.* **2005**, *4*, 1265–1272. [CrossRef]

49. Chakraborty, P.; Kuo, R.; Vervelde, L.; Dutia, B.M.; Kaiser, P.; Smith, J. Macrophages from Susceptible and Resistant Chicken Lines have Different Transcriptomes following Marek's Disease Virus Infection. *Genes* **2019**, *10*, 74. [CrossRef]

50. Tahiri-Alaoui, A.; Smith, L.P.; Kgosana, L.; Petherbridge, L.J.; Nair, V. Identification of a neurovirulence factor from Marek's disease virus. *Avian Dis.* **2013**, *57*, 387–394. [CrossRef] [PubMed]

51. Hong, Y.; Coussens, P.M. Identification of an immediate-early gene in the Marek's disease virus long internal repeat region which encodes a unique 14-kilodalton polypeptide. *J. Virol.* **1994**, *68*, 3593–3603. [PubMed]

52. Tahiri-Alaoui, A.; Smith, L.P.; Baigent, S.; Kgosana, L.; Petherbridge, L.J.; Lambeth, L.S.; James, W.; Nair, V. Identification of an intercistronic internal ribosome entry site in a Marek's disease virus immediate-early gene. *J. Virol.* **2009**, *83*, 5846–5853. [CrossRef]

53. Hearn, C.; Preeyanon, L.; Hunt, H.D.; York, I.A. An MHC class I immune evasion gene of Mareks disease virus. *Virology* **2015**, *475*, 88–95. [CrossRef]

54. Becker, Y.; Asher, Y.; Tabor, E.; Davidson, I.; Malkinson, M. Open reading frames in a 4556 nucleotide sequence within MDV-1 BamHI-D DNA fragment: Evidence for splicing of mRNA from a new viral glycoprotein gene. *Virus Genes* **1994**, *8*, 55–69. [CrossRef]

55. Schippers, T.; Jarosinski, K.; Osterrieder, N. The ORF012 gene of Marek's disease virus type 1 produces a spliced transcript and encodes a novel nuclear phosphoprotein essential for virus growth. *J. Virol.* **2015**, *89*, 1348–1363. [CrossRef] [PubMed]

56. Jarosinski, K.W.; Osterrieder, N. Marek's disease virus expresses multiple UL44 (gC) variants through mRNA splicing that are all required for efficient horizontal transmission. *J. Virol.* **2012**, *86*, 7896–7906. [CrossRef] [PubMed]

57. Liu, J.L.; Lin, S.F.; Xia, L.; Brunovskis, P.; Li, D.; Davidson, I.; Lee, L.F.; Kung, H.J. MEQ and V-IL8: Cellular genes in disguise? *Acta Virol.* **1999**, *43*, 94–101. [PubMed]

58. Tombacz, D.; Csabai, Z.; Szucs, A.; Balazs, Z.; Moldovan, N.; Sharon, D.; Snyder, M.; Boldogkoi, Z. Long-Read Isoform Sequencing Reveals a Hidden Complexity of the Transcriptional Landscape of Herpes Simplex Virus Type 1. *Front. Microbiol.* **2017**, *8*, 1079. [CrossRef] [PubMed]

59. Gimeno, I.; Silva, R.F. Deletion of the Marek's disease virus UL41 gene (vhs) has no measurable effect on latency or pathogenesis. *Virus Genes* **2008**, *36*, 499–507. [CrossRef] [PubMed]

60. Sadek, J.; Read, G.S. The Splicing History of an mRNA Affects Its Level of Translation and Sensitivity to Cleavage by the Virion Host Shutoff Endonuclease during Herpes Simplex Virus Infections. *J. Virol.* **2016**, *90*, 10844–10856. [CrossRef] [PubMed]

61. Mandel, C.R.; Bai, Y.; Tong, L. Protein factors in pre-mRNA 3′-end processing. *Cell. Mol. Life Sci.* **2008**, *65*, 1099–1122. [CrossRef]

62. Proudfoot, N.J. Ending the message: Poly(A) signals then and now. *Genes Dev.* **2011**, *25*, 1770–1782. [CrossRef]

63. Kamil, J.P.; Tischer, B.K.; Trapp, S.; Nair, V.K.; Osterrieder, N.; Kung, H.J. vLIP, a viral lipase homologue, is a virulence factor of Marek's disease virus. *J. Virol.* **2005**, *79*, 6984–6996. [CrossRef]

64. Hardy, W.R.; Sandri-Goldin, R.M. Herpes-Simplex Virus Inhibits Host-Cell Splicing, and Regulatory Protein Icp27 Is Required for This Effect. *J. Virol.* **1994**, *68*, 7790–7799. [PubMed]

65. Zhu, H.; Li, H.; Han, Z.; Shao, Y.; Wang, Y.; Kong, X. Identification of a spliced gene from duck enteritis virus encoding a protein homologous to UL15 of herpes simplex virus 1. *Virol. J.* **2011**, *8*, 156. [CrossRef] [PubMed]

66. Zhu, G.S.; Iwata, A.; Gong, M.; Ueda, S.; Hirai, K. Marek's disease virus type 1-specific phosphorylated proteins pp38 and pp24 with common amino acid termini are encoded from the opposite junction regions between the long unique and inverted repeat sequences of viral genome. *Virology* **1994**, *200*, 816–820. [CrossRef]

67. Lv, H.; Zhang, Y.; Sun, G.; Bao, K.; Gao, Y.; Qi, X.; Cui, H.; Wang, Y.; Li, K.; Gao, L.; et al. Genetic evolution of Gallid herpesvirus 2 isolated in China. *Infect. Genet. Evol.* **2017**, *51*, 263–274. [CrossRef] [PubMed]

68. Morgan, R.W.; Xie, Q.; Cantello, J.L.; Miles, A.M.; Bernberg, E.L.; Kent, J.; Anderson, A. Marek's disease virus latency. *Curr. Top. Microbiol. Immunol.* **2001**, *255*, 223–243.

69. Cantello, J.L.; Parcells, M.S.; Anderson, A.S.; Morgan, R.W. Marek's disease virus latency-associated transcripts belong to a family of spliced RNAs that are antisense to the ICP4 homolog gene. *J. Virol.* **1997**, *71*, 1353–1361. [PubMed]

70. Burnside, J.; Bernberg, E.; Anderson, A.; Lu, C.; Meyers, B.C.; Green, P.J.; Jain, N.; Isaacs, G.; Morgan, R.W. Marek's disease virus encodes MicroRNAs that map to meq and the latency-associated transcript. *J. Virol.* **2006**, *80*, 8778–8786. [CrossRef] [PubMed]

71. Yao, Y.; Zhao, Y.; Xu, H.; Smith, L.P.; Lawrie, C.H.; Watson, M.; Nair, V. MicroRNA profile of Marek's disease virus-transformed T-cell line MSB-1: Predominance of virus-encoded microRNAs. *J. Virol.* **2008**, *82*, 4007–4015. [CrossRef]

72. Jarosinski, K.W.; Schat, K.A. Multiple alternative splicing to exons II and III of viral interleukin-8 (vIL-8) in the Marek's disease virus genome: The importance of vIL-8 exon I. *Virus Genes* **2007**, *34*, 9–22. [CrossRef]

73. Anobile, J.M.; Arumugaswami, V.; Downs, D.; Czymmek, K.; Parcells, M.; Schmidt, C.J. Nuclear localization and dynamic properties of the Marek's disease virus oncogene products Meq and Meq/vIL8. *J. Virol.* **2006**, *80*, 1160–1166. [CrossRef]

74. Okada, T.; Takagi, M.; Murata, S.; Onuma, M.; Ohashi, K. Identification and characterization of a novel spliced form of the meq transcript in lymphoblastoid cell lines derived from Marek's disease tumours. *J. Gen. Virol.* **2007**, *88*, 2111–2120. [CrossRef]

75. Trapp, S.; Parcells, M.S.; Kamil, J.P.; Schumacher, D.; Tischer, B.K.; Kumar, P.M.; Nair, V.K.; Osterrieder, N. A virus-encoded telomerase RNA promotes malignant T cell lymphomagenesis. *J. Exp. Med.* **2006**, *203*, 1307–1317. [CrossRef] [PubMed]

76. Tycowski, K.T.; Guo, Y.E.; Lee, N.; Moss, W.N.; Vallery, T.K.; Xie, M.; Steitz, J.A. Viral noncoding RNAs: More surprises. *Genes Dev.* **2015**, *29*, 567–584. [CrossRef] [PubMed]

77. Zhuang, G.; Sun, A.; Teng, M.; Luo, J. A Tiny RNA that Packs a Big Punch: The Critical Role of a Viral miR-155 Ortholog in Lymphomagenesis in Marek's Disease. *Front. Microbiol.* **2017**, *8*, 1169. [CrossRef] [PubMed]

78. Figueroa, T.; Boumart, I.; Coupeau, D.; Rasschaert, D. Hyperediting by ADAR1 of a new herpesvirus lncRNA during the lytic phase of the oncogenic Marek's disease virus. *J. Gen. Virol.* **2016**, *97*, 2973–2988. [CrossRef]

79. Wyler, E.; Menegatti, J.; Franke, V.; Kocks, C.; Boltengagen, A.; Hennig, T.; Theil, K.; Rutkowski, A.; Ferrai, C.; Baer, L.; et al. Widespread activation of antisense transcription of the host genome during herpes simplex virus 1 infection. *Genome Biol.* **2017**, *18*, 209. [CrossRef] [PubMed]

80. Greninger, A.L.; Knudsen, G.M.; Roychoudhury, P.; Hanson, D.J.; Sedlak, R.H.; Xie, H.; Guan, J.; Nguyen, T.; Peddu, V.; Boeckh, M.; et al. Comparative genomic, transcriptomic, and proteomic reannotation of human herpesvirus 6. *BMC Genom.* **2018**, *19*, 204. [CrossRef]
81. Olah, P.; Tombacz, D.; Poka, N.; Csabai, Z.; Prazsak, I.; Boldogkoi, Z. Characterization of pseudorabies virus transcriptome by Illumina sequencing. *BMC Microbiol.* **2015**, *15*, 130. [CrossRef] [PubMed]
82. Shabalina, S.A.; Ogurtsov, A.Y.; Spiridonov, N.A.; Koonin, E.V. Evolution at protein ends: Major contribution of alternative transcription initiation and termination to the transcriptome and proteome diversity in mammals. *Nucl. Acids Res.* **2014**, *42*, 7132–7144. [CrossRef] [PubMed]

Article

Gram-Positive Bacteria-Like DNA Binding Machineries Involved in Replication Initiation and Termination Mechanisms of Mimivirus

Motohiro Akashi * and Masaharu Takemura

Laboratory of Biology, Department of Liberal Arts, Faculty of Science, Tokyo University of Science, Shinjuku, Tokyo 162-8601, Japan; takemura@rs.kagu.tus.ac.jp
* Correspondence: motoa@rs.tus.ac.jp

Received: 29 January 2019; Accepted: 14 March 2019; Published: 17 March 2019

Abstract: The detailed mechanisms of replication initiation, termination and segregation events were not yet known in *Acanthamoeba polyphaga mimivirus* (APMV). Here, we show detailed bioinformatics-based analyses of chromosomal replication in APMV from initiation to termination mediated by proteins bound to specific DNA sequences. Using GC/AT skew and coding sequence skew analysis, we estimated that the replication origin is located at 382 kb in the APMV genome. We performed homology-modeling analysis of the gamma domain of APMV-FtsK (DNA translocase coordinating chromosome segregation) related to FtsK-orienting polar sequences (KOPS) binding, suggesting that there was an insertion in the gamma domain which maintains the structure of the DNA binding motif. Furthermore, UvrD/Rep-like helicase in APMV was homologous to *Bacillus subtilis* AddA, while the chi-like quartet sequence 5′-CCGC-3′ was frequently found in the estimated *ori* region, suggesting that chromosomal replication of APMV is initiated via chi-like sequence recognition by UvrD/Rep-like helicase. Therefore, the replication initiation, termination and segregation of APMV are presumably mediated by DNA repair machineries derived from gram-positive bacteria. Moreover, the other frequently observed quartet sequence 5′-CGGC-3′ in the *ori* region was homologous to the mitochondrial signal sequence of replication initiation, while the comparison of quartet sequence composition in APMV/*Rickettsia*-genome showed significantly similar values, suggesting that APMV also conserves the mitochondrial replication system acquired from an ancestral genome of mitochondria during eukaryogenesis.

Keywords: DNA replication; *ori*; mitochondria; Rickettsia; gram-positive bacteria; APMV; Mimivirus; giant virus; eukaryogenesis

---

1. Introduction

Understanding the mechanism of genomic replication for all organisms, including the "giant viruses", is an important scientific endeavor. Mimivirus, the first giant virus to be discovered, has a 750-nm-long virion and a 1.2 Mb linear dsDNA genome [1,2]. The method of replication termination for *Acanthamoeba polyphaga mimivirus* (APMV) has been previously hypothesized [3]. The first model suggested that the replication of the lagging strand of APMV's linear genome is mediated by homologous recombination of approximately 617 bp located on both ends of the viral chromosome, similar to T4 phage replication, and is processed with Mimivirus R555 recombinase (Mre11/Rad50 fusion protein) [3,4]. Recently, the second model of replication termination and segregation of APMV was proposed [5]. In this model, the FtsK-like protein (also called packaging ATPase), binds FtsK orienting polar sequences (KOPS) and is localized to both ends of the nucleosome, resulting in chromosome segregation by the recombination of *dif* sequences [5]. The KOPS is the recognition sites of FtsK protein, and this protein controls the chromosome segregation in bacteria [6]. The second model reinforces the first model regarding homologous recombination between chromosomal ends. However, it should be noted that the bacteria do not perfectly conserve KOPS among species [6–8], suggesting that APMV could have KOPS of its own, which might be similar to bacterial KOPS.

The initiation of DNA replication of bacteria is mainly driven by DnaA-DnaA box interaction, which subsequently unwinds dsDNA via DnaB [9]. Rep protein, related DNA helicase such as DnaB, initiates plasmid replication [10]. The Rep protein relates not only to plasmids but also to chromosome replication in *Escherichia coli* [11]. The homologues of these helicases in gram-positive bacteria are PcrA and AddAB; both are involved in DNA repair, and the former is involved in rolling-circle replication [12–14]. Especially, AddAB mediates the homologous recombination with recognizing the five-nucleotide sequence called chi [12], so the recombination is sequence specific. The sequence chi is known as a site-specific recombination site that is catalyzed by the RecBCD pathway in *E. coli*, and RecB is a homologous helicase to AddA [15,16]. Chi sequence varies among bacterial species: *E. coli*: 5′-GCTGGTGG-3′; *Bacillus subtilis*: 5′-AGCGG-3′; *Lactococcus lactis*: 5′-GCGCGTG-3′ [12,15–17]. Based on the model of replication termination in APMV [5], the DNA replication initiation mechanism might also be homologous to that of bacteria. If so, the pair of bacteria-like DNA sequences and their recognition proteins which are related to the DNA replication initiation and segregation could be found in APMV. Plotting the nucleotide composition bias called genomic GC skew is a tool for visualizing the bias of the nucleotide composition on the genome, which is able to determine the origin of replication since the values of GC skew switch across the replication origin and its terminus [18], and which has been determined in bacteria with some improvements of the technique [19–21]. This nucleotide composition bias shaping the genomic polarity is thought to have results of the mutation and selection pressure against the different replication mechanism of leading/lagging strand [22,23]. Although the origin of replication remains putative in Mimivirus, the GC skew analysis against the Mimivirus genome with high resolution may facilitate the detection of the signal sequence of DNA replication initiation. Therefore, using bioinformatics, we analyzed the APMV genome to determine the DNA replication initiation/termination segregation mechanism in detail, starting with the GC skew analysis.

2. Materials and Methods

2.1. GC/AT, Coding Sequence (CDS) Skew Analysis

GC and AT skew of the APMV genome (AY653733.1) was analyzed using a method described previously [20,21]. Each index was calculated using the following formulas: GC skew = [G-C]/[G+C]; AT skew = [A-T]/[A+T] (window size: 10,001 bp; step size: 1000 bp), and the GC/AT skew and cumulative graph of these were plotted. To calculate coding sequence (CDS) skew, we indexed the CDS direction on the APMV genome (direction of the gene (D): positive: +1; negative: -1) and the CDS length (L). Subsequently, the CDS skew index was calculated against every CDS using the following formula: CDS skew = [D] × [L] (Figure 2b). The CDS skew and cumulative CDS skew were plotted with the CDS start positions.

2.2. Correlation Analysis of the CDS Length of Left/Right Side from the Estimated Ori Region

The CDS length of the left or right sides from the estimated origin (380,698 bp), and the CDS length of the positive or negative direction against the APMV genome (AY653733.1), were plotted using a box plot. The statistical differences between each of two groups were calculated by two-sample Kolmogorov–Smirnov (KS) test using "lawstat package" of R software (https://www.r-project.org) with default options.

2.3. Paralogous Gene Localization Analysis

We confirmed the three kinds of paralogous gene locations found on the two GC skew shift points (296,000 bp and 882,000 bp): *ankyrin repeat, serine/threonine protein kinase,* and *collagen triple helix repeat containing protein*. First, we made a gene list of the CDS information annotated as "ankyrin containing protein", "serine/threonine protein kinase", or "collagen triple helix repeat containing protein" in the APMV genome (AY653733.1), containing the data of locus tags, start/end physical positions, gene length and gene direction and GC content (Supplementary Data). Every pair of each of the three genes was listed and the distance between these pairs from the start position was calculated. This pair of directions on the physical position of the APMV genome was plotted using the Circos v0.69-6 software [24]. The pair of gene direction was also labeled (matched direction: forward-forward/reverse-reverse; mismatched direction: forward-reverse/reverse-forward). When drawing the graph with Circos, color coding was used to display the three kinds of genes, and states of matched/mismatched gene directions were displayed with shades of colors (ankyrin containing protein: blue; serine/threonine protein kinase: green; collagen triple helix repeat containing protein: red; matched: light; mismatched: shade). The columns of id, gene1_start_pos, gene2_start_pos, line_colour, and line_thickness in the Supplementary Data without header (Supplementary Data; the pane "Paralogous gene direction") can be used for loading the data to Circos. The distance between every two CDSs in each of the three genes was plotted by the CDS direction-matching patterns on the APMV genome ("match" or "mismatch"). The statistical differences between the two groups were calculated by two-sample KS test using the "lawstat package" in the R software with default options.

2.4. Sequence Alignment Analysis Neighbor-Net Network Analysis

The accessions of analyzed FtsK with positions of motor (ATPase) domain and gamma domain were as follows: APMV: AAV50705.1, 5-215 aa, 216-284 aa; *E. coli*: NP_415410.1, 868-1242 aa, 1268-1324 aa; *L. lactis*: NP_267812.1, 312-668 aa, 695-749 aa; *Pseudomonas aeruginosa*: Q9I0M3, 343-716 aa, 749-803 aa; *B. subtilis*: WP_003231869.1, 346-703 aa, 582-723 aa. The accession numbers of analyzed UvrD/Rep-like helicase were as follows: APMV: AKI80299.1; *E. coli*: YP_026251.1 (Rep), AAA67609.1 (UvrD), NP_417297.1 (RecB); *L. lactis*: WP_003132060.1 (PcrA), WP_010905024.1 (AddA); *B. subtilis*: WP_003233919.1 (PcrA), WP_003233100.1 (AddA). The dataset without the APMV sequence was used for alignment. Poor-quality sequences were masked using Prequal software [25], and sequence alignment analyses was performed using the MAFFT v7.222 software with "—auto" option [26]. Subsequently, the APMV sequence was added and realigned with MAFFT. To determine the alignment of the chi-binding site of AddA with APMV, two sequences of AddA from *L. lactis* and *B. subtilis* (WP_010905024.1 and WP_003233100.1) were recursively added with the APMV sequence. For the Neighbor-Net network analysis, alignment data containing APMV sequence were trimmed using trimAl 1.2rev59 software with "-strictplus" option [27], and the numbers of aligned residues used were: FtsK ATPase domain: 322 aa; FtsK gamma domain: 49 aa; UvrD/Rep-like helicase: 483 aa. The Neighbor-Net network tree was drawn by SplitTree4 (version 4.14.8) with 1000 bootstrap replicates [28,29].

2.5. KOPS Distribution in the Genome

The bacterial KOPS distributions on APMV and the bacterial genomes were plotted using the following genome data: APMV: AY653733.1; *L. lactis* Il1403: NC_002662.1; *E. coli* MG1655: NC_000913.3; *B. subtilis* 168: NC_000964.3. The information on the *ori* and *ter* positions of these bacteria was provided by Genome projector website (http://www.g-language.org/g3/) [30]. KOPS of each bacteria were listed as follows: *L. lactis*: 5'-GAGAAG-3'; *B. subtilis*: 5'-GAGAAGGG-3'; *E. coli*: 5'-GGGNAGGG-3' [6–8].

2.6. Quartet Sequence Composition Analysis

Every quartet sequence compositions on APMV and the bacterial genomes were confirmed with the compseq program of the EMBOSS 6.6.0.0 software [31] with the "-word 4 " option, using the following genome data: APMV: AY653733.1; *L. lactis* Il1403: NC_002662.1; *E. coli* MG1655: NC_000913.3; *B. subtilis* 168: NC_000964.3; *Rickettsia prowazekii*: NC_000963.1; *Homo sapiens* mitochondria (MT): CM001971.1. To compare the compositions between the estimated *ori* region and the whole genome of APMV, sequence composition was also confirmed on the 375–385 kb region, and the ratio of *ori* region per whole genome (quartet nucleotide composition ratio) was calculated as follows: quartet nucleotide composition ratio = [*ori* observed/expected frequency]/[all genome observed/expected frequency]. The Grubbs test was performed on this data to detect the highly observed sequence on the *ori* region with the "outliers" package of the R software (options: type = 11; opposite = TRUE; two.sided = TRUE). The mutual observed frequency of quartet sequence among APMV, APMV *ori* region, and bacteria were plotted, while statistical differences were confirmed by two-sample KS test using the "lawstat package" of the R software with default options.

2.7. Homology Modeling Analysis

Homology modeling analyses of the gamma domain of packaging ATPase (FtsK-like protein, 216-284 aa region of AAV50705.1) were performed using the I-TASSER server [32]. Template structures of *P. aeruginosa* (2J5O and 2VE9) were used for modeling [33,34].

3. Results

3.1. GC/AT Skew and CDS Skew Analyses

The GC skew plot showed the two highest and lowest peaks at the symmetrical position point of the genome (296 kb and 882 kb), and the plot was able to separate the three regions by these positions (Figure 1a). Both end regions (<296 kb, >882 kb) were increasing; however, the former value was negative and the latter value was positive, indicating that the 5′ end contained a C nucleotide rather than G nucleotide, and the 3′ end contained a G nucleotide rather than a C nucleotide. The middle region of the graph (from 296 kb to 882 kb) was almost flat (Figure 1b), and the cumulative GC skew plot of this region was increasing (Figure 1b), indicating that the number of G and C nucleotides in this region was slightly skewed to the G nucleotide. The AT skew graph showed the shift point of the value, which corresponded to the peak of the valley of the cumulative AT skew graph at 382,000 ± 5000 bp region, suggesting that the origin of DNA replication is located in this region (Figure 1). However, the cumulative AC skew did not show any peaks (Figure S1).

CDS skew analysis showed that the shift point and cumulative CDS skew analysis exhibited a V-shaped graph, similar to the AT skew/cumulative AT skew (Figure 2a). Furthermore, the valley of this cumulative CDS skew graph was located at 382,698 bp, and the gene direction faced outward from this peak. Therefore, we concluded that the estimated *ori* region is located at the 382 kb position of APMV genome. The location of the estimated *ori* region is biased toward the 5′ end from the center of the genome; however, there were no significant differences in CDS length between the right and left sides of the estimated *ori* region and between the positive and negative strands (Figure S2). We found the same paralogous genes on the GC skew shift points (296,000 bp, 882,000 bp): ankyrin repeat, serine/threonine protein kinase, and collagen triple helix repeat containing protein. Paralogues of these genes were located on the line of symmetry position in the genome, especially *collagen triple helix repeat containing protein* gene, which exhibited the exact positions of the GC skew shift points with opposite gene direction, suggesting that the nucleotide compositions of these paralogues formed the shift points (Figure 3a). Additionally, we analyzed the gene-to-gene distances between each pair of paralogues for each of the three genes, indicating that the distance between the gene direction matched pair was shorter than that between the mismatched pair (fold change: 2.6, $p < 0.05$, Figure 3b). Thus, APMV forms the double-folding structure and is the cause of paralogue generation by homologous recombination (Figure S3).

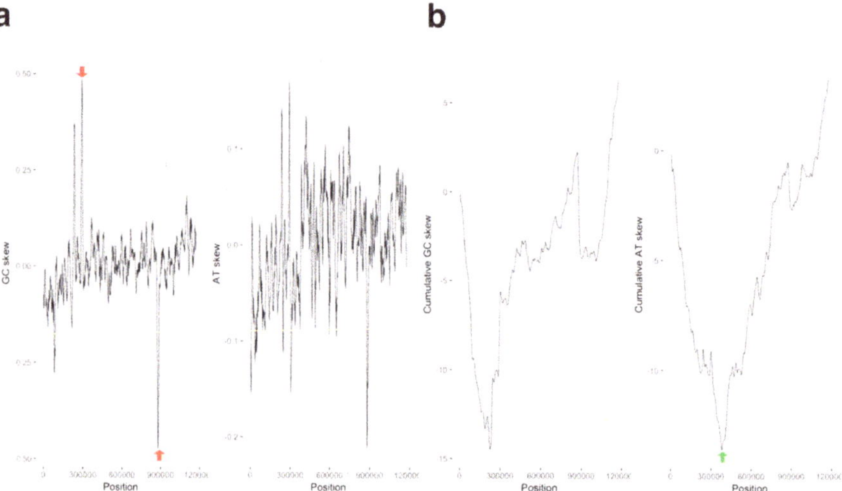

Figure 1. Analyses of the GC/AT skew in the *Acanthamoeba polyphaga mimivirus* (APMV) genome. (**a**) GC and AT skew plot of APMV genome (AY653733.1, step size: 1000 bp; window size: 10,001 bp). Red arrows on the GC skew plot indicate the highest/lowest peaks, which are located at the symmetrical position of the genome (GC skew: 0.488234 at 296,000 bp; -0.472989 at 882,000 bp). (**b**) Cumulative GC and AT skew plots corresponding to the graphs on the panel (a). The green arrow on the graph on the cumulative AT skew plot shows the lowest valley point on this graph, which was estimated in the genomic region as an origin of replication (382,000 ± 5000 bp).

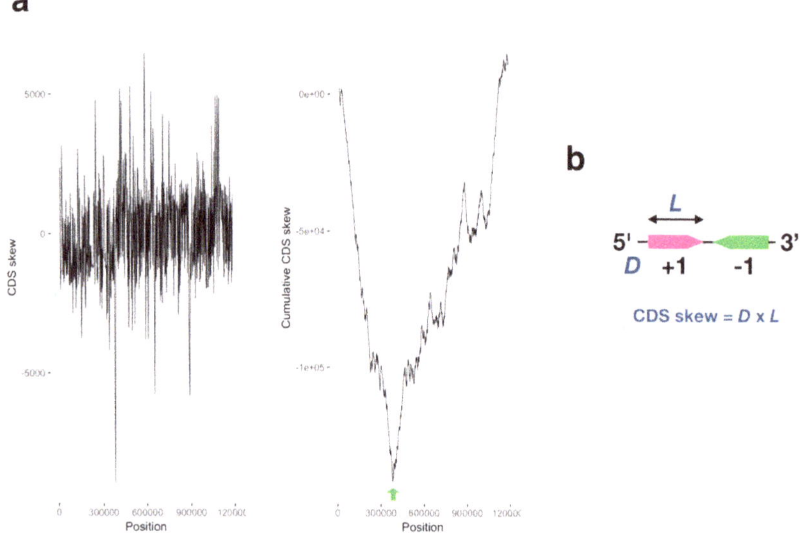

Figure 2. Analyses of the coding sequence (CDS) skew of the APMV genome. (**a**) CDS skew plot (left) and cumulative CDS skew plot (right) of the APMV genome (AY653733.1). The green arrow on the graph on the cumulative CDS skew plot shows the lowest valley point, which was estimated to be genomic region of the origin of 382,698 bp. (**b**) Calculation of the CDS skew index. D: Direction values of the CDS against the APMV genome (AY653733.1); L: CDS length (bp). Pink and green arrows indicate the positive and negative CDS direction on the APMV genome, respectively.

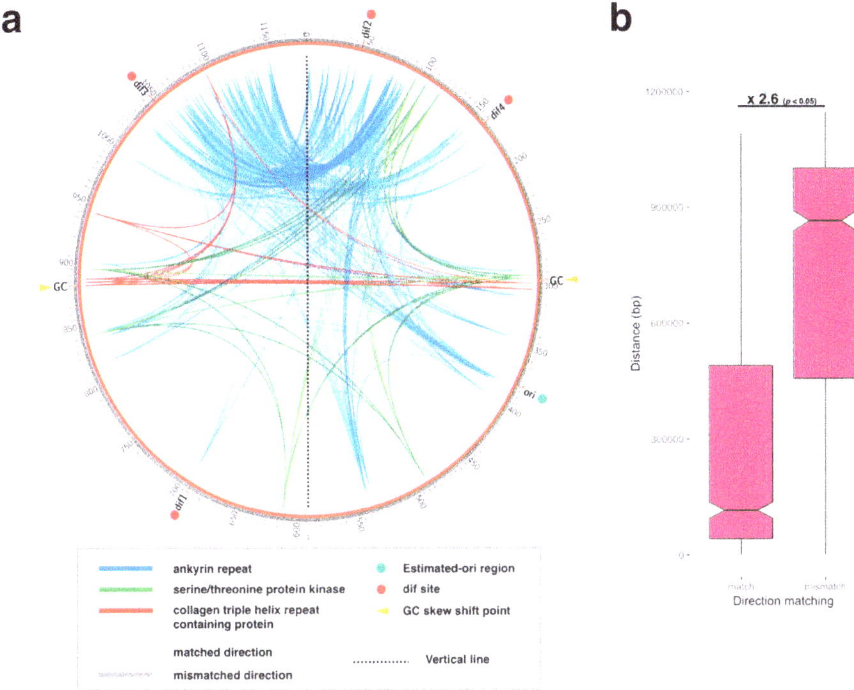

Figure 3. Paralogous gene localization found at the GC skew shift point. (**a**) The three genes found on the GC skew shift point were plotted (ankyrin repeat, serine/threonine kinase, and collagen triple helix repeat containing protein). Four *dif* sites (*dif*1-4) were described previously (see text). Each of the three genes were separately categorized as matching or mismatching the coding sequence (CDS) direction on the *Acanthamoeba polyphaga mimivirus* (APMV) genome (AY653733.1) with shades of colors (light: matched direction; dark: mismatched direction). (**b**) Distances between each pair of CDSs in each of the three genes were plotted by the CDS direction matching patterns on the APMV genome ("match" or "mismatch"). Labels above the box plots indicate the fold change between the two values with the resulting *p* value from KS-tests.

3.2. Initiation of DNA Replication

3.2.1. Sequence Analysis of UvrD/Rep-Like Helicase

Since the replication initiation mechanism had not yet been analyzed in APMV, we sought to determine the protein that participated in the initiation of DNA replication. In doing so, we discovered a similar sequence to Rep helicase that was related to the both chromosome and plasmid replication [10,35]. This protein has already been annotated as "UvrD/REP helicase family protein", suggesting that this protein is possibly an initiator of DNA replication in APMV (accession ID of NCBI protein database: AKI80299.1). We aligned the UvrD/REP helicase family protein of APMV (AKI80299.1) with the bacterial homologues: Rep, UvrD, and RecB of *E. coli*; PcrA and AddA of *B. subtilis*; and *L. lactis* [10,12,13]. The alignment and phylogenetic analysis showed that the UvrD/Rep-like helicase of APMV is a close relative to the AddA rather than Rep (Figure 4a,c). Seven regions that were known to be conserved among AddA and other helicases [36] were also highly conserved in the UvrD/Rep-like helicase (Figure 4a). AddA forms a heterodimer with AddB, which recognizes the chi sequence of *B. subtilis* [37]. Therefore, we compared the homologous region of the chi binding site of AddA with UvrD/Rep-like helicase of APMV. The results showed that there were three regions containing chi binding sites in UvrD/Rep-like helicase of APMV, and two out of seven sites were conserved: Q1115 and I1157 of *Bacillus*-AddA corresponded to the Q922 and I924 of UvrD/Rep-like helicase of APMV; and Y1204 of *Bacillus*-AddA presumably corresponded to the Y973 of UvrD/Rep-like helicase of APMV, which was shifted one amino acid residue to the C-terminus from the homologous site against Y1204 in *B. subtilis* (Figure 4b). Although we have aligned the other region containing four DNA-binding residues between the 1012–1019 aa region of *Bacillus*-AddA with the 610–617 aa region of UvrD/Rep-like helicase, both of which are constructed with polar amino acids, we could not find any conserved sequences of DNA binding sites (K1013, S1015, V1016, and S1017 in *Bacillus* AddA, Figure 4b). These four residues in *Bacillus*-AddA bind the phosphate at the 3′ end of the chi sequence [37], so that the content of polar amino acids in this region, rather than the exact amino acid sequence, is important for *Bacillus*-AddA to bind to the chi sequence. Therefore, we estimated that the 610–617 aa region of UvrD/Rep-like helicase of APMV could also bind the 3′ end of the chi-like sequence in the APMV genome. Altogether, the UvrD/Rep-like helicase of APMV was similar to the AddA of gram-positive bacteria, and thus this protein would presumably recognize the chi sequence.

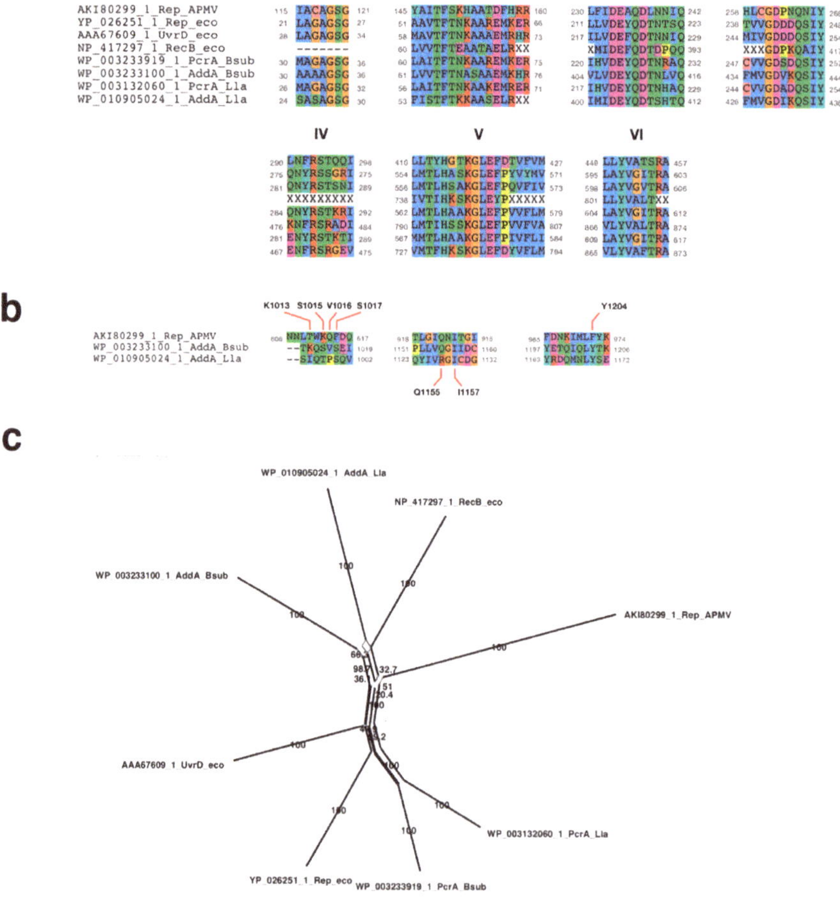

Figure 4. Sequence analyses of UvrD/Rep-like helicase of APMV. (**a**) Sequence alignment of seven conserved regions among APMV and three bacteria. Character "X" on the alignment indicates the

residues masked by the prequel program [25]. Accession number and species name are labeled on the left side of the alignments. Grey colored numbers, which are located on both sides of the alignments, indicate the physical positions of each sequence. (**b**) Sequence alignment of chi sequence binding sites. Accession number and species name are labeled on the left side of the alignment. Grey colored numbers, which are located on both sides of the alignments, indicate the physical positions of each sequence. Labels above/under the alignment indicate the amino acids related to the chi sequence binding sites determined using *B. subtilis*-AddA [37]. (**c**) Neighbor-Net network tree of the UvrD/Rep-like helicase. Accession number and species name corresponding to the sequence alignment are labeled at the end of the branch. Numbers on nearby branches indicate bootstrap test values with 1000 replicates. Scale bar: number of substitutions per site.

3.2.2. Signal Sequence of the Initiation of DNA Replication

Subsequently, we sought to find the sequence that was related to the origin of recognition on the probable *ori* region. To accomplish this, we first calculated the quartet nucleotide composition ratio of the *ori*-containing region per total genome (Figure 5). For details of the calculation of the ratio, see Materials and Methods. The four types of sequences that had a ratio greater than 2 contained 75–100% of GC nucleotides, and two of four sequences, CGGC and CCGC, exhibited significantly different ratios from the other sequences ($p < 0.05$, Figure 5a). Furthermore, two sequences (CCGC and GCGG) were complementary to each other (Figure 5a). This pair of sequences was identical to four out of five nucleotides on the chi sequence, which was recognized by *B. subtilis*-AddA (5′-AGCGG-3′) [12]. Additionally, the sequence CGGC is known to be a part of a complementary sequence that is known as a replication signal sequence of mitochondrial genome in humans, 3′-GGCCG-5′ [38]. Therefore, we analyzed the distribution of these quartet sequence in the APMV genome. The sequence densities of CCGC and GCGG in the APMV genome showed that both sequences had two peaks, and one of each peak was located on the *ori*-containing region, while the others were on the axial symmetric position (about 800,000 bp, Figure 5b). Similar to this, those of CGGC and GCCG exhibited two peaks and the peak on the 5′ side in GCCG located on the *ori*-containing region, although that of the CGGC slightly skewed from the *ori*-containing region to the 5′ end of the genome (Figure 5c). Each type of quartet sequence composition frequencies was calculated in APMV, four kinds of bacteria (*B. subtilis*, *L. lactis* and *E. coli, and R. prowazekii*), and human mitochondrial genome (*H. sapiens* MT). *R. prowazekii* is considered as an ancestor of mitochondria [39], while the mitogenome sequence similarity against the Mimivirus genome has been reported recently [40]. These frequencies were then plotted in every pair of bacteria, mitochondria, APMV, and the *ori*-containing region of APMV (375–385k) (Figure 6). Every pair without APMV-APMV-*ori* and APMV-*R.prowazekii* exhibited a significant difference of sequence composition ($p<0.05$, Figure 6). *R. prowazekii* possesses the most similar sequence composition compared with APMV ($p = 0.2528$). The composition similarity of the *R. prowazekii*-APMV pair was higher than those of the *R. prowazekii*-*H. sapiens* MT and APMV-*H. sapiens* MT pairs (*R. prowazekii*-*H. sapiens* MT: $p = 0.04685$; APMV-*H. sapiens* MT: $p = 0.00146$), suggesting that the APMV conserves the sequence derived from an ancestor of mitochondria rather than from highly evolved mitochondria in humans. Interestingly, the composition of human mitochondria and *L. lactis* also showed a high similarity ($p = 0.3552$) rather than the *R. prowazekii*-*H. sapiens* MT pair, suggesting that the mitochondrial genome still conserves a remnant of bacterial characteristics.

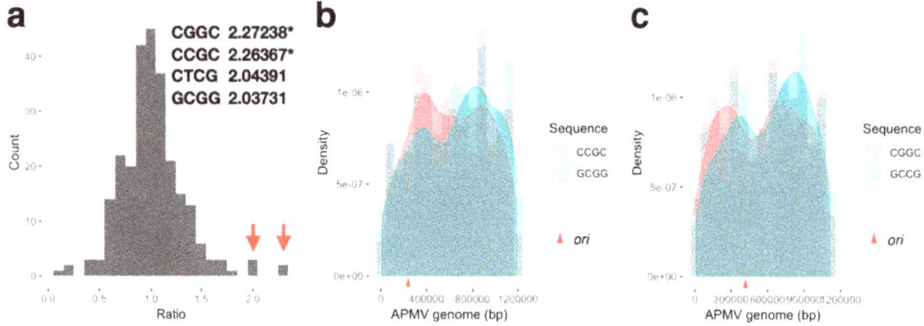

Figure 5. Comparative analysis of the quartet nucleotide composition ratio between the *ori*-containing region and the total genome of APMV. (**a**) Frequency distribution of fold-changes (= [*ori*-containing region]/[whole genome], *ori*-containing region: 375–385 kb of AY653733.1). Red arrows indicate that the fold-changes of the bins are greater than two. Sequences and fold-changes greater than two are listed on the top right of the figure. Asterisks indicate significant outliers from the population calculated by Grubbs test (*p*<0.05). (**b**,**c**) Density graph of the 5′-CCGC-3′/5′-GCGG-3′ and 5′-CGGC-3′/5′-GCCG-3′ on the APMV genome (AY653733.1). Red arrowhead on the scale indicate the estimated location of the *ori* region (382,000 ± 5000 bp).

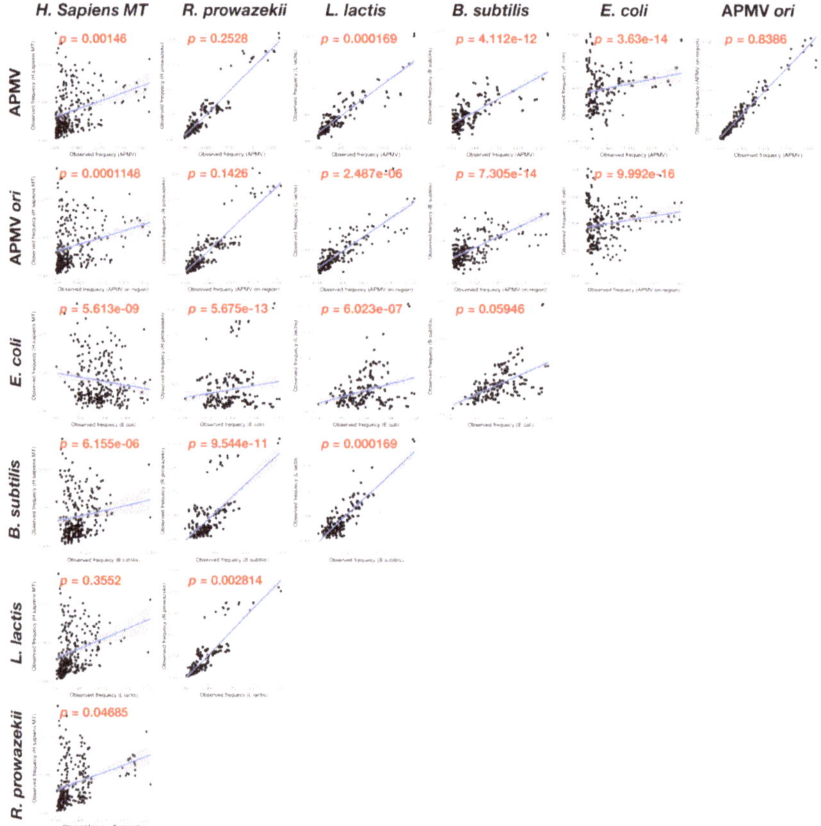

Figure 6. Comparative analysis of quartet nucleotide composition frequencies among APMV, human

mitochondria (*H. sapiens* MT), and four bacterial genomes (*Escherichia Coli, Bacillus subtilis, Lactococcus lactis, Rickettsia prowazekii*). Pairs of every frequency between two species were plotted with an approximate line. "APMV *ori*" indicates the frequencies of sequence compositions in the estimated location of the *ori* region (375–385 kb region of AY653733.1). *p*-values in each top left corner of the graphs are the significant differences calculated between two groups as determined by KS test.

3.3. Termination of DNA Replication and Chromosome Segregation

3.3.1. Sequence Analysis of FtsK-Like Protein in APMV

FtsK mainly conserves the motor domain (alpha/beta domain) and gamma domain, the latter of which is our main target, and recognizes KOPS [5,26]. The DNA binding sites of FtsK has been determined in *P. aeruginosa* [34,41]. We aligned the gamma domain of APMV with *P. aeruginosa* and other bacteria (*L. lactis, B. subtilis,* and *E. coli*), revealing that the DNA binding sites determined in *P. aeruginosa* were conserved to almost the same degree among APMV and bacteria. The phylogenetic analysis of FtsK-gamma domain among these species indicated that the sequence of *L. lactis* was most closely related to the APMV-FtsK gamma domain (Figure 7b). However, in APMV, there was a 10 aa insertion in the region where the DNA binding sites were localized (Figure 7a), suggesting that DNA-binding activities of APMV-FtsK had possibly collapsed. Therefore, we confirmed the three-dimensional structure via homology modeling of the gamma domain in APMV-FtsK, using two models from *P. aeruginosa* as templates [34,41]. As a result, the DNA binding sites of APMV-FtsK were thought to be topologically reconstructed and conserved by this insertion (Figure 8). Estimated DNA binding sites against the DNA backbone were K242, K243, K253, and K256, which corresponded with R770, K771, R778, and R781 in *P. aeruginosa*, respectively, and the binding site for KOPS specifically was N246 at APMV-FtsK, which corresponded to N777 in *P. aeruginosa*. K242, K243, and N246 were placed on the insertion (Figure 7a, Figure 8a) [33]. These four basic amino acids localize in two helix motifs, which bind to the KOPS region and support the recognition of KOPS by asparagine between the two helix motifs (Figure 8a,b) [33]. Next, we confirmed the conservation of the ATP binding sites on the FtsK in APMV. The FtsK conserves the Walker A motif (the ATPase active site), and in *P. aeruginosa*, substitution of amino acid residue in this motif leads to the deactivation of ATPase (K472N) [41]. The sequence alignment showed that the motor domain of APMV conserved lysine on the Walker A motif (K32) as well as other three bacteria (Figure 9a). The other ATP binding sites found in *P. aeruginosa* [41] were partially conserved in the APMV-FtsK (Figure 9a). However, some amino acids were not conserved, even in bacteria (R418 and H675 of *P. aeruginosa*-FtsK, Figure 9a), suggesting that these amino acid residues in binding sites could be replaced with other amino acid residues. Phylogenetic analyses of these motor domains indicated that the sequence of APMV was distinct from bacteria (Figure 9b). Therefore, APMV-FtsK might have speciated from its bacterial group prior to the emergence of bacterial FtsK.

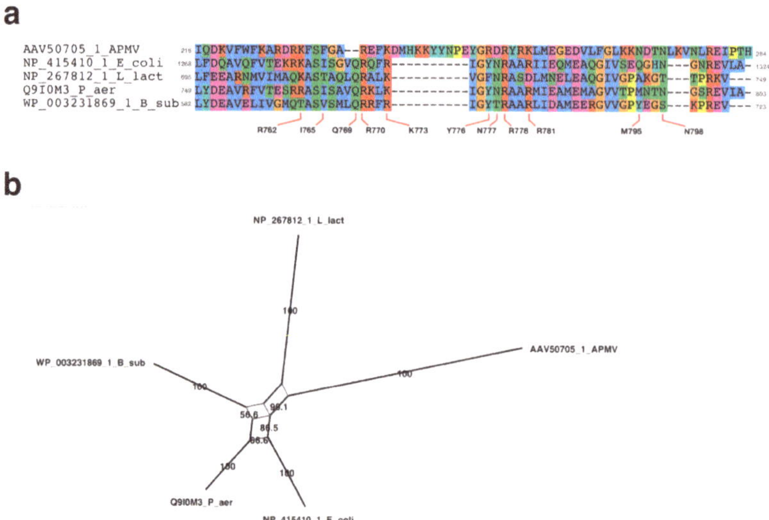

Figure 7. Sequence analyses of APMV-FtsK gamma domain. (**a**) Sequence alignment of FtsK-like protein gamma domain. Accession number and species name are labeled on the left side of each alignment. Grey numbers located on both sides of the alignments indicate the exact positions of each sequence. Amino acids labeled under the alignment indicate the amino acid residues responsible for DNA binding sites of *Pseudomonas aeruginosa* [25]. (**b**) Neighbor-Net network tree of FtsK gamma domains. Accession number and species name corresponding to the sequence alignment are labeled at the end of the branch. Numbers on nearby branches indicate bootstrap test values with 1000 replicates. Scale bar: number of substitutions per site.

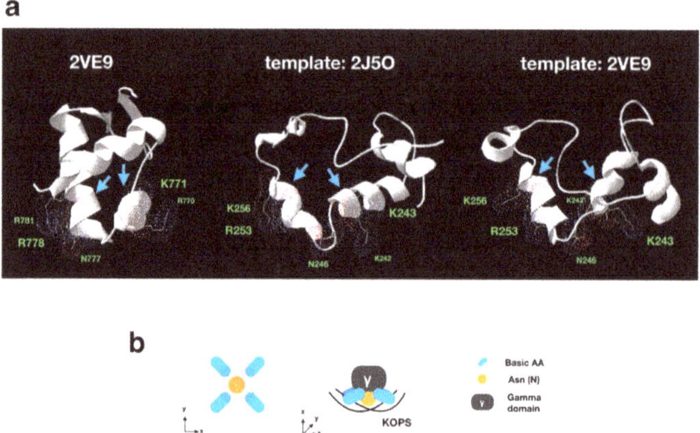

Figure 8. Homology modeling analysis of APMV-FtsK gamma domain. (**a**) Estimated structures of FtsK gamma domain using two different templates of *P. aeruginosa*-FtsK (PDBID; middle: 2J5O; right: 2VE9). The left model is the gamma domain of *P. aeruginosa*-FtsK (PDBID: 2VE9). Side chains on the left model are directly bound to FtsK orienting polar sequences (KOPS), while the other two models are estimated amino acids, which are functionally homologous to 2VE9. Blue arrows indicate two helix motifs harboring KOPS binding residues. (**b**) Model of the gamma domain of APMV-FtsK against KOPS. Four basic amino acids and Asn (N) correspond to the side chain in every model on panel (a).

a

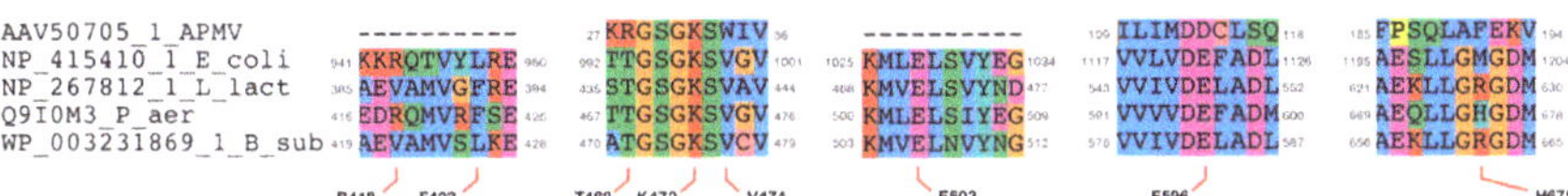

b

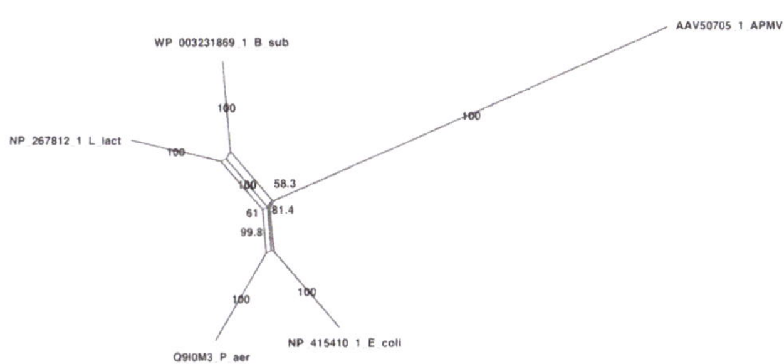

Figure 9. Sequence analyses of APMV-FtsK ATPase domain. (**a**) Motor domain of FtsK-like protein sequence alignment. Accession number and species name are labeled on the left side of alignments. Grey numbers located on both sides of the alignments indicate the exact positions of each sequence. Labels under the alignment indicate the amino acids related to ATPase activities determined in *P. aeruginosa* [41]. (**b**) Neighbor-Net network tree of the gamma domains. Accession number and species name correspond to the sequence alignment labeled at the end of the branch. Numbers on nearby branches indicate bootstrap test values with 1000 replicates. Scale bar: number of substitutions per site.

3.3.2. Distribution Pattern of KOPS on APMV Genome

The most frequently observed type of KOPS sequence was from *L. lactis*, whereas those of *B. subtilis* and *E. coli* were hardly encountered (*L. lactis*, 5'-GAGAAG-3': 225; *B. subtilis*, 5'-GAGAAGGG-3': 5; *E. coli*, 5'-GGGNAGGG-3': 9; Figure 10a). This KOPS distribution pattern was different between the positive and negative strand, and the density graphs showed that these two patterns crossed at the estimated *ori* position (Figure 10b). These KOPS distribution patterns were analyzed in each bacterial genome, and the results suggested that the density of KOPS on the positive/negative strand switched on the exact points of the *ori/ter* region (Figure 10c). This is similar to the distribution pattern of *L. lactis*-derived KOPS in the APMV genome and to KOPS in the bacterial genomes, indicating that *L. lactis* KOPS could be one of the commonly used termination sequences in DNA replication in APMV. It should be noted that we did not have evidence that the other two KOPS, 5'-GAGAAGGG-3' and 5'-GGGNAGGG-3', found in APMV, were inactive (Figure 10a).

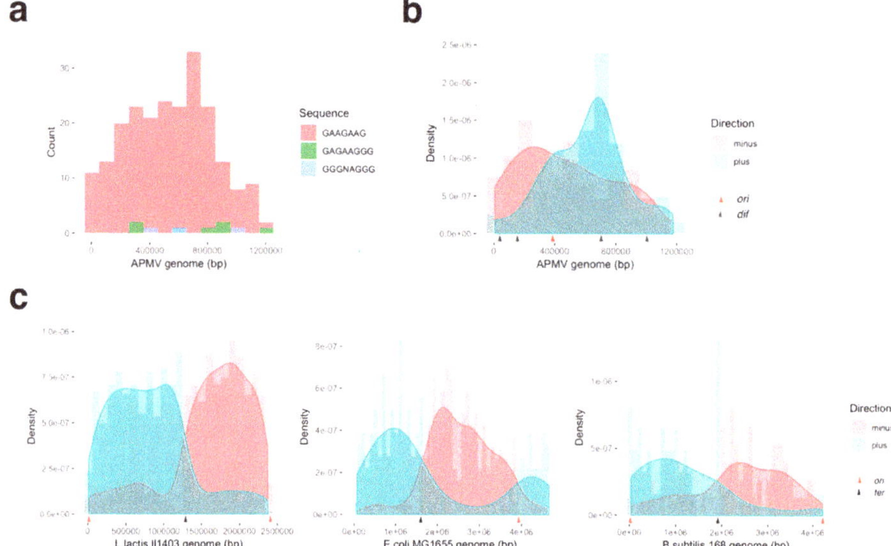

Figure 10. Bacterial KOPS distribution on APMV. (**a**) Frequency distribution of bacterial KOPS distribution in the APMV genome (AY653733.1, *L. lactis*; 5′-GAAGAAG-3′, *B. subtilis*: 5′-GAGAAGGG-3′, *E. coli*: 5′-GGGNAGGG-3′). Complementary sequences were also counted. (**b**) Density graph of the *L. lactis* type KOPS in the APMV genome (AY653733.1). KOPS on the positive and negative strand of the APMV genome were plotted separately. Red arrow: estimated *ori* region; dark gray arrows: *dif* sequence positions (**c**) Bacterial KOPS distributions in bacterial genomes (*L. lactis* Il1403: NC_002662.1, *E. coli* MG1655: NC_000913.3, *B. subtilis* 168: NC_000964.3). Red and black arrows indicate the *ori* and *ter* positions, respectively.

4. Discussion

The GC/AT skew bias is the result of replication bias in bacteria [19], and the replication-transcription conflict causes a high mutation rate in genes, causing genetic transcription and DNA replication to be co-directional [42]. We determined the estimated *ori* region using high-resolution cumulative skew graphs of AT nucleotides and CDSs (382 kb, Figures 1 and 2), which suggests that the transcription and DNA replication of the APMV genome are co-directional. This characteristic is not unique to cellular organisms but is the same in APMV. Both bacteria and viruses replicate faster than eukaryotes. APMV has a 1.2 Mb genome, a size which is similar to that of bacteria. Therefore, its large genomic structure would likely be constructed while facing the selection pressure of DNA replication.

The negative and positive values on both ends of the GC skew plot suggest that the G/C nucleotide composition switches between the 5′ region (<296 kb) and the 3′ region (>882 kb). Both end regions have *dif* sequences, which have been previously estimated [5], suggesting that this nucleotide composition bias is perhaps a key factor for replication termination with directional homologous recombination (Figure S3). Indeed, both regions harbor paralogous genes of ankyrin repeats in opposite directions, indicating that recombination between the 5′ and 3′ ends frequently occurs (Figure 3a). Moreover, the ankyrin repeats are symmetrically placed between *dif* 2 and *dif* 3; thus, these two *dif* sequences might often be used for termination and segregation (Figure 3a). A model for the termination of DNA replication in APMV has been previously described, hypothesizing that the genomic DNA bent symmetrically during DNA replication of the 3′ end [3]. The symmetrical distribution of the sequences and genes in APMV suggest that homologous recombination and/or sequence insertions would occur prior to chromosome segregation. Furthermore, the large, symmetrical inversion between APMV and

Megavirus chillensis (lineage C of *Mimiviridae*) [43] indicates that the topology of the genomic DNA during replication led to the diversification of the viral family *Mimiviridae*.

In bacteria, the *dnaA* box sequence on the *ori* region is recognized by DnaA when DNA replication is initiated [9], and the Rep protein is used when plasmid replication is initiated [44]. Rep is also involved in chromosomal replication, and a lack of Rep function can cause delay in chromosomal replication [11]. In *E. coli*, RecB, which is a homologous helicase to Rep, acts in the reconstruction of a stalled replication fork in the RecBCD pathway, while recognizing chi sequences (5′-GCTGGTGG-3′) [15,16]. The gram-positive bacteria *B. subtilis* encodes the homolog of the RecB, AddA, and recognizes the chi sequence 5′-AGCGG-3′ [12]. Based on our results, UvrD/Rep-like helicase of APMV is more similar to AddA than it is to Rep, and a portion of the gram-positive chi-like sequence (5′-CCGC-3′) is frequently found in the estimated *ori* region of the APMV genome (Figures 4 and 5) This suggests that the initiation of DNA replication is presumably mediated by the interaction between the UvrD/Rep-like helicase and chi-like sequences in APMV. In *B. subtilis*, AddA, Q1155, and I1157, homologous residues of Q922 and I924 in UvrD/Rep-like helicase of APMV are known to recognize the fourth and fifth G nucleotides of the chi sequence (5′-AGCGG-3′) [37]. Therefore, these conservations are appropriate for binding APMV helicase to the GCGG sequence, of which the complementary sequence is frequently found in the estimated *ori* region of the APMV genome (Figures 4b and 5). However, S1015, V1016, S1017, and Y1204 in *B. subtilis* AddA bind phosphate at the 3′ end of the chi sequence [37], suggesting that these residues are not involved in the recognition of a specific chi sequence. Phosphate binding sites were not conserved in UvrD/Rep-like helicase of APMV at the sequence level, however these regions contain polar amino acids similar to *B. subtilis* AddA (Figure 4b), which indicates that the UvrD/Rep-like helicase of APMV could also bind phosphate at the 3′ end of the chi sequence. Furthermore, the other quartet sequence, 5′-CGGC-3′, which was detected on the estimated *ori* region of APMV (Figure 5a,c), is reported to be a part of the DNA replication signal sequence of mitochondria (3′-GGCCG-5′) [38], indicating that the DNA replication of APMV is also initiated by host replication machineries involved in the replication of mitochondrial DNA.

KOPS and FtsK determine the region of DNA replication termination, and FtsK mediates DNA segregation process in bacteria [7,8]. KOPS are different among bacteria [6–8], in that the KOPS recognition mechanism is thought to be defined by the structure of FtsK and the sequence of KOPS itself. Our results showed that, in the APMV genome, the distribution pattern of *L. lactis*-KOPS was similar to those of bacteria (Figure 10b,c). The sequence analysis of the FtsK gamma domain, which interacts with KOPS, also showed a high similarity between *L. Lactis* and APMV (Figure 7), suggesting that the pair of KOPS and the FtsK-gamma domain structure in APMV might be homologous to that of *L. lactis*. Furthermore, we found that there was an insertion in the gamma domain of APMV-FtsK. Interestingly, the three-dimensional structure estimated by homology modeling revealed that this insertion was reconstructed and conserved in the DNA-binding domain (Figure 8). We also showed that the Walker A motif was conserved in the FtsK-motor domain of APMV (Figure 9a). These results suggest that APMV-FtsK might be functionally homologous to that of bacteria. The phylogeny of APMV-FtsK (gamma domain) was also found to be similar to that of *L. lactis* (Figure 7b). Therefore, the mechanisms of interaction between these proteins and specific sequences are presumably homologous between APMV and bacteria such as *L. lactis*, although further molecular biological studies and structural analyses are required to certify this model.

The phylogenetic relations between *Rickettsia* and mitochondria and between mitochondria and Mimivirus were described previously [39,40]. Interestingly, the genome size and GC content of *Rickettsia* are 1.1 Mb and 29%, respectively [39], which is highly similar to APMV (1.2Mb, 28%). Furthermore, the backbone of the Mimivirus genome is reported to be derived from the ancestor of mitochondria [40]. Considering the quartet sequence similarities (Figure 6) and the phylogenetic relation between *Rickettsia* and mitochondria, the ancestor of Mimivirus infected the ancestor of eukaryotic cells (last archaeal common ancestor, LACA) before the endosymbiosis of mitochondrial ancestor to the first eukaryotic common ancestor (FECA), while the ancestor of mitochondria and

Rickettsia also infected to the ancestor of eukaryotic cells. This ancestral virus presumably harbored shorter genomic DNA than the present-day Mimivirus, and it acquired the long genome from ancestor of mitochondria by genomic fusion. The sequence 5′-CGGC-3′ found in the APMV *ori* region might also have been acquired from the ancestor of mitochondria, and is still conserved as a DNA replication signal sequence (3′-GGCCG-5′) in the mitochondrial genome [38]. It has been reported that the comparison between mitochondrial genes and the *Rickettsia* genome shows much higher similarities than that between mitochondrial genes and megaviral genome (poxvirus), while the synteny of these three species are significantly conserved [45]. Thus, according to our results, the genomic remnants of the ancestor of mitochondria may be still conserved in the Mimivirus genome to a greater extent than in the poxvirus genome. Moreover, the co-infection (or preying on) of the LACA cells might have occurred not only in ancestor of mitochondria and Mimivirus but also in other bacteria, and therefore APMV conserves the bacteria-like machineries such as UvrD/Rep-like helicase and APMV-FtsK derived from the co-infected ancestor of gram-positive bacteria by horizontal gene transfer.

5. Conclusions

Here, we presented a proposed model of the initiation and termination of DNA replication and chromosome segregation for APMV. The estimated *ori* region exists at the 382 kb position in the genome, which contains the chi-like sequence recognized by *B. subtilis* AddA, which is homologous to the UvrD/Rep-like helicase of APMV. The other sequence has a homology of a DNA replication signal sequence of mitochondria, indicating that the DNA replication of APMV may initiate with the replication machineries of mitochondria. The KOPS distribution pattern and the structure of APMV-FtsK indicate that the KOPS recognition system by APMV-FtsK is similar to that of *L. lactis*. Consequently, replication initiation, termination and segregation systems of APMV are presumably mediated by DNA repair machineries, similar to that of gram-positive bacteria, such as *L. lactis*. Furthermore, the comparison of quartet sequence compositions shows the similarity between APMV and *Rickettsia,* which may have the closest common ancestor of mitochondria, indicating that Mimivirus has acquired a large bacteria-like genome and its DNA replication machineries from ancestor of mitochondria during the co-infection to the LACA cells. The evolutionary history of APMV remains unclear; however, the further analyses of such a chimeric genome of APMV may illustrate the early stage of evolution of eukaryotic cells and Mimivirus.

Supplementary Materials: The following are available online at http://www.mdpi.com/1999-4915/11/3/267/s1: Figure S1: AC skew analyses of the *Acanthamoeba polyphaga mimivirus* (APMV) genome; Figure S2: Correlation of the CDS length of left/right side from the estimated *ori* region; Figure S3: Modes of the generation of paralogous genes far from each other and their directions on the linear genome; Supplementary Data: All numeric datasets in this paper.

Author Contributions: M.A. designed the research and performed the analyses. M.A. and M.T. wrote the manuscript.

Funding: This research received no external funding.

Acknowledgments: We would like to thank K. Aoki, S. Fukaya, and M. Kurabayashi from our laboratory for helpful discussions. We would also like to thank Editage (www.editage.jp) for English language editing.

Conflicts of Interest: The authors declare no conflict of interest.

References

1. La Scola, B.; Audic, S.; Robert, C.; Jungang, L.; de Lamballerie, X.; Drancourt, M.; Birtles, R.; Claverie, J.M.; Raoult, D. A giant virus in amoebae. *Science* **2003**, *299*, 2033. [CrossRef] [PubMed]
2. Raoult, D.; Audic, S.; Robert, C.; Abergel, C.; Renesto, P.; Ogata, H.; La Scola, B.; Suzan, M.; Claverie, J.M. The 1.2-Megabase Genome Sequence of Mimivirus. *Science* **2004**, *306*, 1344–1350. [CrossRef] [PubMed]
3. Yoshida, T.; Claverie, J.-M.; Ogata, H. Mimivirus reveals Mre11/Rad50 fusion proteins with a sporadic distribution in eukaryotes, bacteria, viruses and plasmids. *Virol. J.* **2011**, *8*, 427. [CrossRef] [PubMed]

4. Kreuzer, K.N. Recombination-dependent DNA replication in phage T4. *Trends Biochem. Sci.* **2000**, *25*, 165–173. [CrossRef]

5. Chelikani VRanjan, T.; Zade, A.; Shukla, A.; Kondabagil, K. Genome segregation and packaging machinery in *Acanthamoeba polyphaga* mimivirus is reminiscent of bacterial apparatus. *J. Virol.* **2014**, *88*, 6069–6075. [CrossRef] [PubMed]

6. Nolivos, S.; Touzain, F.; Pages, C.; Coddeville, M.; Rousseau, P.; El Karoui, M.; Le Bourgeois, P.; Cornet, F. Co-evolution of segregation guide DNA motifs and the FtsK translocase in bacteria: Identification of the atypical *Lactococcus lactis* KOPS motif. *Nucleic Acids Res.* **2012**, *40*, 5535–5545. [CrossRef]

7. Ptacin, J.L.; Nollmann, M.; Becker, E.C.; Cozzarelli, N.R.; Pogliano, K.; Bustamante, C. Sequence-directed DNA export guides chromosome translocation during sporulation in *Bacillus subtilis*. *Nat. Struct. Mol. Biol.* **2008**, *15*, 485–493. [CrossRef] [PubMed]

8. Bigot, S.; Saleh, O.A.; Lesterlin, C.; Pages, C.; El Karoui, M.; Dennis, C.; Grigoriev, M.; Allemand, J.F.; Barre, F.X.; Cornet, F. KOPS: DNA motifs that control *E. coli* chromosome segregation by orienting the FtsK translocase. *EMBO J.* **2005**, *24*, 3770–3780. [CrossRef]

9. Kaguni, J.M. Replication initiation at the *Escherichia coli* chromosomal origin. *Curr. Opin. Chem. Biol.* **2011**, *15*, 606–613. [CrossRef]

10. Lohman, T.M. *Escherichia coli* DNA helicases: Mechanisms of DNA unwinding. *Mol. Microbiol.* **1992**, *6*, 5–14. [CrossRef]

11. Lane, H.E.; Denhardt, D.T. The *rep* mutation. IV. Slower movement of replication forks in *Escherichia coli rep* strains. *J. Mol. Biol.* **1975**, *97*, 99–112. [CrossRef] [PubMed]

12. Chédin, F.; Noirot, P.; Biaudet, V.; Ehrlich, S.D. A five-nucleotide sequence protects DNA from exonucleolytic degradation by AddAB, the RecBCD analogue of *Bacillus subtilis*. *Mol. Microbiol.* **1998**, *29*, 1369–1377. [CrossRef] [PubMed]

13. Velankar, S.S.; Soultanas, P.; Dillingham, M.S.; Subramanya, H.S.; Wigley, D.B. Crystal structures of complexes of PcrA DNA helicase with a DNA substrate indicate an inchworm mechanism. *Cell* **1999**, *97*, 75–84. [CrossRef]

14. Petit, M.A.; Dervyn, E.; Rose, M.; Entian, K.D.; McGovern, S.; Ehrlich, S.D.; Bruand, C. PcrA is an essential DNA helicase of *Bacillus subtilis* fulfilling functions both in repair and rolling-circle replication. *Mol. Microbiol.* **1998**, *29*, 261–273. [CrossRef] [PubMed]

15. Horiuchi, T.; Fujimura, Y. Recombinational rescue of the stalled DNA replication fork: A model based on analysis of an *Escherichia coli* strain with a chromosome region difficult to replicate. *J. Bacteriol.* **1995**, *177*, 783–791. [CrossRef]

16. Amundsen, S.K.; Sharp, J.W.; Smith, G.R. RecBCD Enzyme "Chi Recognition" Mutants Recognize Chi Recombination Hotspots in the Right DNA Context. *Genetics* **2016**, *204*, 139–152. [CrossRef] [PubMed]

17. Biswas, I.; Maguin, E.; Ehrlich, S.D.; Gruss, A. A 7-base-pair sequence protects DNA from exonucleolytic degradation in *Lactococcus lactis*. *Proc. Natl. Acad. Sci. USA* **1995**, *92*, 2244–2248. [CrossRef]

18. Frank, A.C.; Lobry, J.R. Asymmetric substitution patterns: A review of possible underlying mutational or selective mechanisms. *Gene* **1999**, *238*, 65–77. [CrossRef]

19. Arakawa, K.; Tomita, M. The GC skew index: A measure of genomic compositional asymmetry and the degree of replicational selection. *Evol. Bioinform.* **2007**, *3*, 159–168. [CrossRef]

20. Freeman, J.M. Patterns of Genome Organization in Bacteria. *Science* **1998**, *279*, 1827. [CrossRef]

21. Grigoriev, A. Analyzing genomes with cumulative skew diagrams. *Nucleic Acids Res.* **1998**, *26*, 2286–2290. [CrossRef] [PubMed]

22. Lobry, J.R.; Sueoka, N. Asymmetric directional mutation pressures in bacteria. *Genome Biol.* **2002**, *3*, RESEARCH0058.1. [CrossRef]

23. Lobry, J.R.; Louarn, J.-M. Polarisation of prokaryotic chromosomes. *Curr. Opin. Microbiol.* **2003**, *6*, 101–108. [CrossRef]

24. Krzywinski, M.; Schein, J.; Birol, I.; Connors, J.; Gascoyne, R.; Horsman, D.; Jones, S.J.; Marra, M.A. Circos: An information aesthetic for comparative genomics. *Genome Res.* **2009**, *19*, 1639–1645. [CrossRef] [PubMed]

25. Whelan, S.; Irisarri, I.; Burki, F. PREQUAL: Detecting non-homologous characters in sets of unaligned homologous sequences. *Bioinformatics* **2018**, *34*, 3929–3930. [CrossRef] [PubMed]

26. Katoh, K.; Standley, D.M. MAFFT multiple sequence alignment software version 7: Improvements in performance and usability. *Mol. Biol. Evol.* **2013**, *30*, 772–780. [CrossRef]

27. Capella-Gutiérrez, S.; Silla-Martínez, J.M.; Gabaldón, T. trimAl: A tool for automated alignment trimming in large-scale phylogenetic analyses. *Bioinformatics* **2009**, *25*, 1972–1973. [CrossRef] [PubMed]
28. Nakhleh, L. Evolutionary Phylogenetic Networks: Models and Issues. In *Problem Solving Handbook in Computational Biology and Bioinformatics*; Springer: Boston, MA, USA, 2010; pp. 125–158. [CrossRef]
29. Huson, D.H.; Bryant, D. Application of phylogenetic networks in evolutionary studies. *Mol. Biol. Evol.* **2006**, *23*, 254–267. [CrossRef]
30. Arakawa, K.; Tamaki, S.; Kono, N.; Kido, N.; Ikegami, K.; Ogawa, R.; Tomita, M. Genome Projector: zoomable genome map with multiple views. *BMC Bioinform.* **2009**, *10*, 31. [CrossRef]
31. Rice, P.; Longden, I.; Bleasby, A. EMBOSS: The European Molecular Biology Open Software Suite. *Trends Genet.* **2000**, *16*, 276–277. [CrossRef]
32. Yang, J.; Yan, R.; Roy, A.; Xu, D.; Poisson, J.; Zhang, Y. The I-TASSER Suite: Protein structure and function prediction. *Nat. Methods* **2015**, *12*, 7–8. [CrossRef]
33. Löwe, J.; Ellonen, A.; Allen, M.D.; Atkinson, C.; Sherratt, D.J.; Grainge, I. Molecular mechanism of sequence-directed DNA loading and translocation by FtsK. *Mol. Cell* **2008**, *31*, 498–509. [CrossRef] [PubMed]
34. Sivanathan, V.; Allen, M.D.; de Bekker, C.; Baker, R.; Arciszewska, L.K.; Freund, S.M.; Bycroft, M.; Löwe, J.; Sherratt, D.J. The FtsK gamma domain directs oriented DNA translocation by interacting with KOPS. *Nat. Struct. Mol. Biol.* **2006**, *13*, 965–972. [CrossRef] [PubMed]
35. Arai, N.; Arai, K.; Kornberg, A. Complexes of *Rep* protein with ATP and DNA as a basis for helicase action. *J. Biol. Chem.* **1981**, *256*, 5287–5293.
36. Kooistra, J.; Haijema, B.J.; Hesseling-Meinders, A.; Venema, G. A conserved helicase motif of the AddA subunit of the *Bacillus subtilis* ATP-dependent nuclease (AddAB) is essential for DNA repair and recombination. *Mol. Microbiol.* **1997**, *23*, 137–149. [CrossRef]
37. Krajewski, W.W.; Fu, X.; Wilkinson, M.; Cronin, N.B.; Dillingham, M.S.; Wigley, D.B. Structural basis for translocation by AddAB helicase-nuclease and its arrest at χ sites. *Nature* **2014**, *508*, 416–419. [CrossRef] [PubMed]
38. Hixson, J.E.; Wong, T.W.; Clayton, D.A. Both the conserved stem-loop and divergent 5′-flanking sequences are required for initiation at the human mitochondrial origin of light-strand DNA replication. *J. Biol. Chem.* **1986**, *261*, 2384–2390.
39. Andersson, S.G.; Zomorodipour, A.; Andersson, J.O.; Sicheritz-Pontén, T.; Alsmark, U.C.; Podowski, R.M.; Näslund, A.K.; Eriksson, A.S.; Winkler, H.H.; Kurland, C.G. The genome sequence of *Rickettsia prowazekii* and the origin of mitochondria. *Nature* **1998**, *396*, 133–140. [CrossRef] [PubMed]
40. Seligmann, H. Giant viruses as protein-coated amoeban mitochondria? *Virus Res.* **2018**, *253*, 77–86. [CrossRef]
41. Massey, T.H.; Mercogliano, C.P.; Yates, J.; Sherratt, D.J.; Löwe, J. Double-stranded DNA translocation: Structure and mechanism of hexameric FtsK. *Mol. Cell* **2006**, *23*, 457–469. [CrossRef]
42. Paul, S.; Million-Weaver, S.; Chattopadhyay, S.; Sokurenko, E.; Merrikh, H. Accelerated gene evolution through replication-transcription conflicts. *Nature* **2013**, *495*, 512–515. [CrossRef] [PubMed]
43. Arslan, D.; Legendre, M.; Seltzer, V.; Abergel, C.; Claverie, J.-M. Distant Mimivirus relative with a larger genome highlights the fundamental features of Megaviridae. *Proc. Natl. Acad. Sci. USA* **2011**, *108*, 17486–17491. [CrossRef]
44. Han, M.; Yagura, M.; Itoh, T. Specific interaction between the initiator protein (Rep) and origin of plasmid ColE2-P9. *J. Bacteriol.* **2007**, *189*, 1061–1071. [CrossRef] [PubMed]
45. Seligmann, H. Giant viruses: Spore-like missing links between Rickettsia and mitochondria? *Ann. N. Y. Acad. Sci.* **2019**. [CrossRef]

Article

Functional RNA Structures in the 3′UTR of Tick-Borne, Insect-Specific and No-Known-Vector Flaviviruses

Roman Ochsenreiter [1], **Ivo L. Hofacker** [1,2] **and Michael T. Wolfinger** [1,2,*]

[1] Department of Theoretical Chemistry, University of Vienna, Währingerstraße 17, 1090 Vienna, Austria; romanoch@tbi.univie.ac.at (R.O.); ivo@tbi.univie.ac.at (I.L.H.)

[2] Research Group BCB, Faculty of Computer Science, University of Vienna, Währingerstraße 29, 1090 Vienna, Austria

* Correspondence: michael.wolfinger@univie.ac.at

Received: 1 March 2019; Accepted: 20 March 2019; Published: 24 March 2019

Abstract: Untranslated regions (UTRs) of flaviviruses contain a large number of RNA structural elements involved in mediating the viral life cycle, including cyclisation, replication, and encapsidation. Here we report on a comparative genomics approach to characterize evolutionarily conserved RNAs in the 3′UTR of tick-borne, insect-specific and no-known-vector flaviviruses in silico. Our data support the wide distribution of previously experimentally characterized exoribonuclease resistant RNAs (xrRNAs) within tick-borne and no-known-vector flaviviruses and provide evidence for the existence of a cascade of duplicated RNA structures within insect-specific flaviviruses. On a broader scale, our findings indicate that viral 3′UTRs represent a flexible scaffold for evolution to come up with novel xrRNAs.

Keywords: flavivirus; non-coding RNA; secondary structure

---

1. Introduction

Flaviviruses are small, single-stranded positive-sense RNA viruses that are typically transmitted between arthropod vectors and vertebrate hosts. They are endemic in tropic and sub-tropic regions and represent a global health threat, although humans are considered dead end hosts in many cases.

The genus *Flavivirus* within the *Flaviviridae* family comprises more than 70 species, which are organized into four groups, each with a specific host association: Mosquito-borne flaviviruses (MBFVs) and tick-borne flaviviruses (TBFVs) spread between vertebrate (mammals and birds) and invertebrate (mosquitoes and ticks) hosts, whereas insect-specific flaviviruses (ISFVs) replicate specifically in mosquitoes and no-known-vector flaviviruses (NKVs) have only been found in rodents and bats, respectively. This natural host-range-based classification is in good agreement with sequence-based phylogenetic clustering, mainly because all flaviviruses share a common genome organization [1]. Conversely, epidemiology, disease association [2] and transmission cycles [3] are fundamentally different among different flavivirus groups.

Emerging and re-emerging MBFVs such as Dengue virus (DENV), Japanese encephalitis virus (JEV), West Nile virus (WNV), Yellow fever virus (YFV) or Zika virus (ZIKV) are the causative agents of large-scale outbreaks that result in millions of human and veterinary infections every year [4]. Likewise, tick-borne encephalitis virus (TBEV), Powassan virus (POWV) and other members of the tick-borne serocomplex are neuropathogenic agents that cause a large number of infections every year, resulting in a massive incidence increase since the 1970ies [5]. Consequently, much research effort has gone into studying MBFV and TBFV biology, biochemistry and phylogeny [6]. The two remaining groups, ISFVs and NKVs, however, have received limited attention in the research community, mainly because they are generally not associated with human or veterinary disease and therefore are still underrepresented in the literature. The phylogenetic relationship among the four ecological flavivirus groups is shown in Figure 1. Table A1 lists all viral species considered in the present study.

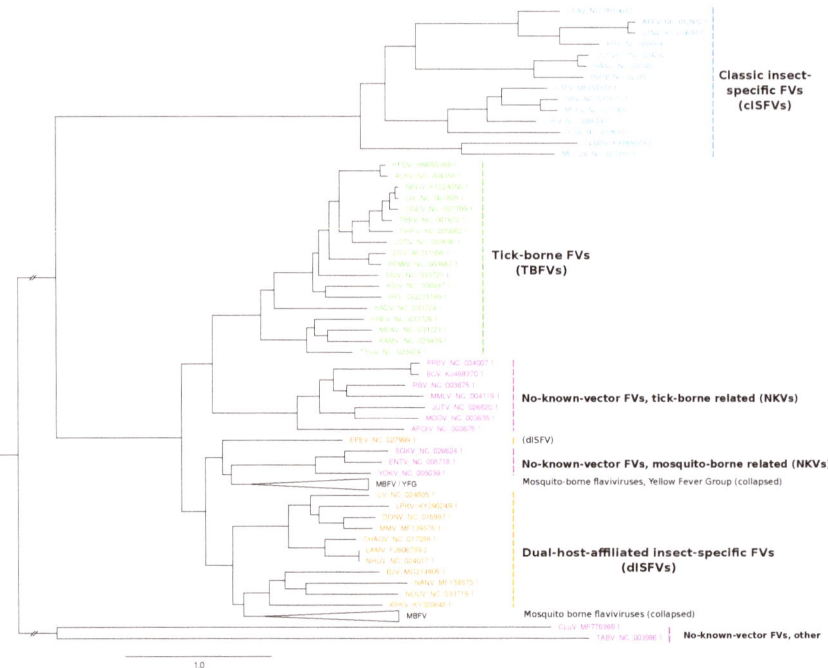

Figure 1. Maximum-likelihood phylogenetic tree of the genus *Flavivirus*, highlighting the major groups ISFVs (blue), dISFVs(orange), TBFVs (green), and NKVs (magenta). The MBFV Yellow Fever virus group (YFG) and the main MBFV branch were not covered in this study and are both collapsed. The tree has been computed from a `MAFFT` alignment of complete polyprotein amino acid sequences with `iq-tree`. Figure rendered with `FigTree`.

TBFVs form a monophyletic group consisting of a single serocomplex, although pathology and clinical manifestations vary among different viruses. They comprise more than a dozen of recognized species and separate into three groups: Mammalian tick-borne flaviviruses (M-TBFV), seabird tick-borne flaviviruses (S-TBFV) and the Kadam virus group. See [7] for a comprehensive review.

ISFVs naturally infect hematophagous *Diptera* and are typically divided into two groups [8]: Classical insect-specific flaviviruses (cISFVs) naturally infect mosquitoes and excursively replicate in mosquito cells in vitro. They form a phylogenetically distinct clade among known flaviviruses, appearing at the root of the MBFV, TBFV and NKV branches. The cISFV group separates into two clades, one associated with *Aedes* spp. mosquitoes and the other associated with *Culex* spp. mosquitoes, respectively [9]. They lack the ability to infect vertebrates and to replicate in vertebrate cell lines and have not been in the research spotlight until very recently. The second group is comprised of arbovirus-related or dual-host affiliated insect-specific flaviviruses (dISFVs), which represent a non-monophyletic group which is phylogenetically and antigenically related to mosquito/vertebrate flaviviruses, although they do not appear to infect vertebrate cells [10]. Insect-specific viruses play a crucial role in the mosquito microbiome and have been shown to modulate the replication of other arboviruses [11]. In this line, they are currently considered as biological control agents and vaccine platforms [12].

NKVs represent an ecologically and phylogenetically diverse set of viruses which have been isolated exclusively from vertebrates (mainly bats and rodents), without evidence for transmission by arthropod vectors. They form a non-monophyletic group among flaviviruses and are typically divided into bat- (B-NKV) and rodent-associated (R-NKV) groups, see Table A1. B-NKVs can be further separated into Entebbe virus group, which is phylogenetically closer to MBFVs, and Rio Bravo virus group, which is a sister clade to TBFVs. Species in the R-NKV group form the Modoc virus group, which is phylogenetically close to the B-NKV Rio Bravo group [13]. While NKVs are poorly characterized they represent a valuable resource to study evolutionary traits related to host-switch capacity mediated by conserved genomic elements.

Conserved RNA Structures Mediate Pathogenesis

Conserved RNA structures in the untranslated regions (UTRs) of RNA viruses are of particular interest because they mediate the viral life cycle by promoting or enhancing replication, as proposed for elements in both 5′UTRs [14] and 3′UTRs [15–18]. Mosquito/vertebrate viruses must operate efficiently in vectors and hosts, phylogenetically distinct organisms with different cellular machineries. This requires a high level of flexibility of viral regulatory elements to evade various antiviral response strategies while assuring proper replication conditions required for maintaining a stable quasispecies population. To achieve this resilience in host adaptation, RNA duplication strategies have been proposed as an evolutionary trait for MBFVs [19]. Tandem RNA structures within DENV 3′UTR are under different selective pressures in alternating hosts, suggesting the idea that duplicated RNA structures differentially evolved to accommodate specific functions in the two hosts [20]. Likewise, there is evidence for evolutionary pressure on maintaining the primary sequence of parts of duplicated RNA elements, as recently shown for flaviviral dumbbell (DB) elements in the context of finding a biophysical model for explaining a possible route for ZIKV-induced neurotropism [21].

Viral RNA genomes are different from procaryotic and eucaryotic mRNA. In addition to coding for and regulating the viral machinery, viral genomic RNA (gRNA) exhibits functional regions that act upon different stages of the viral life cycle. The 10-12kB flaviviral gRNA is capped, but non-polyadenylated and encodes a single open reading frame (ORF). Upon translation, a polyprotein is produced, which is then cleaved by viral and cellular enzymes into structural and non-structural proteins [22]. The ORF is flanked by highly structured untranslated regions (UTRs), which contain evolutionary conserved RNA elements that are crucially related to regulation of the viral life cycle, thereby inducing processes such as genome circularization, viral replication and packaging [23–25].

Upon flavivirus infection, accumulation of both gRNA as well as viral long non-coding RNAs (lncRNAs) is observed. These lncRNAs, which have been referred to as subgenomic flaviviral RNAs (sfRNAs) [26] are stable decay intermediates produced by exploiting the host's mRNA degradation machinery [27] and are associated with viral replication, pathogenesis and cytopathicity [28,29]. The production of sfRNA is induced by partial degradation of viral gRNA by the 5′-3′ exoribonuclease Xrn1, an enzyme associated with the cell's RNA turnover machinery [30,31]. Mechanistically, sfRNAs are generated by stalling Xrn1 at conserved structural elements in the viral 3′UTR, termed xrRNA (exoribonuclease-resistant RNA elements). These structures efficiently stall Xrn1 from progressing through from the 5′ direction, thus protecting the downstream RNA from degradation, while pass-through in the 3′-5′ direction, as required for viral RNA-dependent RNA-polymerase is still possible [32]. In particular, different types of stem-loop (SL) and dumbbell (DB) elements found in many MBFVs and TBFVs have been related to quantitative protection of downstream virus RNA against exoribonuclease degradation [33].

Xrn1 stalling results in dysregulation of cellular function with the aim of promoting viral infections. In this regard, functions of sfRNA in modulating cellular mRNA decay and RNAi pathways [34] as well as modulating anti-viral interferon response [35,36] have been reported.

The genomic architecture of flaviviruses has been extensively studied to understand the molecular principles required for sfRNA production. Chemical and enzymatic probing methods [37], together with x-ray crystallography revealed the 3'UTR structure of the MBFVs WNV [38], YFV [39], DENV [40], Murray Valley encephalitis virus (MVEV) [41], ZIKV [42] and recently different species of the TBFV and NKV groups [33], highlighting the possibility that exoribonuclease resistance might be a pervasive mechanism of the viral world. Interestingly, several conserved RNA structural elements in viral 3'UTRs have been predicted in our group [43–47], some of which have later been attributed to xrRNA functionality [26]. To further expand the set of potential xrRNAs, we report here on a comparative genomics survey aimed at characterization of evolutionary conserved RNA structures in flavivirus 3'UTRs, focusing on TBFVs and the hitherto understudied groups of ISFVs and NKVs. A detailed study on the evolutionary traits of conserved RNAs in MBFV 3'UTRs will be published elsewhere.

2. Materials and Methods

Viral genome data for the present study were obtained from the public National Center for Biotechnology Information (NCBI) refseq (https://www.ncbi.nlm.nih.gov/refseq/) and genbank (https://www.ncbi.nlm.nih.gov/genbank/) databases on 28 May 2018. We downloaded all complete viral genomes under taxonomy ID 11051 (genus *Flavivirus*) and filtered for TBFV, ISFV and NKV species listed in Table A1. Whenever refseq annotation was not available for a species, we selected the longest complete genome from the genbank set as representative sequence. In total, the data set is comprised of 86 ISFV, 275 TBFV and 27 NKV isolates, respectively. The number of isolates with available 3'UTR sequence data per species varies between 1 and 167.

2.1. Phylogeny Reconstruction

The polyprotein/coding sequence (CDS) regions of most flaviviruses can be aligned consistently, however, UTRs typically show large variance both in length and sequence composition, rendering them ill-suited for phylogeny reconstruction. A phylogeny of all members of the genus *Flavivirus* (Figure 1) was therefore reconstructed via a multiple sequence alignment (MSA) of the nucleotide sequences of the CDS regions only. The MSA was computed with MAFFT[48] and subsequent maximum-likelihood tree reconstruction was performed using iq-tree [49] using the GTR+F+R7 substitution model.

2.2. Structural Homology Search with Covariance Models

The present study is centered around structural homology of RNA elements among phylogenetically narrow subgroups. A straightforward approach to finding novel homologous RNA structures is to search RNA sequence databases with Covariance Models (CMs), i.e., statistical models of RNA structure that extend classic Hidden-Markov-Models (HMMs) to simultaneously represent sequence and secondary structure. CMs, as implemented in the infernal package [50] allow for rapid screening of large RNA sequence databases to find even weakly conserved sequence-only or structurally homologous RNAs. We have recently applied this approach to identify novel telomerase RNAs in *Saccharomycetes* [51].

Here, structural multiple sequence alignments of the viral 3'UTR sequences were generated with locARNA [52] and CMs were built for known or experimentally verified xrRNAs [33]. All 3'UTR sequences were then screened and novel candidate sequences were added to perform iterative refinement until convergence. Weak sequence conservation of putative xrRNA elements resulted in initially fragile results, indicating that infernal default parameters are typically not optimal. Adjusting parameters, in particular disabling both heuristic filtering and local end detection, however, allowed our CMs to find homologs with strongly conserved secondary structures in presence of large sequence deviation from the original sequence the CM was built from. Likewise, cmsearch E-values turned out unsuitable for assessing hit quality in case of major sequence divergence. We therefore employed a cutoff approach, requiring a hit to form at least 75% of all base pairs listed in the CM in order to be considered significant.

2.3. De novo Discovery of Conserved RNA Elements

Beside characterization of RNAs with homology to known structurally conserved elements, we aimed at identifying novel elements, considering both thermodynamic stability and sequence covariation as evolutionary traits. In this line, locARNA-generated structural alignments of full UTR sequences were cut manually into blocks corresponding to conserved secondary structures. Alternatively, we employed RNALalifold from the ViennaRNA package [53] to compute locally stable secondary structures for aligned UTR sequences. A CM was built for each structure and searched against all flavivirus 3′UTRs, keeping only CMs that scored well multiple times per UTR. The rationale here is that the occurrence of multiple copies hints towards a possible functional role of a structural element, given that the ability of two or more independently evolving sequences to form a common structure is unlikely.

The above approach is implemented as a set of custom Perl and Python scripts for semi-automatic characterization and annotation of conserved RNAs in viral UTR sequences. Internally, these scripts build on the ViennaRNA scripting language interface for thermodynamics calculations, the ViennaNGS [54] suite for extraction of genomic loci, the RNAaliSplit package [55] for splitting alignments into subparts with common consensus structures (i.e., common structures formed by all individual sequences), R2R [56] for visualization, and the ETE3 framework [57] for tree annotation and visualization.

3. Results

Several flaviviruses have previously been studied in great detail, yielding a varied landscape of repeated RNA sequence and structure elements within the 3′UTRs of these viruses, which are likely to have evolved from numerous duplications [19,58,59]. Many of these studies relied on single sequence predictions, which resulted in a good understanding of both structure and genomic position of conserved elements in individual species. A unified picture of homologous RNAs within the 3′UTRs of flaviviruses, however, has not been available.

The comparative approach applied in the present study outperforms single sequence predictions by considering consensus structures formed by all sequences. This allows us not only to confirm previously described RNA structures but also to elucidate hitherto unrecognized tandem repeats in many species. In this line our results can help in understanding the complex evolution of flavivirus 3′UTRs.

3.1. Construction of Seed Alignments

Based on recent experimental evidence for the existence of xrRNAs in TBFVs, ISFVs and NKVs, and previously characterized conserved RNA elements in flaviviral 3′UTRs, we built seed alignments for initial CMs, which were then refined iteratively, i.e., subjected to multiple rounds of screening and incorporation of best hits into the CM. Likewise, candidate structures from RNALalifold calculations were used as seeds for identification of conserved RNA structures. Figure 2 shows an overview of refined consensus structures for each ecologic group of flaviviruses analyzed here.

The four ecologic groups of flaviviruses show a varied 3′UTR architecture, however, the terminal 3′ stem-loop structure (3′SL, also referred to as 3′ long stable hairpin, 3′LSH) has been shown to be associated with panhandle-formation during virus replication and is therefore present in the terminal region of all flaviviruses [14]. The element is listed in Rfam as RF00185 (Flavivirus 3′UTR cis-acting replication element, Flavi_CRE) and we could use it to consistently identify terminal regions within 3′UTRs. Absence of this element from a UTR sequence is indicative of incomplete or truncated sequence data. The underlying sequences generally form a stable stem-loop structure upon structural alignment and single sequence folding. We built individual 3′SL seed alignments and CMs for each ecologic group, termed T.3SL, N.3SL and I.3SL, respectively.

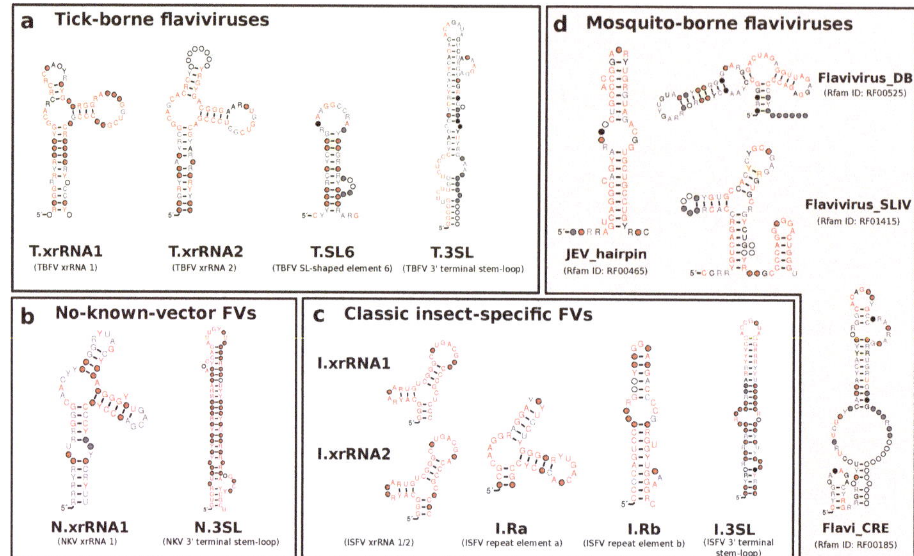

Figure 2. Overview of consensus structures of all CMs used for the annotation of flavivirus 3′UTRs. TBFV, ISFV, and NKV elements were refined from published experimental data (T.xrRNA1/2, I.xrRNA1/2, N.xrRNA) or identified computationally (T.SL6, I.Ra, I.Rb as well as all 3′-terminal stem-loop structures). MBFV elements were obtained from Rfam. Throughout this paper, all CMs are referred to by the name written in bold. References to xrRNA-like structures refer to the generalized xrRNA CM (Section 3.6).

3.2. Tick-Borne Flaviviruses

MacFadden et al. [33] suggested two different exoribonuclease-resistant structures in TBFV 3′UTRs. We used the proposed sequences from TBEV, POWV, Karshi virus (KSIV), Langat virus (LGTV), Louping ill virus (LIV), Omsk hemorrhagic fever virus (OHFV) and Alkhumra hemorrhagic fever virus (ALKV) as templates for a set of initial structural alignments and CMs. These models were then employed to search for high confidence hits within all TBFV 3′UTRs to construct seed alignments of the two exoribonuclease resistant structures in TBFVs, termed T.xrRNA1 and T.xrRNA2. These models allowed us to construct highly specific CMs for both TBFV xrRNAs, which were subsequently used to annotate xrRNA instances in already studied and previously unstudied TBFV species (Figure 3a).

The full structural alignment of the 3′UTRs of selected tick-borne species moreover suggests a short stem-loop element in several species, which is characterized by high sequence heterogeneity but heavily conserved structure supported by multiple covariations. Evidence for this element, termed stem-loop 6 (SL6), has been reported earlier for at least TBEV, LGTV and OHFV [58,59]. We kept this nomenclature and identified the exact position in each TBFV 3′UTR (Figure 3d).

Our data further shows that both TBFV xrRNA CMs (Figure 2a), as well as NKV xrRNA CMs (Figure 2b and Section 3.5) consistently yield plausible hits with a high degree of structure conservation immediately upstream of the strongly conserved terminal stem-loop element. Existence of a Y-shaped element (termed Y1) and putative similarity to NKVs has been proposed earlier based on single sequence structure predictions [58]. Structural `locARNA` alignment and subsequent `RNAalifold` consensus structure prediction indicates strong secondary structure conservation with frequent structure-conserving sequence covariations. Taken together, this suggests good evidence that respective regions in TBFVs harbor a putatively structured and functional xrRNA-like RNA (Y1, Figure 3d).

a Tick-borne flaviviruses

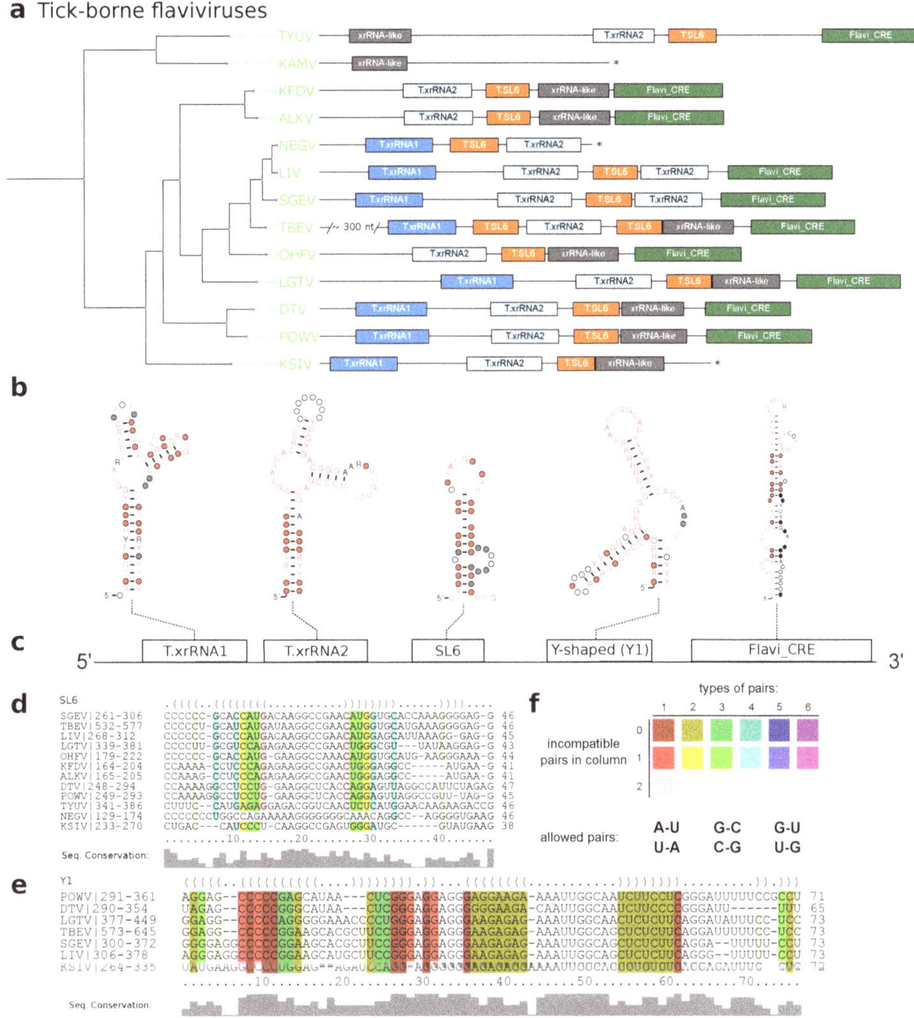

Figure 3. (**a**) Annotated 3′UTRs of TBFVs. The phylogenetic tree on the left has been computed from complete coding sequence nucleotide alignments and corresponds to the TBFV subtree in Figure 1. For each species with available 3′UTR sequence a sketch of the 3′UTR is drawn to scale next to the leaves of the tree. Colored boxes represent conserved RNA structural elements. Identifiers within the boxes indicate the CM which was used to infer homology at this position. Asterisks indicate incomplete 3′UTR sequences. Species without available 3′UTR are not shown. (**b**) Consensus structure plots of CM hits as calculated by *mlocarna*. (**c**) Schematic depiction of the common structural architecture of TBFV 3′UTRs. (**d**,**e**) Structural alignments of elements SL6 and Y1. (**f**) `RNAalifold` coloring scheme for paired columns in alignments. Colors indicate the number of basepair combinations found in pair of columns. Fainter colors indicate that some sequences cannot form a base pair.

Despite the differences in length and sequence composition, the 3′UTRs of most species in the TBFV group share a common architecture. Similar to MBFV SL-elements [19], two copies of xrRNAs are found in almost every species of this ecologic group, generally succeeded by one instance of SL6 and Y1. Likewise, the terminal 3′ stem loop is conserved in all TBFVs and can be reliably annotated by both our CM, T.3SL, and the Rfam Flavi_CRE model, which is used in Figure 3. Among all investigated species, only ALKV, OHFV and Kyasanur forest disease virus (KFDV) do not have a copy of xrRNA1, indicating that these viruses may have previously lost this element. Conversely, the two seabird-associated TBFVs with available 3′UTR data, Tyuleniy virus (TYUV) and Kama virus (KAMV) do not fit into this general scheme. Likewise, we were not able to annotate additional homologous or conserved structures with any CM used in this screen in the variable region of the 3′UTR of TBEV [5], despite the substantially longer UTR (+300 nts).

3.3. Classic Insect-Specific Flaviviruses

Classic insect-specific flaviviruses present diverged 3′UTR architectures, which likely result from the association of different species to *Aedes* spp. and *Culex* spp. vectors, respectively, which are also reflected by clade separation in the ISFV phylogenetic tree. Previous studies employed single sequence predictions to propose a varied set of homologous RNA structures in combination with an unusually large number of duplicated sequence signals [59]. Recent experimental evidence, however, suggests the presence of xrRNAs that have a similar fold to those known from MBFVs in cISFVs. Consequently, we set out to independently characterize conserved RNA elements for different subclades.

3.3.1. Exoribonuclease-Resistant RNAs in *Aedes*-Associated cISFVs

MacFadden et al. [33] used SHAPE structure probing to report the presence of two exoribonuclease-resistant RNAs in Cell fusing agent virus (CFAV), and provided evidence for a duplicated set of homologous structures in Aedes flavivirus (AEFV) and Kamiti river virus (KRV). We constructed initial alignments from the reported sequences in this clade in *Aedes* spp. associated viruses, resulting in two seed alignments, termed I.xrRNA1 and I.xrRNA2 (Figure 2c). For both elements, seed CMs were iteratively built from structural locARNA alignments. Minor manual adjustments to the alignments were required here, since the predicted consensus structures diverged slightly from the published SHAPE-guided prediction. Both models were then employed to search for additional high confidence hits within other isolates of CFAV, AEFV and KRV which were subsequently added to the seed alignments.

Screening the entire set of flavivirus 3′UTRs revealed that both ISFV xrRNA elements, I.xrRNA1 and I.xrRNA2, are only found in CFAV, AEFV and KRV, i.e., species initially used for the construction of the respective CMs. Furthermore, also in terms of pure structural conservation, no reliable hits in any other ISFV species could be obtained with any of these CMs (Figure 4a). This suggests that both ISFV xrRNAs may represent a specialized class of xrRNA elements only present in CFAV, AEFV and KRV. The 3′UTR of KRV is unique among all known flaviviruses because it harbors an additional copy of the terminal 3′ stem-loop element 600 nts upstream of the actual 3′-terminus, supporting previous reports that the KRV 3′UTR has undergone a full duplication during its evolution [60].

3.3.2. Conserved Structures in *Culex*-Associated cISFVs

The second distinct clade of cISFVs includes Culex flavivirus (CxFV), Quang Binh virus (QBV), Mosquito flavivirus (MSFV), Palm Creek virus (PCV), Culex theileri flavivirus (CTFV) as well as a few other species with only partial genome sequence availability [9] and is associated with *Culex* spp. vectors. An interesting observation in this clade is that no other CM from any of the four ecologic flavivirus groups shows a hit, not even with remote sequence or structure conservation. We therefore set out to produce a a high quality structural alignment of the complete 3′UTRs of CxFV, QBV and MSFV. Consensus structure folding of the full alignment revealed each species to harbor 3–4 repeats of two highly conserved elements supported by multiple co-varying base pairs (Figure 4b,c). We termed these "Repeat element a/b", respectively (Ra and Rb). Both elements, while strongly conserving their folds, show highly variable loop regions as well as weak sequence conservation in the case of the Ra element. Structure conservation and occurrence in multiple copies, as typically seen with other exoribonuclease-stalling elements, hints towards possible functional importance. These results complement earlier reports of sequence repeats in the 3′UTR of CxFV [61] with the identification of a quadruplicated pair of conserved structures.

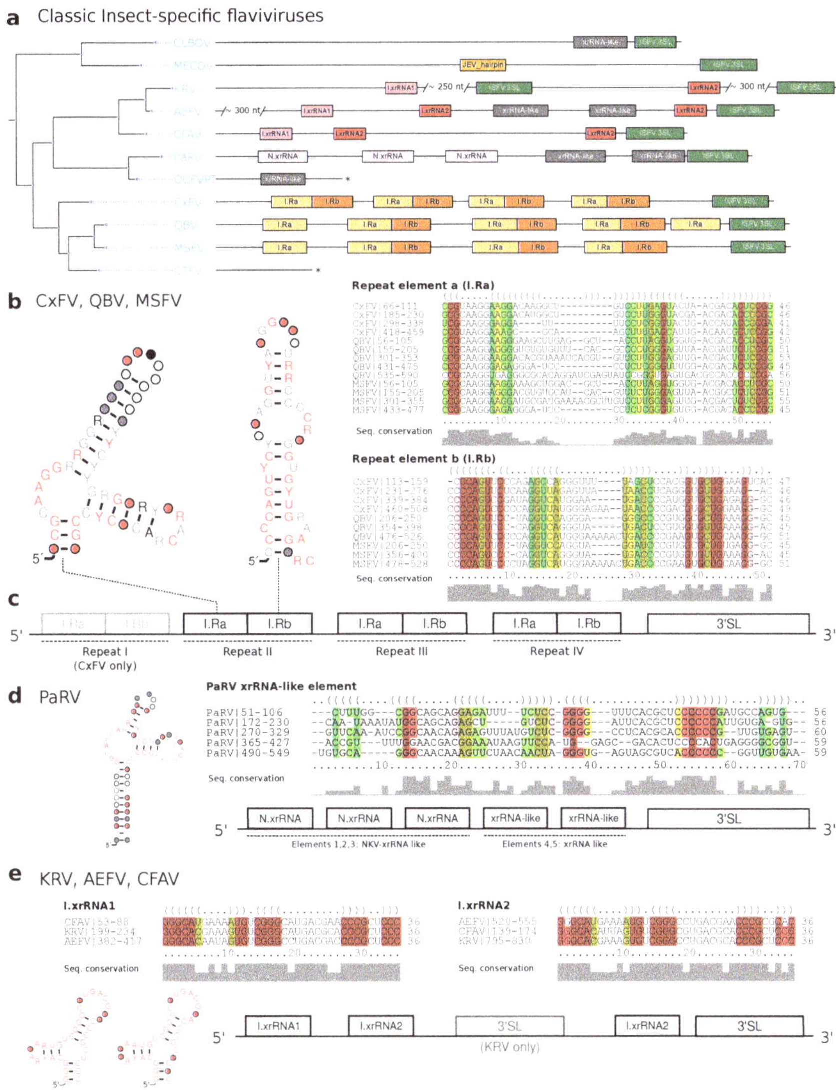

Figure 4. (**a**) Annotated Tree of cISFV 3′UTRs. Asterisks denote incomplete 3′UTR sequences. Species without available 3′UTR are not shown. (**b**) Consensus secondary structure plots and structural alignments of CM hits of Repeat a/b elements in CxFV, QBV, and MSFV. (**c**) Schematic of the common architecture of CxFV, QBV, and MSFV. Element I-IV refers to the respective repeat of elements. (**d**) Consensus structure plot and structural alignment of all CM hits of xrRNA-like elements in PaRV. (**e**) Proposed 3′UTR architecture of KRV, AEFV, and CFAV with consensus structure plots and structural alignments of I.xrRNA1 and I.xrRNA2.

3.3.3. Diverged 3′UTR Architecture in Many cISFVs

Interestingly, a screen of all available CMs in Parramatta River virus (PaRV) revealed five xrRNA-like elements (Figure 4a), with elements 1–3 sharing sequence and structure properties with NKV xrRNAs (Section 3.5), while elements 4 and 5 only conserve N.xrRNAs structure. All five hits can be structurally aligned into a consistent consensus structure (Figure 4d), despite the overall weak sequence consensus.

Conversely, the 3′UTRs of Calbertado virus (CLBOV) and Mercadeo virus (MECDV) appear structurally different from the other cISFVs. A general lack of characteristic CM hits lets these species appear more like an outgroup among cISFVs. In particular, we could only find significant hits for the omnipresent terminal 3′stem-loop structure, a putative xrRNA-like element in CLBOV and a single instance of a structure homologous to the Rfam model RF00465 (Japanese encephalitis virus hairpin structure) in MECDV. Still, limited availability of 3′UTR sequence data renders the characterization of conserved elements and interpretation difficult here.

Our data suggests that the 3′UTRs of cISFVs, in contrast to TBFVs (Section 3.2 and dISFVs (Section 3.4), do not appear to have a consistent architectural organization. In agreement with the cISFV phylogenetic subtree (Figure 1) we constitute three diverged groups with common 3′UTR organization that conform to their respective sub-clades: (i) CFAV-AEFV-KRV, each with two instances of xrRNAs, (ii) CxFV-QBV-MSFV with 3–4 copies of I.Ra/I.Rb elements and (iii) PaRV with 4–5 copies of xrRNA-like structures. Although no full 3′UTR sequences are available for the phylogenetically closest relatives of PaRV, HANV and OCFVPT, an xrRNA-like element in the small available fragment (syntenic to PaRV UTR) of OCFVPT 3′UTR suggests that both viruses might be organized in a similar manner, as supported by earlier reports that these viruses should be classified within the same species [1]. For CLBOV and MECDV, no clear pattern of conserved elements can be identified with our CMs. Both viruses either employ an entirely different class of elements or might not require capability for exoribonuclease stalling at all. The only element shared universally among all cISFVs is the 3′-terminal stem-loop, although cISFVs seem to diverge from other flaviviruses here, indicated by the inability of Rfam model RF00185 (Flavivirus CRE) to reliably annotate any cISFV 3′UTR.

3.4. Dual-Host Affiliated Insect-Specific Flaviviruses

Isolated almost exclusively from mosquitoes, dISFVs do not seem to infect vertebrate cells, despite their phylogenetic proximity to MBFVs (Figure 1). This association is reflected by good hits of the Rfam covariance models RF00525 (Flavivirus DB element) and RF00465 (Japanese encephalitis virus hairpin structure) in all dISFV isolates studied here (Figure 5). Interestingly, we could not find evidence for any sequences or structures homologous to tick-borne or other insect-specific flaviviruses. In this line, our data is in good agreement with the phylogenetic location of these viruses, which share ancestral roots with MBFVs [10].

An unusual species within this group is Ecuador Paraiso Escondido virus (EPEV), which has been isolated from New World sandflies and has been classified as insect-specific virus. EPEV phylogenetically appears at the root of the Entebbe bat virus group (ENTVG), a clade comprised of the three NKVs Entebbe bat virus (ENTV), Sokoluk virus (SOKV) and Yokose virus (YOKV). While all of these viruses contain homologs of conserved stem-loop (SL) and dumbbell (DB) elements found in MBFVs, ENTVG species may have lost their vector dependence [1].

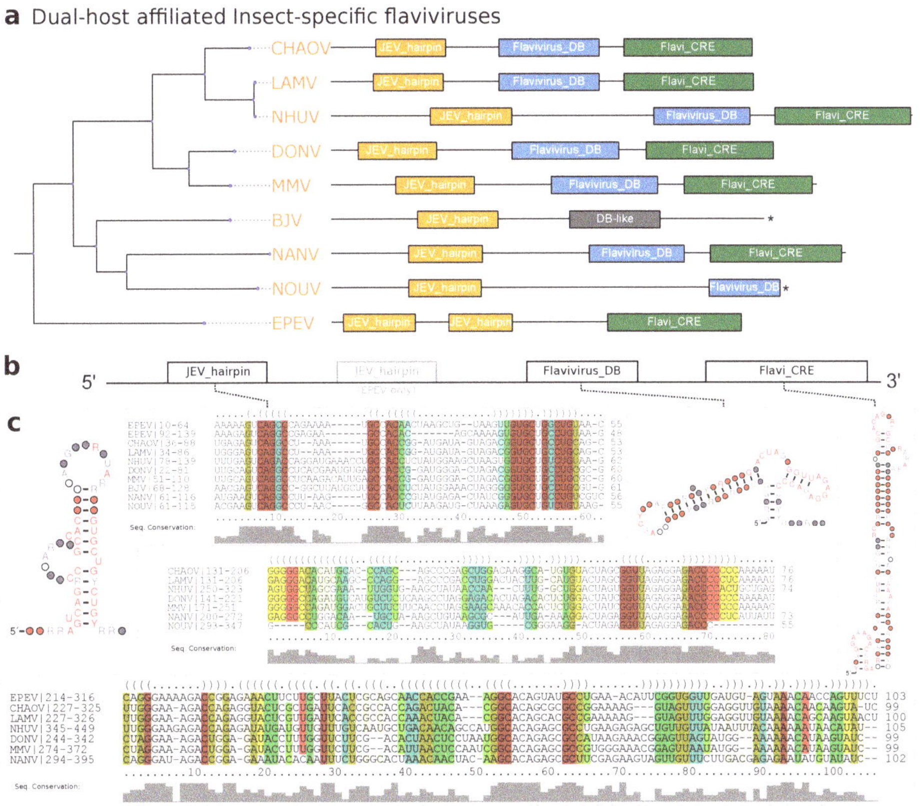

a Dual-host affiliated Insect-specific flaviviruses

Figure 5. (**a**) Annotated Tree of dISFV 3′UTRs. Asterisks denote incomplete 3′UTR sequences. Species without available 3′UTR are not shown. (**b**) Schematic Architecture of the dISFV 3′UTR. (**c**) Structural alignments and consensus structure plots of dISFV elements.

3.5. No-Known-Vector Flaviviruses

Rather than forming a monophyletic group, the no-known-vector flaviviruses can be separated into two distinct lineages, which are closely related to either TBFVs or MBFVs, respectively (Figure 1). Two additional NKVs, Tamana bat virus (TABV) and Cyclopterus lumpus virus (CLuV), are phylogenetically distant and serve as an outgroup to all flaviviruses. In analogy to the procedure outlined above for TBFVs (Section 3.2) and ISFVs (Section 3.3), we built a CM for experimentally verified xrRNAs in tick-borne related NKVs, termed N.xrRNA.

We found multiple hits of this CM at various loci within the 3′UTRs of tick-borne related NKVs, indicating that these species, in contrast to TBFVs, do not conserve a common 3′UTR architecture (Figure 6a). Surprisingly, we could identify several high-quality hits of the Rfam model RF00525 (Flavivirus DB element), an element typically found in MBFVs, in Rio Bravo virus (RBV), Montana myotis leukoencephalitis virus (MMLV) and Modoc virus (MODV). This is in so far remarkable as there is no evidence for conservation of this element in TBFVs, which phylogenetically cluster with this clade of NKVs. This element might have been introduced by an ancestral recombination event. Alternatively, conservation of an MBFV element in NKVs might be indicative of an association with an unknown vector, in agreement with the hypothesis that vector specificity is mediated by characteristic 3′UTR elements [19].

Conversely, there seems to be no generally conserved 3′UTR architecture among members of the mosquito-borne related NKVs (Figure 6c). While sequence data has not been available for Sokoluk virus (SOKV), we could annotate typical MBFV elements in the next relatives Entebbe bat virus (ENTV) and Yokose virus (YOKV), as proposed previously [32].

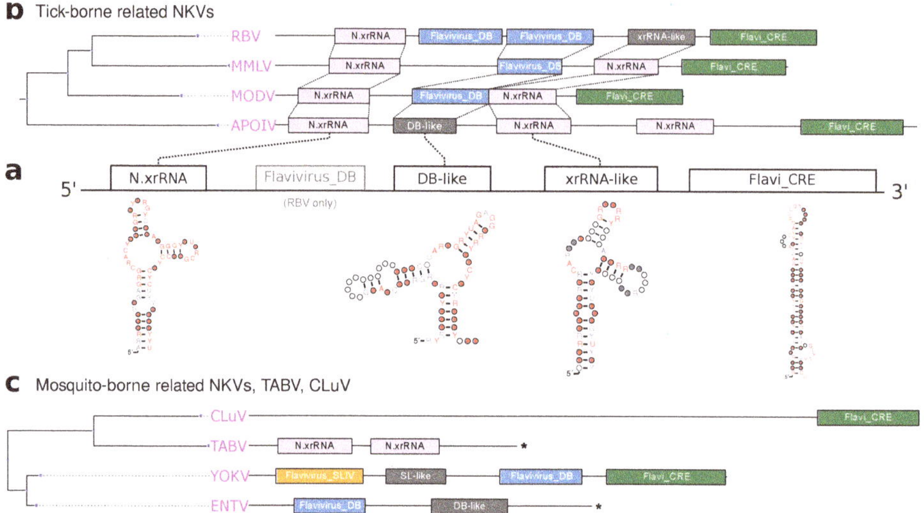

Figure 6. (**a**,**c**) Annotated 3′UTRs of NKVs. Asterisks denote incomplete 3′UTR sequences. (**b**) Schematic of TBFV-associated NKV-FV UTR architecture with consensus structures of NKV structure elements.

3.6. A Generalized xrRNA Structure

Earlier work suggested that xrRNAs from TBFVs and tick-borne related NKVs fall into a more general structural class of xrRNAs [33]. Following this line of reasoning, we investigated whether all high confidence hits obtained with our TBFV and NKV CMs could be assembled into one coherent CM that conserves the xrRNA-typical fold. A further advantage of a generalized CM would be higher sensitivity, allowing for identification of common features and eventually lead to annotation of previously unannotated xrRNAs.

Structural alignment and consensus structure prediction revealed all high confidence hits to fold into a common secondary structure (Figure 7a,b). While most of the consensus structure is characterized by low sequence conservation, stem 3 (S3) and loop 1 (L1) show medium to high degree of sequence conservation. The length of all stems is well conserved, although both major loop regions L2 and L3 show large fluctuations, with the length of L3 being de facto constant and L2 showing a high degree of flexibility.

We further investigated whether any high confidence hits from I.xrRNA1/2 in cISFVs could be aligned to the generalized xrRNA model. Although both cISFV xrRNAs (Figure 2c) bear some similarity to the generalized model, in particular to S3 and L3, we were not able to build a common alignment or consensus structure. Despite seemingly similar shape, our data also suggests that cISFV xrRNAs form a separate xrRNA subclass, unrelated to MBFV xrRNAs. In particular, we could not obtain hits of Rfam CMs (which can be seen as representatives of MBFV elements) in cISFVs, nor could we confirm any hits of cISFV-specific elements in MBFVs.

In addition to learning xrRNA features, a more generalized CM enabled us to detect xrRNA-like structures (indicated as such in all annotation plots) that could not be found previously.

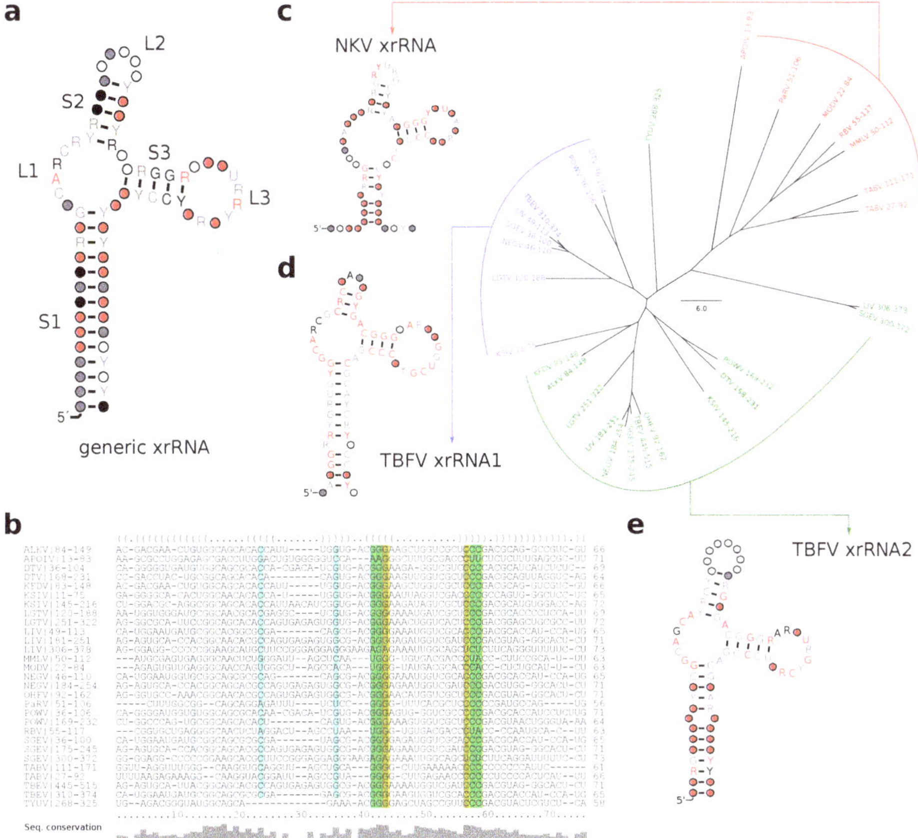

Figure 7. Generalized structure of all high confidence (cmsearch evalue $< 10^{-5}$) hits of T.xrRNA1, T.xrRNA2 and N.xrRNA. (**a**) Consensus structure prediction and (**b**) structural alignment of all high confidence hits. (**c–e**) Neighbor-joining tree of all high confidence hits of N.xrRNA (**c**), T.xrRNA1 (**d**), and T.xrRNA2 (**e**). Leaves are grouped and colored by the CM used for annotation, coordinates correspond to the position in the respective 3′UTR. For each group a separate structural alignment was computed, the consensus structures are shown.

4. Discussion

Mediated gRNA decay in the form of exoribonuclease resistance seems to be a pervasive strategy employed by viruses to circumvent host immune responses. Evidence of sfRNA production following incomplete Xrn1 degradation has not only been observed in different members of the *Flavivirus* genus [62], but also in other species of the *Flaviviridae* family, however, with major differences in xrRNA structure and sfRNA characteristics. While MBFV produce a 300-500nt sfRNA that corresponds to degradation products of the gRNA 3′UTR, hepaciviruses and pestiviruses produce a long subgenomic RNA whose 5′ end is located within the first 130nt of the viral gRNA [63].

Moreover, recent studies have identified xrRNA functionality in several phylogenetically distant RNA viruses, such as animal-infecting, segmented viruses of the *Bunyaviridae* and *Arenaviridae* [64] families, as well as plant-infecting viruses of the *Tombusuviridae* and *Luteoviridae* families [65,66]. The interesting question whether exoribonucleases other than Xrn1 would be blocked as well has recently been answered. MacFadden et al. [33] could show that both RNAse J1 and Dxo1 are stalled by MBFV xrRNAs, thereby demonstrating the general nature of this structure-induced blocking mechanism. These novel findings, together with previous knowledge of Xrn1 stalling in segmented plant viruses [67,68] provide evidence for a convergent evolution scenario where xrRNAs depend on a specific folded RNA structure and form a distinct class of functional RNAs.

Repeated RNA elements appear to be a hallmark of flavivirus 3′UTR architecture. While there seems to be a plethora of conserved structure classes, our data emphasizes the consistent trend that these elements typically do not occur as single copies. Rather, duplicate or even multiple occurrences of these elements hint towards functional relevance. This is further underlined by the evolutionary conservation of both patterns and elements among different species. Since, exoribonuclease stalling is presumably never perfect, it makes sense that viruses might employ multiple copies of such elements.

In this contribution we set out to identify homologs of known exoribonuclease stalling elements and novel conserved structures. To this end, we computationally characterized homologs of experimentally verified xrRNA in tick-borne and no-known vector viruses that seem to form a coherent class of RNA structures with capability to stall exoribonucleases (T.xrRNA1/2, N.xrRNA). Likewise, we identified another class of xrRNAs in classic insect-specific flaviviruses (I.xrRNA1/2) which appears to be only distantly related to the former class. In the same line, we predicted a set of novel conserved elements in cISFVs that appear in quadruples and do not coincide with other insect-specific elements (I.Ra,I.Rb).

While we did not focus on studying the evolutionary history of these groups of elements in detail, our data suggests that many elements share ancestral roots. This is supported by the observation that at least the tick-borne, no-known-vector and *Aedes* spp. associated xrRNAs fold into a similar Y-shaped substructure, although the exact fold varies significantly among individual species.

We compiled a set of covariance models that can be used for rapid screening assays in the identification and characterization of novel flaviviruses. All models are available from GitHub via https://github.com/mtw/ITNFV-Data.

A major problem is the limited availability of diverse 3′UTR sequence data for many viruses analyzed here, particularly within cISFVs. Many novel ISFVs have previously been discovered, but 3′UTR sequences were only available for a subset of them. Future studies are required to shed more light on the evolutionary history of 3′UTR evolution in flaviviruses.

Author Contributions: M.T.W. conceived the study. R.O. and M.T.W. analyzed the data, characterized conserved RNA elements and performed phylogenetic studies. All authors contributed to writing the paper and approved of the submitted version.

Funding: This work was funded in part by the Doktoratskolleg RNA Biology at Univ. Vienna and the Austrian Science Fund (SFB F43, I-1303).

Conflicts of Interest: The authors declare no conflict of interest.

Appendix A. Data Set Analyzed in This Study

Table A1. Viral genomes considered in this study. Flaviviruses are categorized into the groups tick-borne flaviviruses (TBFV), insect-specific flaviviruses (ISFV) and no-known-vector flaviviruses (NKV). The length of the 3′UTR is listed for each isolate. [†] Representative accession number from the `refseq` database. Whenever a `refseq` genome was not available, the isolate with the longest 3′UTR was selected as representative species. [a] Mammalian TBFVs. [b] Seabird TBFVs. [c] Classic ISFVs. [d] Dual-host affiliated ISFVs. [e] Rodent-associated NKVs. [f] Bat-associated NKVs. [N/A] 3′UTR partial or not available in the `refseq` data set.

Group	Accession Number [†]	Acronym	Scientific Name	3′UTR Length (nt)	Isolates
ISFV [c]	NC_012932.1	AEFV	Aedes flavivirus	942	3
ISFV [c]	NC_001564.2	CFAV	Cell fusing agent virus	553	3
ISFV [c]	KX669689.1	CLBOV	Calbertado virus	546	8
ISFV [c]	MF153378.1	CTFV	Culex theileri flavivirus	112	1

Table A1. *Cont.*

Group	Accession Number [†]	Acronym	Scientific Name	3′UTR Length (nt)	Isolates
ISFV [c]	NC_008604.2	CxFV	Culex flavivirus	654	13
ISFV [c]	NC_030401.1	HANV	Hanko virus	N/A	N/A
ISFV [c]	NC_005064.1	KRV	Kamiti River virus	1208	3
ISFV [c]	NC_027819.1	MECDV	Mercadeo virus	638	3
ISFV [c]	NC_021069.1	MSFV	Mosquito flavivirus	674	6
ISFV [c]	NC_034242.1	OCFVPT	Ochlerotatus caspius flavivirus	148	2
ISFV [c]	NC_027817.1	PaRV	Parramatta River virus	629	2
ISFV [c]	NC_033694.1	PCV	Palm Creek virus	N/A	N/A
ISFV [c]	NC_012671.1	QBV	Quang Binh virus	673	2
ISFV [d]	MG214905.1	BJV	Barkedji virus	335	1
ISFV [d]	NC_017086.1	CHAOV	Chaoyang virus	326	2
ISFV [d]	NC_016997.1	DONV	Donggang virus	343	2
ISFV [d]	NC_027999.1	EPEV	Paraiso Escondido virus	316	2
ISFV [d]	NC_024805.1	ILV	Ilomantsi virus	N/A	N/A
ISFV [d]	KY320648.1	KPKV	Kampung Karu virus	N/A	N/A
ISFV [d]	FJ606789.2	LAMV	Lammi virus	326	1
ISFV [d]	KY290249.1	LPKV	Long Pine Key virus	N/A	N/A
ISFV [d]	KY320649.1	LTNV	La Tina virus	N/A	N/A
ISFV [d]	MF139576.1	MMV	Marisma mosquito virus	376	1
ISFV [d]	MF139575.1	NANV	Nanay virus	399	1
ISFV [d]	NC_024017.1	NHUV	Nhumirim virus	451	1
ISFV [d]	NC_033715.1	NOUV	Nounane virus	347	3
NKV [e]	NC_003676.1	APOIV	Apoi virus	576	1
NKV	KJ469370.1	BCV	Batu Cave virus	N/A	N/A
NKV	MF776369.1	CLuV	Cyclopterus lumpus virus	601	1
NKV [f]	NC_008718.1	ENTV	Entebbe bat virus	308	3
NKV [e]	NC_026620.1	JUTV	Jutiapa virus	N/A	N/A
NKV [f]	NC_004119.1	MMLV	Montana myotis leukoencephalitis virus	460	1
NKV [e]	NC_003635.1	MODV	Modoc virus	366	1
NKV [f]	NC_034007.1	PPBV	Phnom Penh bat virus	N/A	N/A
NKV [f]	NC_003675.1	RBV	Rio Bravo virus	486	2
NKV [f]	NC_026624.1	SOKV	Sokoluk virus	N/A	N/A
NKV	NC_003996.1	TABV	Tamana bat virus	241	1
NKV [f]	NC_005039.1	YOKV	Yokose virus	429	1
TBFV [a]	NC_004355.1	ALKV	Alkhumra hemorrhagic fever virus	393	21
TBFV [a]	AF311056.1	DTV	Deer tick virus	459	1
TBFV [a]	NC_033723.1	GGV	Gadgets Gully virus	N/A	N/A
TBFV	NC_033724.1	KADV	Kadam virus	N/A	N/A
TBFV [b]	NC_023439.1	KAMV	Kama virus	282	2
TBFV [a]	HM055369.1	KFDV	Kyasanur forest disease virus	392	6
TBFV [a]	NC_006947.1	KSIV	Karshi virus	381	3
TBFV [a]	NC_003690.1	LGTV	Langat virus	568	5
TBFV [a]	NC_001809.1	LIV	Louping ill virus	500	5
TBFV [b]	NC_033721.1	MEAV	Meaban virus	N/A	N/A
TBFV [a]	KT224355.1	NEGV	Negishi virus	266	1
TBFV [a]	NC_005062.1	OHFV	Omsk hemorrhagic fever virus	410	4
TBFV [a]	NC_003687.1	POWV	Powassan virus	480	23
TBFV	DQ235149.1	RFV	Royal Farm virus	N/A	N/A
TBFV [a]	NC_027709.1	SGEV	Spanish goat encephalitis virus	493	2
TBFV [b]	NC_033726.1	SREV	Saumarez Reef virus	N/A	N/A
TBFV [a]	NC_001672.1	TBEV	Tick-borne encephalitis virus	764	167
TBFV [b]	NC_023424.1	TYUV	Tyuleniy virus	591	3

References

1. Kuno, G.; Chang, G.J.J.; Tsuchiya, K.R.; Karabatsos, N.; Cropp, C.B. Phylogeny of the Genus *Flavivirus*. *J. Virol.* **1998**, *72*, 73–83.
2. Gaunt, M.W.; Sall, A.A.; de Lamballerie, X.; Falconar, A.K.; Dzhivanian, T.I.; Gould, E.A. Phylogenetic relationships of flaviviruses correlate with their epidemiology, disease association and biogeography. *J. Gen. Virol.* **2001**, *82*, 1867–1876. [CrossRef]

3. Vasilakis, N.; Weaver, S.C. Flavivirus transmission focusing on Zika. *Curr. Opin. Virol.* **2017**, *22*, 30–35. [CrossRef]

4. Barrows, N.J.; Campos, R.K.; Liao, K.C.; Prasanth, K.R.; Soto-Acosta, R.; Yeh, S.C.; Schott-Lerner, G.; Pompon, J.; Sessions, O.M.; Bradrick, S.S.; et al. Biochemistry and Molecular Biology of Flaviviruses. *Chem. Rev.* **2018**, *118*, 4448–4482. [CrossRef]

5. Kellman, E.; Offerdahl, D.; Melik, W.; Bloom, M. Viral Determinants of Virulence in Tick-Borne Flaviviruses. *Viruses* **2018**, *10*, 329. [CrossRef] [PubMed]

6. Braack, L.; de Almeida, A.P.G.; Cornel, A.J.; Swanepoel, R.; De Jager, C. Mosquito-borne arboviruses of African origin: Review of key viruses and vectors. *Parasite Vector* **2018**, *11*, 29. [CrossRef] [PubMed]

7. Gritsun, T.; Lashkevich, V.; Gould, E. Tick-borne encephalitis. *Antivir. Res.* **2003**, *57*, 129–146. [CrossRef]

8. Guzman, H.; Contreras-Gutierrez, M.A.; da Rosa, A.P.T.; Nunes, M.R.; Cardoso, J.F.; Popov, V.L.; Young, K.I.; Savit, C.; Wood, T.G.; Widen, S.G.; et al. Characterization of three new insect-specific flaviviruses: Their relationship to the mosquito-borne flavivirus pathogens. *Am. J. Trop. Med. Hyg.* **2018**, *98*, 410–419. [CrossRef] [PubMed]

9. Blitvich, B.; Firth, A. Insect-Specific Flaviviruses: A Systematic Review of Their Discovery, Host Range, Mode of Transmission, Superinfection Exclusion Potential and Genomic Organization. *Viruses* **2015**, *7*, 1927–1959. [CrossRef] [PubMed]

10. Alkan, C.; Zapata, S.; Bichaud, L.; Moureau, G.; Lemey, P.; Firth, A.E.; Gritsun, T.S.; Gould, E.A.; de Lamballerie, X.; Depaquit, J.; et al. Ecuador Paraiso Escondido virus, a new flavivirus isolated from New World sandflies in Ecuador is the first representative of a novel clade in the genus *Flavivirus*. *J. Virol.* **2015**, *89*, 11773–11785. [CrossRef] [PubMed]

11. Halbach, R.; Junglen, S.; van Rij, R.P. Mosquito-specific and mosquito-borne viruses: Evolution, infection, and host defense. *Curr. Opin. Insect. Sci.* **2017**, *22*, 16–27. [CrossRef] [PubMed]

12. Bolling, B.; Weaver, S.; Tesh, R.; Vasilakis, N. Insect-Specific Virus Discovery: Significance for the Arbovirus Community. *Viruses* **2015**, *7*, 4911–4928. [CrossRef]

13. Blitvich, B.; Firth, A. A Review of Flaviviruses that Have No Known Arthropod Vector. *Viruses* **2017**, *9*, 154. [CrossRef] [PubMed]

14. Filomatori, C.V.; Lodeiro, M.F.; Alvarez, D.E.; Samsa, M.M.; Pietrasanta, L.; Gamarnik, A.V. A 5' RNA element promotes dengue virus RNA synthesis on a circular genome. *Gene Dev.* **2006**, *20*, 2238–2249. [CrossRef] [PubMed]

15. Alvarez, D.E.; Ezcurra, A.L.D.L.; Fucito, S.; Gamarnik, A.V. Role of RNA structures present at the 3' UTR of dengue virus on translation, RNA synthesis, and viral replication. *Virology* **2005**, *339*, 200–212. [CrossRef] [PubMed]

16. Manzano, M.; Reichert, E.D.; Polo, S.; Falgout, B.; Kasprzak, W.; Shapiro, B.A.; Padmanabhan, R. Identification of Cis-Acting Elements in the 3'-Untranslated Region of the Dengue Virus Type 2 RNA That Modulate Translation and Replication. *J Biol. Chem.* **2011**, *286*, 22521–22534. [CrossRef]

17. Brinton, M.A.; Basu, M. Functions of the 3' and 5' genome RNA regions of members of the genus Flavivirus. *Virus Res.* **2015**, *206*, 108–119. [CrossRef]

18. Ng, W.; Soto-Acosta, R.; Bradrick, S.; Garcia-Blanco, M.; Ooi, E. The 5' and 3' untranslated regions of the flaviviral genome. *Viruses* **2017**, *9*, 137. [CrossRef] [PubMed]

19. Villordo, S.M.; Carballeda, J.M.; Filomatori, C.V.; Gamarnik, A.V. RNA Structure Duplications and Flavivirus Host Adaptation. *Trends Microbiol.* **2016**, *24*, 270–283. [CrossRef] [PubMed]

20. de Borba, L.; Villordo, S.M.; Marsico, F.L.; Carballeda, J.M.; Filomatori, C.V.; Gebhard, L.G.; Pallarés, H.M.; Lequime, S.; Lambrechts, L.; Vargas, I.S.; et al. RNA Structure Duplication in the Dengue Virus 3' UTR: Redundancy or Host Specificity? *mBio* **2019**, *10*, e02506–18. [CrossRef]

21. de Bernardi Schneider, A.; Wolfinger, M.T. Musashi binding elements in Zika and related Flavivirus 3'UTRs: A comparative study in silico. *bioRxiv* **2018**. [CrossRef]

22. Rice, C.M.; Lenches, E.M.; Eddy, S.R.; Shin, S.; Sheets, R.; Strauss, J. Nucleotide sequence of yellow fever virus: Implications for flavivirus gene expression and evolution. *Science* **1985**, *229*, 726–733. [CrossRef] [PubMed]

23. Hahn, C.S.; Hahn, Y.S.; Rice, C.M.; Lee, E.; Dalgarno, L.; Strauss, E.G.; Strauss, J.H. Conserved elements in the 3' untranslated region of flavivirus RNAs and potential cyclization sequences. *J. Mol. Biol.* **1987**, *198*, 33–41. [CrossRef]

24. Villordo, S.M.; Alvarez, D.E.; Gamarnik, A.V. A balance between circular and linear forms of the dengue virus genome is crucial for viral replication. *RNA* **2010**, *16*, 2325–2335. [CrossRef] [PubMed]

25. de Borba, L.; Villordo, S.M.; Iglesias, N.G.; Filomatori, C.V.; Gebhard, L.G.; Gamarnik, A.V. Overlapping local and long-range RNA-RNA interactions modulate dengue virus genome cyclization and replication. *J. Virol.* **2015**, *89*, 3430–3437. [CrossRef] [PubMed]

26. Pijlman, G.P.; Funk, A.; Kondratieva, N.; Leung, J.; Torres, S.; Van der Aa, L.; Liu, W.J.; Palmenberg, A.C.; Shi, P.Y.; Hall, R.A.; et al. A highly structured, nuclease-resistant, noncoding RNA produced by flaviviruses is required for pathogenicity. *Cell Host Microbe* **2008**, *4*, 579–591. [CrossRef] [PubMed]

27. Akiyama, B.M.; Eiler, D.; Kieft, J.S. Structured RNAs that evade or confound exonucleases: Function follows form. *Curr. Opin. Struct. Biol.* **2016**, *36*, 40–47. [CrossRef] [PubMed]

28. Moon, S.L.; Anderson, J.R.; Kumagai, Y.; Wilusz, C.J.; Akira, S.; Khromykh, A.A.; Wilusz, J. A noncoding RNA produced by arthropod-borne flaviviruses inhibits the cellular exoribonuclease XRN1 and alters host mRNA stability. *RNA* **2012**, *18*, 2029–2040. [CrossRef]

29. Roby, J.A.; Pijlman, G.P.; Wilusz, J.; Khromykh, A.A. Noncoding subgenomic flavivirus RNA: Multiple functions in West Nile virus pathogenesis and modulation of host responses. *Viruses* **2014**, *6*, 404–427. [CrossRef]

30. Jones, C.I.; Zabolotskaya, M.V.; Newbury, S.F. The 5′-> 3′ exoribonuclease Xrn1/Pacman and its functions in cellular processes and development. *Wiley Interdiscip. Rev. RNA* **2012**, *3*, 455–468. [CrossRef]

31. Antic, S.; Wolfinger, M.T.; Skucha, A.; Hosiner, S.; Dorner, S. General and miRNA-mediated mRNA degradation occurs on ribosome complexes in Drosophila cells. *Mol. Cell Biol.* **2015**, doi:10.1128/MCB.01346-14. [CrossRef]

32. Kieft, J.S.; Rabe, J.L.; Chapman, E.G. New hypotheses derived from the structure of a flaviviral Xrn1-resistant RNA: Conservation, folding, and host adaptation. *RNA Biol.* **2015**, *12*, 1169–1177. [CrossRef]

33. MacFadden, A.; O'Donoghue, Z.; Silva, P.A.; Chapman, E.G.; Olsthoorn, R.C.; Sterken, M.G.; Pijlman, G.P.; Bredenbeek, P.J.; Kieft, J.S. Mechanism and structural diversity of exoribonuclease-resistant RNA structures in flaviviral RNAs. *Nat. Commun.* **2018**, *9*, 119. [CrossRef]

34. Hussain, M.; Torres, S.; Schnettler, E.; Funk, A.; Grundhoff, A.; Pijlman, G.P.; Khromykh, A.A.; Asgari, S. West Nile virus encodes a microRNA-like small RNA in the 3′ untranslated region which up-regulates GATA4 mRNA and facilitates virus replication in mosquito cells. *Nucleic Acids Res.* **2012**, *40*, 2210–2223. [CrossRef] [PubMed]

35. Schuessler, A.; Funk, A.; Lazear, H.M.; Cooper, D.A.; Torres, S.; Daffis, S.; Jha, B.K.; Kumagai, Y.; Takeuchi, O.; Hertzog, P.; et al. West Nile virus noncoding subgenomic RNA contributes to viral evasion of the type I interferon-mediated antiviral response. *J. Virol.* **2012**, *86*, 5708–5718. [CrossRef] [PubMed]

36. Bidet, K.; Dadlani, D.; Garcia-Blanco, M.A. G3BP1, G3BP2 and CAPRIN1 are required for translation of interferon stimulated mRNAs and are targeted by a dengue virus non-coding RNA. *PLoS Pathog* **2014**, *10*, e1004242. [CrossRef] [PubMed]

37. Lorenz, R.; Wolfinger, M.T.; Tanzer, A.; Hofacker, I.L. Predicting RNA Structures from Sequence and Probing Data. *Methods* **2016**, *103*, 86–98. [CrossRef] [PubMed]

38. Funk, A.; Truong, K.; Nagasaki, T.; Torres, S.; Floden, N.; Melian, E.B.; Edmonds, J.; Dong, H.; Shi, P.Y.; Khromykh, A.A. RNA structures required for production of subgenomic flavivirus RNA. *J. Virol.* **2010**, *84*, 11407–11417. [CrossRef]

39. Silva, P.A.; Pereira, C.F.; Dalebout, T.J.; Spaan, W.J.; Bredenbeek, P.J. An RNA pseudoknot is required for production of yellow fever virus subgenomic RNA by the host nuclease XRN1. *J. Virol.* **2010**, *84*, 11395–11406. [CrossRef] [PubMed]

40. Sztuba-Solinska, J.; Teramoto, T.; Rausch, J.W.; Shapiro, B.A.; Padmanabhan, R.; Le Grice, S.F. Structural complexity of Dengue virus untranslated regions: cis-acting RNA motifs and pseudoknot interactions modulating functionality of the viral genome. *Nucleic Acids Res.* **2013**, *41*, 5075–5089. [CrossRef] [PubMed]

41. Chapman, E.G.; Costantino, D.A.; Rabe, J.L.; Moon, S.L.; Wilusz, J.; Nix, J.C.; Kieft, J.S. The structural basis of pathogenic subgenomic flavivirus RNA (sfRNA) production. *Science* **2014**, *344*, 307–310. [CrossRef] [PubMed]

42. Akiyama, B.M.; Laurence, H.M.; Massey, A.R.; Costantino, D.A.; Xie, X.; Yang, Y.; Shi, P.Y.; Nix, J.C.; Beckham, J.D.; Kieft, J.S. Zika virus produces noncoding RNAs using a multi-pseudoknot structure that confounds a cellular exonuclease. *Science* **2016**, *354*, 1148–1152. [CrossRef]

43. Rauscher, S.; Flamm, C.; Mandl, C.W.; Heinz, F.X.; Stadler, P.F. Secondary structure of the 3′-noncoding region of flavivirus genomes: comparative analysis of base pairing probabilities. *RNA* **1997**, *3*, 779–791. [PubMed]

44. Hofacker, I.L.; Fekete, M.; Flamm, C.; Huynen, M.A.; Rauscher, S.; Stolorz, P.E.; Stadler, P.F. Automatic detection of conserved RNA structure elements in complete RNA virus genomes. *Nucleic Acid Res.* **1998**, *26*, 3825–3836. [CrossRef] [PubMed]

45. Witwer, C.; Rauscher, S.; Hofacker, I.L.; Stadler, P.F. Conserved RNA secondary structures in Picornaviridae genomes. *Nucleic Acid Res.* **2001**, *29*, 5079–5089. [CrossRef]

46. Hofacker, I.L.; Stadler, P.F.; Stocsits, R.R. Conserved RNA secondary structures in viral genomes: A survey. *Bioinformatics* **2004**, *20*, 1495–1499. [CrossRef]

47. Thurner, C.; Witwer, C.; Hofacker, I.L.; Stadler, P.F. Conserved RNA secondary structures in Flaviviridae genomes. *J. Gen. Virol.* **2004**, *85*, 1113–1124. [CrossRef]

48. Katoh, K.; Standley, D.M. MAFFT multiple sequence alignment software version 7: Improvements in performance and usability. *Mol. Biol. Evol.* **2013**, *30*, 772–780. [CrossRef]

49. Nguyen, L.T.; Schmidt, H.A.; von Haeseler, A.; Minh, B.Q. IQ-TREE: A fast and effective stochastic algorithm for estimating maximum-likelihood phylogenies. *Mol. Biol. Evol.* **2014**, *32*, 268–274. [CrossRef] [PubMed]

50. Nawrocki, E.P.; Eddy, S.R. Infernal 1.1: 100-fold Faster RNA Homology Searches. *Bioinformatics* **2013**, *29*, 2933–2935. [CrossRef]

51. Waldl, M.; Thiel, B.C.; Ochsenreiter, R.; Holzenleiter, A.; de Araujo Oliveira, J.V.; Walter, M.E.M.; Wolfinger, M.T.; Stadler, P.F. TERribly Difficult: Searching for Telomerase RNAs in Saccharomycetes. *Genes* **2018**, *9*, 372. [CrossRef] [PubMed]

52. Will, S.; Reiche, K.; Hofacker, I.L.; Stadler, P.F.; Backofen, R. Inferring noncoding RNA families and classes by means of genome-scale structure-based clustering. *PLoS Comp. Biol.* **2007**, *3*, e65. [CrossRef] [PubMed]

53. Lorenz, R.; Bernhart, S.H.; Zu Siederdissen, C.H.; Tafer, H.; Flamm, C.; Stadler, P.F.; Hofacker, I.L. ViennaRNA Package 2.0. *Algorithm Mol. Biol.* **2011**, *6*, 26. [CrossRef] [PubMed]

54. Wolfinger, M.T.; Fallmann, J.; Eggenhofer, F.; Amman, F. ViennaNGS: A toolbox for building efficient next-generation sequencing analysis pipelines. *F1000Research* **2015**, *4*. [CrossRef]

55. Wolfinger, M.T. Bio::RNA::RNAaliSplit 0.09. 2019. Available online: https://github.com/mtw/Bio-RNA-RNAaliSplit (accessed on 23 March 2019).

56. Weinberg, Z.; Breaker, R.R. R2R-software to speed the depiction of aesthetic consensus RNA secondary structures. *BMC Bioinform.* **2011**, *12*, 3. [CrossRef]

57. Huerta-Cepas, J.; Serra, F.; Bork, P. ETE 3: Reconstruction, analysis, and visualization of phylogenomic data. *Mol. Biol. Evol.* **2016**, *33*, 1635–1638. [CrossRef] [PubMed]

58. Gritsun, T.; Gould, E. Origin and Evolution of 3′UTR of Flaviviruses: Long Direct Repeats as A Basis for the Formation of Secondary Structures and Their Significance for Virus Transmission. *Adv. Virus Res.* **2006**, *69*, 203–248.

59. Gritsun, D.J.; Jones, I.M.; Gould, E.A.; Gritsun, T.S. Molecular Archaeology of Flaviviridae Untranslated Regions: Duplicated RNA Structures in the Replication Enhancer of Flaviviruses and Pestiviruses Emerged via Convergent Evolution. *PLoS ONE* **2014**, *9*, e92056. [CrossRef]

60. Gritsun, T.; Gould, E. The 3′ untranslated regions of Kamiti River virus and Cell fusing agent virus originated by self-duplication. *J. Gen. Virol.* **2006**, *87*, 2615–2619. [CrossRef]

61. Hoshino, K.; Isawa, H.; Tsuda, Y.; Yano, K.; Sasaki, T.; Yuda, M.; Takasaki, T.; Kobayashi, M.; Sawabe, K. Genetic characterization of a new insect flavivirus isolated from *Culex pipiens* mosquito in Japan. *Virology* **2007**, *359*, 405–414. [CrossRef]

62. Clarke, B.; Roby, J.; Slonchak, A.; Khromykh, A. Functional non-coding RNAs derived from the flavivirus 3′ untranslated region. *Virus Res.* **2015**, *206*, 53–61. [CrossRef]

63. Moon, S.L.; Blackinton, J.G.; Anderson, J.R.; Dozier, M.K.; Dodd, B.J.; Keene, J.D.; Wilusz, C.J.; Bradrick, S.S.; Wilusz, J. X1 Stalling in the 5′UTR of Hepatitis C Virus and Bovine Viral Diarrhea Virus Is Associated with Dysregulated Host mRNA Stability. *PLoS Pathog.* **2015**, *11*, e1004708. [CrossRef]

64. Charley, P.A.; Wilusz, C.J.; Wilusz, J. Identification of Phlebovirus and Arenavirus RNA sequences that stall and repress the exoribonuclease XRN1. *J. Biol. Chem.* **2018**, *293*, 285–295. [CrossRef] [PubMed]

65. Steckelberg, A.L.; Akiyama, B.M.; Costantino, D.A.; Sit, T.L.; Nix, J.C.; Kieft, J.S. A folded viral noncoding RNA blocks host cell exoribonucleases through a conformationally dynamic RNA structure. *Proc. Natl. Acad. Sci. USA* **2018**, *115*, 6404–6409. [CrossRef] [PubMed]

66. Steckelberg, A.L.; Vicens, Q.; Kieft, J.S. Exoribonuclease-Resistant RNAs Exist within both Coding and Noncoding Subgenomic RNAs. *mBio* **2018**, *9*, e02461-18. [CrossRef] [PubMed]

67. Iwakawa, H.o.; Mizumoto, H.; Nagano, H.; Imoto, Y.; Takigawa, K.; Sarawaneeyaruk, S.; Kaido, M.; Mise, K.; Okuno, T. A viral noncoding RNA generated by cis-element-mediated protection against 5′- 3′ RNA decay represses both cap-independent and cap-dependent translation. *J. Virol.* **2008**, *82*, 10162–10174. [CrossRef] [PubMed]

68. Flobinus, A.; Hleibieh, K.; Klein, E.; Ratti, C.; Bouzoubaa, S.; Gilmer, D. A viral noncoding RNA complements a weakened viral RNA silencing suppressor and promotes efficient systemic host infection. *Viruses* **2016**, *8*, 272. [CrossRef] [PubMed]

Article

Protein Structure-Guided Hidden Markov Models (HMMs) as A Powerful Method in the Detection of Ancestral Endogenous Viral Elements

Heleri Kirsip [1,*] and Aare Abroi [2,*]

[1]   Department of Bioinformatics, University of Tartu, Tartu, 51010, Riia 23, Estonia
[2]   Institute of Technology, University of Tartu, Tartu, 50411, Nooruse 1, Estonia
*   Correspondence: heleri16@ut.ee (H.K.); abroi@ut.ee (A.A.)

Received: 29 January 2019; Accepted: 27 March 2019; Published: 2 April 2019

Abstract: It has been believed for a long time that the transfer and fixation of genetic material from RNA viruses to eukaryote genomes is very unlikely. However, during the last decade, there have been several cases in which "virus-to-host" gene transfer from various viral families into various eukaryotic phyla have been described. These transfers have been identified by sequence similarity, which may disappear very quickly, especially in the case of RNA viruses. However, compared to sequences, protein structure is known to be more conserved. Applying protein structure-guided protein domain-specific Hidden Markov Models, we detected homologues of the Virgaviridae capsid protein in Schizophora flies. Further data analysis supported "virus-to-host" transfer into Schizophora ancestors as a single transfer event. This transfer was not identifiable by BLAST or by other methods we applied. Our data show that structure-guided Hidden Markov Models should be used to detect ancestral virus-to-host transfers.

Keywords: endogenous viral elements; bioinformatics; horizontal gene transfer; virus-to-host gene transfer; HMM; tobacco mosaic virus; *Drosophila*; capsid protein

---

1. Introduction

Viruses are one of the most abundant and prevalent biological entities on Earth and thus are an important and integral part of the biosphere (for viral importance in environments, see [1–4]). The importance of viruses for humans and the biosphere is illustrated by the fact that, in our bodies, there are 10 times more bacteria than human cells (~10^{13} cells), and the number of viruses is even higher than the number of bacteria by an order of magnitude [5]. The same prevalence also applies to viruses in other biotopes; for example, there are approximately 10^8 viruses in one litre of water [6] and approximately 10^6–10^7 viruses per m^3 in the atmosphere [7]. In most biotopes, the number of viruses exceeds the number of prokaryotes by an order of magnitude [4]. The overall virus–host interaction is an important aspect driving viral evolution and the role of viruses in evolution. These roles are not restricted to the host–parasite arms race. It has been known for a long time that bacteriophages mediate transduction in bacteria. It is also known that retroviruses integrate into their hosts' genomes and thus can affect the genomic organisation of the vertebrate's cells [8]. However, Retroviridae is one viral family among approximately one hundred viral families that infect eukaryotes, and vertebrates constitute only a small fraction of eukaryotes. What is less known is that other, non-retroviral viruses are able to do the same, i.e., integrate viral genes into the chromosomes of infected cells, both in vertebrates and in other eukaryotic cells, of course not in such direct way as retroviruses. [9,10]. The viral genome elements that have integrated into eukaryotic genomes and have become fixed are called endogenous viral elements (EVEs) [9]. It has been shown that both cytoplasm- and nucleus-replicating non-retroviral viruses with different replication strategies ((+)ssRNA, (−)ssRNA, dsRNA, ssDNA and dsDNA viruses) can integrate into animal [11,12], plant [13] and fungal [14] genomes (for a thorough review, see [15]). Thus, the integration of viral genes into eukaryotic genomes is quite widespread. In contrast, the function and expression of retroviral EVEs have only been thoroughly researched in model organisms [16], and for non-retroviral EVEs, additional analyses of EVE function and expression have been performed only for human and squirrel endogenous bornavirus-like nucleoprotein (EBLN) elements [17–19] and tobacco plant geminivirus-related DNA sequences (GRDs) [9,20]. The exact pathway by which viral genes integrate and become fixed in eukaryotic organisms is not clear. However, to be effectively spread over the population, EVEs must also benefit the organism (although spreading via genetic drift, founder effects and population bottlenecks cannot be excluded).

The main strategy for detecting EVEs in organisms was formalised by Kondo et al. [21]. The detection protocol for EVEs consists of three main stages: (a) EVE detection with bioinformatics tools, mostly with BLAST searches in different databases; (b) EVE detection and confirmation in experiments, mostly with PCR and DNA sequencing to confirm integration; and (c) EVE phylogeny construction combined with viral sequences to identify EVEs closest relatives and to evaluate the integration time.

Following this protocol, scientists have discovered a large amount of viral-like sequences in different eukaryotic genomes. However, there is a problem with detecting distant homologues, which was noted by Brenner et al.; BLAST does not work well with sequences under 30% of identity [22]. Park et al. [23] showed that Intermediate Sequence Search (ISS) method, Hidden Markov Models (HMMs) and PSI-BLAST work considerably more effectively than pairwise sequence comparison methods (e.g., BLAST), especially when the sequences have changed greatly over time. Profile-HMMs are probabilistic models that can be used to describe the patterns shared by sets of protein/domain sequences (HMMs are generated from constructed multiple sequence alignments (MSAs), and the generated model covers both the diversity and conservation within the input MSA). It is possible to search for a sequence that is the most similar to the model, which is especially important when dealing with viral sequences because the virus mutation speed is much higher, especially in RNA viruses (approximately 10^{-2} to 10^{-7} substitutions per position per year compared to about 10^{-9} substitutions per position per year in mammals [24–26]), and confident pairwise similarity between homologous sequences may disappear. Thus, profile-HMMs should detect distant homologous sequences more efficiently.

Advantages of other methods in comparison with BLAST in the case of viruses have been shown in practise. Kuchibhatla et al. started from the assumptions above and used sequence-profile comparisons (such as HMMER3) and profile–profile comparisons (HHpred) in addition to BLAST analyses [27]. They showed that sequences that were previously classified as "ORFans" (more accurately "taxonomically restricted to only one taxon") according to BLAST analyses, had distant homologues in various viruses and that this approach could also be applied to organisms. Thus, considering the fast evolution of viruses, profile-based algorithms that can detect distant homologues should be more suitable for detecting (distant) homologues of viral proteins [28,29].

It is also known that protein structure is more conserved than sequence [30,31]. Therefore, the protein structure is also important to consider when detecting distant homologous sequences. When dealing with the identification of similarities between evolutionarily distant, but homologous sequences, taking protein structure-based information into account may improve the results. Challis and Schmidler showed that, when including structural information, the phylogenetic inference for distant relationships improves [32]. Additionally, Herman et al. showed that structure contains more information that can be obtained from sequences; hence, it vastly reduces the alignment and phylogenetic tree topology uncertainty [33]. This finding is especially important in the case of viruses, which lose sequence similarity within a relatively short time, especially compared to organisms. Thus, structural information should be included when studying deep evolutionary relationships.

In this study, we tested whether a complex method, such as profile-HMM together with protein structure-guided information, is a suitable and effective first step in EVE detection and whether it should be implemented in the EVE-detection protocol. To complete this aim, a comparison between BLAST and profile-HMM search results was made with the objective of finding whether profile-HMMs could detect more distant EVEs and how many more EVEs are identified when implementing a structure-guided profile-HMM search. This work shows that using profile-HMMs with additional structure-guided information enables the detection of more distant homologous sequences; in this case, endogenous *Tobamovirus* (family Virgaviridae; +ssRNA viruses) coat protein-like (eTCPL) elements in different fly genomes that have not been detected previously by using pairwise sequence comparison method BLAST.

2. Materials and Methods

2.1. EVE Detection Using BLAST Analysis

To detect eukaryotic endogenous *Tobamovirus* coat protein-like (eTCPL) elements, the BLAST method was used (Figure 1). Different *Tobamovirus* (family Virgaviridae, +ssRNA viruses), *Hordeivirus* (family Virgaviridae, +ssRNA viruses), *Tobravirus* (family Virgaviridae, +ssRNA viruses) and *Pecluvirus* (family Virgaviridae, +ssRNA viruses) coat proteins (CPs) were used as queries (the sequences used are shown in Table S1) in NCBI BLASTp and tBLASTn (protein sequence against the nucleotide database translated into 6 reading frames) (BLAST+ 2.2.30+, updated 6.10.2014; using the NCBI BLAST server http://blast.ncbi.nlm.nih.gov/ [34]) searches in eukaryotes (NCBI Taxonomy ID 2759). The search was performed against nucleotide collection (nt/nr), Transcriptome Shotgun Assembly (TSA), expressed sequence tag (Est) and whole-genome shotgun contig (WGS) databases using default parameters with an additional implemented E-value threshold of 1×10^{-5} (May 2017) (results in Table S2). additionally, BLASTp (protein sequence against protein database) was run using the same query sequences and search parameters (including an E-value threshold of 1×10^{-5}) against a non-redundant protein database ("nr", which also includes most of the UniProtKB sequences). Searches and databases used are visualised in Figure 1.

2.2. EVE Detection Using the Profile-HMM Method

As an alternative method, searches were also performed with the HMMER3 package (Figure 1) (version 1.8; using the web server http://www.ebi.ac.uk/Tools/hmmer/ [35]). The searches were performed using the same *Tobamovirus*, *Hordeivirus*, *Tobravirus* and *Pecluvirus* CP sequences as described above for BLAST analyses with taxonomic restriction to Eukaryotes (NCBI Taxonomy ID 2759) using the default parameters with an additional E-value threshold of 1×10^{-7}.

In the "phmmer" search type (protein sequence against protein sequence database), the databases used were UniProtKB (version v.2017_05), Swiss-Prot (version v.2017_05), Ensemble All (version v.88) and Ensemble Genomes (version v.35) because no other databases were available during the analyses (May 2017) (results in Table S3).

In "hmmscan" (protein sequence against the profile-HMM database), viral sequences were used to search against protein families in the Pfam-A [36] and SUPERFAMILY [37] databases (April 2017). Both resources contain protein domain-specific HMM profiles and assign these profiles to available protein sequences in different databases. The "hmmscan" search results give the protein domain-specific HMM profile ID. Further, it is also possible to analyse these HMM profiles and the sequences that are assigned to the profile. Eukaryotic sequences assigned to these profiles can be analysed as possible eTCPLs. The E-value threshold for belonging to the respective domain family was set to 1×10^{-7} (results in Table S4).

For the "hmmsearch" search (protein alignment/profile-HMM against protein sequence database), the *Tobamovirus* CP MSA and the *Tobamovirus-Pecluvirus-Tobravirus-Hordeivirus* CP MSA were constructed using the MUSCLE algorithm (with default parameters) within MEGA (version 7.0.20; [38]).

The search was performed in the UniProtKB and Swiss-Prot databases (April 2017) (results in Table S5). Next, "hmmsearch" was performed using previous "hmmscan" results (predefined protein family models from the Pfam-A/SUPERFAMILY databases) as queries (May 2017) (results in Table S6).

2.3. Testing for False Positive Hits with Alternative Methods

As stated by Pearson, "When a scientifically unexpected alignment appears to be statistically significant, investigators should consider alternate strategies for estimating statistical significance" [39]; we applied alternative methods to confirm the homology of high scoring hits (Figure 2). First, to test for annotation artefacts, database annotations of sequence's scaffold/contig/chromosome (their origin), overall length, sequence location and surrounding area were scanned. Data confirmation was needed to exclude the possibility of sample contamination that could have occurred during cloning or sequence assembly and/or a misannotation of the protein itself (for example, nucleic acid is isolated from tissues infected with a virus, but all sequences are still annotated as host sequences).

Second, to exclude false positive hits (sequences that are more similar to other sequences or domains), a reciprocal sequence similarity search was performed with default parameters using tBLASTn for protein sequences (against the NCBI nr/nt database) and BLASTn (nucleotide sequence against the nucleotide database) for nucleotide sequences (against the NCBI nr/nt database). When the reciprocal search gives the most significant hit to primary query sequences or their close relatives (in viruses in the same viral genus or family), it is noted as such in all of the tables and in other cases, the best hit is described; additionally, "no hits" are marked with "−". Other viral hits that were not closely related to Virgaviridae viruses were classified as false positives for eTCPL but will remain possible positive hits for other EVEs and will be analysed in the future. Additionally, a "hmmscan" (protein sequence against the profile-HMM database) search was performed on the HMMER webpage to test whether the sequence belongs to another protein domain family with much higher confidence.

Third, to determine possible false positive results with alternative methods, protein structure prediction was performed with the LOMETS [40] meta-server for all remaining positive hits from both the BLAST and HMMER3 analyses. Additionally, this step was important for the positive hits (both eukaryotic and viral hits) obtained from the SUPERFAMILY and Pfam-A databases to confirm that profile-HMM models have not falsely assigned sequences to the models. The LOMETS meta-server simultaneously uses many different algorithms to predict possible protein structures from sequences using the protein threading method. If the sequence is a real homologue, the server should recognise the tobamoviral coat protein or its structural relatives in SCOP, i.e., the protein domains with "SCOP concise classification string" (SCCS) (also named "SCOP superfamily identifier") staring with "a.24.5" as a template for high-confidence models. Additionally, the confidence score, the alignment length and the template coverage should be high, and at least one algorithm not based on HMMs should give a high confidence score. In addition, the Z-score was considered, for which a higher score is considered to be better than a lower score. All possible eTCPL hits that were not classified as false positives or non-determined (ND) were used in the phylogenetic analyses.

In general, all the sequences that reciprocal BLAST classified as similar to tobamoviruses or other Virgaviridae but only had RNA data available were classified as ND because we cannot distinguish between actual viral sequences and new EVEs.

2.4. Phylogenetic Analysis for eTCPL

For the phylogenetic analysis, a joint list of viral coat protein sequences and eukaryotic EVE sequences were used. Viral sequences were taken from the NCBI full genomes database, belonging to tobamoviruses, pecluviruses, hordeiviruses, tobraviruses and goraviruses (Table S1).

The eukaryotic sequences used in this study are from two datasets. The first dataset consists of eukaryotic sequences from the SUPERFAMILY database that have been assigned to eTCPL elements (Table S7). Sequences that were included in the phylogeny were those that remained after the false positives were excluded. For these sequences, only the eTCPL region that was assigned to the SUPERFAMILY HMM model was used. The second dataset (search results) consists of the HMMER analysis results (Table S5) that were obtained when using Virgaviridae viral MSAs as queries. One additional sequence (*Lasioglossum albipes*; NCBI accession ID: ANOB01025386.1), which was the one positive hit from the tBLASTn analyses using the viral sequence as a query, was added to the second dataset. All phylogenetic analyses were performed using the protein domain part of the sequences (*Drosophila melanogaster*, FlyBase gene ID FBgn0029799, UniProt ID Q9W483, protein region 122–259). To determine the domain part of the sequences (where the domain part was not previously designated), the two datasets were merged and aligned within MEGA using ClustalW and were edited using the first eukaryotic dataset as a baseline to cut out the regions that were outside of the eTCPL protein domain. Additionally, Jalview [41] was used to remove identical sequences (remove redundancy threshold of 100%).

All the results were additionally filtered to ensure data quality because most of the partial domains in proteins are alignment and/or annotation artefacts [42]. Protein sequences shorter than 130 AA were removed and not used in the MSA. All sequences were aligned using the Mafft alignment algorithm (version 7) with the default parameters [43]. To select the best model for phylogeny construction, the ProtTest server was used [44]. The best model was selected using the Akaike information criterion framework: LG + I + G + F. The phylogenetic tree was constructed using the MEGA program package with maximum likelihood methods using LG with frequencies (+F) model along with the gamma distribution with invariant sites (G + I). The number of discrete gamma categories was 5. Gaps and missing data were dealt with via partial deletion using a site coverage cut-off of 95%. The initial tree was performed automatically (using the NJ/BioNJ method). For the phylogeny test, the bootstrap method (500 replications) was used.

2.5. Additional BLAST Analysis to Broaden the Search to Databases that Are not Accessible in HMMER3

For the final step in detecting EVEs, an additional BLAST analysis was performed to identify potential EVEs not yet annotated as proteins (Figure 1). This analysis was done because the HMMER3 web server has few directly connected databases, none of which are nucleotide databases. Hence, performing an additional BLAST analysis enables us to broaden the phylogenomic coverage to incomplete genomes and/or nonannotated proteins. Eukaryotic sequences (Table S8), obtained through protein domain databases (SUPERFAMILY), were used as queries to search for eukaryotes (NCBI Taxonomy ID 2759) using tBLASTn with the same parameters described previously and with E-value 1×10^{-7}.

2.6. eTCPL Synteny in Complete Fly Genomes

To determine the possible integration events in further detail, completely sequenced and annotated genomes that harbour potential eTCPLs were explored in more detail. From all the results, only *Drosophila* species were fully sequenced; thus, they were used in the synteny analysis. All the work was performed using the FlyBase database [45] and its genome browser (June 2017). eTCPL genes in corresponding fly genomes were determined, and the gene regions surrounding it on both sides were analysed. Additionally, the copy number of the eTCPLs, the localisation of the surrounding genes and the mobile element existing near the eTCPL site were included in the analysis.

3. Results

3.1. Searching for EVEs Using "Sequence versus Sequence" Search Algorithms (BLAST and "Phmmer")

First, to identify potential homologues of *Tobamovirus* coat proteins (tobamo-CPs) from eukaryotic genomes, a BLAST analysis [34], as a standard protocol for detecting EVEs, was used. In general, the viral protein sequences were used as queries in BLAST analyses (a list of viral sequences used can be found in Table S1), but the analyses gave no informative *Tobamovirus*-related eukaryotic results (Figure 1 and Table S2). In BLASTp (protein vs. protein search), one result (UniProt ID: P93362_TOBAC) had a significant E-value and bit-score; however, closer examination of the annotation showed that its viral sequences have been annotated as a part of the *Nicotiana tabacum* plant genome.

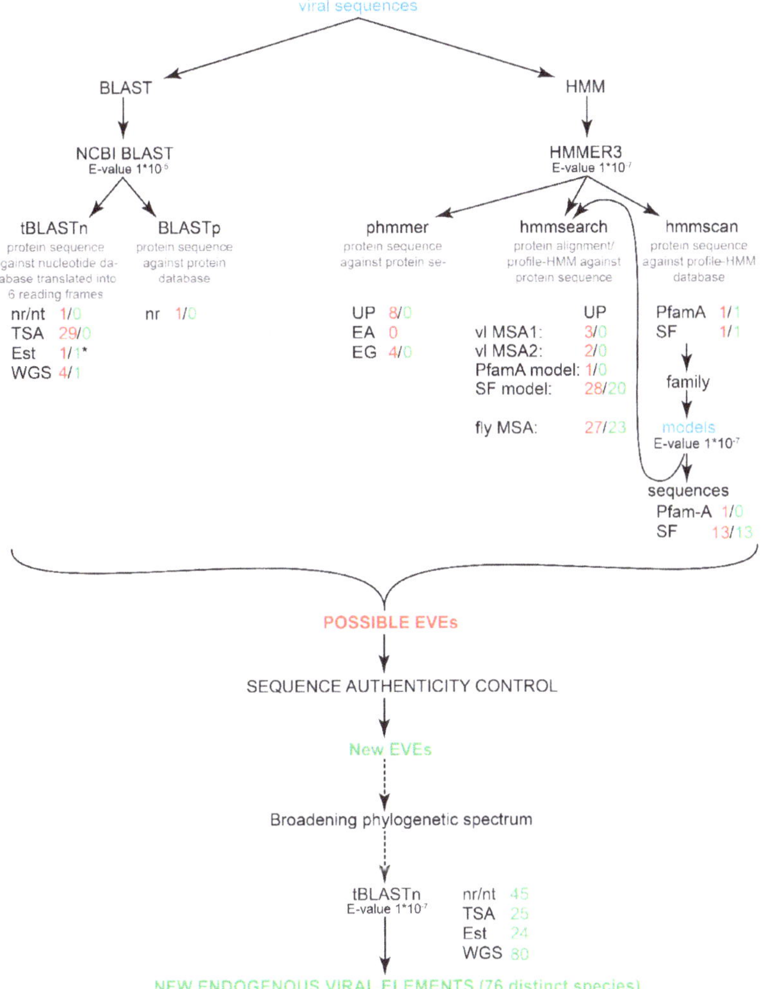

Figure 1. Workflow of the identification of new EVEs. The first step is the search against eukaryotic sequences from different databases (the Swiss-Prot database results are left out because this database gave no hits) using viral sequences as a query. All the search inputs (sequences and models) are indicated in blue script. All the hits are then examined to determine false positives and mis-annotations. All the numbers after the databases (written in red) indicate the distinct hit IDs that were collected from the databases. The second number (written in green) indicates (after all the noted sequence authenticity controls) how many possible EVEs were identified from the databases. The "*" denotes one recombinant sequence that was found in the analyses. "vl MSA1" denotes *Tobamovirus*-specific MSA, "vl MSA2" denotes *Tobamovirus-Hordeivirus-Pecluvirus-Tobravirus* specific MSA and "fly MSA" denotes an MSA that was constructed using 13 *Drosophila* fly sequences from the SUPERFAMILY database. Abbreviations: UniProtKB, UP; Ensembl All, EA; Ensembl Genomes, EG; SUPERFAMILY, SF. Note that "nr" database in BLASTp search include also UniProtKB. The sequence authenticity control workflow is described in detail in Figure 2.

We extended our search to tBLASTn (protein sequence as a query against nucleotide database translated into 6 reading frames) to find EVEs not yet annotated or not coding a protein. Using tobamo-CP as a query in tBLASTn against the nr/nt database, no true positives were found. However, against WGS (Whole Genome Sequence) database, four significant hits were obtained (Figure 1 and Table S2). In reciprocal BLAST analyses, one sequence was highly similar to Polydnaviridae, and two of the sequences seemed to be relatives of the tobamo-CP. However, since the contig length of the DNA was similar or shorter than the known *Tobamovirus* genome length, these sequences were classified as ND (non-determined because of a lack of information to confidently classify this sequence as part of the eukaryotic genome). One of these, *Lasioglossum albipes* (bee) sequence ANOB01025384.1 (NCBI accession ID) in the WGS database, however, did have a DNA scaffold over 40 Mbp (40,000 bp) in length. Protein structure prediction of the region that harbours the potential EVE indicated that the tobamoviral coat protein was the best match. However, the database annotations also state that DNA was extracted from the whole body; hence, it is currently not possible to determine whether the scaffold is part of the actual bee genome or is from some other organism found in the bee's holobiont.

We also performed a similar tBLASTn search against the TSA (Transcription Shotgun Assembly) and Est (Expressed Sequenced Tags) databases. No true positive hits were obtained because, while sequencing the whole-body RNA, no evidence was given that the RNA was transcribed from the eukaryotic genome (Figure 1 and Table S2).

To determine whether using a more complex method (while still using the protein sequence as a query) can give us more information than a pairwise comparison method, additional analyses were performed. Hidden Markov Models have been shown to identify distant homologous sequences more successfully than BLAST. HMMER3 [35] was used for this purpose because it has been shown to be more sensitive while not losing computational speed. When using the same *Tobamovirus* coat protein sequences as queries in HMMER3 "phmmer" search against the protein sequence databases, no new significant results were found (Table S3), but this result may be due to the small number of databases that can be searched against (UniProt, SwissProt and Ensemble). None of the 18 primary hits from this analysis met the other criteria described in the Materials and Methods Section.

Thus, sequence versus sequence searches gave one possible true hit (*Lasioglossum albipes*; NCBI accession ID: ANOB01025384.1). Despite expanding the search queries to include *Hordeivirus* (family Virgaviridae; +ssRNA viruses), *Pecluvirus* (family Virgaviridae; +ssRNA viruses) and *Tobravirus* (family Virgaviridae; +ssRNA viruses) capsid proteins (sequences listed in Table S1), no additional EVEs were found.

3.2. Searching for EVEs Using "MSA versus Sequence" search Algorithms ("hmmsearch")

As noted above, it has been known for a long time that profile-based methods work better than sequence-based methods, especially in distant homologous sequence detection. We used *Tobamovirus* and *Tobamovirus-Hordeivirus-Pecluvirus-Tobravirus* capsid proteins (sequences listed in Table S1) to generate MSAs and the corresponding profiles, and then we searched the sequence databases using these profiles. All these analyses gave hits (Table S5) that had been previously classified as false positives, whether by viral contamination or by belonging to other protein domains. Hence, the "MSA versus sequence" search did not result in any new EVEs (Figure 1).

3.3. Identification of EVEs Using Profile-HMMs from the Pfam-A Database

Additionally, when searching single sequences against databases that use protein domain profile-HMMs (e.g., Pfam-A, TIGRFAM, and SUPERFAMILY), HMMER3 "hmmscan" gave highly reliable results when using the tobamo-CP as a query (Table S4). The Pfam-A database gave the result of protein family "Virus coat protein (TMV like)" (Pfam-A code PF00721.20). This Pfam-A family consists of sequences of *Tobamovirus, Hordeivirus, Pecluvirus, Tobravirus* (Virgaviridae, +ssRNA virus), *Furovirus* (Virgaviridae, +ssRNA virus), *Pomovirus* (Virgaviridae, +ssRNA virus) and *Benyvirus* (Benyviridae, +ssRNA virus) coat proteins. However, this Pfam-A family does not contain any eukaryotic sequences (neither in Pfam-A version 28 nor in versions 29, 30 and 31), except for one *Nicotiana tabacum* protein (UniProt accession ID: P93362_TOBAC) that has previously identified as a viral sequence. Pfam-A version 31 is based on UniProt release 2016_10 and does not include sequences deposited after October 2016. To overcome this limitation, we used Pfam-A's own profile-HMM model as a search query against the UniProt database ("hmmsearch") on the HMMER webpage (https://www.ebi.ac.uk/Tools/hmmer/), which uses a monthly updated version of the UniProt. However, no new results were found (Table S6).

3.4. Searching for EVEs Using Structure Guided Profile-HMMs from the SUPERFAMILY Database

SUPERFAMILY is the other profile-HMM database, where the protein domain family is presented as a profile-HMM [37]. It identified "TMV-like viral coat proteins" (SUPERFAMILY accession ID: 47195 or SF_47195) as a positive hit (Table S4). SUPERFAMILY is a resource of genomic assignment of protein structural domains. This database constructs profile-HMMs for protein structural domains at the SCOP [46] superfamily level (throughout this article, we used term "superfamily" as it is defined in SCOP) and then uses completely sequenced and annotated genomes to identify distant homologous sequences that belong to the same structural superfamily. According to the SCOP authors, a superfamily level groups protein domains with a common ancestor.

According to the SCOP classification, the superfamily "TMV-like viral coat protein" (SCOP concise classification string "sccs" a.24.5) is composed of four *Tobamovirus* coat protein structures (PDB ID: 1EI7, 1VTM, 1RMV, 1CGM). Obviously, the *Tobamovirus* capsid sequences (22 genomes) are assigned to HMM models of this superfamily. Additionally, the capsid sequences of tobraviruses (three genomes), hordeiviruses (one genome) and pecluviruses (two genomes) are assigned to the "TMV-like viral coat protein" superfamily. Additionally, three sequences from Bymoviruses (Potyviridae, +ssRNA virus), each from different genomes, gave highly significant hits. However, these sequences are not capsid proteins but are from a RNA2-encoded polyprotein (P2). Recall that SUPERFAMILY resources use non-redundant, complete viral genomes from NCBI as a source for viral sequences.

In addition to viruses, thirteen *Drosophila* fly species are also assigned to "TMV-like viral coat protein" superfamily (Table S7). All thirteen *Drosophila* species were assigned with very high confidence (an *E*-value less than 1×10^{-33}) and with an almost full alignment length. Thus, these sequences are most likely homologous to tobamo-CPs. Although this database uses completely sequenced and annotated genomes and the profile-HMMs are trained to be highly specific, false positives classifications should not be overlooked; hence, protein structure prediction was performed to confirm the SUPERFAMILY database results. Additionally, when using HMM models of the SF_47195 SCOP classification from the SUPERFAMILY database as queries in the "hmmsearch" on the HMMER webpage with up-to-date databases, several hits were found (Table S6). Most of these species belonged to different *Drosophila* species, and others belonged to different fly species from the Schizophora section (*Bactrocera* species, *Ceratitis capitata*, *Lucilia cuprina*, *Musca domestica*, *Stomoxys calcitrans* and *Glossina* species). Similar results were also observed when using the SUPERFAMILY *Drosophila* sequence (protein domain part)-based MSA as the "hmmsearch" query (Table S5).

3.5. Confirming SUPERFAMILY Hits as True Homologues to TMV-CP

To exclude false positives, the LOMETS metaserver [40] was chosen for protein structure prediction (Figure 2). This program uses protein threading, which attempts to fit the given sequence to previously identified protein structures, aligns each amino acid accordingly, and then it evaluates how well the target fits the template. The results of the LOMETS metaserver for these *Drosophila* sequences show that the best modelling templates, according to several different prediction algorithms (including those that do not use HMMs), belong to tobamo-CP structures. This result confirms that *Drosophila* fly sequences are highly likely to have structures very similar to tobamo-CPs and thus have a high likelihood of being real homologues.

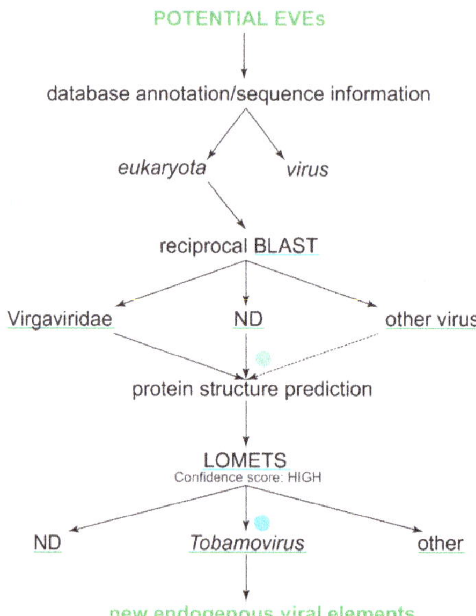

Figure 2. A detailed sequence authenticity control workflow. Each step describes what kind of results can be obtained from each step and what results can be used in further analyses. The programs used in the analyses are underlined with blue line and the results of the analyses are underlined with green line. The green dot shows that we cannot distinguish between viruses and EVEs when only RNA-seq data are available. Similarly, the blue dot shows that, even when protein structure prediction may indicate that the sequence is homologous to a certain virus, this sequence may have a close relative whose structure is not yet solved. However, the protein structure prediction still indicates that the potential EVE sequence most likely belongs to the respective structural superfamily (in terms of the SCOP).

When looking at the surrounding region of the eTCPL in *Drosophila* (and in other true positive hits for SUPERFAMILY HMMs) in more detail (i.e., when analysing contig length and the presence of other viral domains), no other indication of a misannotated viral origin was found. Thus, the *Drosophila* eTCPL is coded by the real part of the genome and not by viral sequences accidently annotated as a part of the genome. The eTCPL is found in several *Drosophila* species, which significantly reduces the probability of sequencing artefacts. In conclusion, the SUPERFAMILY HMMs found several true positives among Diptera.

3.6. Phylogenetic Analysis of Virgaviridae CP and Diptera eTCPLs

The above analysis shows that *Drosophila* and other flies contain the eTCPL domain. This result by itself does not address transfer direction. To evaluate the phylogenetic relationship of eTCPLs and exogenous viruses, we performed a comprehensive phylogenetic analysis using CP sequences of corresponding Virgaviridae viruses (tobamoviruses, tobraviruses, pecluviruses, hordeiviruses and goraviruses) (Table S1); and the integrated eTCPLs in eukaryotes (only high-confidence eTCPLs were used).

As shown in the phylogenetic tree in Figure 3, all viral genera (tobamoviruses, hordeiviruses, pecluviruses and tobraviruses) have been clearly separated according to their known phylogeny. The overall distribution of viral genera is very similar to the known divergence of these viruses, with hordeiviruses and pecluviruses being very closely related, as determined based on both replication proteins and capsid protein phylogenies of Virgaviridae [47]. These data show that the phylogenetic analysis is biologically significant and meaningful.

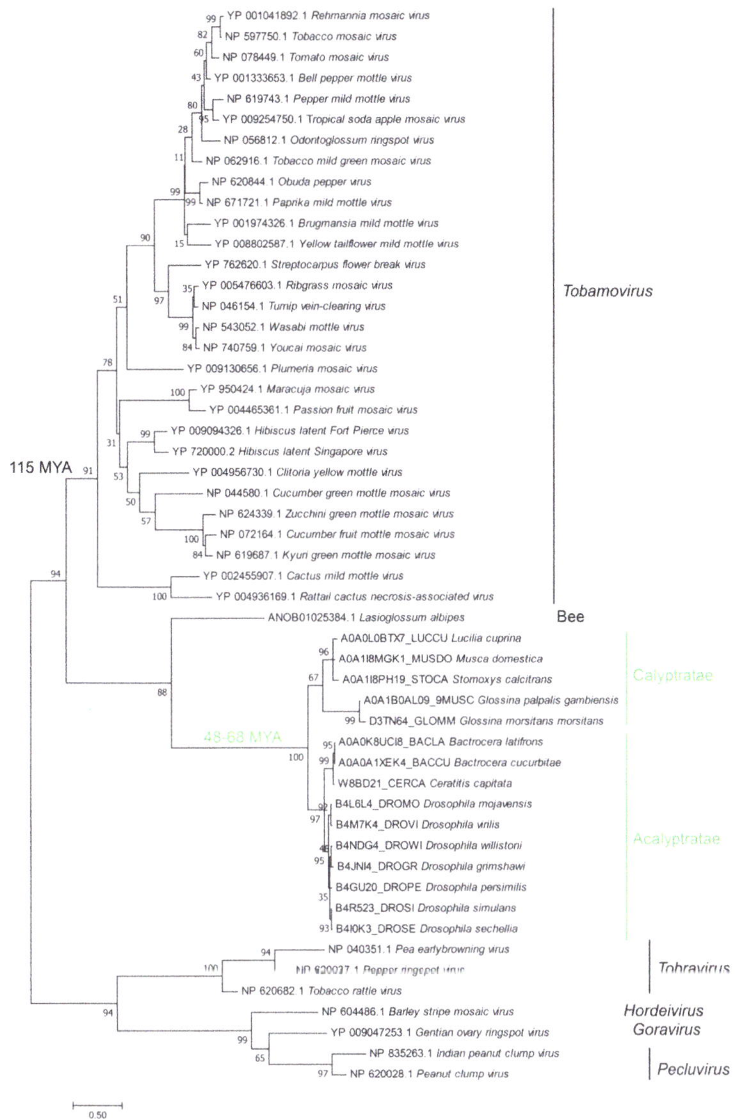

Figure 3. Phylogeny of eTCPLs and the corresponding viral sequences. The tree was constructed using the maximum likelihood method (LG + I + G + F model) using the MEGA program package [48]. All viral genera are monophyletically grouped according to their phylogeny. All the eukaryotic sequences are also grouped, and the pattern follows known fly phylogeny. The Acalyptratae-Calyptratae divergence from other flies [49–52] and *Tobamovirus* divergence from other Virgaviridae [53,54] have been noted.

All flies are distributed to two main groups based on the two subsections of Schizophora flies: Acalyptratae and Calyptratae. All the eTCPLs from different flies cluster together with very high confidence and form a sister clade with extant viruses. The phylogenetic pattern inside the eTCPL clade is consistent with the divergence of flies in the Calyptratae and Acalyptratae subsections and resembles the phylogeny of the flies itself [55].

It has been shown that the Acalyptratae-Calyptratae split occurred approximately 63–68 MYA according to three independent studies [49–51]. A more focused and through study showed that the split occurred approximately 48 MYA [52]. On the other hand, the ancestral split of *Tobamovirus* from other Virgaviridae has been shown to have occurred approximately 115 MYA [53,54]. Overall, when the phylogenetic tree is supplemented with available divergence dates, it provides strong support for the virus-to-host transfer.

3.7. eTCPL Synteny in Completely Sequenced Fly Genomes

In this work, eTCPLs have been found in many different fly genomes. Of these, only a few genomes have been fully sequenced, assembled and deposited into the FlyBase database [45]. All of them are from *Drosophila* species (Table S7). Using their assembled genomes, the eTCPLs and the surrounding region were analysed. All the eTCPLs are present in a single copy, and the protein domain itself is coded by a single exon (in *Drosophila willistoni* and *Drosophila ananassae*, the respective genes have two exons, but the eTCPL domain is coded by a single exon). The genes surrounding *Drosophila melanogaster* eTCPL are IntS6, CG4078, CG15771 and lin-52 (the eTCPL gene overlaps with the last gene but not in the coding sequence). The same trend is apparent in all of the *Drosophila* species (Figure 4). However, some changes have occurred in either *Drosophila* or the *Sophophora* subgenus. For *Drosophila virilis* and *Drosophila grimshawi*, the genes IntS6 and CG4078 are located further away and are located in different scaffolds for *Drosophila mojavensis*. Additionally, *Drosophila sechellia* acquired a new gene (GM19534) that has no known orthologues. Henceforth, a possible viral gene could have integrated in this locus before the *Drosophila* fly split.

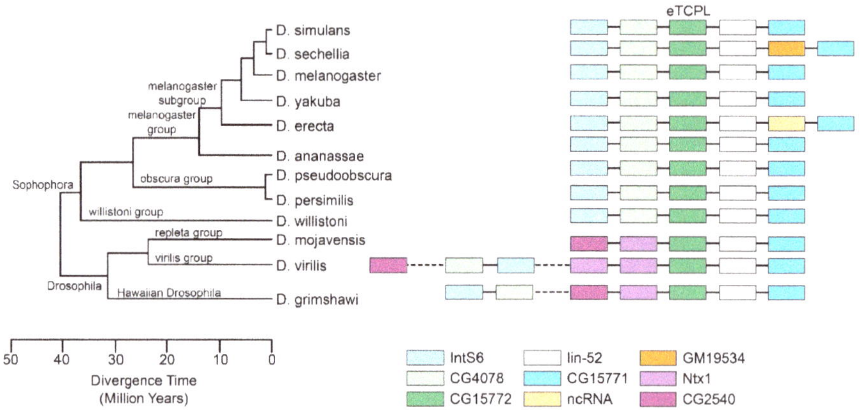

Figure 4. eTCPL gene synteny comparisons on fully sequenced genomes from *Drosophila* flies. The eTCPL is noted in dark green and was taken as an observation centre. *Drosophila melanogaster*, the most well-researched fly, was taken as a baseline and two to three genes surrounding the eTCPL were also considered. All data were collected from the FlyBase database [45], including the tree picture. In some flies whose genes surrounding the eTCPL were different, previously noted genes (IntS6 and CH4078) were also located for this analysis.

3.8. EVE Detection from Incompletely Sequenced Genomes

True positive eukaryotic hits of the SF_47195 from the SUPERFAMILY database (which uses the complete genomes and UniProt) indicated flies are hosts to eTCPLs. The next step in the analysis was an additional BLAST analysis to evaluate their phylogenomic distribution in uncomplete genomes, whose proteins may not yet be annotated.

In our approach, HMMs can recognise deeper evolutionary relationships, and BLAST will give a better phylogenomic coverage since the HMM search was limited to a few protein databases and

BLAST has a much wider possible database usage range and allows searching against nucleotide databases (including subsets that do not have annotated translations). Using the fly eTCPLs as queries, 76 different species (including 13 species from SUPERFAMILY) were identified as containing true positive eTCPLs in their genome (Table S8). The results mostly consist of both Acalyptratae and Calyptratae species (65 species belonging to the Schizophora section) covering both of their phylogenies as best as possible with the data available.

Most of the species (in the Schizophora group) also have the same genes surrounding the eTCPLs as in *Drosophila melanogaster* (genes IntS6, CG15771 and CG4078 located on the same contig as the eTCPL) with the exception of a few species (data not shown), most likely because those species are not sequenced as fully as others. This also indicates that eTCPLs' synteny shown for *Drosophilidae* (Figure 4) is also extended to Calyptratae and thus for all Schizophora.

We can use protein structure-guided profile-HMM analyses with an additional tBLASTn search to cover the un-annotated coding sequences. With this approach, it is possible to detect eTCPL sequences in a large number of eukaryotic genomes, which was not possible when using only a BLAST search.

4. Discussion

During the last decade, many EVEs have been detected in eukaryotes. According to the published literature, the number of viral families that have donated EVEs to eukaryotes is at least 34 for identified nucleic acid transfers, including at least 21 viral families where the coding potential of integrated viral sequences have remained (data collected from the literature by H. Kirsip). The recipient eukaryotic organisms have been fungi, plants, vertebrates, arthropods and others. Even when using sequence-based EVE search strategies, the impact of non-retroviral viruses in cellular genomes/proteomes is (and has been) significant. Currently applied methods do not take into account that viruses have very fast evolutionary rates, and thus, sequence similarity may not be detectable using pairwise sequence comparison methods.

The data presented in this study show that, as expected from the literature, there exist methods that are more sensitive than BLAST to detect "virus-to-host" transfers of genetic material. It should be noted that the same method could also be applied to the "host-to-virus" approach. According to the literature, HMM-based searches should be more sensitive than BLAST, and structure-guided HMMs should be even more sensitive than sequence-based HMM searches in the detection of remote homologues. In our case, BLAST found one possible homologue in eukaryotes for *Tobamovirus* capsid proteins. Additionally, using different HMMER3 search capabilities ("phmmer" and the viral MSA-based "hmmsearch", both based on HMMs) did not give any new reliable results. At the same time, structure-guided HMMs from SUPERFAMILY found distant homologues of tobamoviral proteins in eukaryotes. Our data show that, as a "proof of principle", structure-guided HMMs can be used to detect V2H transfers that are not detectable by other methods.

In modern high throughput times, the annotations of the data in different databases vary, particularly from the point of view that relevant aspects of annotations vary for different researchers (and for their scientific questions). In our case, the authenticity of sequences (their origin) must be carefully tested because very often the taxonomic annotations of the origin of a nucleic acid correspond to major species in holobionts and not to their real origin. This phenomenon has also been used by scientists in a positive way, extracting, for example, the viruses (real viruses, not EVEs) infecting insects from the data with taxonomic annotation to insects [56]. If the sequence data are not linked to publications (or if the methods of how the probe was isolated and prepared have not been properly described), it is very hard to evaluate the real origin of a sequenced nucleic acid. Some types of data, by definition, do not allow us to distinguish between viruses and EVEs (for example, RNA-seq data). Curated datasets such as complete proteomes or complete genomes seem to have fewer un-authentic sequences compared to WGS. However, using WGS and similar databases are very useful for acquiring taxonomic coverage that is as wide as possible. Of course, every method may have false positive hits;

therefore, we analysed all the search hits with multiple alternative methods to obtain more confidence results. Unfortunately, these methods are also not as "high throughput".

Structure-guided HMMs help to detect remote homology; however, detecting homology is not enough to classify the sequence as an EVE. In addition to vertical inheritance from primordial worlds, there are two processes that may lead to homologous proteins in viruses and organisms: "virus-to-host" transfer and "host-to-virus" transfer. To detect EVEs, the transfer direction of homologues found in viruses and cells must be determined. Determination of the transfer direction for more ancestral transfers is not as trivial as in the case of more recent events. In the latest case, it is possible to use the highly confident outgroup sequences to root the phylogenetic tree and polarised it on time. At the same time, when dealing with more older events, determining a good outgroup is a challenge. Studying the evolutionary history of a single viral protein (or a protein domain) selection of outgroups is complicated by rampant horizontal gene transfer between viruses. Even such an unlikely event as gene transfer between ssDNA viruses and RNA viruses has been reported [10,57,58]. Known divergence times of cellular and viral taxa would also help to polarise the phylogenetic tree. Unfortunately, very few virus group divergence times have been determined. Therefore, to determine the transfer direction, different independent approaches should be used.

To determine the transfer direction of proteins under study, we combined several approaches. First, we tested the synteny of eTCPLs in well-annotated species in the FlyBase database. The eTCPL and its surrounding area in fully sequenced genomes were analysed. The data in Figure 4 show the synteny of the eTCPL in these species. In different *Drosophila* fly genomes, the eTCPL is in the same locus, is coded by a single exon, and the surrounding genes are similar. The exception is between *Drosophila* and the *Sophophora* subgenus, where some gene region reorganisation has occurred in one of them. Respective species (where the eTCPL has the same location according to FlyBase) diverged approximately 48 MYA (the known divergence time for *Drosophila* species) [52], and at least for these species, the eTCPL shows a clear monophyletic origin with a single transfer event according to the proposed phylogeny of the *Drosophila*/*Sophophora* common ancestor. We also reconstitute the phylogenetic tree of viral tobamo-CP and cellular eTCPL sequences together. This tree (and several other trees with different alignment and tree building algorithms [59]) shows very strong bootstrap confidence intervals for viral genera as well as for the Calyptratae/Acalyptratae split (from an alignment with a length of ~130 AA), indicating the biological significance of the tree. Seeing as both Schizophora subsections (Acalyptratae and Calyptratae) have eTCPL in their genomes and the eTCPL phylogeny is consistent with known fly phylogeny [55], it can be concluded that the integration of viral elements occurred before these two groups diverged. One group of flies, Aschiza, diverged before the Acalyptratae/Calyptratae divergence and are closest relatives of Schizophora. In this group of flies, there are three organisms with fully sequenced genomes (according to the NCBI Genomes database): *Megaselia abdita*, *Megaselia scalaris* and *Eristalis dimidiata*. None of the organisms could be identified as carrying the eTCPL gene; however, the neighbouring genes of eTCPL were detected using BLAST [59]. Therefore, it could be concluded that the main integration of the tobamo-CP gene was into the ancestor of the Schizophora flies. Supplementing the phylogenomic tree with known divergence times of tobamoviruses, the Calyptratae/Acalyptratae split, and synteny analysis, tobamoviruses are much older than eTCPLs in Schizophora. Therefore, taking all this information into account, the most likely single transfer event took place between 115 and 48 MYA with a direction from viruses to insects.

The presence of plant virus genes in insect genomes may seem very unlikely, but we must take into account that insects are common plant virus vectors and otherwise have very intimate contact with plants. According to that, the described transfer is no longer unexpected. Additionally, single V2H transfer events in one organism must be somehow beneficial for hosts to be fixed and to spread in populations and maintained over millions of years. The detected EVE seems to be biologically active, as the respective mRNA has been observed in several fly species (Table S8). In *Drosophila melanogaster*, the expression is observed in different developmental stages and different tissues (our in silico analysis of preexisting data deposited in different databanks, data not shown). In addition, the peptide corresponding to amino acids 229–246 in UniProt Q9W483 (a product of gene CD15772, i.e., The *D. melanogaster* eTCPL) have been observed in adult *Drosophila melanogaster* heads according to a search in the PeptideAtlas database (www.peptideatlas.org).

However, more detailed experiments need to be performed to determine the full function of the eTCPLs in flies.

5. Conclusions

This work showed that fast evolving viral protein homologues can be identified in eukaryotic genomes using structure-guided HMM searches, even when the integration event itself is very ancient.

Supplementary Materials: The following are available online at http://www.mdpi.com/1999-4915/11/4/320/s1, Supplementary File 1; containing Tables S1–S8.

Author Contributions: Conceptualisation, A.A.; methodology, A.A.; software, H.K.; validation, H.K., A.A.; formal analysis, H.K. and A.A.; investigation, A.A. and H.K.; data curation, H.K.; writing—original draft preparation, H.K.; writing—review and editing, A.A. and H.K.; visualisation, H.K.; and supervision, A.A.

Funding: The authors received no SPECIFIC funding for this work. However, the work of AA was supported by grant PRG198 from Estonian Research Council to Prof. Mart Ustav and by the Center of Excellence in Molecular Cell Engineering (2014-2020.4.01.15-0013) through the European Regional Development Fund. The work of HK was partially supported by grant IUT20-27 from the Estonian Research Council to Prof. Andres Merits.

Acknowledgments: The authors are grateful to Lauri Saag for their support with phylogenetic trees.

Conflicts of Interest: The authors declare no conflicts of interest.

References

1. Thingstad, T.F.; Lignell, R. Theoretical models for the control of bacterial growth rate, abundance, diversity and carbon demand. *Aquat Microb. Ecol.* **1997**, *13*, 19–27. [CrossRef]
2. Middelboe, M.; Lyck, P.G. Regeneration of dissolved organic matter by viral lysis in marine microbial communities. *Aquat Microb. Ecol.* **2002**, *27*, 187–194. [CrossRef]
3. Suttle, C.A. Marine viruses–major players in the global ecosystem. *Nat. Rev. Microbiol.* **2007**, *5*, 801–812. [CrossRef]
4. Cobián Güemes, A.G.; Youle, M.; Cantú, V.A.; Felts, B.; Nulton, J.; Rohwer, F. Viruses as Winners in the Game of Life. *Annu. Rev. Virol.* **2016**, *3*, 197–214. [CrossRef] [PubMed]
5. Mokili, J.L.; Rohwer, F.; Dutilh, B.E. Metagenomics and future perspectives in virus discovery. *Curr. Opin. Virol.* **2012**, 63–77. [CrossRef]
6. Bergh, Ø.; Børsheim, K.Y.; Bratbak, G.; Heldal, M. High abundance of viruses found in aquatic environments. *Nature* **1989**, *340*, 467–468. [CrossRef] [PubMed]
7. Whon, T.W.; Kim, M.-S.; Roh, S.W.; Shin, N.-R.; Lee, H.-W.; Bae, J.-W. Metagenomic Characterization of Airborne Viral DNA Diversity in the Near-Surface Atmosphere. *J. Virol.* **2012**, *86*, 8221–8231. [CrossRef]
8. Lander, E.S.; Linton, L.M.; Birren, B.; Nusbaum, C.; Zody, M.C.; Baldwin, J.; Devon, K.; Dewar, K.; Doyle, M.; Fitzhugh, W.; et al. Initial sequencing and analysis of the human genome. *Nature* **2001**, *409*, 860–921. [CrossRef]
9. Bejarano, E.R.; Khashoggi, A.; Witty, M.; Lichtenstein, C. Integration of multiple repeats of geminiviral DNA into the nuclear genome of tobacco during evolution. *Proc. Natl. Acad. Sci. USA* **1996**, *93*, 759–764. [CrossRef]
10. Stedman, K.M. Deep Recombination: RNA and ssDNA Virus Genes in DNA Virus and Host Genomes. *Annu. Rev. Virol.* **2015**, *2*, 203–217. [CrossRef] [PubMed]
11. Katzourakis, A. Endogenous Viral Elements in Animal Genomes. *PLoS Genet.* **2010**. [CrossRef] [PubMed]
12. Horie, M.; Tomonaga, K. Non-retroviral fossils in vertebrate genomes. *Viruses* **2011**, *3*, 1836–1848. [CrossRef] [PubMed]
13. Chu, H.; Jo, Y.; Cho, W.K. Evolution of endogenous non-retroviral genes integrated into plant genomes. *Curr. Plant Biol.* **2014**, 55–59. [CrossRef]
14. Frank, A.C.; Wolfe, K.H. Evolutionary capture of viral and plasmid DNA by yeast nuclear Chromosomes. *Eukaryot. Cell* **2009**. [CrossRef]
15. Aiewsakun, P.; Katzourakis, A. Endogenous viruses: Connecting recent and ancient viral evolution. *Virology* **2015**. [CrossRef] [PubMed]

16. Frank, J.A.; Feschotte, C. Co-option of endogenous viral sequences for host cell function. *Curr. Opin. Virol.* **2017**, 81–89. [CrossRef] [PubMed]

17. Myers, K.N.; Barone, G.; Ganesh, A.; Staples, C.J.; Howard, A.E.; Beveridge, R.D.; Maslen, S.; Skehel, J.M.; Collis, S.J. The bornavirus-derived human protein EBLN1 promotes efficient cell cycle transit, microtubule organisation and genome stability. *Sci. Rep.* **2016**, *6*. [CrossRef]

18. He, P.; Sun, L.; Zhu, D.; Zhang, H.; Zhang, L.; Guo, Y.; Liu, S.; Zhou, J.; Xu, X.; Xie, P. Knock-down of endogenous bornavirus-like nucleoprotein 1 inhibits cell growth and induces apoptosis in human oligodendroglia cells. *Int. J. Mol. Sci.* **2016**, *17*, 435. [CrossRef]

19. Fujino, K.; Horie, M.; Honda, T.; Merriman, D.K.; Tomonaga, K. Inhibition of Borna disease virus replication by an endogenous bornavirus-like element in the ground squirrel genome. *Proc. Natl. Acad. Sci. USA* **2014**, *111*, 13175–13180. [CrossRef]

20. Murad, L.; Bielawski, J.P.; Matyasek, R.; Kovařík, A.; Nichols, R.A.; Leitch, A.R.; Lichtenstein, C.P. The origin and evolution of geminivirus-related DNA sequences in Nicotiana. *Heredity* **2004**, *92*, 352–358. [CrossRef]

21. Kondo, H.; Chiba, S.; Suzuki, N. Detection and analysis of non-retroviral RNA virus-like elements in plant, fungal, and insect genomes. *Methods Mol. Biol.* **2015**, *1236*, 73–88. [CrossRef]

22. Brenner, S.E.; Chothia, C.; Hubbard, T.J. Assessing sequence comparison methods with reliable structurally identified distant evolutionary relationships. *Proc. Natl. Acad. Sci. USA.* **1998**, *95*, 6073–6078. [CrossRef]

23. Park, J.; Karplus, K.; Barrett, C.; Hughey, R.; Haussler, D.; Hubbard, T.; Chothia, C. Sequence comparisons using multiple sequences detect three times as many remote homologues as pairwise methods. *J. Mol. Biol.* **1998**, *284*, 1201–1210. [CrossRef] [PubMed]

24. Abroi, A.; Gough, J. Are viruses a source of new protein folds for organisms?—Virosphere structure space and evolution. *BioEssays* **2011**, *33*, 626–635. [CrossRef] [PubMed]

25. Aiewsakun, P.; Katzourakis, A. Time-Dependent Rate Phenomenon in Viruses. *J. Virol.* **2016**, *90*, 7184–7195. [CrossRef] [PubMed]

26. Duffy, S.; Shackelton, L.A.; Holmes, E.C. Rates of evolutionary change in viruses: Patterns and determinants. *Nat. Rev. Genet.* **2008**, *9*, 267–276. [CrossRef] [PubMed]

27. Kuchibhatla, D.B.; Sherman, W.A.; Chung, B.Y.W.; Cook, S.; Schneider, G.; Eisenhaber, B.; Karlin, D.G. Powerful Sequence Similarity Search Methods and In-Depth Manual Analyses Can Identify Remote Homologs in Many Apparently "Orphan" Viral Proteins. *J. Virol.* **2014**, *88*, 10–20. [CrossRef]

28. Dunbrack, R.L. Sequence comparison and protein structure prediction. *Curr. Opin. Struct. Biol.* **2006**, 374–384. [CrossRef] [PubMed]

29. Karlin, D.; Belshaw, R. Detecting remote sequence homology in disordered proteins: Discovery of conserved motifs in the N-termini of mononegavirales phosphoproteins. *PLoS ONE* **2012**, *7*. [CrossRef] [PubMed]

30. Illergård, K.; Ardell, D.H.; Elofsson, A. Structure is three to ten times more conserved than sequence—A study of structural response in protein cores. *Proteins Struct. Funct. Bioinform.* **2009**, *77*, 499–508. [CrossRef]

31. Chothia, C.; Lesk, M. The relation between the divergence of sequence and structure in proteins. *EMBO J.* **1986**, *5*, 823–826. [CrossRef] [PubMed]

32. Challis, C.J.; Schmidler, S.C. A stochastic evolutionary model for protein structure alignment and phylogeny. *Mol. Biol. Evol.* **2012**, *29*, 3575–3587. [CrossRef] [PubMed]

33. Herman, J.L.; Challis, C.J.; Novák, A.; Hein, J.; Schmidler, S.C. Simultaneous Bayesian Estimation of Alignment and Phylogeny under a Joint Model of Protein Sequence and Structure. *Mol. Biol. Evol.* **2014**, *31*, 2251–2266. [CrossRef] [PubMed]

34. Altschul, S.F.; Madden, T.L.; Schäffer, A.A.; Zhang, J.; Zhang, Z.; Miller, W.; Lipman, D.J. Gapped BLAST and PSI-BLAST: A new generation of protein database search programs. *Nucleic Acids Res.* **1997**, *25*, 3389–3402. [CrossRef]

35. Finn, R.D.; Clements, J.; Eddy, S.R. HMMER web server: Interactive sequence similarity searching. *Nucleic Acids Res.* **2011**, *39*, 29–37. [CrossRef]

36. Finn, R.D.; Coggill, P.; Eberhardt, R.Y.; Eddy, S.R.; Mistry, J.; Mitchell, A.L.; Potter, S.C.; Punta, M.; Qureshi, M.; Sangrador-Vegas, A.; et al. The Pfam protein families database: Towards a more sustainable future. *Nucleic Acids Res.* **2016**, *44*, D279–D285. [CrossRef] [PubMed]

37. Gough, J.; Karplus, K.; Hughey, R.; Chothia, C. Assignment of homology to genome sequences using a library of hidden Markov models that represent all proteins of known structure. *J. Mol. Biol.* **2001**, *313*, 903–919. [CrossRef] [PubMed]

38. Kumar, S.; Stecher, G.; Tamura, K. MEGA7: Molecular Evolutionary Genetics Analysis Version 7.0 for Bigger Datasets. *Mol. Biol. Evol.* **2016**, *33*, 1870–1874. [CrossRef] [PubMed]
39. Pearson, W.R. An introduction to sequence similarity ("homology") searching. *Curr. Protoc. Bioinform.* **2013**. [CrossRef]
40. Wu, S.; Zhang, Y. LOMETS: A local meta-threading-server for protein structure prediction. *Nucleic Acids Res.* **2007**, *35*, 3375–3382. [CrossRef]
41. Waterhouse, A.M.; Procter, J.B.; Martin, D.M.; Clamp, M.; Barton, G.J. Jalview Version 2-A multiple sequence alignment editor and analysis workbench. *Bioinformatics* **2009**, *25*, 1189–1191. [CrossRef] [PubMed]
42. Triant, D.A.; Pearson, W.R. Most partial domains in proteins are alignment and annotation artifacts. *Genome Biol.* **2015**, *16*, 1–12. [CrossRef] [PubMed]
43. Katoh, K.; Rozewicki, J.; Yamada, K.D. MAFFT online service: Multiple sequence alignment, interactive sequence choice and visualization. *Brief Bioinform.* **2017**. [CrossRef] [PubMed]
44. Abascal, F.; Zardoya, R.; Posada, D. ProtTest: Selection of best-fit models of protein evolution What can I use ProtTest for?—Introduction The program: Using ProtTest. *Bioinformatics* **2005**, *21*, 1–17. [CrossRef]
45. Attrill, H.; Falls, K.; Goodman, J.L.; Millburn, G.H.; Antonazzo, G.; Rey, A.J.; Marygold, S.J. FlyBase Consortium. Flybase: Establishing a gene group resource for Drosophila melanogaster. *Nucleic Acids Res.* **2016**, *44*, D786–D792. [CrossRef] [PubMed]
46. Hubbard, T.J.P.; Ailey, B.; Brenner, S.E.; Murzin, A.G.; Chothia, C. SCOP: A structural classification of proteins database. *Nucleic Acids Res.* **1999**, 254–256. [CrossRef]
47. Adams, M.J.; Antoniw, J.F.; Kreuze, J. Virgaviridae: A new family of rod-shaped plant viruses. *Arch Virol.* **2009**, *154*, 1967–1972. [CrossRef] [PubMed]
48. Tamura, K.; Stecher, G.; Peterson, D.; Filipski, A.; Kumar, S. MEGA6: Molecular Evolutionary Genetics Analysis Version 6.0. *Mol. Biol. Evol.* **2013**, *30*, 2725–2729. [CrossRef] [PubMed]
49. Junqueira, A.C.M.; Azeredo-Espin, A.M.L.; Paulo, D.F.; Marinho, M.A.T.; Tomsho, L.P.; Drautz-Moses, D.I.; Purbojati, R.W.; Ratan, A.; Schuster, S.C. Large-scale mitogenomics enables insights into Schizophora (Diptera) radiation and population diversity. *Sci. Rep.* **2016**, *6*. [CrossRef]
50. Thomas, J.A.; Trueman, J.W.H.; Rambaut, A.; Welch, J.J. Relaxed phylogenetics and the palaeoptera problem: Resolving deep ancestral splits in the insect phylogeny. *Syst. Biol.* **2013**, *62*, 285–297. [CrossRef]
51. Ding, S.; Li, X.; Wang, N.; Cameron, S.L.; Mao, M.; Wang, Y.; Xi, Y.; Yang, D. The phylogeny and evolutionary timescale of muscoidea (Diptera: Brachycera: Calyptratae) inferred from mitochondrial genomes. *PLoS ONE* **2015**, *10*, e0134170. [CrossRef] [PubMed]
52. Wiegmann, B.M.; Yeates, D.K.; Thorne, J.L.; Kishino, H. Time Flies, a New Molecular Time-Scale for Brachyceran Fly Evolution Without a Clock. *Syst. Biol.* **2003**, *52*, 745–756. [CrossRef]
53. Gibbs, A.J.; Wood, J.; Garcia-Arenal, F.; Ohshima, K.; Armstrong, J.S. Tobamoviruses have probably co-diverged with their eudicotyledonous hosts for at least 110 million years. *Virus Evol.* **2015**, *1*, vev019. [CrossRef] [PubMed]
54. Stobbe, A.H.; Melcher, U.; Palmer, M.W.; Roossinck, M.J.; Shen, G. Co-divergence and host-switching in the evolution of tobamoviruses. *J. Gen. Virol.* **2012**, *93*. [CrossRef] [PubMed]
55. Wiegmann, B.M.; Trautwein, M.D.; Winkler, I.S.; Barr, N.B.; Kim, J.-W.; Lambkin, C.; Bertone, M.A.; Cassel, B.K.; Bayless, K.M.; Heimberg, A.M.; et al. Episodic radiations in the fly tree of life. *Proc. Natl. Acad. Sci. USA* **2011**, *108*, 5690–5695. [CrossRef] [PubMed]
56. Kondo, H.; Chiba, S.; Maruyama, K.; Andika, I.B.; Suzuki, N. A novel insect-infecting virga/nege-like virus group and its pervasive endogenization into insect genomes. *Virus Res.* **2017**. [CrossRef] [PubMed]
57. Roux, S.; Enault, F.; Bronner, G.; Vaulot, D.; Forterre, P.; Krupovic, M. Chimeric viruses blur the borders between the major groups of eukaryotic single-stranded DNA viruses. *Nat. Commun.* **2013**, *4*. [CrossRef]
58. Diemer, G.S.; Stedman, K.M. A novel virus genome discovered in an extreme environment suggests recombination between unrelated groups of RNA and DNA viruses. *Biol. Direct.* **2012**, *7*. [CrossRef]
59. Kirsip, H.; University of Tartu, Tartu, Estonia. Phylogenetic and synteny analyses for the eTCPL in Diptera genomes, 2018.

Article

Interpreting Viral Deep Sequencing Data with GLUE

Joshua B. Singer [1,*], Emma C. Thomson [1], Joseph Hughes [1], Elihu Aranday-Cortes [1],
John McLauchlan [1], Ana da Silva Filipe [1], Lily Tong [1], Carmen F. Manso [2], Robert J. Gifford [1],
David L. Robertson[1], Eleanor Barnes [3], M. Azim Ansari [3], Jean L. Mbisa [2], David F. Bibby [2]Daniel
Bradshaw [2] and David Smith [3]

[1]   MRC-University of Glasgow Centre for Virus Research, Glasgow G61 1QH, UK;
      Emma.Thomson@glasgow.ac.uk (E.C.T.); joseph.hughes@glasgow.ac.uk (J.H.);
      Elihu.Aranday-Cortes@glasgow.ac.uk (E.A.-C.); John.McLauchlan@glasgow.ac.uk (J.M.);
      Ana.daSilvaFilipe@glasgow.ac.uk (A.d.S.F.); lily.tong@glasgow.ac.uk (L.T.);
      Robert.Gifford@glasgow.ac.uk (R.J.G.); David.L.Robertson@glasgow.ac.uk (D.L.R.)
[2]   Virus Reference Department, National Infection Service, Public Health England, Colindale,
      London NW9 5EQ, UK; Carmen.Manso@phe.gov.uk (C.F.M.); Tamyo.Mbisa@phe.gov.uk (J.L.M.);
      David.Bibby@phe.gov.uk (D.F.B.); Daniel.Bradshaw@phe.gov.uk (D.B.)
[3]   Peter Medawar Building for Pathogen Research, Nuffield Department of Medicine, University of Oxford,
      Oxford OX1 3SY, UK; ellie.barnes@ndm.ox.ac.uk (E.B.); ansari@well.ox.ac.uk (M.A.A.);
      david.smith@stcatz.ox.ac.uk (D.S.)
*   Correspondence: josh.singer@glasgow.ac.uk

Received: 28 February 2019; Accepted: 14 March 2019; Published: 3 Apri 2019

Abstract: Using deep sequencing technologies such as Illumina's platform, it is possible to obtain reads from the viral RNA population revealing the viral genome diversity within a single host. A range of software tools and pipelines can transform raw deep sequencing reads into Sequence Alignment Mapping (SAM) files. We propose that interpretation tools should process these SAM files, directly translating individual reads to amino acids in order to extract statistics of interest such as the proportion of different amino acid residues at specific sites. This preserves per-read linkage between nucleotide variants at different positions within a codon location. The samReporter is a subsystem of the GLUE software toolkit which follows this direct read translation approach in its processing of SAM files. We test samReporter on a deep sequencing dataset obtained from a cohort of 241 UK HCV patients for whom prior treatment with direct-acting antivirals has failed; deep sequencing and resistance testing have been suggested to be of clinical use in this context. We compared the polymorphism interpretation results of the samReporter against an approach that does not preserve per-read linkage. We found that the samReporter was able to properly interpret the sequence data at resistance-associated locations in nine patients where the alternative approach was equivocal. In three cases, the samReporter confirmed that resistance or an atypical substitution was present at NS5A position 30. In three further cases, it confirmed that the sofosbuvir-resistant NS5B substitution S282T was absent. This suggests the direct read translation approach implemented is of value for interpreting viral deep sequencing data.

Keywords: deep sequencing; virus genomics; hepatitis C virus; variant calling; sequence interpretation; drug resistance; bioinformatics

---

1. Introduction

For some virus species, their highly error-prone replication mechanism produces a population of related genomic variants of the virus within a single infected host individual [1]. Sequencing systems such as Illumina's platform produce short, relatively accurate nucleotide sections of viral genome, often generating thousands of reads for a given genomic location from a single sample [2]. Such deep sequencing technologies therefore offer methods for understanding the nature of viral intra-host diversity. Whole genome and deep sequencing of virus genomes has been widely applied in basic virology research but has also found applications in clinical contexts such as the detection of drug resistance [3].

Various bioinformatics stages must be applied in the interpretation of viral deep sequencing data. Reads unrelated to the virus genome are removed and low-quality reads removed or trimmed. Following this, we must then construct an alignment: how the reads are arranged relative to each other within the virus genome, accounting for sequence homology. Reference-based alignment or mapping methods such as Bowtie 2 [4], BWA [5], MOSAIK [6], Stampy [7] or Tanoti [8] use one or more reference sequences to guide the alignment of reads. In contrast, de novo assembly approaches such as SPAdes [9] and VICUNA [10] use associations derived purely from the read data itself to propose large genome fragments, avoiding the biases arising from the choice of reference sequence. A drawback of de novo methods is that they may not accurately capture the full genomic structure or diversity, thus, for well-known viruses with high levels of genomic diversity, combinations of de novo assembly and reference-based alignment methods, such as shiver [11], are often used. One aspect common to almost all recent methods in this area is that they output their results in the form of a sequence alignment mapping (SAM) file. The SAM format [12] integrates nucleotide, read quality and alignment data in a single file. It was standardised at an early point in the adoption of deep sequencing, allowing diverse methods to be compared with each other and integrated into processing pipelines for a broad set of applications.

A range of variant-calling methods have been developed to analyse genomic heterogeneity within deep sequencing data. Error rates in short read technologies such as Illumina are low but it can be challenging to distinguish errors from real single nucleotide variants (SNVs) occurring at frequencies comparable to the error rate. Therefore, variant-calling methods such as LoFreq [13] and V-Phaser [14,15] apply statistical techniques to the aligned read data to identify probable SNVs occurring even at very low frequency in the presence of sequencing errors. The focus on low frequency SNVs is critical in applications such as cancer genomics where somatic deviations from the consensus are both rare and of high consequence [16]. However, virus bioinformatics has distinct priorities from fields focused on eukaryote or bacterial organisms with higher replication fidelity [17]. Since viruses typically have a low replication fidelity, there is a higher level of diversity within an infected host and the viral population can be expected to contain many variants.

For reasons of clarity within the research community, virus genome locations can be defined in terms of a standard virus strain with a well-established "master reference" genome. In hepatitis C virus (HCV) for example, codons within viral proteins are numbered relative to the H77 strain [18]. Polymorphisms at these standardised locations are reported with phenotypic associations established experimentally or in clinical trials. The advent of deep sequencing data prompts questions such as: What are the relative proportions of different amino acid residues at a given genome location? What proportion of reads support the presence of a certain sequence motif? What proportion of reads indicate a deletion? However, it is challenging to answer these questions within deep sequencing data, since read alignments do not in general use the reference coordinate space, and a mapping between the two spaces must be established and applied.

The genomes of the virus population may contain multiple nucleotide base variants at different positions within a single codon location. Both Verbist et al. [19] and Döring et al. [20] pointed out that linkage between nucleotide positions is lost when variants are called as SNVs. This linkage must be retained within datasets in order to accurately predict the amino acid residues arising from protein translation. Suppose for example we observe significant levels of both adenine (A) and thymine (T) at the first position of a particular codon location. At the second position, we observe cytosine (C) and guanine (G). With cytosine at the third position, what amino acids are the genomes in the virus population coding for? Without retaining linkage, these observations are consistent with a mix of Threonine (codon ACC) and Cysteine (TGC), or alternatively with purely Serine (AGC/TCC), or with any combination of these amino acids. By retaining linkage, we can accurately select between these interpretations. Haplotype reconstruction methods aim to determine linkage by associating sub-populations of reads as haplotypes. Schirmer et al. [21] found that this was exceptionally unreliable for viral deep sequencing data. However, full haplotype reconstruction may not be necessary for practical applications where the variants of interest are linked within the span of a single read or read pair.

We present a subsystem of the GLUE software package [22] called samReporter, focused on the analysis of aligned deep sequencing viral genome data. It directly processes the SAM file format produced by most methods, and can also process the more compact Binary Alignment Map (BAM) format. The samReporter can be instantiated within an existing GLUE project containing reference sequences and alignments for a given virus. This allows the software to establish a reading frame for coding region reads within a SAM file, and map between the read alignment coordinate space and standardised genome locations. In turn, this facilitates the scanning of reads directly for different classes of sequence pattern such as codons, amino acid residues, indels and motifs. This approach of scanning reads directly has the advantage of retaining linkage and we can report how often combinations of variations appear together on the same viral RNA, certainly within a codon location but also further, at least as far as paired-end read data allows.

We demonstrate the benefits of applying the GLUE samReporter to hepatitis C virus (HCV) deep sequencing data. HCV is a positive-sense single-stranded RNA virus of the family Flaviviridae. Its genome of about 9000 bases codes for a single polyprotein that produces 10 mature viral proteins. HCV affects over 100 million people worldwide and can cause liver disease and cirrhosis. The infection can be treated with a range of direct-acting antiviral (DAA) drugs which inhibit three of the mature proteins: NS3, NS5A and NS5B. Such therapies produce a sustained virological response (SVR) in the vast majority of patients, clearing the virus in around 95% of cases [23]. Notwithstanding this therapeutic success, HCV is proving difficult to treat in certain categories of patients, including "retreatment" patients: those for whom prior DAA treatment has failed. It has been shown both in vitro and in vivo that certain resistance-associated substitutions (RASs) in the viral genome confer resistance to DAA drugs [24].

Vermehren et al. suggested that retreatment patients have RASs in multiple drug target genes and that therefore "genomic resistance testing may be useful to select the optimal combination and treatment duration" for subsequent rounds of drug therapy [23]. Recent guidance suggests that, if deep sequencing is used, observing a RAS in 15% of the virus population may be clinically relevant [25]. Tools aimed at HCV resistance testing such as geno2pheno[ngs-freq] [20] suggest frequencies of 2%, 10% and 15%. Thus, while HCV RAS testing can benefit from deep sequencing methods, moderately-low rather than ultra-low frequencies are of most interest. RAS testing for retreatment patients therefore provides a good case study for a deep sequencing data interpretation system.

2. Results

We analysed viral genome diversity within a group of 241 HCV retreatment patients sampled within the United Kingdom. A range of genotypes were represented: Gt1 $n = 115$, Gt2 $n = 5$, Gt3 $n = 104$, Gt4 $n = 14$ and Gt6 $n = 3$. The five most frequent subtypes: were 1a $n = 98$, 1b $n = 13$, 3a $n = 95$, 3b $n = 6$ and 4r $n = 6$. Fourteen other subtypes were represented each by three or fewer patients. In three cases, a subtype could not be assigned.

The samReporter scans aligned reads directly, retaining linkage within reads with the intention of more accurate detection of specific variants. To test the benefits of this approach, we also contrived an alternative method that attempts to capture within-host diversity without retaining linkage.

Besides the four concrete bases A, C, G and T/U, IUPAC notation for nucleic acids, used in the FASTA sequence file format, contains 11 ambiguity codes, covering all possible combinations of more than one base [26]. For example, code S can represent a combination of C and G. Any method which calls SNVs can encode these variants within a FASTA file. This aspect of the encoding is used to capture minor nucleotide variants. Web-based HCV drug resistance interpretation systems such as HCV-GLUE [27] and geno2pheno[hcv] [28] do attempt to interpret ambiguity codes if they appear in the input data. We produced FASTA files with ambiguity codes for each sample using the samReporter `nucleotide-consensus` command, which produces one IUPAC code for each nucleotide position in the SAM reference coordinate space. Read bases with a Phred quality score of less than 25 were excluded. A "concrete" base (A, C, G or T) was encoded at a given position if it appeared both in at least five individual reads and in 5% of the quality-filtered reads at that location. IUPAC ambiguity codes are then used if multiple concrete bases are to be encoded. We found that FASTA files for all but two samples contained at least one ambiguity code. On average, files contained ambiguity codes which represent two bases at 1.02% of nucleotide positions (std. dev. 1.3%) and codes which represent three bases at 0.0161% of nucleotide positions (std. dev. 0.0475%).

A triplet of concrete bases, i.e. a codon, specifies a single amino acid. If ambiguity codes occur within nucleotide data for a given codon location, multiple distinct codons are present in the underlying data at that location; the precise composition is unknown. For a given ambiguous triplet (possibly containing ambiguity codes), there is set of "possible" amino acids comprising any residues coded by one or more of the possible codons. For example, for the ambiguous triplet YTM, the set of possible amino acids is Leucine (L) and Phenylalanine (F) because the set of codons and their corresponding amino acids are CTA (L), CTC (L), TTA (L) and TTC (F). Additionally, there is a (possibly empty) subset of "definite" amino acid residues, i.e. those that must be coded by at least some of the underlying codons, whatever the composition, under reasonable assumptions. For the ambiguous triplet YTM, every combination of codons which produces the ambiguity codes contains at least some codons for Leucine (L); this is the single definite amino acid. In general, if there is a single ambiguity code encoding two bases within the triplet, there will be one or two definite amino acids, and these will also be the only possible amino acids.

The FASTA files were analysed for "ambiguous" codon locations where the definite and possible amino acid sets were different. This typically occurs when there are two ambiguity codes within a triplet. Such locations present a challenge for drug resistance interpretation systems based on FASTA file inputs. Whereas amino acid residues in the definite set can be inferred to be present in the virus population, the status of amino acids in the possible set but not the definite set cannot be established clearly from FASTA data. We excluded from the analysis degenerate locations i.e. those where the possible set, excluding stops, contained more than five amino acids or where the read depth for the whole codon location was less than 10.

In total, 435 ambiguous locations were found in patients, within all ten viral proteins: Core $n = 62$, E1 $n = 24$, E2 $n = 205$, p7 $n = 5$, NS2 $n = 14$, NS3 $n = 30$, NS4A $n = 1$, NS4B $n = 9$, NS5A $n = 38$ and NS5B $n = 47$. The full set of ambiguous locations is given in the Supplementary Materials. Scaling by the length of each region, this implies that the E2 and Core proteins had a higher rate of ambiguous locations. The drug target proteins NS3, NS5A and NS5B have rates in the lower part of the range.

The current version of HCV-GLUE [27] documents 44 locations associated in the literature with resistance to six DAA drugs in current use: 18 in NS3, 15 in NS5A and 11 in NS5B. These are listed in the Supplementary Materials. Within the FASTA data, we found 10 ambiguous resistance-associated locations in nine patients, six in NS5A and four in NS5B. We resolved these locations using the GLUE samReporter, calculating the frequencies of codons and amino acids by directly analysing reads. Codons were excluded if any Phred base quality was below 25. Amino acid residues were deemed to be present if 5% or more of filtered read codons at the location coded for the residue. The 10 locations are shown in Table 1. HCV-GLUE classifies an amino acid as typical at a location for a given subtype if 10% or more of the GenBank sequences of that subtype contain the residue, these are also shown in the table. In all cases except one (R25, NS5B position 159) the definite residues set (not shown) was empty.

Table 1. Ambiguous resistance-associated locations resolved using GLUE samReporter.

Sequencing Facility	Sample ID	Subtype	Virus Protein	Codon Location	Ambiguous Triplet	Typical Residue (s)	Possible Residues Set	Confirmed Residues Set
Glasgow	HCV294	3b	NS5B	282	WSY	S	CST	S
Glasgow	HCV300	3a	NS5A	30	RMG	A	AEKT	AK
PHE	R127	1a	NS5A	24	RSG	K	AGRT	GT
PHE	R164	3a	NS5A	30	RMG	A	AEKT	AK
PHE	R25	4r	NS5B	159	YTM	L	FL	L
PHE	R25	4r	NS5B	282	WSC	S	CST	S
PHE	R36	4r	NS5B	282	WSC	S	CST	S
PHE	R67	1a	NS5A	30	YAW	Q	HQY	QY
PHE	R91	1a	NS5A	28	RYG	M	AMTV	MV
Oxford	7444	3a	NS5A	62	SYA	ST	ALPV	AL

In the cases shown in Table 1, samReporter was able to eliminate many possible residues. Whereas the possible set contained up to four residues, samReporter confirmed that only one or two were actually present at the 5% level. Two resistance locations occur three times each and merit a discussion. For NS5A position 30, in subtype 1a the typical residue is Glutamine (Q). In sample R67, samReporter found Tyrosine (Y) at around 74%, which has not been documented as a RAS but is atypical for the subtype. In samples HCV300 and R164 (subtype 3a) the typical residue is Alanine (A); samReporter found Lysine (K), a well-documented RAS, at levels of 92% and 82% respectively. Thus, in these three cases samReporter confirmed the presence of a RAS or atypical substitution. Substitutions at NS5B position 282 have been strongly associated with resistance to sofosbuvir, particularly the substitution of the typical Serine (S) with Threonine (T). In contrast with NS5A position 30, samReporter was able to eliminate this resistant residue and the other possible atypical residue Cysteine (C); in these three ambiguous cases, only Serine is present, but is actually coded by significantly distinct codons in each case. For sample HCV294, the codons were TCT at 54%, AGT at 25% and AGC at 20%, for sample R25, AGC at 85% and TCC at 15%, and, for sample R36, TCC at 92% and AGC at 7%. The effect of the presence of these diverse codons is to create ambiguity at the nucleotide level. One possible explanation for the codon diversity is that Threonine codons became frequent in the viral population during sofosbuvir treatment, and that following the end of the treatment course the descendants of these virions reverted to Serine, but now coded using diverse alternative codons.

The HCV296 sample is typical in the sense that the size of the BAM file (24.6 MB) was closest to the mean for this dataset, it contains ≈282,000 paired-end reads with an average depth across the HCV polyprotein of ≈3700. To evaluate performance, we ran some samReporter commands on this file using a 2014 MacBook Pro with a 2.5 GHz Intel Core i7 processor and 16 GB of RAM. The samReporter was configured to use up to four CPU cores. The `amino-acid` command was run to translate reads for the whole polyprotein, producing amino acid residue frequencies at each codon location, without any read filters. Using the auto-align feature with a known target reference sequence, this command took 7.5 s. Using the max-likelihood-placer feature, the command took 29.2 s, with most of the extra time spent in the RAxML-EPA step. See Section 3.2 for details of the samReporter design.

3. Materials and Methods

3.1. Sequencing Data

Deep whole genome HCV sequencing data was derived from blood samples collected from 241 patients resident in the United Kingdom, who had not achieved virological clearance after previous courses of antiviral therapy. Sequencing was performed using target enrichment on Illumina sequencers at three different institutes: the MRC-University of Glasgow Centre for Virus Research ($n = 56$), the University of Oxford Nuffield Department of Medicine ($n = 25$) and the Virus Reference Department at Public Health England (PHE) ($n = 160$).

The Glasgow library preparation protocol was as follows. RNA was isolated from 200 μL of plasma using the RNAdvance Blood extraction kit (Beckman Coulter, Brea, CA, United States) and collected in 27 µL of water. Following conversion of RNA to double-stranded DNA, libraries were prepared for Illumina sequencing using the KAPA DNA LTP Library Preparation Kit (Roche, Basel, Switzerland), and NEBNext Multiplex Oligos for Illumina (New England Biolabs, Ipswich, MA, United States). Libraries were quantified using Qubit dsDNA HS Assay Kit (Invitrogen, Carlsbad, CA, United States) and size distribution assessed using Agilent TapeStation with D1K High Sensitivity Kit (Agilent, Santa Clara, CA, United States); libraries were normalised according to viral load and mass. A 500 ng aliquot of the pooled library was enriched using SeqCap EZ Developer Probes (Roche), following the manufacturer's protocol. Following a 14 cycle post-enrichment PCR, the cleaned pool was sequenced with 151-base paired-end reads on a NextSeq cartridge (Illumina, San Diego, CA, United States).

The Oxford libraries were prepared for Illumina sequencing using the NEBNext Ultra Directional RNA Library Prep Kit (New England Biolabs) with 8 µL of RNA extracted from plasma using NUCLISENS easyMAG (bioMérieux, Marcy-l'Étoile, France) and previously published modifications of the manufacturer's guidelines (v2.0) [29]: omission of heat fragmentation, omission of Actinomycin D at first-strand reverse transcription, library amplification for 15 PCR cycles using custom indexed primers [30], and post-PCR clean-up with 0.85x volume Ampure XP (Beckman Coulter). Libraries were quantified using Quant-iT PicoGreen dsDNA Assay Kit (Invitrogen) and size distribution analysed using Agilent TapeStation D1K High Sensitivity kit. A 500 ng aliquot of the pooled libraries (96 plex) was enriched using the xGen Lockdown protocol (Rapid Protocol for DNA Probe Hybridization and Target Capture Using an Illumina TruSeq or Ion Torrent Library v1.0, Integrated DNA Technologies, Coralville, IA, United States) with equimolar-pooled 120 nt DNA oligonucleotide probes (Integrated DNA Technologies) followed by a 12-cycle on-bead post-enrichment PCR. The cleaned post-enrichment ve-Seq library was quantified by qPCR with the KAPA SYBR FAST qPCR Kit (Roche) and sequenced with 150b paired-end reads on a single run of the Illumina MiSeq.

The PHE library preparation protocol is the laboratory component of a pipeline aimed at clinical use; a manuscript describing the full pipeline is in preparation. RNA was extracted from 350 µL of plasma using the NUCLISENS easyMAG system (bioMérieux). Total eluates were subjected to Turbo DNAse treatment (Thermo Fisher, Waltham, MA, United States) followed by library preparation using KAPA RNA HyperPrep kit (Roche). Libraries were pooled based upon DNA concentration and HCV quantity, assessed using the Quant-iT kit on the Glomax platform (Promega, Madison, WI, United States) and the Qiagen QuantiTect kit with primers and probes from Davalieva et al. [31]. Pools were enriched by hybridisation to a biotinylated probe set (Integrated DNA Technologies, described by Bonsall et al. [32]) followed by further PCR cycles depending upon HCV quantity. The two pools were pooled, again by concentration and HCV quantity. The final pool was quantified using the KAPA SYBR FASTA qPCR kit (Roche) on a PRISM 7500 (Applied BioSystems, Foster City, CA, United States) before being sequenced on a MiSeq using Reagent kit v2 (Illumina).

The Illumina read data were processed into SAM files using different bioinformatics pipelines at the different institutions. At Glasgow, reads were trimmed and filtered using TrimGalore [33] with quality threshold 30 and minimum read length 75. The most appropriate HCV reference sequence was identified via a *k*-mer-based approach, using *k*-mers unique to each genotype [34]. SAM files were generated by mapping against the best-matching HCV reference using Tanoti [8]. At Oxford, de-multiplexed sequence read-pairs were trimmed of low-quality bases using QUASR v7.01 [35] and adapter sequences with CutAdapt version 1.7.1 [36] and subsequently discarded if either read had fewer than 50 bases in its remaining sequence or if both reads matched the human reference sequence using Bowtie version 2.2.4 [4]. Remaining reads were mapped using BWA mem [5] and Stampy [7] against a database of reference sequences, both to choose an appropriate reference and to select those reads which formed a majority population for de novo assembly using VICUNA [10] and finishing with V-FAT [37]. The reads were then mapped back to this assembly using MOSAIK [6]. At PHE, human reads were filtered out from trimmed FASTQ files using SMALT [38], remaining reads were then assembled using VICUNA de novo assembly [10]. Contigs were matched to HCV reference genomes using BLAST [39] and gaps filled using LASTZ [40] to generate a draft assembly. Reads were then mapped to the draft assembly with BWA [5].

The deep sequencing data used in this study has been deposited in the NCBI Sequence Read Archive (http://www.ncbi.nlm.nih.gov/sra), under BioProject accession number PRJNA527067 and experiment accession numbers SRX5528430 to SRX5528670.

3.2. GLUE samReporter Design

The GLUE samReporter aims to provide a convenient tool for interpreting viral deep sequencing data. As part of the wider GLUE system [22], it can be used interactively in the command line interpreter or within bioinformatics scripts. Instantiated within a GLUE project for a specific category of viruses such as HCV-GLUE, it can take advantage of certain data objects within that project.

When interpreting viral deep sequencing data, one obstacle is mapping the SAM file coordinate space to a standard codon numbering system. Within the HCV-GLUE project the H77 strain (RefSeq accession NC_004102) is defined as the "master" reference sequence object. The precursor polyprotein and the 10 mature proteins are defined as coding features and their locations are specified on the H77 sequence. A wider set of reference sequence objects is also defined within HCV-GLUE; there are currently over 200 of these, based on the ICTV HCV resource [41]. HCV-GLUE then specifies an unconstrained "master" alignment object containing all these reference sequences, which is used to map their locations to those on the H77 sequence. HCV-GLUE also contains a reference phylogeny of the same set of sequences, computed using RAxML [42].

SAM files for HCV typically map each read to a single coordinate space. To interpret individual reads the samReporter must infer sequence homology (i.e., pairwise alignment) between this SAM file coordinate space and one of the reference sequences defined within GLUE—the "target" reference. The simplest method, to specify that the SAM file coordinate space is identical to that of a specific target reference sequence, is appropriate if one of the project's reference sequences was used for the SAM file coordinate space. A more flexible "auto-align" method allows GLUE to generate a codon-aware pairwise alignment between the consensus of the SAM file and a selected target reference, using techniques based on BLAST+ [39]. This is appropriate if the SAM virus strain is closely related to the target reference, but importantly, it allows the method producing the SAM file, which may have a de novo element, to construct a coordinate space appropriate for the viral reads. The final and most general and robust method is "max-likelihood-placer". This allows GLUE to select the target reference itself, by feeding the consensus of the SAM file into the first two stages of the GLUE genotyping pipeline. This consists of incorporating it into the master alignment using MAFFT [43], placing it in the reference phylogeny using RAxML-EPA [42] and selecting as the target the reference sequence with the lowest patristic distance from the SAM consensus. The auto-align method is then used to generate the homology. The master alignment will also typically act as the "linking" alignment, providing a mapping between the target reference and the master reference. The result of this process is then a chain of pairwise homology relationships, as shown in Figure 1, from each individual read to the SAM file coordinate space, to the target reference sequence and finally, via the linking alignment, to the master reference sequence.

The samReporter offers a range of GLUE commands for interpreting SAM files (Table 2). These each accept similar arguments for specifying the coordinate homology and genome region. The "variation scan" command scans each read for the presence or absence of sequence patterns defined by GLUE *Variation* objects [22]. If paired-end read data are supplied, the reads in each pair are processed together. *Variation* objects can encapsulate insertions, deletions, regular expressions and combinations at the nucleotide or amino acid level. In HCV drug resistance this capability may become important. For example, the Magellan-1 trial of the drug pibrentasvir found that the combination of a Methionine at NS5A position 28 with the deletion of the residue at NS5A position 32 was associated with resistance to the drug for HCV subtype 1b [44]. It remains to be seen whether such deletions and combinations occurring as minority variants are clinically relevant but if so, the samReporter offers a means of detecting these.

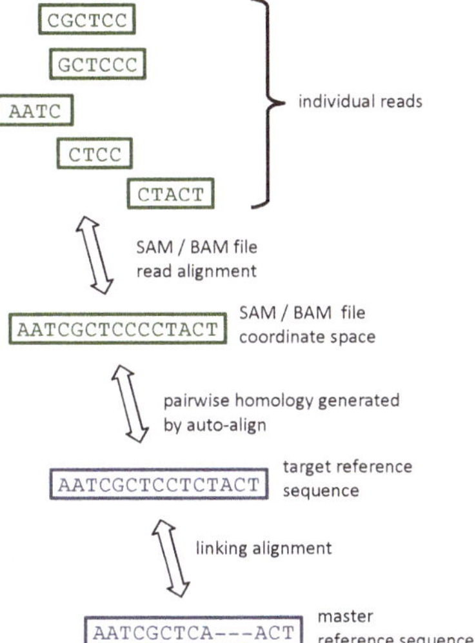

Figure 1. The chain of pairwise homology relationships between reads and the master reference sequence (H77 for HCV), established during the operation of GLUE samReporter.

Table 2. GLUE samReporter commands.

Command	Description
`nucleotide`	Generate a table of nucleotide frequencies within a specific genome region.
`depth`	Generate a table of read depths within a specific genome region.
`nucleotide-consensus`	Generate a FASTA consensus file, optionally using ambiguity codes.
`amino-acid`	Generate a table of amino acid residue frequencies within a specific protein-coding region.
`codon-triplets`	Generate a table of codon frequencies within a specific protein-coding region.
`variation scan`	Scan for the presence or absence of GLUE *Variations* within reads.
`export nucleotide-alignment`	Export a specific part of the SAM alignment as a FASTA file.

The commands also allow simple, optional filtering based on Phred base qualities, MAPQ mapping quality and depth. In command outputs, codon numbering is based on the system proposed by Kuiken et al. [18]; nucleotide coordinates both within the SAM file and the mapped location on the master reference are also given. Individual input files may be processed more quickly using parallelisation of command operations across multiple processors. Finally, for paired-end read data, regions where paired reads overlap are not counted twice in command outputs. As part of the GLUE engine, the SAM reporter is implemented in Java, using the Htsjdk library [45] to interpret the SAM format.

The samReporter is delivered as part of the GLUE software package. This study used GLUE version 1.1.33, HCV-GLUE project version 0.1.51 with PHE-HCV-DRUG-RESISTANCE extension version 0.1.21. GLUE is licensed under the open source GNU Affero General Public License version 3.0. and may be installed on Mac OSX, Windows or Linux systems. Documentation specific to the samReporter may be found at: http://tools.glue.cvr.ac.uk/#/deepSequencingData. Documentation for other aspects of GLUE and links to the source code repository can be found on other pages within the same web site.

4. Discussion

The results show that, within virus genomes of HCV retreatment patients, linkage between nucleotide variants within a codon location is not a purely theoretical issue. In a small number of cases among UK retreatment patients, such linkage did occur at sites critical for drug resistance. Approaches that do not preserve linkage, such as those encoding variants as ambiguity codes, cannot correctly resolve these cases. How would such a system deal with amino acid residues in the possible set but not in the definite set? If the system is configured not to report such residues, the result is false negative detection of a substitution at NS5A position 30 in three patients, obscuring A30K RAS in two cases. Conversely, if the system reports these residues, the result is false positive detection of the NS5B RAS S282T in three other patients. As far as we are aware, the current study is novel in terms of quantifying the effect of such linkage on resistance detection in real HCV patient data.

The current HCV-GLUE database [27] documents many RASs that combine substitutions at locations within the span of a typical Illumina read. The samReporter can report the presence or absence of these on any read (or read pair for paired-end data) that covers the relevant locations. Future work might consider whether detection of these "combination" RASs at a minority level are of clinical relevance. It would also be of interest to incorporate existing low frequency variant-calling mechanisms from the literature into GLUE. The samReporter and GLUE generally are intended to be useful in both research and clinical contexts. However, even once a SAM file has been generated from a sample, the samReporter only represents one part of the process in terms of drug resistance analysis. The HCV-GLUE system is currently being developed to provide a comprehensive drug resistance report, using samReporter to interpret a SAM file.

Other software, for example the VirVarSeq system [19], calls variants at the codon level but is focused on very low frequency variants. DiversiTools [46] provides frequencies of amino acids on a per read basis but does not link to a standardised coordinate system as is available in the GLUE framework. The geno2pheno[ngs-freq] system [20] directly interprets drug resistance in deep sequencing data. Users must transform their data into a table of nucleotide or codon frequencies and a web-based system then performs interpretation on this table, using a user-defined frequency threshold. In comparison with samReporter, this design facilitates fast transfer over a network since the frequency table is much more compact than a typical SAM or BAM file. However, some information is necessarily lost in the processing, for example the codon frequency table cannot encode linkage beyond a codon location which would be required for example to detect combination RASs.

While the current study applied samReporter to HCV, it can also be used to analyse deep sequencing data for other viruses. In many simple cases, the prerequisites would simply be a nucleotide alignment of alternative target reference sequences and a master reference sequence with coding region annotations. In more complex cases, for example where virus genomes contain ambisense genomes or RNA editing, GLUE and samReporter would need to take account of this. The GLUE samReporter shows that a simple, pragmatic software design can conveniently answer some common questions concerning within-host variation in viral deep sequencing data.

Supplementary Materials: The following are available online at http://www.mdpi.com/1999-4915/11/4/323/s1. The complete set of "ambiguous" codon locations that were found is supplied in a tab-delimited supplementary data file `unclearCodonLocations.txt`. Resistance associated locations are supplied in a tab-delimited supplementary data file `resistanceAssociatedLocations.txt`.

Author Contributions: Conceptualisation, J.B.S., E.C.T., J.L.M., J.H., J.M., E.B., D.L.R. and R.J.G.; methodology, J.B.S., J.H., D.S., D.F.B. and M.A.A.; software, formal analysis, and validation, JBS; data curation, E.A.-C., E.C.T., D.B., D.S., D.F.B. and M.A.A.; resources, A.d.S.F., L.T., C.F.M., D.F.B., E.A.C. and M.A.A.; writing—original draft preparation, J.B.S.; writing—review and editing, J.H., E.C.T., J.L.M., R.J.G., E.B., D.L.R. and J.M.; and supervision, R.J.G., D.L.R., E.C.T., J.M., E.B. and J.L.M.

Funding: This work was funded by the Medical Research Council (MRC) of the United Kingdom, award number MC_UU_12014/12. JBS was also part funded by a MRC Confidence in Concept award to the University of Glasgow, MC_PC_16045. ECT was funded by the Wellcome Trust (102789/Z/13/Z). EB is funded by the MRC, the Oxford National Institute for Health Research (NIHR) Biomedical Research Centre and is an NIHR Senior

Investigator. The views expressed in this article are those of the authors and not necessarily those of the National Health Service (NHS), the NIHR, or the Department of Health.

Acknowledgments: The authors would like to thank the participants and clinicians at NHS sites who contributed samples, and Will Irving and Emma Hudson for their contributions, particularly for sample collection and project management.

Conflicts of Interest: The authors declare no conflict of interest.

Abbreviations

The following abbreviations are used in this manuscript:

BAM	Binary alignment mapping
DAA	Direct-acting Antiviral
DNA	Deoxyribonucleic Acid
ICTV	International Committee for the Taxonomy of Viruses
HCV	Hepatitis C virus
RAS	Resistance-associated Substitution
RNA	Ribonucleic Acid
SAM	Sequence alignment mapping
SNV	Single nucleotide variant
SVR	Sustained virological response

References

1. Holmes, E.C. *The Evolution and Emergence of RNA Viruses*; Oxford University Press, Oxford, UK, 2009.
2. Goodwin, S.; McPherson, J.D.; McCombie, W.R. Coming of age: Ten years of next-generation sequencing technologies. *Nat. Rev. Genet.* **2016**, *17*, 333–351. [CrossRef] [PubMed]
3. Houldcroft, C.J.; Beale, M.A.; Breuer, J. Clinical and biological insights from viral genome sequencing. *Nat. Rev. Microbiol.* **2017**, *15*, 183–192. [CrossRef] [PubMed]
4. Langmead, B.; Salzberg, S.L. Fast gapped-read alignment with Bowtie 2. *Nat. Methods* **2012**, *9*, 357–359. [CrossRef]
5. Li, H.; Durbin, R. Fast and accurate short read alignment with Burrows–Wheeler transform. *Bioinformatics* **2009**, *25*, 1754–1760. [CrossRef]
6. Lee, W.P.; Stromberg, M.P.; Ward, A.; Stewart, C.; Garrison, E.P.; Marth, G.T. MOSAIK: A Hash-Based Algorithm for Accurate Next-Generation Sequencing Short-Read Mapping. *PLoS ONE* **2014**, *9*, e90581. [CrossRef]
7. Lunter, G.; Goodson, M. Stampy: A statistical algorithm for sensitive and fast mapping of Illumina sequence reads. *Genome Res.* **2011**, *21*, 936–939. [CrossRef]
8. Tanoti: A BLAST-Guided Reference-Based Short Read Aligner. Available online: http://www.bioinformatics.cvr.ac.uk/tanoti.php (accessed on 24 February 2019).
9. Bankevich, A.; Nurk, S.; Antipov, D.; Gurevich, A.A.; Dvorkin, M.; Kulikov, A.S.; Lesin, V.M.; Nikolenko, S.I.; Pham, S.; Prjibelski, A.D.; et al. SPAdes: A New Genome Assembly Algorithm and Its Applications to Single-Cell Sequencing. *J. Comput. Biol.* **2012**, *19*, 455–477. [CrossRef]
10. Yang, X.; Charlebois, P.; Gnerre, S.; Coole, M.G.; Lennon, N.J.; Levin, J.Z.; Qu, J.; Ryan, E.M.; Zody, M.C.; Henn, M.R. De novo assembly of highly diverse viral populations. *BMC Genom.* **2012**, *13*, 475. [CrossRef]
11. Wymant, C.; Fraser, C.; Hall, M.; Golubchik, T.; Bannert, N.; Fellay, J.; Fransen, K.; Porter, K.; Gourlay, A.; Grabowski, M.K.; et al. Easy and accurate reconstruction of whole HIV genomes from short-read sequence data with shiver. *Virus Evol.* **2018**, *4*, vey007. [CrossRef]
12. Li, H.; Wysoker, A.; Handsaker, B.; Marth, G.; Abecasis, G.; Ruan, J.; Homer, N.; Durbin, R.; Fennell, T.; The 1000 Genome Project Data Processing Subgroup. The Sequence Alignment/Map format and SAMtools. *Bioinformatics* **2009**, *25*, 2078–2079. [CrossRef] [PubMed]
13. Wilm, A.; Aw, P.P.K.; Bertrand, D.; Yeo, G.H.T.; Ong, S.H.; Wong, C.H.; Khor, C.C.; Petric, R.; Hibberd, M.L.; Nagarajan, N. LoFreq: A sequence-quality aware, ultra-sensitive variant caller for uncovering cell-population heterogeneity from high-throughput sequencing datasets. *Nucleic Acids Res.* **2012**, *40*, 11189–11201. [CrossRef]

14. Macalalad, A.R.; Zody, M.C.; Charlebois, P.; Lennon, N.J.; Newman, R.M.; Malboeuf, C.M.; Ryan, E.M.; Boutwell, C.L.; Power, K.A.; Brackney, D.E.; et al. Highly Sensitive and Specific Detection of Rare Variants in Mixed Viral Populations from Massively Parallel Sequence Data. *PLoS Comput. Biol.* **2012**, *8*, e1002417. [CrossRef] [PubMed]

15. Yang, X.; Patrick, C.; Macalalad, A.; Henn, M.R.; Zody, M.C. V-Phaser 2: Variant inference for viral populations. *BMC Genom.* **2013**, *14*, 674. [CrossRef]

16. Sandmann, S.; de Graaf, A.O.; Karimi, M.; van der Reijden, B.A.; Hellström-Lindberg, E.; Jansen, J.H.; Dugas, M. Evaluating Variant Calling Tools for Non-Matched Next-Generation Sequencing Data. *Sci. Rep.* **2017**, *7*, 43169. [CrossRef]

17. Ibrahim, B.; McMahon, D.P.; Hufsky, F.; Beer, M.; Deng, L.; Mercier, P.L.; Palmarini, M.; Thiel, V.; Marz, M. A new era of virus bioinformatics. *Virus Res.* **2018**, *251*, 86–90. [CrossRef] [PubMed]

18. Kuiken, C.; Combet, C.; Bukh, J.; Shin-I, T.; Deleage, G.; Mizokami, M.; Richardson, R.; Sablon, E.; Yusim, K.; Pawlotsky, J.M.; et al. A comprehensive system for consistent numbering of HCV sequences, proteins and epitopes. *Hepatology* **2006**, *44*, 1355–1361. [CrossRef]

19. Verbist, B.M.; Aerssens, J.; Reumers, J.; Thys, K.; Van der Borght, K.; Clement, L.; Thas, O.; Talloen, W.; Wetzels, Y. VirVarSeq: A low-frequency virus variant detection pipeline for Illumina sequencing using adaptive base-calling accuracy filtering. *Bioinformatics* **2014**, *31*, 94–101. [CrossRef] [PubMed]

20. Döring, M.; Büch, J.; Friedrich, G.; Pironti, A.; Kalaghatgi, P.; Knops, E.; Heger, E.; Obermeier, M.; Däumer, M.; Thielen, A.; et al. geno2pheno[ngs-freq]: a genotypic interpretation system for identifying viral drug resistance using next-generation sequencing data. *Nucleic Acids Res.* **2018**, *46*, W271–W277. [CrossRef] [PubMed]

21. Schirmer, M.; Sloan, W.T.; Quince, C. Benchmarking of viral haplotype reconstruction programmes: An overview of the capacities and limitations of currently available programmes. *Brief. Bioinform.* **2012**, *15*, 431–442. [CrossRef]

22. Singer, J.B.; Thomson, E.C.; McLauchlan, J.; Hughes, J.; Gifford, R.J. GLUE: A flexible software system for virus sequence data. *BMC Bioinform.* **2018**, *19*, 532. [CrossRef] [PubMed]

23. Vermehren, J.; Park, J.S.; Jacobson, I.M.; Zeuzem, S. Challenges and perspectives of direct antivirals for the treatment of hepatitis C virus infection. *J. Hepatol.* **2018**, *69*, 1178–1187. [CrossRef] [PubMed]

24. Sorbo, M.C.; Cento, V.; Maio, V.C.D.; Howe, A.Y.; Garcia, F.; Perno, C.F.; Ceccherini-Silberstein, F. Hepatitis C virus drug resistance associated substitutions and their clinical relevance: Update 2018. *Drug Resist. Updates* **2018**, *37*, 17–39. [CrossRef] [PubMed]

25. European Association for the Study of the Liver. Recommendations on Treatment of Hepatitis C 2016. *J. Hepatol.* **2017**, *66*, 153–194. [CrossRef] [PubMed]

26. IUPAC-IUB Commission on Biochemical Nomenclature (CBN). Abbreviations and symbols for nucleic acids, polynucleotides and their constituents. Recommendations 1970. *Biochem. J.* **1970**, *120*, 449–454. [CrossRef]

27. HCV-GLUE: A Sequence Data Resource for Hepatitis C Virus. Available online: http://hcv.glue.cvr.ac.uk (accessed on 24 February 2019).

28. Kalaghatgi, P.; Sikorski, A.M.; Knops, E.; Rupp, D.; Sierra, S.; Heger, E.; Neumann-Fraune, M.; Beggel, B.; Walker, A.; Timm, J.; et al. Geno2pheno[HCV]—A Web-based Interpretation System to Support Hepatitis C Treatment Decisions in the Era of Direct-Acting Antiviral Agents. *PLoS ONE* **2016**, *11*, e0155869. [CrossRef]

29. Batty, E.M.; Wong, T.H.N.; Trebes, A.; Argoud, K.; Attar, M.; Buck, D.; Ip, C.L.C.; Golubchik, T.; Cule, M.; Bowden, R.; et al. A Modified RNA-Seq Approach for Whole Genome Sequencing of RNA Viruses from Faecal and Blood Samples. *PLoS ONE* **2013**, *8*, e66129. [CrossRef] [PubMed]

30. Lamble, S.; Batty, E.; Attar, M.; Buck, D.; Bowden, R.; Lunter, G.; Crook, D.; El-Fahmawi, B.; Piazza, P. Improved workflows for high throughput library preparation using the transposome-based Nextera system. *BMC Biotechnol.* **2013**, *13*, 104. [CrossRef] [PubMed]

31. Davalieva, K.; Kiprijanovska, S.; Plaseska-Karanfilska, D. Fast, reliable and low cost user-developed protocol for detection, quantification and genotyping of hepatitis C virus. *J. Virol. Methods* **2014**, *196*, 104–112. [CrossRef]

32. Bonsall, D.; Ansari, M.; Ip, C.; Trebes, A.; Brown, A.; Klenerman, P.; Buck, D.; STOP-HCV Consortium.; Piazza, P.; Barnes, E.; Bowden, R. ve-SEQ: Robust, unbiased enrichment for streamlined detection and whole-genome sequencing of HCV and other highly diverse pathogens [version 1; referees: 2 approved, 1 approved with reservations]. *F1000Research* **2015**, *4*, 1062. [CrossRef] [PubMed]

33. TrimGalore: A Wrapper around Cutadapt and FastQC to Consistently Apply Adapter and Quality Trimming to FastQ Files, with Extra Functionality for RRBS Data. Available online: https://github.com/FelixKrueger/TrimGalore (accessed on 24 February 2019).

34. Sreenu, V.B. (MRC-University of Glasgow Centre for Virus Research, Glasgow, UK). Personal Communication, 2019.

35. Gaidatzis, D.; Lerch, A.; Hahne, F.; Stadler, M.B. QuasR: Quantification and annotation of short reads in R. *Bioinformatics* **2014**, *31*, 1130–1132. [CrossRef]

36. Martin, M. Cutadapt removes adapter sequences from high-throughput sequencing reads. *EMBnet J.* **2011**, *17*, 10–12. [CrossRef]

37. V-FAT: A Post-Assembly Pipeline for the Finishing and Annotation of Viral Genomes. Available online: https://www.broadinstitute.org/viral-genomics/v-fat (accessed on 24 February 2019).

38. SMALT: A Mapper for DNA Sequencing Reads. Available online: https://www.sanger.ac.uk/science/tools/smalt-0 (accessed on 24 February 2019).

39. Camacho, C.; Coulouris, G.; Avagyan, V.; Ma, N.; Papadopoulos, J.; Bealer, K.; Madden, T.L. BLAST+: Architecture and applications. *BMC Bioinform.* **2009**, *10*, 421. [CrossRef] [PubMed]

40. Harris, R.S. Improved Pairwise Alignment of Genomic DNA. Ph.D. Thesis, Pennsylvania State University, State College, PA, USA, 2007.

41. Smith, D.B.; Bukh, J.; Kuiken, C.; Muerhoff, A.S.; Rice, C.M.; Stapleton, J.T.; Simmonds, P. Expanded classification of hepatitis C virus into 7 genotypes and 67 subtypes: Updated criteria and genotype assignment web resource. *Hepatology* **2014**, *59*, 318–327. [CrossRef]

42. Stamatakis, A. RAxML version 8: A tool for phylogenetic analysis and post-analysis of large phylogenies. *Bioinformatics* **2014**, *30*, 1312–1313. [CrossRef] [PubMed]

43. Katoh, K.; Standley, D.M. MAFFT Multiple Sequence Alignment Software Version 7: Improvements in Performance and Usability. *Mol. Biol. Evol.* **2013**, *30*, 772–780. [CrossRef] [PubMed]

44. Poordad, F.; Pol, S.; Asatryan, A.; Buti, M.; Shaw, D.; Hézode, C.; Felizarta, F.; Reindollar, R.W.; Gordon, S.C.; Pianko, S.; et al. Glecaprevir/Pibrentasvir in patients with hepatitis C virus genotype 1 or 4 and past direct-acting antiviral treatment failure. *Hepatology* **2018**, *67*, 1253–1260. [CrossRef] [PubMed]

45. A Java API for High-Throughput Sequencing Data (HTS) Formats. Available online: http://samtools.github.io/htsjdk/ (accessed on 24 February 2019).

46. DiversiTools: Tool for Analysing Viral Diversity from HTS. Available online: http://josephhughes.github.io/DiversiTools/ (accessed on 24 February 2019).

Article

The Utility of Data Transformation for Alignment, De Novo Assembly and Classification of Short Read Virus Sequences

Avraam Tapinos [1,*], Bede Constantinides [1,2], My V. T. Phan [3], Samaneh Kouchaki [1,4], Matthew Cotten [3,5,6] and David L. Robertson [1,5]

[1] School of Biological Sciences, The University of Manchester, Manchester M13 9PT, UK; bede.constantinides@manchester.ac.uk (B.C.); samaneh.kouchaki@eng.ox.ac.uk (S.K.); david.l.robertson@glasgow.ac.uk (D.L.R.)
[2] Modernising Medical Microbiology Consortium, Nuffield Department of Clinical Medicine, John Radcliffe Hospital, University of Oxford, Oxford OX3 9DU, UK
[3] Department of Viroscience, Erasmus Medical Centre, Doctor Molewaterplein 40, 3015 GD Rotterdam, The Netherlands; v.t.m.phan@erasmusmc.nl (M.V.T.P.); mlcotten13@gmail.com (M.C.)
[4] Department of Engineering Science, Institute of Biomedical Engineering, University of Oxford, Oxford OX3 7DQ, UK
[5] MRC-University of Glasgow Centre for Virus Research, Glasgow G61 1QH, UK
[6] MRC/UVRI & LSHTM Uganda Research Unit Entebbe, P.O. Box 49 Entebbe, Uganda
[*] Correspondence: avraam.tapinos@manchester.ac.uk; Tel.: +44-(0)-161-701-7563

Received: 30 March 2019; Accepted: 22 April 2019; Published: 26 April 2019

Abstract: Advances in DNA sequencing technology are facilitating genomic analyses of unprecedented scope and scale, widening the gap between our abilities to generate and fully exploit biological sequence data. Comparable analytical challenges are encountered in other data-intensive fields involving sequential data, such as signal processing, in which dimensionality reduction (i.e., compression) methods are routinely used to lessen the computational burden of analyses. In this work, we explored the application of dimensionality reduction methods to numerically represent high-throughput sequence data for three important biological applications of virus sequence data: reference-based mapping, short sequence classification and de novo assembly. Leveraging highly compressed sequence transformations to accelerate sequence comparison, our approach yielded comparable accuracy to existing approaches, further demonstrating its suitability for sequences originating from diverse virus populations. We assessed the application of our methodology using both synthetic and real viral pathogen sequences. Our results show that the use of highly compressed sequence approximations can provide accurate results, with analytical performance retained and even enhanced through appropriate dimensionality reduction of sequence data.

Keywords: alignment; assembly; taxonomic classification; time series; data transformation; DWT; DFT; PAA; data compression; compressive genomics

---

1. Introduction

Next-generation sequencing (NGS) enables massively parallel determination of nucleotide order within genetic material, making it possible to rapidly sequence the genomes of individuals, populations and metagenomic samples [1–5]. However, the sequences generated by these instruments are almost always considerably shorter in length than the genomic regions studied. Genomic analyses often begin with the process of sequence assembly, where sequence fragments (reads) are reconstructed into the larger sequences from which they originated. Computational methods play a vital role in the assembly of short reads, and a variety of assemblers and related tools have been developed in tandem with emerging sequencing platforms [6]. All subsequent analyses and investigations depend upon the quality, accuracy and speed of this crucial sequence assembly process.

There are many computational methods to generate consensus sequences representing the genomes of species in a sample. Such approaches include seed-and-extend alignment methods using suffix array derivatives, such as the Burrows-Wheeler Transform (BWT) for aligning short reads informed by a known reference sequence [7,8], graph-based methods employing Overlap Layout Consensus (OLC) [9,10] and de Bruijn graphs of k-mers [11–13] for reference-free de novo sequence assembly. However, for sequencing projects to characterise genetic variation within populations (deep sequencing), metagenomics and pathogen discovery, the effectiveness of the aforementioned approaches varies considerably [14].

Samples with mixed viral infections, especially those comprising divergent variants, present a number of analytical and computational problems. The use of a reference sequence, even the use of a data specific generated sequence, can lead to valuable read information being discarded during the alignment process [15]. On the other hand, while de novo approaches require little a priori knowledge of target sequence composition, the methods are computationally intensive, and their performance scales poorly with datasets of increasing size [9]. Aggressive heuristics must be employed, to traverse graphs and deal with mismatches, reduce the running time of de novo assemblers, which, in turn, can compromise assembly quality. Indexing structures such as the BWT and its relatives are widely used to reduce the burden of pairwise sequence comparison, for both reference-based mapping and de novo assembly. However, they cannot process mismatches within reads, necessitating the use of computationally expensive heuristics to establish relationships between divergent sequences. Increasing sequence length further affects the performance of these approaches [16].

A major challenge in working with NGS data from metagenomic studies is the high levels of diversity present, particularly for the virus genetic material. Also, the number of sequences generated challenge many computational systems for a feasible working solution in terms of time and the computational resources typically available in biological laboratories. For biologists working on outbreak responses or pathogen discovery, both the accuracy of the assembly results and the speed of sequence analyses (e.g., assembly, alignment and pathogen classification) are crucial for crisis response and management. The ability to run analyses in the field on portable computer systems without internet connectivity is also important. Here, we explore the utility of data transform methods to extract major features from viral NGS sequence data and use the features to analyse data in a lower dimensional space.

Similar analytical challenges involving high dimensional sequential data are encountered in other data-intensive fields, such as signal and image processing, and time series analysis, where data transforms and approximation techniques are used for data dimensionality reduction. Data transform/approximation techniques include the discrete Fourier transform (DFT) [17], the discrete wavelet transform (DWT) [18,19] and piece-wise aggregate approximation (PAA) [20,21]. The DFT or DWT are used to transform data to their frequency domains, allowing feature extraction [22], and PAA is used as a data approximation approach. In data-intensive fields, data transformations/approximations are commonly used as dimensionality reduction approaches for obtaining fast approximate solutions for a given problem. Due to the ordered nature of genetic data, many of these transformation approaches can be applied to sequences of nucleotides [23] or amino acids [24]. An example of a successful implementation of a Fourier transform in computational biology is the multiple sequence alignment based on fast Fourier transform alignment algorithm MAFFT [25] where the physiochemical properties of amino acids are used to represent sequences for fast matching of homologous sequence regions for alignment. Since most transformation approaches are suitable only for numerical sequences, the strings of letters representing genetic sequences must be mapped into numerical space using a numerical sequence representation method [26].

In addition to the DFT, the DWT and PAA, suitable methods for measuring the pairwise similarity of sequential data or transformations include the Lp-norms [27], dynamic time warping (DTW) [28], longest common subsequence (LCS) [29], and alignment approaches, such as the Needleman-Wunsch and

Smith-Waterman algorithms. Euclidean distance is arguably the most widely used Lp-norm method for sequential data comparison but can only be used on sequences of the same length. Furthermore, Lp-norm methods do not accommodate shifts in the *x*-axis (time or position) and are thus limited in their ability to identify similar features within offset data. Elastic similarity/dissimilarity methods, such as LCS, unbounded DTW and various alignment algorithms, permit comparison of data with different dimensions and tolerate shifts in the *x*-axis. These properties of elastic similarity methods can be very useful in the analysis of speech signals, for example, but can be computationally expensive [30]. Several approaches have been proposed to permit fast searching with DTW, including the introduction of global constraints (wrapping path) or the use of lower bounding techniques, such as LB_keogh [28].

While pairwise comparison methods may be used for clustering, classification and similarity searches, they are very time consuming for large datasets ($O(n^2)$ time complexity). Indexing structures, such as the *R**-tree, *KD*-tree, *VP*-tree and *MVP*-tree have significantly lower time complexity ($O(n\ log(n))$) for similarity search [31] and are more appropriate for efficient analysis of large datasets. The *R**-tree [32,33] and *KD*-tree [34] indexing structures are very accurate for low dimensional datasets. However, their performance deteriorates significantly in high dimensional space [31], a phenomenon known as the 'curse of dimensionality' [35,36]. Metric trees, such as the *VP*-tree [37] and *MVP*-tree [38], are less prone to this limitation. Metric space indexing structures make use of geometric properties for partitioning data and work efficiently on both low and high dimensional data [39]. The curse of dimensionality can be further mitigated using data approximations, such as the DFT, the DWT and the PAA, to partition a dataset in an approximated space without loss of generality [21].

Here, we investigate the performance of three established dimensionality reduction techniques on three common analysis tasks involving viral short read sequence data: classification, reference-based mapping/alignment and de novo assembly. We benchmarked the accuracy of our proposed methodology against existing tools, and demonstrate the applicability of time series and signal processing data mining techniques for the analysis of viral NGS data.

2. Materials and Methods

2.1. Symbolic to Numeric Sequence Representations

Various numeric sequence representation methods can be used for symbolising a nucleotide sequence to a numerical space (see 51). Depending on the chosen numerical representation, each nucleotide is associated with a specific numerical value or vector. The specific values are assigned to the position of each nucleotide indicating the presence of a nucleotide at each sequence position (Equation 1). R_i is the indicator for a specific nucleotide in the ith position of the sequence S with a length of n nucleotides. Values $v_1 \ldots v_5$ correspond to the numerical value or numerical vector associated with each nucleotide.

$$R = \begin{cases} v_1 \; if \; i = A \\ v_2 \; if \; i = T \\ v_3 \; if \; i = C \quad \forall i \in S \\ v_4 \; if \; i = G \\ v_5 \; otherwise \end{cases} \tag{1}$$

Methods, such as the electron-ion interaction pseudopotentials (EIIP) [40] and the atomic representation approach [41], aim to mimic the biochemical properties of nucleic acids but introduce some mathematical bias that does not exist in reality [26]. Other methods, like the Voss indicator [42] and the Tetrahedron approach, do not introduce internucleotide mathematical bias, meaning the pairwise distances between each non-identical transformed nucleotide are the same (for example, the distance between A and T is equal to the distances between A and C as well as A and G). Furthermore, the cumulative sum of a numerical representation R can be used to indicate the trajectory of a sequence in nucleotide space. Table 1 indicates the values used for different representation methods [26].

Table 1. Numerical nucleotide sequence representation methods.

Method	Numerical Representation
Integer number	$A = 1,\ C = -1,\ G = 2,\ T = -2,\ N = 0$
Real number	$A = -1.5,\ C = 0.5, G = -0.5,\ T = 1.5,\ N = 0.0$
EIIP	$A = 0.1260,\ C = 0.1340,\ G = 0.0806,\ T = 0.1335,\ N = 0$
Atomic	$A = 70,\ C = 58,\ G = 78,\ T = 66,\ N = 0$
Pair	$A\ or\ T = 1,\ C\ or\ G = -1,\ N = 0$
Complex number	$A = 1 + 1i,\ C = -1 + 1i,\ G = -1 - 1i,\ T = 1 - 1i,\ N = 0 + 0i$
DNA Walk	$A = [1,0],\ C = [0,1],\ G = [0,-1],\ T = [-1,0],\ N = [0,0]$
Tetrahedron	$A = [0,0,1],\ C = \left[-\frac{\sqrt{2}}{3}, -\frac{\sqrt{6}}{3}, \frac{1}{3}\right],$ $G = \left[-\frac{\sqrt{2}}{3}, -\frac{\sqrt{6}}{3}, -\frac{1}{3}\right],\ T = \left[2 \times \frac{\sqrt{2}}{3}, 0, -\frac{1}{3}\right],\ N = [0,0,0]$
Voss indicator	$A = [0,0,1,0],\ C = [1,0,0,0],\ G = [0,1,0,0],\ T = [0,0,0,1],\ N = [0,0,0,0]$

2.2. Sequence Transformation

Effective methods for transforming/approximating sequential data should: (i) accurately transform/approximate data without loss of useful information, (ii) have low computational overheads, (iii) facilitate rapid comparison of data and (iv) provide lower bounding—where the distance between data representations is always less than or equal to that of the original data—precluding false negative results [43]. The lower bounding property guarantees that if two data points are nearby in their original space, they will remain so in their transformed/approximate space. We employ the DFT and the DWT transformation methods and the PAA approximation method as they satisfy the above requirements, and these are widely used for analysing discrete signals [44] and can be used to transform/approximate nucleotide sequence numerical representations to different levels of resolution, permitting reduced dimensionality sequence analysis.

Figure 1A illustrates an example of the DFT and DWT transformations and PAA approximation of a short nucleotide sequence. The DFT and the fast Fourier transform (FFT) convert data from their original domain into the frequency domain. In principle, the DFT decomposes a numerically represented nucleotide sequence with n positions (dimensions) into a series of n frequency components ordered by their frequency. A subset of the resulting Fourier frequencies are used to approximate the original sequence in a lower dimensional space [17], and the tradeoff between analytical speed and accuracy can be varied according to the number of frequencies considered [45].

The DWT transforms data into the time-frequency domain, capturing both frequency and temporal location information [18,46,47], in contrast to DFT, which only provides frequency information. DWT is a set of averaging and differencing functions that may be used recursively to represent sequential data at different resolutions, and each resolution can be used as an approximation of the original data. Figure 1B depicts DWT transformations of a short nucleotide sequence.

In PAA, a numerical sequence is divided into n equally sized windows, the mean values of which together form a compressed sequence representation [20,21]. The selection of n determines the resolution of the compressed or approximate representation. While PAA is faster and easier to implement than the DFT and the DWT, unlike these two methods, PAA is irreversible, meaning that the original sequence cannot be recovered from the approximation. Figure 1C depicts an example of the PAA transformations of a short nucleotide sequence.

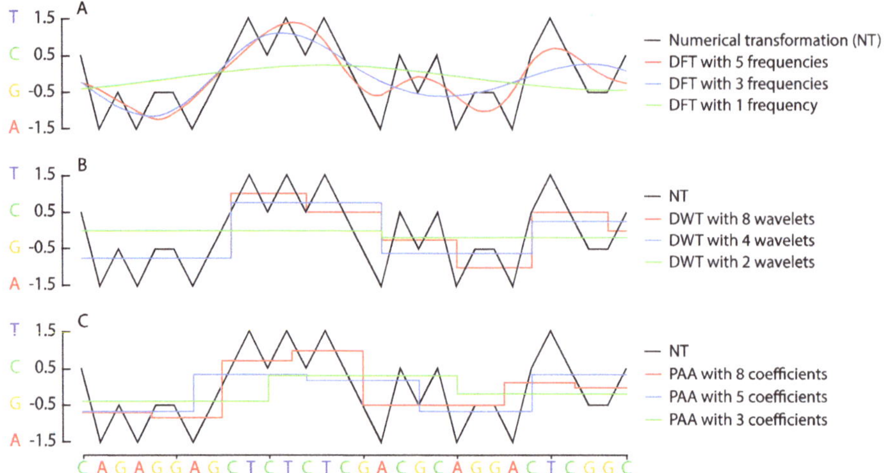

Figure 1. A numerically represented DNA sequence transformed at various levels of spatial resolution using the discrete Fourier transform (DFT) of the whole sequence (**A**), the Haar discrete wavelet transform (DWT) (**B**) and piecewise aggregate approximation (PAA) (**C**). A 30 nucleotide sequence (*x*-axis) is represented as a numerical sequence (black lines) using the real number representation method (*y*-axis where T = 1.5, C = 0.5, G = −0.5 and A = −1.5) for DFT approximations of the sequence with 5 (red), 3 (blue) and 1 (green) Fourier frequencies (**A**); DWT approximations of the same sequence with 8 level wavelets (red), 4 level wavelets (blue) and 2 level wavelets (green) (**B**); PAA approximations of the same sequence with 8 (red), 5 (blue) and 3 (green) coefficients (**C**).

2.3. Similarity Search Approaches for Sequential Data

Here, we adopt the Euclidian distance and *VP*-tree index to perform a fast *k*-nearest neighbour (*k*-NN) similarity search for aligning the reads to a reference genome.

In a *VP*-tree indexing structure, data is segregated using the distance between data points, thus implementing data partitioning in a metric space. A data point to use as a vantage point is selected (either randomly or by applying some heuristic to find and use the furthest point in the dataset [37]), and the rest of the data points are partitioned into two nodes based on their distance to that point. Data found to be closer to the vantage point than a given threshold (the median distance between all the data points and the vantage point) are assigned to the same node, and the rest of the data points to a different node. This function is repeated recursively in order to complete the partitioning process. The resulting indexing structure can then be used for fast identification of a *k*-nearest neighbour (*k*-NN) search. A *k*-NN-search returns the data points that are closest to a query *q*. Initially, the distance between the query *q* and the vantage point in the top node is calculated. If the distance between *q* and the vantage point satisfies a set of given conditions (the distance is smaller or larger than a given threshold – this threshold being the median distance between the vantage point and other data points within the node), a decision is made to visit either one or both of the child nodes. This process is repeated until the entire tree has been traversed. The *k* data points—in this case, reads—found closest to our query are the *k*-nearest neighbours to the query *q*.

2.4. Proposed Short Reads Processing Methodology

Our methodology for taxonomic classification, reference-based mapping and de novo assembly of short reads used time series and digital signal processing data transformation techniques. Figure 2 illustrates the fundamental concept of our approach. The short reads and reference genomes are mapped to a numerical space using an appropriate method from Table 1. Subsequently, lower dimensional approximations were generated for all data using the appropriate data transformation method, such as DFT, DWT and PAA. A *VP*-tree was constructed to allow fast data comparison. Depending on the application, the *VP*-tree was constructed either by using *k*-mer transformations obtained from the reference genomes or by using the short reads' transformations. Consequently, the best matches for our short reads' transformations were identified using a *k*-NN search approach on the *VP*-tree. As a final step, the results obtained from the *k*-NN search were re-evaluated in the original space to remove potential false positive results.

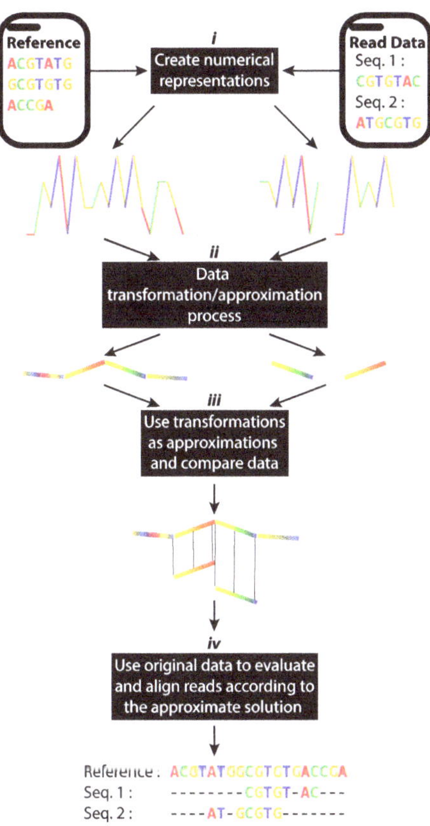

Figure 2. Overview of our proposed methodology using time series transformation/approximation methods: (*i*) Creation of numerical representations of input sequences. (*ii*) Application of an appropriate signal decomposition method to transform sequences into their feature space. (*iii*) Use of approximated transformations to perform rapid data analysis in lower dimensional space. (*iv*) Validation of inferences against original, full-resolution input sequences. In the case of reference-based alignment and taxonomic classification, approximated read transformations were compared with a reference sequence. In our de novo implementation, pairwise comparisons were performed between all of the approximated read transformations.

2.5. Data

The implementations of our proposed methodologies were assessed with both simulated and real virus datasets. The simulated datasets were generated using CuReSim [48] and WGSIM (https://github.com/lh3/wgsim). Simulated data included information, such as the reference genome used, the alignment position and alignment direction, for each read, enabling rigorous evaluation of the proposed techniques. We used two simulators to examine our approach in a variety of use cases. CuReSim is highly customisable, allowing the user to control the type of variation (insertion, deletion and substitution) to simulate. WGSIM can simulate genomes with uniform insertion, deletion and substitution variation.

CuReSim was used to generate 16 HIV-1 HXB2 simulated datasets with different levels and types of variation. WGSIM was used to generate 4 mixed virus datasets with different levels of variation. Each simulation contained 200,000 reads generated using 5 Norovirus, 5 Ebola virus and 5 Respiratory syncytial virus (RSV) genomes, with various types and extents of simulated variation. HXB2 and simulated mixed virus datasets and corresponding reference genomes used to simulate them are deposited on GitHub (https://github.com/Avramis/Supporting-data/tree/master/Simulated%20Data). Table 2 contains detailed information about the simulated datasets.

Table 2. Simulated read data. Each row contains details for each simulated dataset (i.e., virus family, virus, GenBank ID, variation type, variation level, number of reads and simulator used to generate data). Abbreviations: Ins, insertions; Del, deletions and Sub, substitutions.

Family	Virus	GenBank Genome ID	Variation Type (%)			Reads	Simulator
			Ins	Del	Sub		
HIV	HXB2	K03455	0.0	0.0	0.0	2133	CuReSim
HIV	HXB2	K03455	0.0	0.0	1.0	2133	CuReSim
HIV	HXB2	K03455	0.0	0.0	2.0	2133	CuReSim
HIV	HXB2	K03455	0.0	0.0	3.0	2133	CuReSim
HIV	HXB2	K03455	0.0	0.0	4.0	2133	CuReSim
HIV	HXB2	K03455	0.0	0.0	5.0	2133	CuReSim
HIV	HXB2	K03455	0.5	0.5	0.0	2133	CuReSim
HIV	HXB2	K03455	1.0	1.0	0.0	2133	CuReSim
HIV	HXB2	K03455	1.5	1.5	0.0	2133	CuReSim
HIV	HXB2	K03455	2.0	2.0	0.0	2133	CuReSim
HIV	HXB2	K03455	2.5	2.5	0.0	2133	CuReSim
HIV	HXB2	K03455	0.5	0.5	1.0	2133	CuReSim
HIV	HXB2	K03455	1.0	1.0	2.0	2133	CuReSim
HIV	HXB2	K03455	1.5	1.5	3.0	2133	CuReSim
HIV	HXB2	K03455	2.0	2.0	4.0	2133	CuReSim
HIV	HXB2	K03455	2.5	2.5	5.0	2133	CuReSim
Mixed Viruses: Caliciviridae, Filoviridae, Pneumoviridae	Norovirus, Ebola virus, RSV	KM198529, KM198528, KM198511, KM198500, KM198486, KU296608, KU296553, KU296549, KU296528, KU296416, KP317952, KP317946, KP317934, KP317923, KP317922	0.0	0.0	0.0	200,000	WGSIM
Mixed Viruses: Caliciviridae, Filoviridae, Pneumoviridae	Norovirus, Ebola virus, RSV	KM198529, KM198528, KM198511, KM198500, KM198486, KU296608, KU296553, KU296549, KU296528, KU296416, KP317952, KP317946, KP317934, KP317923, KP317922	1.0	1.0	1.0	200,000	WGSIM
Mixed Viruses, Caliciviridae, Filoviridae, Pneumoviridae	Norovirus, Ebola virus, RSV	KM198529, KM198528, KM198511, KM198500, KM198486, KU296608, KU296553, KU296549, KU296528, KU296416, KP317952, KP317946, KP317934, KP317923, KP317922	3.33	3.33	3.33	100,000	WGSIM
Mixed Viruses, Caliciviridae, Filoviridae, Pneumoviridae	Norovirus, Ebola virus, RSV	KM198529, KM198528, KM198511, KM198500, KM198486, KU296608, KU296553, KU296549, KU296528, KU296416, KP317952, KP317946, KP317934, KP317923, KP317922	6.66	6.66	6.66	200,000	WGSIM

Furthermore, 15 publicly available real virus datasets were used for the evaluation of our methodology. The real datasets comprise 5 Norovirus, 5 Ebola virus and 5 human respiratory syncytial virus (RSV) short read datasets. Norovirus NGS datasets (ERR225628, ERR225629, ERR225631, ERR225632, ERR225633) were generated from diarrhoeal patients in Vietnam [49]. Group A rotavirus datasets were obtained from human and pig samples from Vietnam [50]. Human coronavirus NL63 datasets were obtained from Kenya [51]. The Ebola virus datasets (SRR3107337, SRR3107338, SRR3107340, SRR3107342, SRR3107343) were retrieved from the bioproject PRJNA309162, generated during the outbreaks in West Africa in 2013–2016 [52]. The human respiratory syncytial virus (RSV) datasets (ERR303259, ERR303260, ERR303261, ERR303262, ERR303263) [53] were generated from humans in Kenya. All 15 datasets are publicly available. The accession numbers of Sequence Read Archive (SRA) and European Nucleotide Archive (ENA) can be found in Table 3.

Table 3. Real short reads data. Rows contain information for each real reads' dataset (i.e., virus family, virus, genome strain GenBank ID, SRA project ID, number of reads and technology used to sequence data). SRA: Sequence Read Archive; ENA: European Nucleotide Archive.

Family	Virus	Amplicon/Random Primer	GenBank Genome ID	ENA/SRA_ID	Reads	Sequencing Technology
Caliciviridae	Norovirus	Amplicon	KM198486	ERR225628	2126502	Illumina MiSeq
Caliciviridae	Norovirus	Amplicon	KM198500	ERR225629	3037674	Illumina MiSeq
Caliciviridae	Norovirus	Amplicon	KM198511	ERR225631	3285078	Illumina MiSeq
Caliciviridae	Norovirus	Amplicon	KM198528	ERR225632	4361884	Illumina MiSeq
Caliciviridae	Norovirus	Amplicon	KM198529	ERR225633	5187234	Illumina MiSeq
Filoviridae	Ebola virus	Amplicon	KU296608	SRR3107337	522968	Ion Torrent PGM
Filoviridae	Ebola virus	Amplicon	KU296549	SRR3107338	771031	Ion Torrent PGM
Filoviridae	Ebola virus	Amplicon	KU296416	SRR3107340	186657	Ion Torrent PGM
Filoviridae	Ebola virus	Amplicon	KU296553	SRR3107342	478346	Ion Torrent PGM
Filoviridae	Ebola virus	Amplicon	KU296528	SRR3107343	42410	Ion Torrent PGM
Pneumoviridae	RSV	Amplicon	KP317934	ERR303259	7275032	Illumina MiSeq
Pneumoviridae	RSV	Amplicon	KP317922	ERR303260	9278070	Illumina MiSeq
Pneumoviridae	RSV	Amplicon	KP317946	ERR303261	11111114	Illumina MiSeq
Pneumoviridae	RSV	Amplicon	KP317923	ERR303262	13293226	Illumina MiSeq
Pneumoviridae	RSV	Amplicon	KP317952	ERR303263	15237848	Illumina MiSeq

The HIV-1 HXB2 genome (K03455) was used as a reference index to align and/or run the taxonomic classification analysis for the HXB2 simulated dataset. The Norovirus genome KM198509, the Ebola virus genome KM034562 and the RSV genome KP317934 were used as a reference index to align and/or run the taxonomic classification analysis for the mixed virus datasets. The Norovirus genome KM198509 was used to run the taxonomic classification analysis on the real Norovirus datasets, the Ebola virus genome KM034562 was used to run the taxonomic classification analysis on the real Ebola datasets and the RSV genome KP317934 was used to perform the taxonomic classification analysis on the real RSV datasets. All reference genomes used in this study are available from the NCBI (https://www.ncbi.nlm.nih.gov/genome), and accession numbers can be found in Table 4.

Table 4. Reference genomes used during classification and reference-based alignment.

Family	Virus	GenBank ID:	Length (nt)
Retroviridae	Human immunodeficiency virus 1 (HXB2)	K03455	9179
Caliciviridae	Norovirus	KM198509.1	7425
Filoviridae	Zaire ebolavirus	KM034562.1	18957
Pneumoviridae	Human orthopneumovirus (Respiratory Syncytial Virus)	KP317934.1	15233

2.6. Classification and Alignment Evaluation

The accuracy of a classification and an alignment tool can be quantified in terms of the *F*-measure [48], a balanced measure of precision and recall, with precision = true positive/true positive + false positive, recall = true positive/true positive + false negative and the *F*-measure = 2 × (precision *x* recall)/(precision + recall) [48]. In the case of simulated data, information concerning the position of the read on the reference and alignment direction can be used to establish the correctness of alignment, and thereby provide a more informative *F*-measure score. Unclassified reads are considered a false negative result. Any reported match to the correct region of the genome in the correct direction is considered a true positive result. However, if the alignment position or direction information is unavailable, the *F*-measure can be calculated from the number of hits reported for a read, or the absence of a hit. Again, unclassified reads are considered false negative results, and classified reads are considered true positive results. In the case of mixed genome data, the *F*-measure score can be calculated by taking into consideration the number of hits that are reported for a read, as well as if a read is assigned to a reference genome from the same family. If a read is assigned to a genome from a different virus family, it is considered a false positive result, while unclassified reads are considered a false negative result.

3. Results

3.1. Classification by Numbers (CBN)

For the taxonomic classification analysis, a classification tool was implemented in C++ (https://github.com/Avramis/ClassificationByNumbers). The implementation was developed to evaluate our methodology but was not optimised for speed. Users might specify parameters, such as the representation method, transformation method, search stringency and the *k*-mer length. A *VP*-tree indexing structure classified reads using a given set of genomic references. *VP*-tree construction began with the extraction of all unique *k*-mers, of a user-specified length *k*, from the set of supplied reference genomes. Each unique *k*-mer was represented in numerical sequence and then transformed into a lower dimensional space. The transformed data were then used to generate the *VP*-tree indexing structure. Subsequently, each short read from a query set was converted into numerical space, transformed to a lower dimensional space and evaluated against the *VP*-tree. The approximate solution arising from this was then evaluated using the original data to identify false positive matches. The CBN algorithm generated two output files. The first output was a text file providing detailed information on all of the classification matches generated for each read, including the reference name, the direction in which the query read was aligned to the reference, the start and end position of the query on the reference, the alignment score, the CIGAR string describing how the read aligns with the reference and the actual alignment of the query read on the reference genome. The second tabular output file provided a brief overview of the alignment. Each line contained the name of the read, the number of classifications generated for that particular read, the highest classification score obtained, the name of the reference, which provided the highest classification score, the alignment direction and starting position on the reference.

The CBN tool was evaluated against NCBI-BLAST 2.8.1 BLASTn [54] and Kaiju 1.6.3 [55] classifier tools. BLASTn performs the analysis in nucleotide space, whereas Kaiju translates nucleotide sequences from every possible reading frame and performs the analysis in protein sequence space. Figures 3–5 illustrate the results of the classification evaluation process. Both BLASTn and Kaiju were evaluated using their default parameters. CBN was evaluated using *k*-mers of 100, 150, 200, 250 and 300 for the HXB2 simulated reads and 50, 100 and 150 for the mixed virus and real datasets. For the DFT and PAA methods, we evaluated the use of transformation/approximations with 2, 4, 6, 8, 10 and 12 Fourier frequencies or PAA coefficients, respectively. For the DWT variant, we tested the cases of 2, 4, 8, 16 and 32 wavelets.

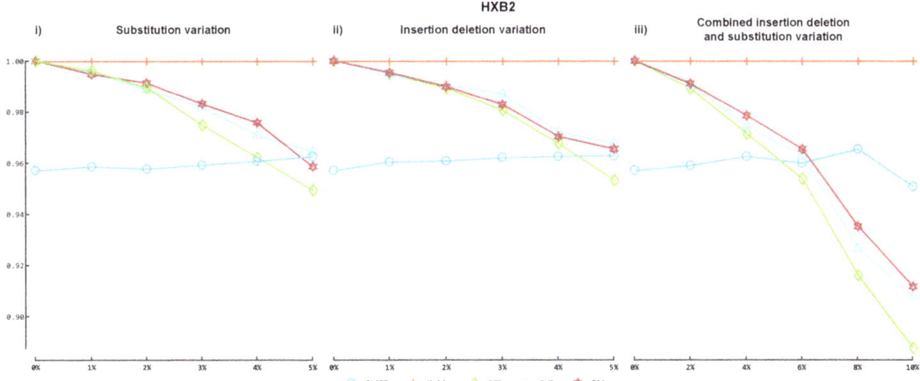

Figure 3. Accuracy of our prototype classification implementation and two established tools on HIV-1 HXB2 simulated datasets. All plots illustrate the *F*-measures obtained on the 16 different HIV datasets. The y-axis indicates the *F*-measure score, and the x-axis depicts the reads data files. Plot 3-i depicts the *F*-measures obtained for each classifier on the simulations with 0% to 5% of substitution variation rate. Plot 3-ii illustrates the *F*-measures obtained for each classifier on the simulations with 0% to 5% uniform insertion/deletion variation, and plot 3-iii illustrates the *F*-measures obtained for each tool on simulations of uniform 0% to 10% insertion/deletion and substitution variation.

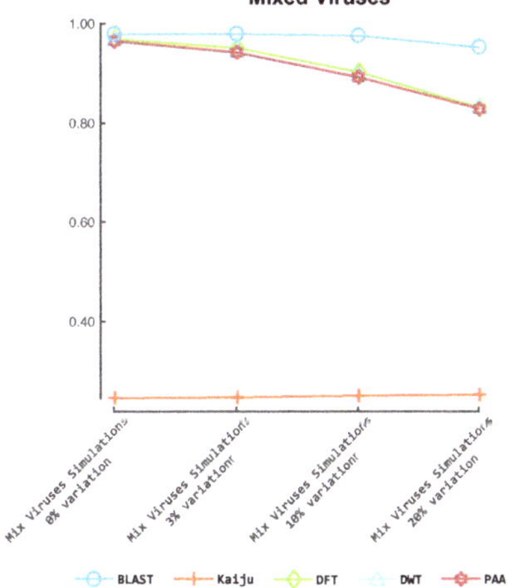

Figure 4. Accuracy of our prototype classification implementation and two established tools on mixed viruses simulated datasets. The *y*-axis indicates the *F*-measure score, and the *x*-axis depicts the reads data files. The plot depicts the *F*-measures obtained for each classifier on the mixed virus simulations. DFT: discrete Fourier transform; DWT: discrete wavelet transform; PAA: piece-wise aggregate approximation.

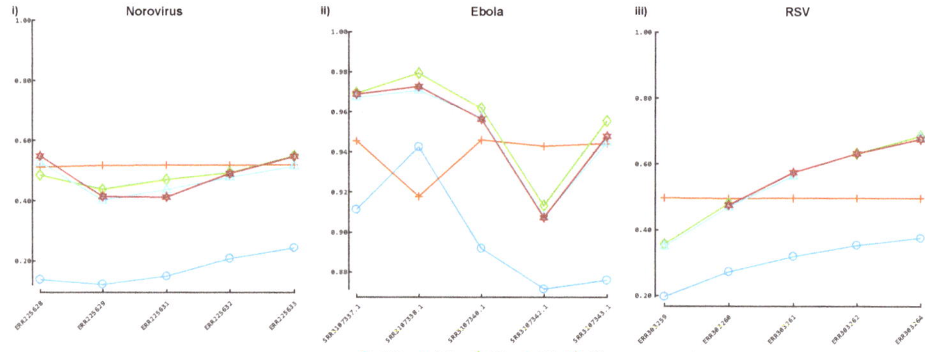

Figure 5. Accuracy of our prototype classification implementation and two established tools on real sequences. The *y*-axis indicates the *F*-measure score, and the *x*-axis depicts the reads data files. Plot 5-i depicts the *F*-measures obtained for each classifier on the Norovirus sequences data. Plot 5-ii illustrates the *F*-measures obtained for each classifier on the Ebola sequence data. Plot 5-iii illustrates the *F*-measures obtained for each tool on Respiratory syncytial virus (RSV) sequence data. DFT: discrete Fourier transform; DWT: discrete wavelet transform; PAA: piece-wise aggregate approximation.

Figure 3 shows the results obtained from the classification process on HIV-1 HXB2 data. Figure 4 illustrates the results of the mixed virus datasets. Figure 5 illustrates the results obtained from the real data. For taxonomic classification of HIV-1 HXB2 simulated reads, where the short reads were classified against the genome used to generate them, Kaiju reported the highest accuracy scores. CBN outperformed BLASTn in most cases, falling behind in terms of accuracy only on datasets with high variation rates. For the mixed virus simulated datasets, where reads were classified against species strains related to those used to generate reads, BLASTn correctly assigned the most species, followed closely by CBN and finally Kaiju. In the evaluation of the tools on the real data, where reads were classified using a publicly available species-specific reference sequence, CBN generated more accurate results than other tools, followed by Kaiju and BLASTn.

3.2. Alignment by Numbers (ALBN)

To test the applicability of sequential data transformations and feature selection for read alignment, we implemented a prototype *k*-NN read aligner (Figure 6) in C++ (available at https://github.com/Avramis/Alignment_by_numbers). As with the CBN classification analysis, the ALBN code was not optimised for speed. Users might specify parameters, such as the representation method, transformation method, search stringency and the *k*-mer length used for seeding alignments. The algorithm's output was used to construct gapped alignments in the widely used Sequence Alignment/Map (SAM) file format.

1) Represent short reads and reference genome as numerical sequences.
2) Select a *k*-mer length.
3) Create transformations of each reference sequence *k*-mer, build *VP*-tree, and create transformations of the initial *k*-mer of each short read.
4) Identify candidate alignments using data transformations.
for each read *i*
    candidate_alignments[*i***] =** *VPtree.k***-NNSearch(query** *i***)**
end
5) Align approximate results with original data using the Smith-Waterman (SW) algorithm:
for each read *i*
    best_score = null
   best_aln = []
    for each k neighbour in candidate_alignments[*i***]**
       if SW_score(*k* **neighbour, read** *i***)**
         best_score = SW_score(*k* **neighbour, read** *i***)**
         best_aln = SW_aln(*k* **neighbour, read** *i***)**
       end
    end
end
6) Output alignment in Sequence Alignment/Map (SAM) format.

Figure 6. Pseudocode for the alignment procedure.

The ALBN tool was evaluated against a set of well-established, widely used, state-of-the-art tools, such as Bowtie2 (version 2.3.1) [56], BWA-MEM (version 0.7.16) [7], GraphMap (version 0.5.2) [57] and Segmehl (version 0.3.4) [58]. Existing state-of-the-art tools were evaluated with default settings. ALBN was evaluated using *k*-mer lengths of 100, 150, 200, 250 and 300 for the HXB2 simulated reads and 50, 100 and 150 for the mixed virus and real datasets. For the DFT and PAA variants, we evaluated the use of transformation/approximations with 2, 4, 6, 8, 10 and 12 frequencies and PAA coefficients accordingly. For the DWT variant, we tested the cases of 2, 4, 8, 16 and 32 wavelets.

Each aligner's accuracy was quantified in terms of the *F*-measure [48]. CuReSim provides information, such as the simulated read's origin on the reference genome and its alignment direction, enabling evaluation of each aligner's output and calculation of alignment accuracy in terms of the *F*-measure. For mixed virus datasets, tool performance was evaluated in terms of ability to match and align reads to the appropriate virus reference genome. For the real data, *F*-measures were calculated according to the number of reads aligned to the given genome or otherwise.

Figures 7–9 illustrate the *F*-measures obtained by evaluating alignments from each aligner. Figure 7 illustrates alignment performance for each of the 16 datasets simulated using the K03455 HIV-1 HXB2 reference genome. Figure 8 illustrates the alignment performance for virus reads simulated with Norovirus genome KM198509.1, Ebola genome KM034562.1 and the RSV genome KP317934.1. Figure 9i–iii illustrate alignment performance (*F*-measure) for alignments of real Norovirus, Ebola virus and RSV sequences against the same reference genomes as those used for simulation.

ALBN provided accurate results in all scenarios tested. Regarding the HIV-1 HXB2 data, where short reads were aligned to the genome used to generate them, ALBN provided the most accurate results in all 16 cases, followed by Bowtie2. This was also the case for the mixed virus datasets, where reads were aligned to reference strains related to those used to generate the dataset. In both cases, GraphMap and BWA-MEM were third and fourth in terms of accuracy, respectively. ALBN also generated the most accurate alignment results using real data, where reads were aligned to species-specific reference genomes.

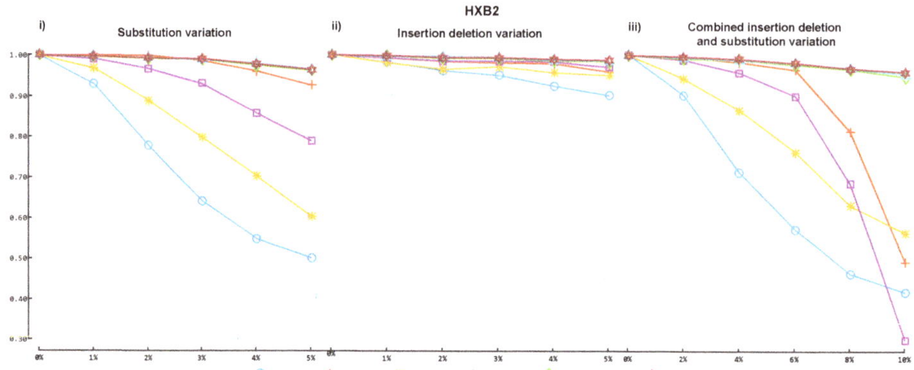

Figure 7. Accuracy of our prototype reference alignment implementation and four established tools on HIV-1 HXB2 simulated datasets. This Figure illustrates the *F*-measures obtained on the 16 different HIV datasets. Plot 6-(**i**) depicts the *F*-measures obtained for each aligner on the simulations with 0% to 5% of substitution variation rate. Plot 6-(**ii**) illustrates the *F*-measures obtained for each aligner on the simulations with 0% to 5% uniform insertion/deletion variation, and plot 6-(**iii**) illustrates the *F*-measures obtained for each tool on simulations of uniform 0% to 10% insertion/deletion and substitution variation. DFT: discrete Fourier transform; DWT: discrete wavelet transform; PAA: piece-wise aggregate approximation.

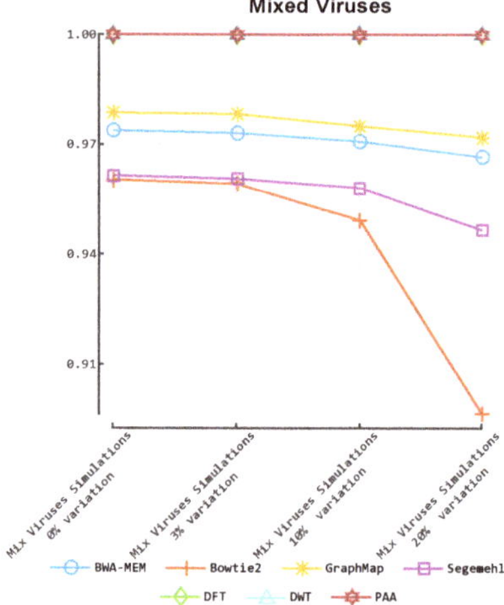

Figure 8. Accuracy of our prototype aligner implementation and four established tools on mixed viruses simulated datasets. The *y*-axis indicates the *F*-measure score, and the *x*-axis depicts the reads data files. The plot depicts the *F*-measures obtained for each aligner on the mixed virus simulations. DFT: discrete Fourier transform; DWT: discrete wavelet transform; PAA: piece-wise aggregate approximation.

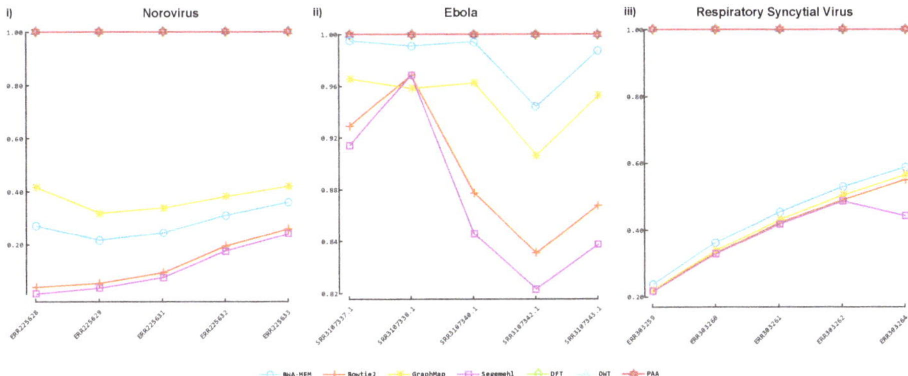

Figure 9. Accuracy of our prototype aligner implementation and four established tools on real sequences datasets. The y-axis indicates the F-measure score, and the x-axis depicts the reads data files. Plot 8-(**i**) depicts the F-measures obtained for each aligner on the Norovirus sequences data. Plot 8-(**ii**) illustrates the F-measures obtained for each aligner on the Ebola sequences data. Plot 8-(**iii**) illustrates the F-measures obtained for each tool on the Respiratory syncytial virus (RSV) sequences data. DFT: discrete Fourier transform; DWT: discrete wavelet transform; PAA: piece-wise aggregate approximation.

3.3. De novo Assembly by Numbers

Lastly, to test the applicability of this approach to the de novo assembly of short reads, we implemented assembly by numbers (ASBN), a prototype algorithm for all-against-all k-mer comparison, using data transformations/approximation. Note, preliminary results have been presented as a conference paper [59]. Figure 10 illustrates the main concept of our de novo assembly approach. For the ASBN tool, reads are represented as numerical sequences using an appropriate numerical representation method (Table 1). Here, we used the tetrahedron numerical representation. Every k-mer of each numerically represented read was identified and transformed to lower dimensional space using the chosen transformation method. All k-mers' transformations were used to build a VP-tree, to allow for fast data comparison. Afterwards, all k-mers were compared to the rest of the data using the VP-tree index. Information from the data comparison was used to construct a weighted graph similar to that shown in Figure 10A. The shortest path on the weighted graph was identified with a breadth-first search (BFS) (Figure 10B). Reads overlaps were used to generate an OLC alignment of short reads (Figure 10C).

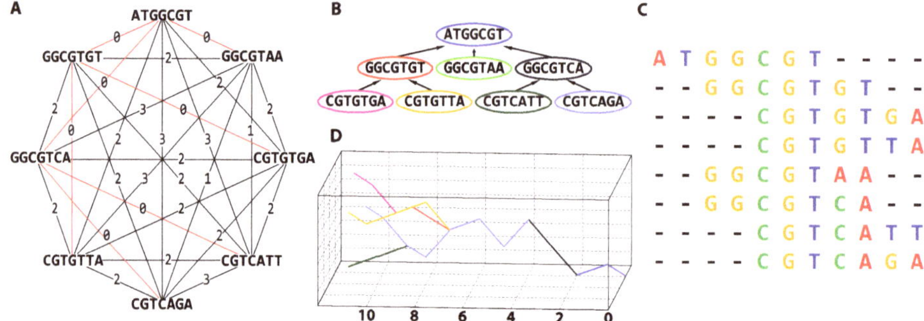

Figure 10. A de novo assembly methodology for numerically represented nucleotide reads. All-against-all sequence comparison (**A**) enables the construction of a read graph with weighted edges. The weight assigned to each edge is the smallest pairwise distance obtained between every possible *k*-mer representation of the two reads. In this example, a 5-mer was used. The smallest distance between every possible *k*-mer can be obtained by either using a sliding window approach or break reads every possible subsequence with length *k*. (**B**) The shortest path in the graph is identified with a breadth-first search algorithm (red coloured edges) thereby (**C**) enabling read alignment. A DNA walk representation of the overlapped reads (**D**) may subsequently be used as a three-dimensional graphical portrayal of the reads, illustrating alignment characteristics.

The ASBN assembler was compared with Megahit (version 1.1.3) [60] and SPAdes (version 3.13.0) [61] de novo assemblers on the HIV-1 HXB2, and mixed virus simulated datasets accordingly. Megahit, SPAdes, BLASTn and Kaiju were evaluated using default parameters. ASBN was evaluated using *k*-mer lengths 100, 150, 200, 250 and 300 for the HXB2 simulated reads and 50, 100 and 150 for the mixed virus datasets. For the DFT and PAA variants, we evaluated the use of transformation/approximations with 2, 4, 6, 8, 10 and 12 frequencies and PAA coefficients accordingly. For the DWT variant, we tested the cases of 2, 4, 8, 16 and 32 wavelets.

The derived contigs from each assembler were evaluated against the reference genomes used to generate the data simulations with BLASTn [54]. From the BLASTn output, information about the contigs' alignment position on the genome and the length of the alignment were obtained. Subsequently, a measure of assembly contiguity and the sum of gaps/mismatches were calculated and plotted on an X-Y matrix (similar to Figures 11 and 12) with *x* being the total coverage of the genomes generated and *y* being the total number of gaps in the coverage. A perfect assembly would have $x =$ full genome length and $y = 0$, indicating that the contig is identical to the genome in terms of length and nucleotide composition. For the HIV-1 HXB2 datasets, the contigs were evaluated against the K03455 genome, and the contigs obtained from the mixed virus datasets were evaluated against the 15 different genomes: KM198529, KM198528, KM198511, KM198500, KM198486, KU296608, KU296553, KU296549, KU296528, KU296416, KP317952, KP317946, KP317934, KP317923 and KP317922.

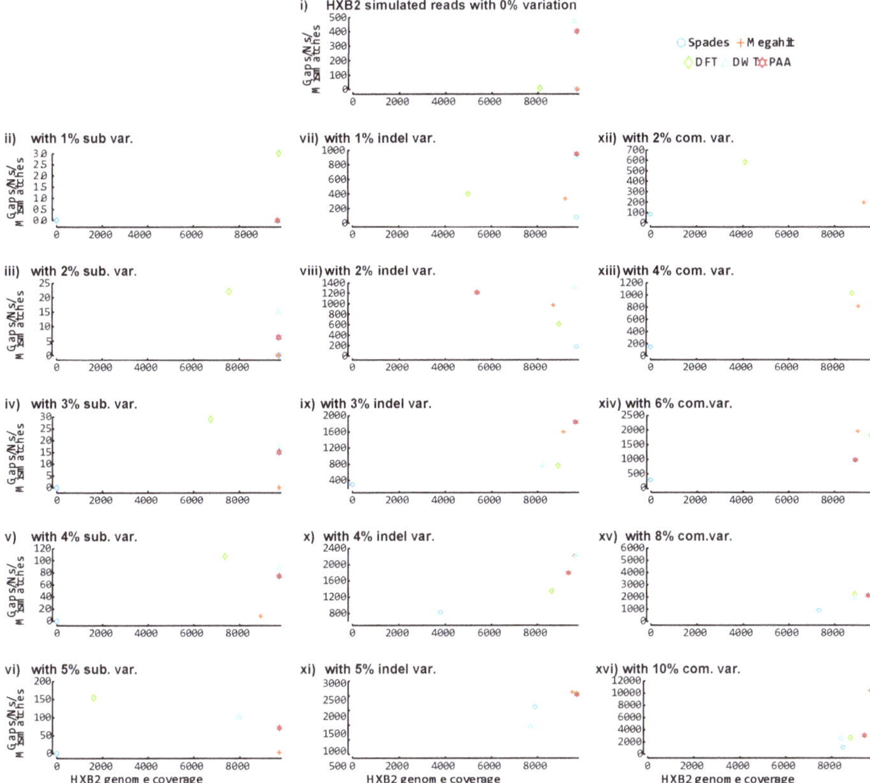

Figure 11. Accuracy of our prototype de novo assembly implementation and two established tools on HIV-1 HXB2 simulated datasets. The contigs obtained for each assembler were evaluated against the reference genome used to generate the simulated data. BLASTn was used to align all contigs to an HIV-1 HXB2 reference genome and determine genome coverage. The *y*-axis indicates the number of gaps and mismatches that exist in the contigs obtained for each tool, and the *x*-axis depicts the length of the genome the reported contigs cover. The contigs obtained from the assembly of the HIV-1 HXB2 simulated short read data were evaluated against the K03455 reference genome. Plot 10-**i** illustrates results obtained from all assemblers on variation-free data. Plots 10-**ii** to 10-**vi** illustrate results obtained from all assemblers on data with different levels of substitution variation. Plots 10-**vii** to 10-**xi** illustrate results obtained from all assemblers on data with different levels of insertion/deletion variation. Plots 10-**xii** to 10-**xvi** illustrate results obtained from all assemblers on data with different levels of combined insertion/deletion and substitution variation.

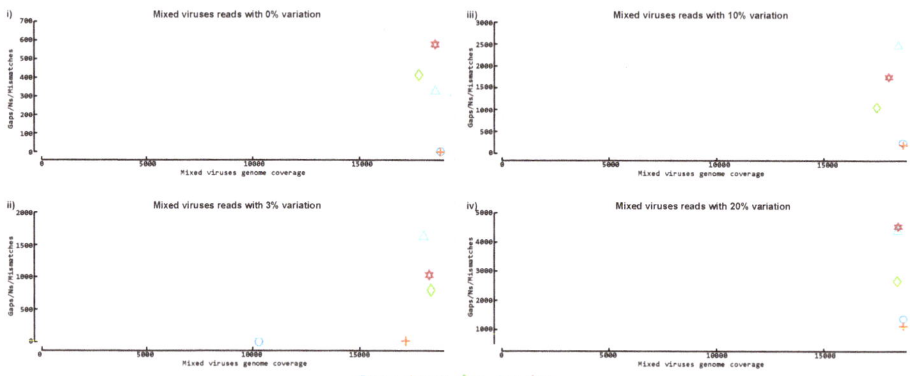

Figure 12. Accuracy of our prototype de novo assembly implementation and two established tools on mixed viruses simulated datasets. The contigs obtained for each assembler were evaluated against the reference genome that was used to generate the simulated data. BLASTn was used to align all contigs to an HIV-1 HXB2 reference genome and determine how much of the particular genome they cover. The y-axis indicates the number of gaps and mismatches that exist in the contigs obtained for each tool, and the x-axis depicts the length of the genome the reported contigs cover. The contigs obtained from the mixed virus simulated dataset were evaluated against the KM198529, KM198528, KM198511, KM198500, KM198486, KU296608, KU296553, KU296549, KU296528, KU296416, KP317952, KP317946, KP317934, KP317923 and KP317922 references genomes. Plots 11-**i** to 11-**iv** illustrate results obtained from all assemblers on data with 0%, 3%, 10% and 20% variation levels accordingly.

Figure 11 illustrates the assembly results of SPAdes, Megahit and all three variants of ASBN on the 16 simulated HIV-1 HXB2 datasets. Figure 12 illustrates the assembly results on the mixed virus simulated databases. Although ASBN processes data and assembles short reads in a lower dimensional space, it nevertheless generated contigs that collectively cover the expected genome length and provided comparable results to both existing state of the art de novo assemblers tested in this experiment (Figures 11 and 12). In all cases, ASBN generated contigs spanning the whole genomes of their respective viral species.

4. Discussion

Although well-established data compression methods for reversible compression of one-dimensional and multivariate signals, images, text and binary exist [62–64], there are very few examples of their application to biological sequence data. We have developed algorithms incorporating signal compression methods for three common biological sequence analysis problems: classification, alignment and de novo assembly of NGS short read virus data. Our results in Figures 3–12 show that this approach permits accurate classification of de novo assembly and reference alignment in spite of high rates of sequence variation or the use of a divergent reference genome. Data approximation/summarisation techniques, such as the DFT, the DWT and the PAA, can be used to extract major features of sequence data and to suppress noise or low-level variation. This allows sequence comparison exploiting the major characteristics of the data, thus enabling the identification of similarities that might otherwise be concealed by minor variation or sequencing error/noise.

Collectively, our results demonstrate that complete nucleotide-level sequence resolution is not a prerequisite of accurate sequence analysis and that analytical performance can be preserved and even enhanced through appropriate dimensionality reduction (compression) of sequences. While our implementations use k-mers, the nature of the transformation/compression methods used shows that optimal k-mer length selection is far less important than the conventional exact k-mer matching methods. The inherent error tolerance of the approach also permits the use of longer k-mers than typically used in conventional sequence comparison algorithms, reducing the computational burden of pairwise comparison, and thus, in de novo assembly specifically, the complexity of building and searching an assembly graph.

Efficient mining of terabase-scale biological sequence datasets requires looking beyond substring-indexing algorithms [65] towards more versatile methods of compression for both data storage and analysis. The use of probabilistic data structures can considerably reduce the computer memory required for in-memory sequence lookups at the expense of a few false positives, and Bloom filters and related data structures have seen broad application in *k*-mer centric tasks, such as error correction [66], in silico read normalisation [67] and de novo assembly [68,69]. However, while these hash-based approaches perform well on datasets with high sequence redundancy, for large datasets with many distinct *k*-mers, large amounts of memory are still necessary [67]. Lower bounding transformations and approximation methods (such as the DFT, the DWT and PAA) can exhibit the same attractive one-sided error offered by these probabilistic data structures, but instead of hash tables, use concrete and reusable sequence representations.

Furthermore, transformations allow compression of standalone sequence composition, enabling flexible reduction of sequence resolution according to analytical requirements, so that redundant sequence precision need not hinder analysis. While the problem of read alignment to a known reference sequence is largely considered solved, assembly of large genomes remains a formidable problem in computing. Moreover, consideration of the metagenomic composition of mixed biological samples, as demonstrated, further extends the scope and scale of the assembly problem beyond what is tractable using conventional sequence comparison approaches. By implementing a reference-based aligner and de novo assembler, we have demonstrated that using compressed numerical representations offers a versatile approach for reconstructing genomes and metagenomes sequenced with short reads.

Emerging long read sequencing technologies bring new challenges for sequence data analysis. Whilst the error rate of Oxford Nanopore sequencing platform, for example, has decreased considerably since the technology's introduction [70,71], the relatively high error rate still limits the scope of downstream analyses [72]. Efficient algorithmic approaches are needed to (1) identify sequence identity/infer homology in spite of abundant insertion/deletion errors associated with the platform, which are problematic for approaches dependent on exact subsequence matching and (2) to overcome issues relating to high data dimensionality and the curse of dimensionality [73]. Both in terms of the raw electric current traces generated by DNA translocation through a nanopore and the corresponding base-called sequences, the resemblance between long reads and time series data from other fields is striking, such that the various transformations/approximations we have implemented will be directly applicable.

In conclusion, nucleotide sequences may be effectively represented as numerical series, enabling the application of existing analytical methods from a variety of mathematical and engineering fields for the purposes of sequence alignment and assembly. By applying established signal decomposition methods, compressed representations of nucleotide sequences can be created, permitting reductions in the spatiotemporal complexity of their analysis, without necessarily compromising analytical accuracy.

Author Contributions: A.T. designed and wrote the methods and software, and performed the data analysis with the help from B.C., M.V.T.P., S.K. and M.C., B.C. and M.C. generated the simulated data, and B.C., M.V.T.P., and M.C. helped perform the data evaluation. A.T. and D.L.R. conceived the study. A.T. and B.C. wrote the manuscript with comments from M.V.T.P., M.C. and D.L.R. All authors read and approved the final manuscript.

Funding: This work has been supported by the Welcome Trust [097820/Z/11/B], the BBSRC [BB/H012419/1, BB/M001121/1 and BC by a BBSRC DTP studentship] and the VIROGENESIS project, which receives funding from the European Union's Horizon 2020 research and innovation programme under grant agreement No 634650.

References

1. Margulies, M.; Egholm, M.; Altman, W.E.; Attiya, S.; Bader, J.S.; Bemben, L.A.; Berka, J.; Braverman, M.S.; Chen, Y.-J.; Chen, Z. Genome sequencing in microfabricated high-density picolitre reactors. *Nature* **2005**, *437*, 376–380. [CrossRef] [PubMed]
2. Bentley, D.R.; Balasubramanian, S.; Swerdlow, H.P.; Smith, G.P.; Milton, J.; Brown, C.G.; Hall, K.P.; Evers, D.J.; Barnes, C.L.; Bignell, H.R. Accurate whole human genome sequencing using reversible terminator chemistry. *Nature* **2008**, *456*, 53–59. [CrossRef]
3. Rothberg, J.M.; Hinz, W.; Rearick, T.M.; Schultz, J.; Mileski, W.; Davey, M.; Leamon, J.H.; Johnson, K.; Milgrew, M.J.; Edwards, M. An integrated semiconductor device enabling non-optical genome sequencing. *Nature* **2011**, *475*, 348–352. [CrossRef]
4. Eid, J.; Fehr, A.; Gray, J.; Luong, K.; Lyle, J.; Otto, G.; Peluso, P.; Rank, D.; Baybayan, P.; Bettman, B. Real-time DNA sequencing from single polymerase molecules. *Science* **2009**, *323*, 133–138. [CrossRef]
5. Salipante, S.J.; Roach, D.J.; Kitzman, J.O.; Snyder, M.W.; Stackhouse, B.; Butler-Wu, S.M.; Lee, C.; Cookson, B.T.; Shendure, J. Large-scale genomic sequencing of extraintestinal pathogenic Escherichia coli strains. *Genome Res.* **2015**, *25*, 119–128. [CrossRef] [PubMed]
6. Rose, R.; Constantinides, B.; Tapinos, A.; Robertson, D.L.; Prosperi, M. Challenges in the analysis of viral metagenomes. *Virus Evol.* **2016**, *2*. [CrossRef] [PubMed]
7. Li, H.; Durbin, R. Fast and accurate short read alignment with Burrows–Wheeler transform. *Bioinformatics* **2009**, *25*, 1754–1760. [CrossRef] [PubMed]
8. Shrestha, A.M.S.; Frith, M.C.; Horton, P. A bioinformatician's guide to the forefront of suffix array construction algorithms. *Brief. Bioinform.* **2014**, *15*, 138–154. [CrossRef]
9. Myers, E.W. Toward simplifying and accurately formulating fragment assembly. *J. Comput. Biol.* **1995**, *2*, 275–290. [CrossRef]
10. Kececioglu, J.D.; Myers, E.W. Combinatorial algorithms for DNA sequence assembly. *Algorithmica* **1995**, *13*, 7–51. [CrossRef]
11. Earl, D.; Bradnam, K.; John, J.S.; Darling, A.; Lin, D.; Fass, J.; Yu, H.O.K.; Buffalo, V.; Zerbino, D.R.; Diekhans, M. Assemblathon 1: A competitive assessment of de novo short read assembly methods. *Genome Res.* **2011**, *21*, 2224–2241. [CrossRef] [PubMed]
12. Iqbal, Z.; Caccamo, M.; Turner, I.; Flicek, P.; McVean, G. De novo assembly and genotyping of variants using colored de Bruijn graphs. *Nat. Genet.* **2012**, *44*, 226–232. [CrossRef]
13. Pevzner, P.A.; Tang, H.; Waterman, M.S. An Eulerian path approach to DNA fragment assembly. *Proc. Natl. Acad. Sci. USA* **2001**, *98*, 9748–9753. [CrossRef]
14. Bradnam, K.R.; Fass, J.N.; Alexandrov, A.; Baranay, P.; Bechner, M.; Birol, I.; Boisvert, S.; Chapman, J.A.; Chapuis, G.; Chikhi, R. Assemblathon 2: Evaluating de novo methods of genome assembly in three vertebrate species. *Gigascience* **2013**, *2*, 1–31. [CrossRef] [PubMed]
15. Archer, J.; Rambaut, A.; Taillon, B.E.; Harrigan, P.R.; Lewis, M.; Robertson, D.L. The evolutionary analysis of emerging low frequency HIV-1 CXCR4 using variants through time—An ultra-deep approach. *PLoS Comput. Biol.* **2010**, *6*, e1001022. [CrossRef] [PubMed]
16. Clement, N.L.; Thompson, L.P.; Miranker, D.P. ADaM: Augmenting existing approximate fast matching algorithms with efficient and exact range queries. *BMC Bioinform.* **2014**, *15*, S1. [CrossRef] [PubMed]
17. Agrawal, R.; Faloutsos, C.; Swami, A. Efficient similarity search in sequence databases. In Proceedings of the 4th International Conference on Foundations of Data Organization and Algorithms, Chicago, IL, USA, 13–15 October 1993.
18. Chan, K.-P.; Fu, A.-C. Efficient time series matching by wavelets. In Proceedings of the 15th International Conference on Data Engineering, Sydney, Australia, 23–26 March 1999; pp. 126–133.
19. Woodward, A.M.; Rowland, J.J.; Kell, D.B. Fast automatic registration of images using the phase of a complex wavelet transform: Application to proteome gels. *Analyst* **2004**, *129*, 542–552. [CrossRef] [PubMed]
20. Geurts, P. Pattern extraction for time series classification. In Proceedings of the 5th European Conference on Principles of Data Mining and Knowledge Discovery, Freiburg, Germany, 3–7 September 2001; pp. 115–127.
21. Keogh, E.; Chakrabarti, K.; Pazzani, M.; Mehrotra, S. Locally adaptive dimensionality reduction for indexing large time series databases. *ACM SIGMOD Record* **2001**, *30*, 151–162. [CrossRef]

22. Shumway, R.H.; Stoffer, D.S.; Stoffer, D.S. *Time Series Analysis and Its Applications with R examples*, 2nd ed.; Springer: New York, NY, USA, 2006.

23. Silverman, B.; Linsker, R. A measure of DNA periodicity. *J. Theor. Biol.* **1986**, *118*, 295–300. [CrossRef]

24. Cheever, E.; Searls, D.; Karunaratne, W.; Overton, G. Using signal processing techniques for DNA sequence comparison. In Proceedings of the Fifteenth Annual Northeast Bioengineering Conference, Boston, MA, USA, 27–28 March 1989; pp. 173–174.

25. Katoh, K.; Misawa, K.; Kuma, K.i.; Miyata, T. MAFFT: A novel method for rapid multiple sequence alignment based on fast Fourier transform. *Nucleic Acids Res.* **2002**, *30*, 3059–3066. [CrossRef] [PubMed]

26. Kwan, H.K.; Arniker, S.B. Numerical representation of DNA sequences. In Proceedings of the 2009 IEEE International Conference on Electro/Information Technology, Windsor, ON, Canada, 7–9 June 2009; pp. 307–310.

27. Yi, B.-K.; Faloutsos, C. Fast time sequence indexing for arbitrary Lp norms. In Proceedings of the 26th roceedings of 26th International Conference on Very Large Data Bases, Cairo, Egypt, 10–14 September 2000; pp. 385–394.

28. Keogh, E.; Ratanamahatana, C.A. Exact indexing of dynamic time warping. *Knowl. Inf. Syst.* **2005**, *7*, 358–386. [CrossRef]

29. Vlachos, M.; Kollios, G.; Gunopulos, D. Discovering similar multidimensional trajectories. In Proceedings of the 18th International Conference on Data Engineering, San Jose, CA, USA, 26 February–1 March 2002; pp. 673–684.

30. Kotsakos, D.; Trajcevski, G.; Gunopulos, D.; Aggarwal, C.C. *In Data Clustering: Algorithms and Applications*; Aggarwal, C.C., Reddy, C., Eds.; CRC Press: Boca Raton, FL, USA, 2013; Chapter 15; pp. 357–379.

31. Chávez, E.; Navarro, G.; Baeza-Yates, R.; Marroquín, J.L. Searching in metric spaces. *ACM Comput. Surv. (CSUR)* **2001**, *33*, 273–321. [CrossRef]

32. Beckmann, N.; Kriegel, H.-P.; Schneider, R.; Seeger, B. The R*-tree: An efficient and robust access method for points and rectangles. *SIGMOD Rec.* **1990**, *19*, 322–331. [CrossRef]

33. Agrawal, R.; Lin, K.; Sawhney, H.S.; Shim, K. Fast similarity search in the presence of noise, scaling, and translation in time-series databases. In Proceedings of the 21th International Conference on Very Large Data Bases, Zurich, Switzerland, 11–15 September 1995; pp. 490–501.

34. Bingham, S.; Kot, M. Multidimensional trees, range searching, and a correlation dimension algorithm of reduced complexity. *Phys. Lett. A* **1989**, *140*, 327–330. [CrossRef]

35. Bellman, R. *Adaptive Control Processes: A Guided Tour*; Princeton University Press: London, UK, 1961; Volume 4.

36. Verleysen, M.; François, D. The Curse of Dimensionality in Data Mining and Time Series Prediction. In Proceedings of the 8th International Work-Conference on Artificial Neural Networks, Barcelona, Spain, 8–10 June 2005; pp. 758–770.

37. Yianilos, P.N. Data structures and algorithms for nearest neighbor search in general metric spaces. In Proceedings of the 4th annual ACM-SIAM Symposium on Discrete Algorithms, Austin, TX, USA; 1993; pp. 311–321.

38. Bozkaya, T.; Ozsoyoglu, M. Indexing large metric spaces for similarity search queries. *ACM Trans. Database Syst. (TODS)* **1999**, *24*, 361–404. [CrossRef]

39. Uhlmann, J.K. Satisfying general proximity/similarity queries with metric trees. *Inf. Process. Lett.* **1991**, *40*, 175–179. [CrossRef]

40. Nair, A.S.; Sreenadhan, S.P. A coding measure scheme employing electron-ion interaction pseudopotential (EIIP). *Bioinformation* **2006**, *1*, 197. [PubMed]

41. Holden, T.; Subramaniam, R.; Sullivan, R.; Cheung, E.; Schneider, C.; Tremberger, G.; Flamholz, A.; Lieberman, D.H.; Cheung, T.D. ATCG nucleotide fluctuation of Deinococcus radiodurans radiation genes. In Proceedings of the Instruments, Methods, and Missions for Astrobiology X, San Diego, CA, USA, 1 October 2007.

42. Voss, R.F. Evolution of long-range fractal correlations and 1/f noise in DNA base sequences. *Phys. Rev. Lett.* **1992**, *68*, 3805. [CrossRef] [PubMed]

43. Faloutsos, C.; Ranganathan, M.; Manolopoulos, Y. Fast subsequence matching in time-series databases. In Proceedings of the 1994 ACM SIGMOD International Conference on Management of data, Minneapolis, MN, USA, 24–27 May 1994.

44. Mitsa, T. *Temporal Data Mining*; CRC Press: New York, NY, USA, 2010.
45. Mörchen, F. *Time Series Feature Extraction for Data Mining Using DWT and DFT*; Technical Report 3; Departement of Mathematics and Computer Science Philipps-University Marburg: Marburg, Germany, 2003; pp. 735–739.
46. Jensen, A.; la Cour-Harbo, A. *Ripples in Mathematics: The Discrete Wavelet Transform*; Springer: Berlin, Germany, 2001.
47. Wu, Y.-L.; Agrawal, D.; El Abbadi, A. A comparison of DFT and DWT based similarity search in time-series databases. In Proceedings of the 9th International Conference on Information and Knowledge Management, Washington, DC, USA, 6–11 November 2000; pp. 488–495.
48. Caboche, S.; Audebert, C.; Lemoine, Y.; Hot, D. Comparison of mapping algorithms used in high-throughput sequencing: Application to Ion Torrent data. *BMC Genom.* **2014**, *15*, 264. [CrossRef] [PubMed]
49. Cotten, M.; Petrova, V.; Phan, M.V.; Rabaa, M.A.; Watson, S.J.; Ong, S.H.; Kellam, P.; Baker, S. Deep sequencing of norovirus genomes defines evolutionary patterns in an urban tropical setting. *J. Virol.* **2014**, *88*, 11056–11069. [CrossRef] [PubMed]
50. Phan, M.V.; Anh, P.H.; Cuong, N.V.; Munnink, B.B.O.; van der Hoek, L.; My, P.T.; Tri, T.N.; Bryant, J.E.; Baker, S.; Thwaites, G. Unbiased whole-genome deep sequencing of human and porcine stool samples reveals circulation of multiple groups of rotaviruses and a putative zoonotic infection. *Virus Evol.* **2016**, *2*. [CrossRef]
51. Kiyuka, P.K.; Agoti, C.N.; Munywoki, P.K.; Njeru, R.; Bett, A.; Otieno, J.R.; Otieno, G.P.; Kamau, E.; Clark, T.G.; van der Hoek, L. Human Coronavirus NL63 Molecular Epidemiology and Evolutionary Patterns in Rural Coastal Kenya. *J. Infect. Dis.* **2018**, *217*, 1728–1739. [CrossRef] [PubMed]
52. Arias, A.; Watson, S.J.; Asogun, D.; Tobin, E.A.; Lu, J.; Phan, M.V.; Jah, U.; Wadoum, R.E.G.; Meredith, L.; Thorne, L. Rapid outbreak sequencing of Ebola virus in Sierra Leone identifies transmission chains linked to sporadic cases. *Virus Evol.* **2016**, *2*. [CrossRef] [PubMed]
53. Agoti, C.N.; Otieno, J.R.; Munywoki, P.K.; Mwihuri, A.G.; Cane, P.A.; Nokes, D.J.; Kellam, P.; Cotten, M. Local evolutionary patterns of human respiratory syncytial virus derived from whole-genome sequencing. *J. Virol.* **2015**, *89*, 3444–3454. [CrossRef] [PubMed]
54. Altschul, S.F.; Gish, W.; Miller, W.; Myers, E.W.; Lipman, D.J. Basic local alignment search tool. *J. Mol. Biol.* **1990**, *215*, 403–410. [CrossRef]
55. Menzel, P.; Ng, K.L.; Krogh, A. Fast and sensitive taxonomic classification for metagenomics with Kaiju. *Nat. Commun.* **2016**, *7*, 11257. [CrossRef] [PubMed]
56. Langmead, B.; Salzberg, S.L. Fast gapped-read alignment with Bowtie 2. *Nat. Methods* **2012**, *9*, 357. [CrossRef] [PubMed]
57. Sović, I.; Šikić, M.; Wilm, A.; Fenlon, S.N.; Chen, S.; Nagarajan, N. Fast and sensitive mapping of nanopore sequencing reads with GraphMap. *Nat. Commun.* **2016**, *7*, 11307. [CrossRef] [PubMed]
58. Otto, C.; Stadler, P.F.; Hoffmann, S. Lacking alignments? The next-generation sequencing mapper segemehl revisited. *Bioinform.* **2014**, *30*, 1837–1843. [CrossRef] [PubMed]
59. Tapinos, A.; Robertson, D.L. De novo assembly of nucleotide sequences in a compressed feature space. In Proceedings of the 2017 IEEE Conference on Computational Intelligence in Bioinformatics and Computational Biology (CIBCB), Manchester, UK, 23–25 August 2017; pp. 1–7.
60. Li, D.; Liu, C.-M.; Luo, R.; Sadakane, K.; Lam, T.-W. MEGAHIT: An ultra-fast single-node solution for large and complex metagenomics assembly via succinct de Bruijn graph. *Bioinformatics* **2015**, *31*, 1674–1676. [CrossRef]
61. Anton, B.; Sergey, N.; Dmitry, A.; Alexey, A.; Mikhail, D.; Alexander, S.; Valery, M.; Sergey, I.; Son, P.; Andrey, D. SPAdes: A new genome assembly algorithm and its applications to single-cell sequencing. *J. Comput. Biol.* **2012**, *19*, 455.
62. Tapinos, A.; Mendes, P. A method for comparing multivariate time series with different dimensions. *PloS ONE* **2013**, *8*, e54201. [CrossRef]
63. Sheybani, E.O. An Algorithm for Real-Time Blind Image Quality Comparison and Assessment. *Int. J. Electr. Comput. Eng. (IJECE)* **2011**, *2*, 120–129. [CrossRef]
64. Hendriks, R.C.; Gerkmann, T.; Jensen, J. DFT-domain based single-microphone noise reduction for speech enhancement: A survey of the state of the art. In *Synthesis Lectures on Speech and Audio Processing*; Morgan & Claypool: San Rafael, CA, USA, 2013; pp. 1–80.
65. Kouchaki, S.; Tapinos, A.; Robertson, D.L. A signal processing method for alignment-free metagenomic binning: Multi-resolution genomic binary patterns. *Sci. Rep.* **2019**, *9*, 2159. [CrossRef]

66. Shi, H.; Schmidt, B.; Liu, W.; Müller-Wittig, W. A Parallel Algorithm for Error Correction in High-Throughput Short-Read Data on CUDA-Enabled Graphics Hardware. *J. Comput. Biol.* **2010**, *17*, 603–615. [CrossRef]
67. Zhang, Q.; Pell, J.; Canino-Koning, R.; Howe, A.C.; Brown, C.T. These Are Not the K-mers You Are Looking For: Efficient Online K-mer Counting Using a Probabilistic Data Structure. *PLoS ONE* **2014**, *9*, e101271. [CrossRef]
68. Salikhov, K.; Sacomoto, G.; Kucherov, G. Using cascading Bloom filters to improve the memory usage for de Brujin graphs. *Algorithms Mol. Biol.* **2014**, *9*, 364–376. [CrossRef]
69. Berlin, K.; Koren, S.; Chin, C.-S.; Drake, J.P.; Landolin, J.M.; Phillippy, A.M. Assembling large genomes with single-molecule sequencing and locality-sensitive hashing. *Nat. Biotechnol.* **2015**, *33*, 623–630. [CrossRef]
70. Laver, T.; Harrison, J.; O'neill, P.; Moore, K.; Farbos, A.; Paszkiewicz, K.; Studholme, D.J. Assessing the performance of the oxford nanopore technologies minion. *Biomol. Detect. Quantif.* **2015**, *3*, 1–8. [CrossRef]
71. Fu, S.; Wang, A.; Au, K.F. A comparative evaluation of hybrid error correction methods for error-prone long reads. *Genome Biol.* **2019**, *20*, 26. [CrossRef]
72. Watson, M.; Warr, A. Errors in long-read assemblies can critically affect protein prediction. *Nature Biotechnol.* **2019**, *37*, 124. [CrossRef]
73. Radovanović, M.; Nanopoulos, A.; Ivanović, M. Time-series classification in many intrinsic dimensions. In Proceedings of the 2010 SIAM International Conference on Data Mining, Columbus, OH, USA, 29 April–1 May 2010; pp. 677–688.

Article

RNAseq Analysis Reveals Virus Diversity within Hawaiian Apiary Insect Communities

Laura E. Brettell [1,2,*], Declan C. Schroeder [3,4] and Stephen J. Martin [2]

[1] Hawkesbury Institute for the Environment, Western Sydney University; Locked bag 1797, Penrith 2751, NSW, Australia
[2] School of Environment and life Sciences, University of Salford, Manchester, M5 4WT, UK; s.j.martin@salford.ac.uk
[3] School of Biological Sciences, University of Reading, Reading RG6 6LA, UK; dcschroe@umn.edu
[4] Veterinary Population Medicine, College of Veterinary Medicine, University of Minnesota, St Paul, MN 55108, USA
*   Correspondence: l.brettell@westernsydney.edu.au

Received: 15 March 2019; Accepted: 24 April 2019; Published: 27 April 2019

Abstract: Deformed wing virus (DWV) is the most abundant viral pathogen of honey bees and has been associated with large-scale colony losses. DWV and other bee-associated RNA viruses are generalists capable of infecting diverse hosts. Here, we used RNAseq analysis to test the hypothesis that due to the frequency of interactions, a range of apiary pest species would become infected with DWV and/or other honey bee-associated viruses. We confirmed that DWV-A was the most prevalent virus in the apiary, with genetically similar sequences circulating in the apiary pests, suggesting frequent inter-species transmission. In addition, different proportions of the three DWV master variants as indicated by BLAST analysis and genome coverage plots revealed interesting DWV-species groupings. We also observed that new genomic recombinants were formed by the DWV master variants, which are likely adapted to replicate in different host species. Species groupings also applied when considering other viruses, many of which were widespread in the apiaries. In social wasps, samples were grouped further by site, which potentially also influenced viral load. Thus, the apiary invertebrate community has the potential to act as reservoirs of honey bee-associated viruses, highlighting the importance of considering the wider community in the apiary when considering honey bee health.

Keywords: RNAseq; honey bees; deformed wing virus; quasispecies; apiary pests; recombination

---

1. Introduction

Across the globe, emerging infectious diseases (EIDs) pose a significant threat to biodiversity and health [1]. This has been clearly demonstrated in recent years both by the recent catastrophic decline of amphibians caused by pathogenic fungi [2] and by the cases of large-scale honey bee colony losses, a major factor in which is the spread of pathogenic viruses [3].

EIDs often occur as a consequence of human-mediated translocations of infected hosts and/or parasites and due to the close proximity of wild and domesticated hosts [1]. As such the honey bee, which over the last century has been spread across the globe by humans for honey production and pollination services [4] and shares complex communities with a wide range of insect taxa [5–8], is a prime candidate to facilitate the spread of EIDs into new insect hosts. Pollinators and other insects with which they share environments are of particular interest due to their value in terms of economy (including pollination services) and biodiversity and are currently already experiencing a number of pressures, e.g., from habitat loss and pesticides [9–11]. The combination of multiple pressures can have the additional effect of decreasing immunity and thus potentially increasing susceptibility to pathogens [12], although further studies are needed in this area [13].

Most host switching results in a dead end or a limited low-level outbreak; however, on rare occasions the transmission can result in sustained outbreaks or major epidemics. This can happen when there is sufficient increased exposure or the evolution of new variants in the original or new host allows successful replication and efficient spread between members of the new host species [14]. RNA viruses with their high mutation rates, diverse sequences, and often very high population sizes [15] are thus prime candidates for emergence or re-emergence in novel hosts. This is of particular concern when considering honey bee populations, which host abundant and diverse RNA viruses, are frequently transported en masse, and come into frequent contact with other arthropods both in the hive and when foraging.

Deformed wing virus (DWV) is now well known to be one of the major factors responsible for honey bee colony losses across the world [3,4,16]. This single virus has come to dominate the virome of honey bee populations due to its spread and amplification in the host being aided by the ectoparasitic mite *Varroa destructor* (referred to as varroa from now on) [17,18]. Additionally, the new mite–bee transmission route serves to reduce viral genotypic diversity and select for the amplification of virulent strains [16,17]. Although initially described as a honey bee virus, it has since become apparent that DWV is a generalist insect virus capable of infecting 64 species from eight orders of arthropods [19]. However, the extent of the generality and capacity for emergence in new hosts is still the subject of contention [20]. Worryingly, recent studies have found that viral pathogens circulating in managed pollinators may be driving infections in wild species [21,22], and in the Hawaiian system, it has recently been shown using RT-PCR-based studies that the presence of varroa in honey bee populations has resulted in a dramatic increase in the prevalence of DWV in species of wasps (*Polistes* sp.) and solitary bees (*Ceratina smaragdula*) [23]. Furthermore, in the yellowjacket wasp (*Vespula pensylvanica*) on the Big Island, Hawaii, DWV prevalence has increased along with a decrease in strain diversity [24], mirroring the situation seen in honey bees [17].

Although gene targeted RT-PCR and RT-qPCR have provided very useful insights into the virome of insect communities, RNAseq allows entire genomes of RNA viruses to be sequenced easily and thus provides much deeper anaylsis. More specifically, using oligo dT-derived RNAseq on field collected samples in Hawaiian honey bee apiaries in which varroa is established, we aimed to identify the extent to which +ssRNA viruses (the overwhelming majority of viruses found associated with honey bees are +ss polyadenylated RNA viruses [10]) were present in the wider arthropod community and to characterize any DWV master variant genomes present in the different species. Common pest species were targeted as they would have the most frequent contact with honey bees that are known to harbor high DWV loads in the study locations [17,25]. We hypothesized that the frequency of interactions would result in common DWV genotypes circling in the apiary environment due to repeated viral transmission events between species. Furthermore, we aimed to determine whether other viruses commonly found in honey bees are also found in apiary pests, and if so, whether certain viruses are associated with particular hosts. The pest species investigated were varroa mites (*Varroa destuctor*), small hive beetles (*Aethina tumida*) yellowjacket wasps (*Vespula pensylvanica*), and two species of ant, big headed ants (*Pheidole megacephala*) and ghost ants (*Tapinoma melanocephalum*)—all widely distributed invasive pests known to interact directly with honey bees in Hawaii.

2. Materials and Methods

2.1. Site and Species Selection

In November and December 2012, opportunistic sampling was carried out in managed apiaries on the islands of Oahu and the Big Island, Hawaii (Figure 1), and where found, common apiary pests (yellowjacket wasps, small hive beetles, big headed ants, and ghost ants) were collected from within brood boxes or at hive entrances. Additionally, reference samples of asymptomatic honey bees

(50–100 individuals per hive) were also collected. Varroa populations had been established for at least 3–5 years at all locations sampled, and honey bee populations were known to harbor high DWV levels [17,25]. All samples were kept on ice for transportation to the laboratory, where they were stored at –80 °C. Additional samples of small hive beetles had been collected in the same way earlier in the year from colonies in an advanced stage of collapse, i.e., "slime out phase".

2.2. RNA Extraction and Next Generation Sequencing

Pools of 30 asymptomatic honey bees were taken and checked for the presence of varroa mites, which, where found, were removed and stored separately. The pools of whole bees were then ground in liquid nitrogen using a sterile pestle and mortar to produce a fine homogenous powder, 30 mg of which was used for RNA extraction. Varroa mites that had been removed from bees were homogenized in pools of 10 using a mini pestle in a 1.5 mL eppendorf tube. Small hive beetles were homogenized either individually or in pools of six. Yellowjacket wasps were homogenized individually, and all ants were crushed in pools of 20–40 due to their small size. RNA extractions proceeded for all samples in the same way using the RNeasy mini kit (Qiagen, Manchetser, UK) following the manufacturers' instructions, eluting in 30 µL nuclease free water followed by DNase treatment using DNase I (Promega, Southampton, UK). Initial screening for the presence of DWV was then carried out using RT-PCR as per [17,26]. A selection of positive samples was then chosen to be analyzed in greater depth using total RNA sequencing (RNAseq). The list of samples used for RNAseq analysis along with the species name, site from which they were sampled, and symbol used to denote them in subsequent figures is given in Table S1. These samples were transported to The BBSRC Earlham Institute (Norwich), where cDNA libraries were prepared using oligo (dT) priming. Resulting libraries were then run on the HiSeq 2000 (The Earlham Institute, Norwich). Raw data were deposited in the National Centre for Biotechnology Information (NCBI) Sequence Read Archive (SRA) under BioProject accession number: PRJNA531527. Sample V_des_2 was previously deposited in the European Nucleotide Archive (ENA) as sample "V_S48" under the Study Accession PRJEB8112.

2.3. Bioinformatic Analysis

Initially, quality control of generated reads was performed using FASTQC (http://www. bioinformatics.babraham.ac.uk/projects/fastqc/). Then, a pipeline initially described by [25] was applied to identify DWV-like reads in individual samples. Briefly, this involved taking all reads that passed QC and performing a nucleotide BLAST against a custom database (E value of 10^{-5}) containing DWV types A, B. and C (type A: NC_004830.2 and NC_005876.1 (Kakugo virus), type B: AY251269.2 (VDV-1), and type C: CEND01000001.1). Custom perl, sed, and awk scripts were used to parse the data, take the top BLAST hit, remove empty lines, and remove reads which did not map to the database. BLAST top hit analysis was then used to quantify the numbers of reads belonging to each of the three master variants by identifying which of the reference genomes each read matched to most closely. This was used to produce pie charts showing the proportions of DWV type A, B, and C reads in each sample.

To assemble contigs representing the full diversity of DWV sequences present in each sample, DWV-like reads were identified by BLAST (read1 files for each sample in fasta format), and the corresponding read 2 files were selected. These were then used to generate de novo assemblies using VICUNA, an assembler specifically designed to accommodate the highly variable sequence data typical of RNA viruses [27].

To investigate DWV diversity within the apiary insect community, all VICUNA-generated contigs >300 bp for each sample were imported into Geneious (Version 7.04, Biomatters), and the map to reference tool was then used with a MUSCLE alignment to competitively map contigs to the DWV-A (NC_004830.2), -B (AY251269.2), and -C (CEND01000001.1) reference genomes (mapping ambiguous reads to all). The resulting alignments containing all the DWV contigs were then trimmed to use a 507 bp region of the *RdRp* gene (nucleotide positions 8016–8522 on NC_004830.2, 7989–8495 on AY251269.2, and 7999–8505 on CEND01000001.1 [4]). Alignments were visually inspected, and contigs showing recombination breakpoints were removed so as to comply with the assumptions of the phylogenetic construction, and a Baysian phylogenetic analysis was constructed using the MrBayes v3.2.6 plugin in Geneious. [28], using a GTR substitution model with gamma rate variation. We ran four chains for 1.1×10^6 MCMC generations, sampling trees every 200 generations, and a consensus tree was then created using a 10% burn in and 50% support threshold.

Furthermore, to understand the diversity of DWV-A in Hawaiian apiaries in a global context, a second phylogeny was generated using a 410 bp region of the *RdRp* gene, as used by [4] in their global DWV-A phylogeny along with 10 of their DWV-A sequences of geographically diverse origins from both honey bees and varroa mites (Table S2). DWV-A contigs generated in this study were aligned with the 410 bp sequences from [4] in Geneious (MUSCLE alignment), and this time, no recombinant contigs were observed in the alignment. A Bayesian phylogeny was then created using the same parameters as described above. The corresponding region of the DWV-C reference genome (CEND01000001.1) was used as the outgroup.

To determine whether recombinants were dominating in the samples, competitive alignment plots were created to look at the DWV reference genomes to which reads preferentially mapped across the length of the genome. This used all reads passing QC (read 1, in fasta format) and the "Map to Reference" tool in Geneious, using DWV types A, B and C (NC_004830.2, AY251269.2 and CEND01000001.1) reference genomes, discarding all ambiguous reads. Recombinants were observed to be present if preferential coverage switched from one master variant to another along the length of the genome. To provide additional evidence for the presence of recombinants, two samples were chosen for which competitive alignment plots had revealed recombinants to be present (V_pen_8 and A_tum_5), and their de novo assembled DWV contigs were aligned using MUSCLE and visually inspected to identify recombinant contigs.

Finally, to investigate whether other honey bee-associated viruses were circulating in the apiary community, a custom BLAST database was created using DWV types A, B, and C along with seven common honey bee-associated viruses and an additional two viruses (Moku and Milolii viruses [29,30]) previously identified from RNAseq data generated from Hawaiian apiary insects (data from individual insects were also used in this study) (Table S3), and BLASTn was used to identify reads belonging to each virus (E value 10^{-5}). Read counts were expressed as reads per kilobase million (RPKM).

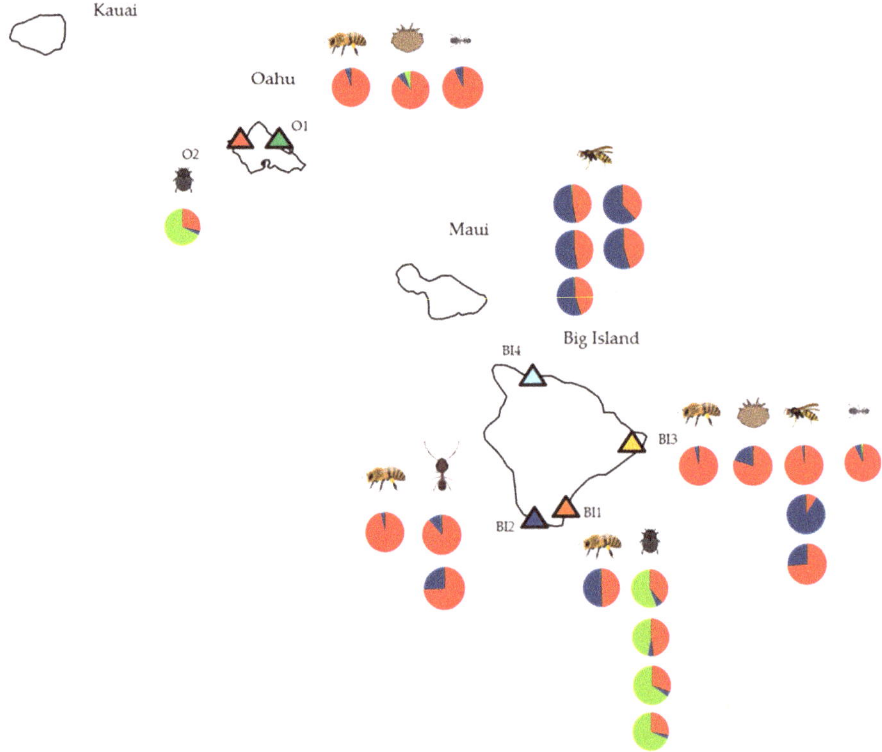

Figure 1. *Cont.*

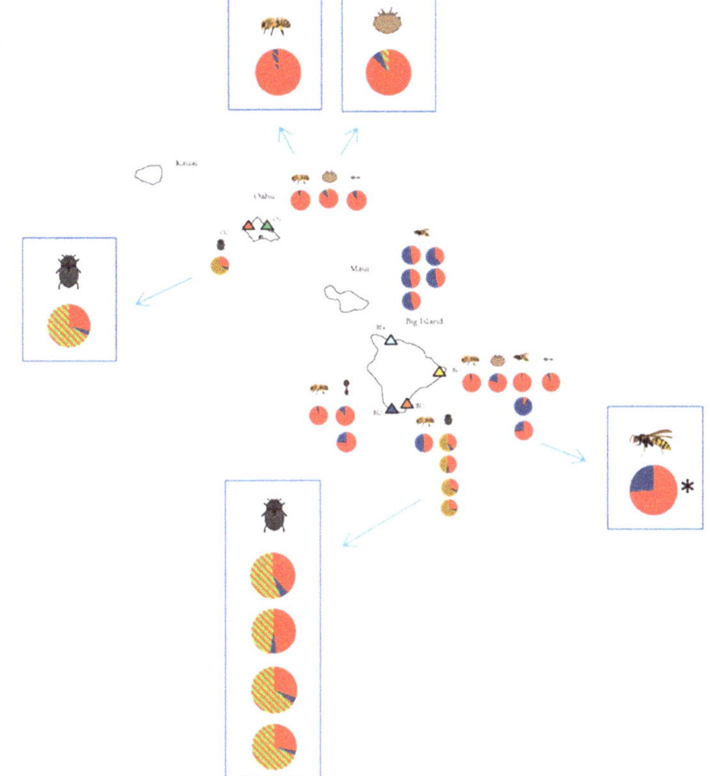

Figure 1. Proportions of deformed wing virus (DWV)-like reads in individual samples from each species mapping to DWV type A (red), type B (blue), and type C (green) along with the site from which they were collected. (**a**) Results using BLAST top hit analysis only (read numbers provided in Table S4); (**b**) where results were amended after considering additional Geneious mapping analysis (Figure S1). Green and red stripes represent that DWV-A/DWV-C recombinants were present (rather that true full-length DWV-C,), red and blue stripes denote a DWV-A/DWV-B recombinant (no full-length DWV-B), and the asterisk represents an additional sample identified as containing recombinant(s). Results for all other samples remain unchanged. Insect images are from BioRender [23,31].

2.4. Statistical Analysis

Redundancy analysis (RDA) was carried out in R (version 3.3.2) to investigate whether the response variables of read count data (RPKM) for each virus were associated with explanatory variables of insect species and site. Dummy variables were randomly assigned as varroa for the species category and B4 for site.

3. Results

3.1. Deformed Wing Virus

RNAseq data was generated from samples of honey bees (A_mel_1-A_mel_4), small hive beetles (A_tum_1-A_tum_5), yellowjacket wasps (V_pen_1-V_pen_8), big headed ants (P_meg_1 and P_meg_2) and ghost ants (T_mel_1 and T_mel_2). As expected from our previous work [17,25], BLAST top hit analysis revealed DWV to dominate sequence reads in all honey bee and varroa mite samples, reaching as high as 91.25% of total reads in the varroa sample V_des_2 from apiary O1. DWV was ubiquitous, being detected in all ant, wasp and beetle samples, although read counts were highly variable (Table S4). The proportions of DWV-types A, B and C present also differed considerably and appeared to group by species (Figure 1). Of the DWV reads, DWV-A reads dominated the majority of honey bee, varroa and ant samples; however the small hive beetle samples were all dominated by type C and the wasps usually contained relatively similar proportions of types A and B. Interestingly, the wasp samples showed further separation by location.

DWV genome coverage plots were generated for each sample, and all alignments showed the typical 3′ bias resulting from the oligo (dT) priming method of cDNA synthesis in library preparation [32]. Similarly, the samples containing type C all showed a spike in the Helicase region caused by the presence of a poly-(A) region of the genome, irrespective of location (Figure 2; Figure S1). Plots showed that the majority of samples were dominated by either full-length DWV-A or DWV-B sequences (or both), as demonstrated by the consistent coverage depth across the genome, and were consistent with the results of the BLAST top hit analysis (Figure 1a). Again, the yellowjacket wasp data appeared to separate by site, with samples V_pen_1–V_pen_5 from site B4 all showing identical coverage plots, with coverage being restricted to the 3′ end indicative of low virus levels in the samples and showing almost identical coverage for DWV-A and DWV-B. The samples V_pens_6 to V_pen_8 from site B3, however, were more variable and contained greater coverage depths, again, in keeping with the BLAST top hit analysis.

Although the majority of samples were dominated by full-length genomes, recombinants were detected in this study as evidenced by the competitive alignment plots (Figure 2) and associated alignments of assembled contigs (Figure S1). The dominant master variants present were the same using both BLASTn and mapping to DWV reference genomes (Geneious). However, the mapping (competitive alignment plots) revealed additional information regarding the presence of recombinants. These additional data were therefore used to create amended pie charts showing variant proportions (Figure 1b) for samples where recombinants were detected. Interestingly all beetles were dominated by a DWV-A/DWV-C recombinant (Figure 1b, Figure 2b, Figures S1 and S2) with a breakpoint in the 5′ UTR immediately upstream of the open reading frame, albeit at low levels, irrespective of location. These beetles also contained lower level full-length DWV-A, as evidenced by type A coverage across the whole genome. The same recombinant was also present in the honey bee sample A_mel_4 (Figure S1), which was collected from a different location from any of the beetles, although the recombinant did not dominate in the honey bee sample, which was instead dominated by full-length DWV-A. Finally, one yellowjacket wasp sample, V_pen_8, showed a distinct coverage pattern not seen in any other sample in this study (Figure 2c). This sample showed three recombination breakpoints—one in the 5′ UTR and one at either end of the helicase gene—and showed coverage of both DWV-A and DWV-B across the full length of the genome. As such, these data cannot confirm the makeup or proportions of recombinant(s) and full-length genomes in this sample.

A Bayesian phylogeny created using a 507 bp fragment of the *RdRp* gene assembled from all samples showed sequences to cluster according to master variant (DWV-A, B, and C), as expected (Figure 3). The origin of all the DWV-C sequences came from a single DWV-C variant, as all the sequences obtained from this region were almost identical to one another and to the reference genome. Within the DWV-A and DWV-B clades, samples did not cluster by either species or location and showed often very similar sequences, indicating that common variants are circling in the apiary.

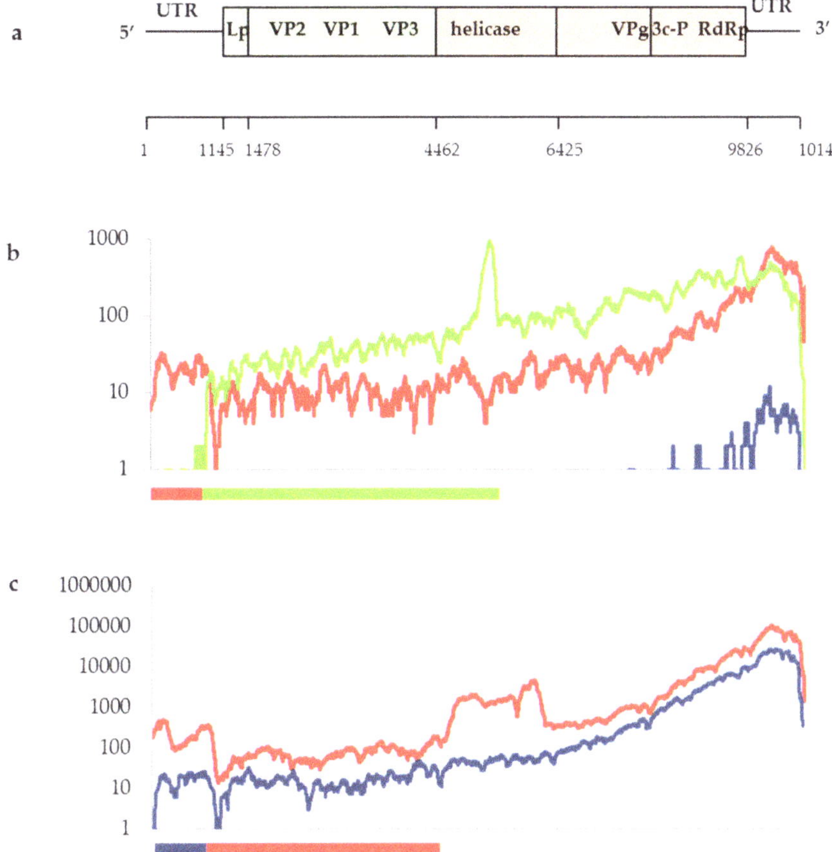

Figure 2. DWV genome coverage plots showing read depth of DWV-A (red), DWV-B (blue), and DWV-C (green), across the length of the genome, along with representations of VICUNA-assembled contigs showing recombination breakpoints; (**a**) genome organization with the structural region in green and the non-structural region in orange, showing nucleotide positions below (adapted from [33,34]); (**b**) small hive beetle sample A_tum_5; (**c**) yellowjacket wasp sample V_pen_8. Plots for all samples are shown in Figure S1, and nucleotide alignments showing the recombinant contigs for samples A_tum_5 and V_pens_8 are shown in Figure S2.

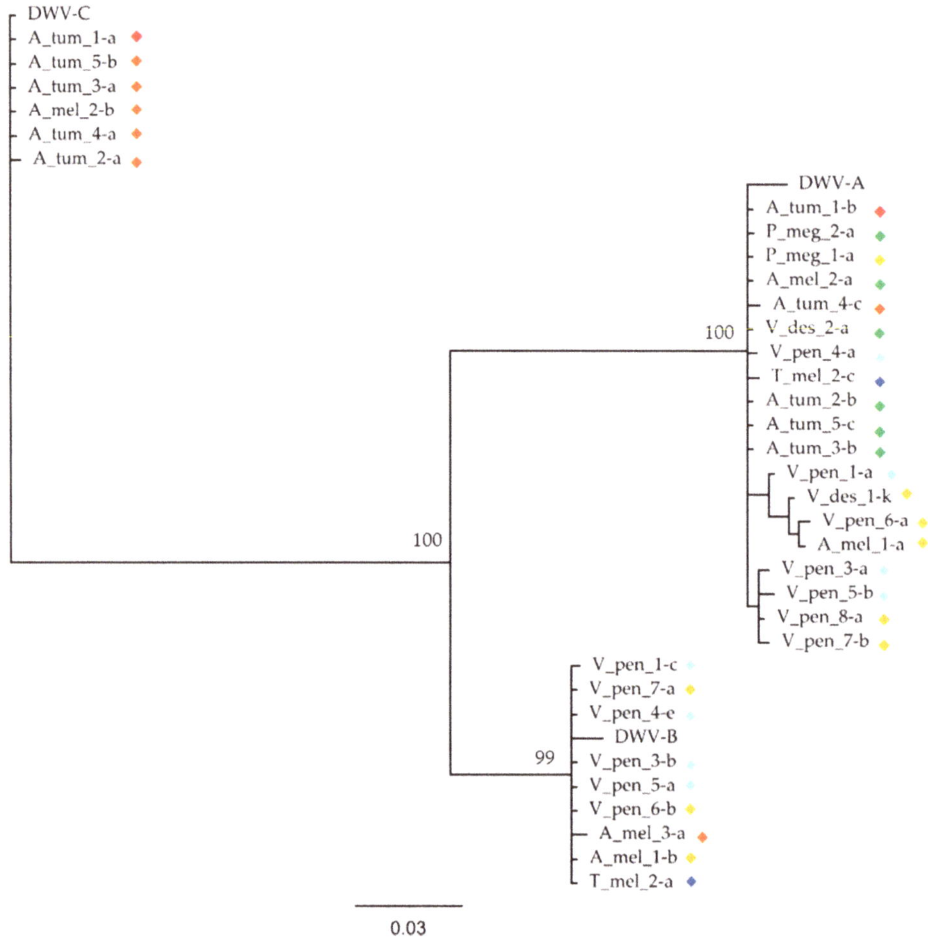

0.03

Figure 3. A Bayesian phylogeny of a 507 bp region of the DWV *RdRp* gene using trimmed contigs assembled for each sample individually using VICUNA, along with the corresponding regions from the DWV-A, B, and C reference genomes (NC_004830.2, AY251269.2, and CEND01000001.1). Colored diamonds represent the sample collection location: O1 = green, O2 = red, B1 = orange, B2 = dark blue, B3 = yellow, and B4 = light blue. Sample name suffixes represent the contig name. The bar represents the number of nucleotide substitutions per site, and the consensus support (%) is shown for selected branches.

To investigate how the DWV-A sequences in the Hawaiian apiary insects compared with DWV-A sequences on a broader scale, we constructed a phylogeny using the same 410 bp *RdRp* region used by [4] in their global phylogeny, along with some of their samples (Figure 4). This revealed that all samples in the current study fell within those detected elsewhere and clustered together with little sequence variation in this region of the genome.

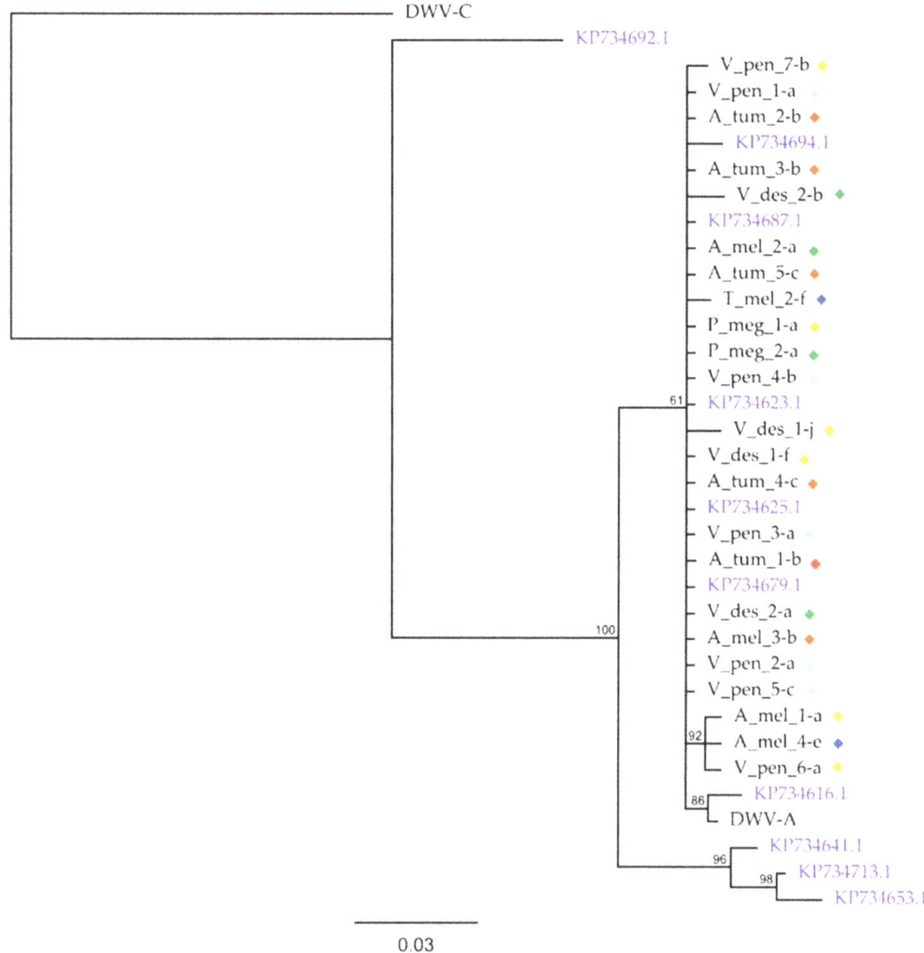

0.03

Figure 4. A Bayesian phylogeny of a 410 bp region of the DWV-A *RdRp* gene using trimmed contigs assembled for each sample individually using VICUNA, along with 10 additional sequences, as featured in the global DWV-A phylogeny of [4] (shown in blue text). Colored diamonds represent the site from which the sample was collected as follows: O1 = green, O2 = red, B1 = orange, B2 = dark blue, B3 = yellow, and B4 = light blue. Sample name suffixes represent the contig name. DWV-C is used as the outgroup, and the bar represents the number of nucleotide substitutions per site. Consensus support values (%) are shown.

3.2. Other Honey Bee-Associated Viruses

In addition to generally grouping in terms of DWV types A, B, and C proportions and coverage, geographical and taxa groupings were also observed when considering the other honey bee-associated viruses they harbored. BLASTn was used to identify reads belonging to eight of the most common honey bee viruses and two additional viruses known from our previous work to be prevalent in Hawaiian insects. As expected, honey bees (with the exception of sample A_mel_3) and varroa were dominated by DWV reads, as were big headed ants. We found ghost ants to be dominated by Milolii virus, yellowjacket wasps to be dominated by Moku virus, and small hive beetles to contain relatively few virus reads (Figure 5). The wasp samples further separated into two groups, with samples

V_pen_1–V_pen_5 (from site B4) containing high numbers of Moku virus reads and also consistent amounts of SBPV, whereas V_pen_6–V_pen_8 (from site B3) showed increased DWV, more variable Moku virus, and no notable SBPV. The honey bees were the only samples to contain any notable BQCV.

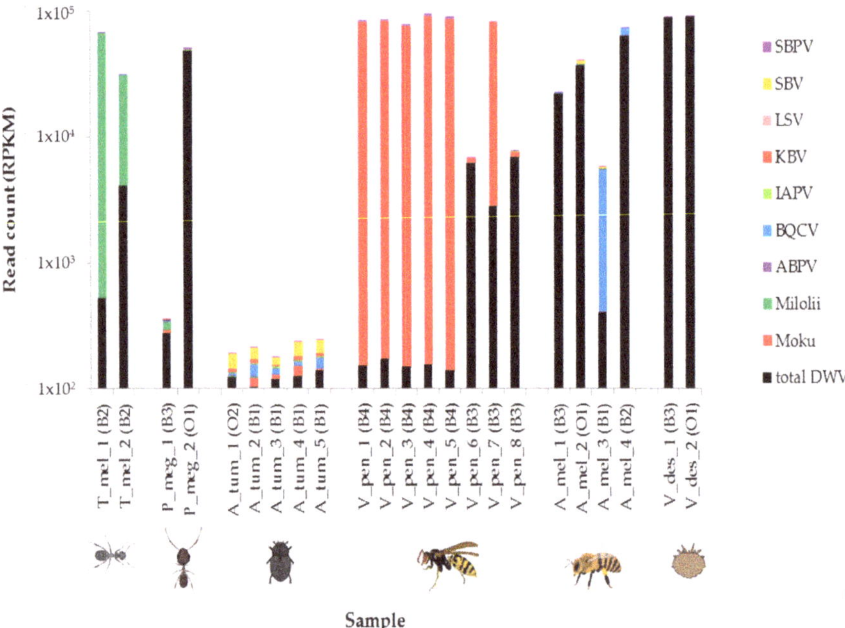

Figure 5. Histograms represent numbers of reads mapping to each of 10 commonly found insect viruses (slow bee paralysis virus: SBPV, sacbrood virus; SBV, Lake Sinai virus; LSV, Kashmir bee virus; KBV, Israeli acute paralysis virus; IAPV, black queen cell virus; BQCV, acute bee paralysis virus; ABPV, Milolii virus, Moku virus and deformed wing virus; DWV) as determined using BLASTn. Read numbers are expressed as reads per kilobase million (RPKM) and are shown on a log-10 scale starting 1×10^2. The samples are positioned according to taxa: ghost ants (T_mel_1–2), big headed ants (P_meg_1–2), small hive beetles (A_tum_1–5), yellowjacket wasps (V_pens_1–8), honey bees (A_mel_1–4), and varroa (V_des_1–2). Apiary locations from where the samples were taken are given after each sample name.

Redundancy analysis was conducted ($p < 0.001$) to further investigate the variation in viral read count data (log-10) for each sample in the context of species and site explanatory variables. RDA1 and RDA2 together explained 54.8% of the variation, and all variables (sites and species) were significantly different from the dummy variables ($p < 0.05$) (Figure 6). The RDA plot shows groupings by site (colors), as well as demonstrating separation by species, especially "honey bee", "small hive beetle", and "yellowjacket wasp".

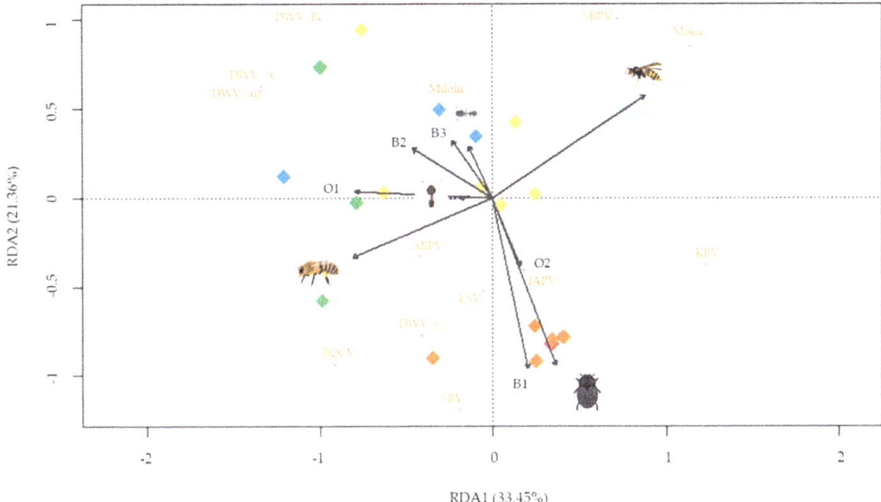

Figure 6. Redundancy Analysis (RDA) plot showing the variation in virus read count data (RPKM log-10) for each sample correlated with environmental variables: site and species. Dummy variables were randomly assigned to varroa (species) and B4 (site). Sites are colored: O1 = green, O2 = red, B1 = orange, B2 = dark blue, B3 = yellow, and B4 = light blue.

4. Discussion

This study revealed that common apiary pests have the potential to act as reservoirs for, or could be impacted by, a number of honey bee-associated viruses. Although recent studies have highlighted the ability of a number of viruses initially described as honey bee pathogens to infect a range of taxa and in some cases cause pathogenic effects (reviewed by [19]), this is the first study to assess the viral burden of taxonomically diverse common apiary pests.

Of 10 common +ssRNA viruses surveyed, DWV was the most common virus in honey bees and varroa, as expected, because it is the most prevalent virus in honey bees across the world [19] and was the most prevalent virus in the apiary with full-genome coverage achieved from samples of each species. DWV-A was the most common variant detected, although DWV-B was also widespread, correlating with the recent finding by [35] that DWV-B dominance is replacing DWV-A on the mainland United States of America (USA). Although the BLAST top hit analysis was reliable when considering full-length virus genomes, such as when detecting a number of different viruses (Figure 5), or when only full length master variants are present, e.g., considering DWV diversity in all ant samples (Figure 1 and Figure S2), methods that identify all reads independently, irrespective of genome location, mean that it is impossible to discriminate between full genomes and recombinant forms. This is evidenced by the beetle data that BLAST top hit analysis suggested was dominated by DWV-C; however closer inspection using competitive alignments revealed this to, in fact, be a DWV-A/DWV-C recombinant. Thus, the two analyses (BLAST top hit results and competitive alignment plots) gave consistent results when only full-length genomes were present but differed when recombinants were detected.

Interestingly, beetles sampled from across two locations were the only ones to be dominated by a DWV-A/DWV-C recombinant. These all had very low read counts but were very consistent. All beetle samples were collected earlier in the year from collapsing honey bee colonies suffering from extremely heavy small hive beetle infestations (slime out conditions), and unfortunately, no corresponding bee samples were taken at this time. As such it may be possible that the same recombinant was present at high levels in the dying bees at that time, and the high amounts of virus in the hives resulted in passive contamination of the beetles, which would explain why DWV is only very rarely detected in beetles [36]. In this case, contamination during library preparation or during the Illumina run cannot be ruled out, because it is a particular issue for RNAseq analysis. This can occur with invertebrate viruses due to the extreme dominance of viral reads in heavily infected samples. These beetle samples had previously tested positive for DWV using RT-PCR prior to selection for RNAseq analysis and a recent RT-PCR-based study showed that stingless bees (*Melipona subnitida*) from Brazil also harbored DWV-C but at higher viral levels [37], confirming this DWV variant also has the ability to infect diverse hosts.

Investigating viral reads generated by oligo (dT), RNAseq is not the only method to assess whether samples are harboring true infections [38,39]. Nonetheless, this method will reveal high read counts, full-genome coverage, and assembled full genomes, suggesting that DWV is replicating in at least some samples from each species. Furthermore, the unique dominance of recombinant(s) in the wasp sample V_pens_8 that was not seen in any other sample provides further evidence that there are specific DWV variants/recombinants present in the quasispecies that are able to successfully replicate in different species.

Considering both the phylogenies together, it appears that common DWV variants are circling in the apiary insect community, suggesting frequent interspecies transmission events. The nature of these transmission events remains unclear, with trophylaxis (between small hive beetles) [20], fecal–oral routes and predation on bees, and consumption of contaminated hive materials, pollen, and nectar all being implicated as routes by which viruses can spread both between and within species [40].

The DWV-A phylogeny, which in the original publication by [4] only contained honey bee- and varroa-derived sequences (Figure 5) showed all sequences from this study to fall within the greater diversity and all cluster within one clade. When considering the phylogeny of DWV sequences from the current study only (Figure 4), again all DWV-A clustered together as was also true of DWV-B sequences. The limitations both in terms of sampling and in using only one gene region means that is not possible to draw strong conclusions on the evolutionary history of these sequences. Furthermore, when considering the read count data and genome coverage plots for wasps only, they appear to separate by location with samples V_pen_1–5 from B4 all showing almost identical results in terms of their DWV variant composition, whereas V_pen_6–8 are very different (Figure 1 and Figure S1). These findings are in keeping with the results of [24] who found, using RT-PCR and Sanger sequencing, that DWV sequences in yellowjacket wasps (also from Hawaii) tracked strain changes observed in the honey bees [17] on which they were feeding, i.e., DWV diversity decreased in honey bees as varroa became established on Oahu and then on the Big Island, and this change was detected in yellowjackets on the Big Island. Although the current study is limited to eight individuals across two locations, it is clear that DWV is present in the wasp samples, that sequences are diverse, and that the presence and potential level of DWV variants are affected by location.

This is the first study of arthropods in the wider apiary community to consider the newly described Moku [29] and Milolii [30] viruses. Although these viruses only dominated in the samples in which they were first described, low numbers of reads were also found in other varroa, beetle, wasp, and ant samples. The wider +ssRNA virus detection study revealed that all of the common honey bee-associated viruses tested for were present at some level in the Hawaiian apiary insects, although IAPV, KBV, ABPV, and LSV were only present in very low levels (as detected through BLAST), and therefore, true infections cannot be confirmed in any sample. It is interesting to note that, similar to the findings of the DWV data, when considering the other honey bee-associated virus data, species and location groupings were seen. In the case of species groupings, this may be related to different species having different virus susceptibilities and, in the case of site groupings, may be more related to variation in particular virus levels at different sites.

5. Conclusions

Although this pilot study has limitations, namely unbalanced sampling design, we have shown that several common honey bee-associated +ssRNA viruses are common in taxonomically diverse apiary pests. We showed that DWV was the most prevalent virus and that DWV infections grouped between species in terms of dominant variants and recombinants, but that common variants (predominantly type A) were circling between all species, suggesting repeated transmission events between species. Within the wasps, DWV was further separated by location. Species also grouped in terms of which other honey bee-associated viruses they harbored, i.e., particular viruses are associated with particular hosts. Therefore, this study highlights the need to consider the wider arthropod community as potential reservoirs of viral pathogens in the apiary.

Supplementary Materials: The following are available online at http://www.mdpi.com/1999-4915/11/5/397/s1, Figure S1: DWV genome coverage plots for individual samples created using Geneious. Read depths are shown on a log-10 scale and represent DWV-A (red), -B (blue), and -C(yellow) along the ~10.1 kb genomes. Figure S2: DWV alignments (MUSCLE) created using Geneious showing de novo assembled contigs from samples A_tum-5 and V_pens-8, which contain recombination breakpoints. (a) Contig A_tum-5-a aligned with DWV-A (NC_004830.2) and DWV-C (CEND01000001.1) reference genomes, (b) contig V_pen_8-b aligned with DWV-A (NC_004830.2) and DWV-B (AY251269.2), and (c) a second contig from sample V_pen_8; V_pen_8-t also aligned with DWV-A (NC_004830.2) and DWV-B (AY251269.2). All alignments show disagreements with the assembled contigs highlighted in black, and recombinant contigs are shaded red where they map most closely to DWV-A, blue to DWV-B, and green to DWV-C. Table S1: Samples used in this study. Sample names are given along with the site from which they were sampled, species name, and the symbol used to denote them in Figures 1 and 5. Table S2: DWV-A *RdRp* sequences originally from [4] and used in this study in the construction of the DWV-A phylogeny in Figure 5. Table S3: Viruses commonly found in bees used for BLAST analysis along with accession numbers. Table S4: Numbers of reads mapping to DWV types A, B, and C using Blast top hit analysis for each sample, along with total numbers of reads passing QC (read1.fasta).

Author Contributions: Conceptualization, L.E.B. and S.J.M.; methodology, L.E.B. and S.J.M.; software, L.E.B.; validation, L.E.B.; formal analysis, L.E.B. and S.J.M.; investigation, L.E.B., S.J.M. and D.C.S; resources, S.J.M. and D.C.S.; data curation, L.E.B.; original draft preparation, L.E.B.; review and editing, L.E.B., S.J.M. and D.C.S.; visualization, L.E.B.; S.J.M. and D.C.S.; and funding acquisition, S.J.M.

Funding: This research was partly funded by Apis-M.

Acknowledgments: We would like to think E. Villolobos and S. Nikaido for their assistance in organizing and carrying out the sampling, C. Ahrens for assistance with statistical analyses using R, J. Finch for assistance with the phylogenetic reconstructions and all the beekeepers in Hawaii who provided access to their bees and apiaries. We also would like to thank the anonymous reviewers, whose comments helped to improve this manuscript.

Conflicts of Interest: The authors declare no conflict of interest

References

1. Daszak, P.; Cunningham, A.A.; Hyatt, A.D. Emerging infectious diseases of wildlife–threats to biodiversity and human health. *Science* **2000**, *287*, 443. [CrossRef] [PubMed]
2. Blaustein, A.R.; Johnson, P.T.J. Conservation biology: When an infection turns lethal. *Nature* **2010**, *465*, 881–882.
3. Schroeder, D.C.; Martin, S.J. Deformed wing virus: The main suspect in unexplained honeybee deaths worldwide. *Virulence* **2012**, *3*, 589–591. [CrossRef] [PubMed]
4. Wilfert, L.; Long, G.; Leggett, H.C.; Schmid-Hempel, P.; Butlin, R.; Martin, S.J.; Boots, M. Deformed wing virus is a recent global epidemic in honeybees driven by Varroa mites. *Science* **2016**, *351*, 594–597. [CrossRef]
5. Potts, S.G.; Vulliamy, B.; Dafni, A.; Ne'eman, G.; Willmer, P. Linking bees and flowers: How do floral communities structure pollinator communities? *Ecology* **2003**, *84*, 2628–2642.
6. Steffan-Dewenter, I.; Tscharntke, T. Resource overlap and possible competition between honeybees and wild bees in central Europe. *Oecologia* **2000**, *122*, 288–296. [CrossRef]
7. Steffan-Dewenter, I.; Tscharntke, T. Insect communities and biotic interactions on fragmented calcareous grasslands—a mini review. *Biol. Cons.* **2002**, *104*, 275–284.
8. Thomson, D. Competitive interactions between the invasive European honeybee and native bumble bees. *Ecology* **2004**, *85*, 458–470. [CrossRef]
9. Potts, S.G.; Biesmeijer, J.C.; Kremen, C.; Neumann, P.; Schweiger, O.; Kunin, W.E. Global pollinator declines: Trends, impacts and drivers. *Trends Ecol. Evol.* **2010**, *25*, 345–353. [CrossRef]
10. Brown, M.J.; Dicks, L.V.; Paxton, R.J.; Baldock, K.C.; Barron, A.B.; Chauzat, M.P.; Freitas, B.M.; Goulson, D.; Jepsen, S.; Kremen, C.; et al. A horizon scan of future threats and opportunities for pollinators and pollination. *PeerJ* **2016**, *4*, e2249. [CrossRef] [PubMed]

11. Vanbergen, A.J. Threats to an ecosystem service: Pressures on pollinators. *Front. Ecol. Environ.* **2013**, *11*, 251–259. [CrossRef]

12. Cornman, R.S.; Tarpy, D.R.; Chen, Y.; Jeffreys, L.; Lopez, D.; Pettis, J.S.; Evans, J.D. Pathogen webs in collapsing honeybee colonies. *PLoS ONE* **2012**. [CrossRef]

13. Collison, E.; Hird, H.; Cresswell, J.; Tyler, C. Interactive effects of pesticide exposure and pathogen infection on bee health—A critical analysis. *Biol. Rev. Camb. Philos. Soc.* **2016**, *91*, 1006–1019. [CrossRef]

14. Parrish, C.R.; Holmes, E.C.; Morens, D.M.; Park, E.C.; Burke, D.S.; Calisher, C.H.; Laughlin, C.A.; Saif, L.J.; Daszak, P. Cross-species virus transmission and the emergence of new epidemic diseases. *Microbiol. Mol. Biol. Rev.* **2008**, *72*, 457–470. [CrossRef] [PubMed]

15. Domingo, E.J.; Holland, J.J. RNA virus mutations and fitness for survival. *Annu. Rev. Microbiol.* **1997**, *51*, 151–178. [CrossRef] [PubMed]

16. Ryabov, E.V.; Wood, G.R.; Fannon, J.M.; Moore, J.D.; Bull, J.C.; Chandler, D.; Mead, A.; Burroughs, N.; Evans, D.J. A virulent strain of Deformed wing virus (DWV) of honeybees (*Apis mellifera*) prevails after *Varroa destructor*-mediated, or in vitro, transmission. *PLoS Pathog.* **2014**, *10*, 1–21. [CrossRef] [PubMed]

17. Martin, S.J.; Highfield, A.C.; Brettell, L.; Villalobos, E.M.; Budge, G.E.; Powell, M.; Nikaido, S.; Schroeder, D.C. Global honeybee viral landscape altered by a parasitic mite. *Science* **2012**, *336*, 1304–1306. [CrossRef]

18. Mondet, F.; de Miranda, J.R.; Kretzschmar, A.; Le Conte, Y.; Mercer, A.R. On the front line: Quantitative virus dynamics in honeybee (*Apis mellifera* L.) colonies along a new expansion front of the parasite *Varroa destructor*. *PLoS Pathog.* **2014**, *10*, e1004323. [CrossRef]

19. Martin, S.J.; Brettell, L.E. Deformed wing virus in Honeybees and Other Insects. *Annu. Rev. Virol.* **2019**, *6*. [CrossRef]

20. Eyer, M.; Chen, Y.P.; Schäfer, M.O.; Pettis, J.; Neumann, P. Small hive beetle, *Aethina tumida*, as a potential biological vector of honeybee viruses. *Apidologie* **2009**, *40*, 419–428. [CrossRef]

21. Fürst, M.A.; McMahon, D.P.; Osborne, J.L.; Paxton, R.J.; Brown, M.J.F. Disease associations between honeybees and bumblebees as a threat to wild pollinators. *Nature* **2014**, *506*, 364–366. [CrossRef]

22. McMahon, D.P.; Fürst, M.A.; Caspar, J.; Theodorou, P.; Brown, M.J.F.; Paxton, R.J. A sting in the spit: Widespread cross-infection of multiple RNA viruses across wild and managed bees. *J. Anim. Ecol.* **2015**, *84*, 615–624. [CrossRef]

23. Santamaria, J.; Villalobos, E.M.; Brettell, L.E.; Nikaido, S.; Graham, J.R.; Martin, S. Evidence of Varroa-mediated deformed wing virus spillover in Hawaii. *J. Invertebr. Pathol.* **2018**, *151*, 126–130. [CrossRef]

24. Loope, K.J.; Baty, J.W.; Lester, P.J.; Wilson Rankin, E.E. Pathogen shifts in a honeybee predator following the arrival of the Varroa mite. *Proc. R. Soc. B* **2019**, *286*. [CrossRef]

25. Mordecai, G.J.; Brettell, L.E.; Martin, S.J.; Dixon, D.; Jones, I.M.; Schroeder, D.C. Superinfection exclusion and the long-term survival of honeybees in Varroa-infested colonies. *Isme J.* **2016**, *10*, 1182–1191. [CrossRef]

26. Brettell, L.E.; University of Salford, Manchester, UK. Results of screening for Deformed wing virus in common apiary pest insects using RT-PCR. 2012.

27. Yang, X.; Charlebois, P.; Gnerre, S.; Coole, M.G.; Lennon, N.J.; Levin, J.Z.; Qu, J.; Ryan, E.M.; Zody, M.C.; Henn, M.R. De novo assembly of highly diverse viral populations. *BMC Genomics* **2012**, *13*, 475. [CrossRef]

28. Huelsenbeck, J.P.; Ronquist, F. MRBAYES: Bayesian inference of phylogenetic trees. *Bioinformatics* **2001**, *17*, 754–755. [CrossRef]

29. Mordecai, G.J.; Brettell, L.E.; Pachori, P.; Villalobos, E.M.; Martin, S.J.; Jones, I.M.; Schroeder, D.C. Moku virus; a new Iflavirus found in wasps, honeybees and Varroa. *Sci. Rep.* **2016**, *6*, 34983. [CrossRef]

30. Brettell, L.E.; Mordecai, G.; Pachori, P.; Martin, S. Novel RNA virus genome discovered in Ghost ants (*Tapinoma melanocephalum*) from Hawaii. *Genome Announc.* **2017**, *5*, e00669-17. [CrossRef]

31. Villalobos, E.M. The mite that jumped, the bee that traveled, the disease that followed. *Science* **2016**, *351*, 554–556. [CrossRef]

32. Brooks, E.M.; Sheflin, L.G.; Spaulding, S.W. Secondary structure in the 3′UTR of EGF and the choice of reverse transcriptases affect the detection of message diversity by RT-PCR. *Biotechniques* **1995**, *19*, 806–812.

33. Lanzi, G.; de Miranda, J.R.; Boniotti, M.B.; Cameron, C.E.; Lavazza, A.; Capucci, L.; Carmazine, S.M.; Rossi, C. Molecular and Biological Characterization of Deformed Wing Virus of Honeybees (*Apis mellifera* L.). *J. Virol.* **2006**, *80*, 4998–5009. [CrossRef]

34. Wood, G.R.; Burroughs, N.J.; Evans, D.J.; Ryabov, E.V. Error correction and diversity analysis of population mixtures determined by NGS. *PeerJ* **2014**, *2*, e645. [CrossRef] [PubMed]

35. Kevill, J.L.; de Souza, F.; Sharples, C.; Schroeder, D.; Martin, S.J. DWV-A lethal to honey bees (*Apis mellifera*): A colony level survey of DWV variants (A, B & C) in England, Wales and 32 states across the US. *Viruses* **2019**. (submitted).

36. Villalobos, E.M.; (University of Hawaii, Hawaii, USA). Personal communication, 2017.

37. De Souza, F.S.; Kevill, J.L.; Correia-Oliveira, M.E.; de Carvalho, C.A.; Martin, S.J. Occurrence of deformed wing virus variants in the stingless bee *Melipona subnitida* and honey bee *Apis mellifera* populations in Brazil. *J. Gen. Virol.* **2019**, *100*, 289–294. [CrossRef]

38. Boncristiani, H.F.; Di Prisco, G.; Pettis, J.S.; Hamilton, M.; Chen, Y.P. Molecular approaches to the analysis of deformed wing virus replication and pathogenesis in the honey bee, *Apis mellifera*. *Virol. J.* **2009**, *6*, 221. [CrossRef]

39. Fung, E.; Hill, K.; Hogendoorn, K.; Glatz, R.V.; Napier, K.R.; Bellgard, M.I.; Barrero, R.A. *De novo* assembly of honey bee RNA viral genomes by tapping into the innate insect antiviral response pathway. *J. Invertebr. Pathol.* **2018**, *152*, 38–47. [CrossRef]

40. Singh, R.; Levitt, A.L.; Rajotte, E.G.; Holmes, E.C.; Ostiguy, N.; Lipkin, W.I.; Toth, A.L.; Cox-Foster, D.L. RNA viruses in Hymenopteran pollinators: Evidence of inter-taxa virus transmission via pollen and potential impact on non-*Apis* hymenopteran species. *PLoS ONE* **2010**, *5*, e14357. [CrossRef]

 *viruses* 

Article

Conserved Secondary Structures in Viral mRNAs

Michael Kiening [1], **Roman Ochsenreiter** [2], **Hans-Jörg Hellinger** [4], **Thomas Rattei** [4], **Ivo Hofacker** [2,3] **and Dmitrij Frishman** [1,5,*]

[1]   Department of Bioinformatics, Wissenschaftszentrum Weihenstephan, Technische Universität München, Maximus-von-Imhof-Forum 3, D-85354 Freising, Germany; m.kiening@wzw.tum.de

[2]   Research Group Bioinformatics and Computational Biology, Faculty of Computer Science, University of Vienna, Währingerstr. 29, 1090 Vienna, Austria; romanoch@tbi.univie.ac.at (R.O.); ivo@tbi.univie.ac.at (I.H.)

[3]   Department of Theoretical Chemistry, Faculty of Chemistry, University of Vienna, Währingerstrasse 17, 1090 Vienna, Austria

[4]   Division of Computational Systems Biology, Department of Microbiology and Ecosystem Science, University of Vienna, Althanstraße 14, 1090 Vienna, Austria; joerg.hellinger@univie.ac.at (H.-J.H.); thomas.rattei@univie.ac.at (T.R.)

[5]   St. Petersburg State Polytechnic University, St. Petersburg 195251, Russia

*   Correspondence: d.frishman@wzw.tum.de

Received: 30 March 2019; Accepted: 26 April 2019; Published: 29 April 2019

Abstract: RNA secondary structure in untranslated and protein coding regions has been shown to play an important role in regulatory processes and the viral replication cycle. While structures in non-coding regions have been investigated extensively, a thorough overview of the structural repertoire of protein coding mRNAs, especially for viruses, is lacking. Secondary structure prediction of large molecules, such as long mRNAs remains a challenging task, as the contingent of structures a sequence can theoretically fold into grows exponentially with sequence length. We applied a structure prediction pipeline to Viral Orthologous Groups that first identifies the local boundaries of potentially structured regions and subsequently predicts their functional importance. Using this procedure, the orthologous groups were split into structurally homogenous subgroups, which we call subVOGs. This is the first compilation of potentially functional conserved RNA structures in viral coding regions, covering the complete RefSeq viral database. We were able to recover structural elements from previous studies and discovered a variety of novel structured regions. The subVOGs are available through our web resource RNASIV (RNA structure in viruses).

Keywords: mRNA structure; structure database; secondary structure; viral mRNA; subVOG; structurally related; RNA structure; structurally homogenous; structurally related; mRNA families

---

1. Introduction

Secondary structures formed in single-stranded mRNA molecules through complementary self-interactions, both in the untranslated (UTR) and coding (CDS) regions of mRNAs, have been implicated in a variety of regulatory functions [1]. For example, riboswitches modulate gene expression through conformational changes in response to various stimuli [2]. Translation initiation, elongation, and termination as well as translation efficiency depend on higher order mRNA secondary structures in non-coding regions [3,4]. CDS hairpins have also been suggested to play a role in the regulation of translation [5], in particular by causing ribosomal stalling and modulating translational efficiency [6]. The relationship between mRNA structure in the CDS and gene expression has been demonstrated both computationally and experimentally [7–11]. In particular, reduced mRNA stability near the start codon has been observed in a wide range of species, probably as a mechanism to facilitate ribosome binding or start codon recognition by initiator-tRNA [12]. Structured elements within CDS directly influence mRNA abundance [13]. Computational studies show that native mRNAs have lower folding energies and are thus more stable than codon-randomized ones [5]. The three mRNA functional domains—5′UTR, CDS, and 3′UTR—form largely independent folding units, while base pairing across domain borders is rare [14]. The ability of viruses to persist in their host in a genus-specific manner is influenced by the interplay between local structural motifs and genome-scale ordered RNA structures (GORS) [15], which impose additional restraints on the RNA sequence space. Evolutionarily conserved local secondary structures have been identified in CDSs [16] and shown to be functional [17]. An indirect indication of the global importance of RNA structures in the coding regions comes from the recent study of Fricke et al. who identified selection favoring specific pairing patterns between synonymous codons within RNA hairpins [18].

Increasing evidence suggests that secondary structural elements in the CDSs of viral RNAs also constitute a previously underappreciated, evolutionarily conserved level of functional organization of viruses. A large number of conserved secondary structural motifs were computationally identified in the Flavivirus genomes [19–21], predicted to restrain sequence variability [22] and experimentally shown to regulate important biological processes, such as replication and infection [21]. Multiple secondary structures were described in the coding regions of the (+) sense RNA of the Influenza A virus [23]. Another example is a secondary structural element within the coding region of the Dengue virus type 2, which is essential for its replication [24]. More recently, using a comparative genomics approach, Goz and Tuller identified a large number of potentially functionally important regions in the coding regions of Dengue viruses, in which the RNA folding strength is conserved independently of sequence conservation and compositional bias [25]. Specific regions in the HIV structural genes were reported to be under strong selection for stable secondary structures [26]. Recent research shows that mechanisms of translational control by RNA structures can be shared between viruses and cellular organisms [27].

Given the important role played by RNA structures in shaping the evolutionary dynamics of viruses and modulating their interaction with the host, a large-scale investigation of RNA motifs in viruses would be warranted. However, there are two major challenges that need to be addressed before embarking on such an investigation. First, accurate structure prediction for long RNA molecules, such as mRNAs, is generally out of reach for the existing computational methods. Second, conserved stem-loop structures can only be derived from a collection of high-quality alignments of orthologous viral transcripts, which are difficult to obtain, given the rapid pace of viral evolution and the ensuing poor sequence conservation, even between closely related species.

Here, we propose a computational approach to explore the RNA structurome of the viral coding regions, in which local structure predictions are applied to VOG (Viral Orthologous Groups, http://vogdb.org), the first comprehensive collection of orthologous groups derived for all viral proteins contained in the RefSeq [28] database. We utilize RNALalifold [29] to scan long input sequences for locally optimal secondary structures. The identified structural boundaries are more accurate than those derived from using a sliding window of fixed length. Functional importance of structured regions is assessed by RNAz [30]. We present a novel database, RNASIV (RNA structure in viruses; http://rnasiv.bio.wzw.tum.de), which contains the largest currently available collection of predicted RNA structures in viruses. It provides access to 201,708 viral mRNA sequences clustered into 42,293 structurally homogenous groups and is intended to become a useful tool for exploring structure–function relationships in virus families.

2. Materials and Methods

2.1. Viral Orthologous Groups (VOGs)

All genome sequences and their annotations were retrieved from the RefSeq viral database release 79 [31] and grouped into phages and non-phages, based on the available taxonomic information. Assemblies containing inconsistently annotated or completely unannotated polyproteins were identified based on the manually curated information provided by ViralZone [32] and excluded from consideration. Phage and non-phage protein sequences were clustered into phage and non-phage preVOGs, using the NCBI's COG software package with all default settings.

For all phage and non-phage preVOGs, multiple sequence alignments were constructed with Clustal Omega v1.2.4 [33] and used to build HMM-profiles using HMMer 3 [34]. The profiles were subsequently aligned against each other, using HHalign from the HHsuite toolkit [35]. The number of aligned HMM columns was used as an alignment score. All scores for alignments with HHalign probability >85, HHalign *e*-Value < 10^{-5}, and more than 70% of aligned columns between the query and the match HMM were stored as an all-against-all matrix. This matrix was clustered into 21,200 VOGs, using the MCL (Markov Clustering) method [36]. Based on the manual inspection of the homogeneity of the protein function descriptions in the resulting clusters, we selected the inflation value of 2.0 for the MCL clustering. For all VOG member proteins, we determined the closest homolog in the UniProt database [37] from BLAST [38] hits with *E*-values better than 10^{-5} and a minimal query coverage of 90%. Functional descriptions of VOGs were automatically derived based on the most frequent protein description found in the UniProt entries or, if not available, in the RefSeq annotation [31]. The complete VOG dataset, which was used in this study, and supplementary files are available for download at http://vogdb.org.

2.2. Mapping VOG Sequences to Specific Hosts

We used Virus-Host DB [39] to assign host information to VOG proteins. For 20757 VOGs, we were able to map all contained sequences to a specific host, while 428 VOGs contain proteins from at least one viral species for which we could not find host annotation. Most VOGs include viruses infecting hosts from only one domain of life, i.e., bacteria (~72%), eukaryotes (~22%), or archaea (4%), while only 2% of VOGs are taxonomically mixed (Figure 1). Only 12 VOGs contain viruses that infect hosts from all three domains of life. The VOG sizes range from 15 proteins of 12 distinct species, up to 265 proteins belonging to 261 different species (on average, 104 proteins from 95 different species). These VOGs mostly harbor highly conserved core enzymes of double-stranded DNA viruses, such as kinases, ligases, methylases, helicases, hydrolases, and synthases [40]. The other two VOGs additionally contain proteins from viruses belonging to the order of Caudovirales, which belong to the bacteriophages, which are not classified as double-stranded DNA viruses, according to the NCBI taxonomy. We excluded from consideration 15 VOGs containing satellite viruses infecting other viruses.

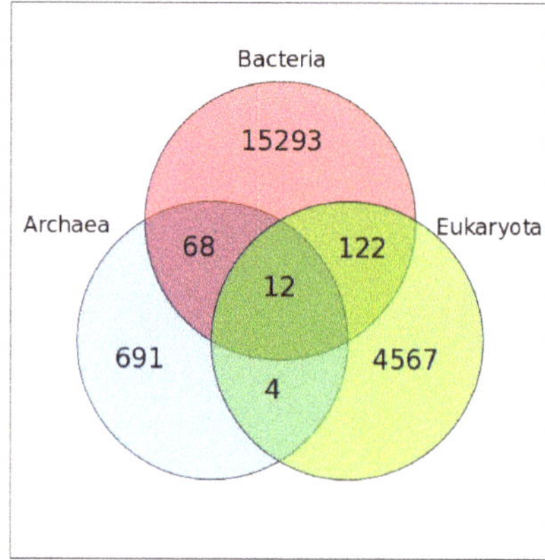

Figure 1. Venn diagram showing the taxonomy of the host organisms within all viral orthologous groups (VOGs). Only those VOGs are included for which host annotation for all viruses is available in the Virus-Host DB.

2.3. Distance Trees of VOG Proteins

Expectedly, we found that RNA structure conservation within VOGs decreases with increasing VOG size. Most VOGs (66%) consist of at least three sequences (size distribution shown in Figure 2) and can therefore potentially be split into smaller groups containing structures that are not conserved across the entire VOG. We therefore utilized distance trees derived by the neighbor-joining algorithm [41] to identify structurally homogeneous subsets of VOGs (subVOGs). All-against-all pairwise alignments of protein sequences were calculated using Clustal Omega and then converted to the nucleotide alphabet. The distance matrices were derived from pairwise sequence identity values, and the trees were created from the matrices using neighbor joining, as implemented in the BioPerl toolkit [42]. The inner nodes of the sequence trees represent possible subVOG candidates, potentially containing structurally homogenous sequences.

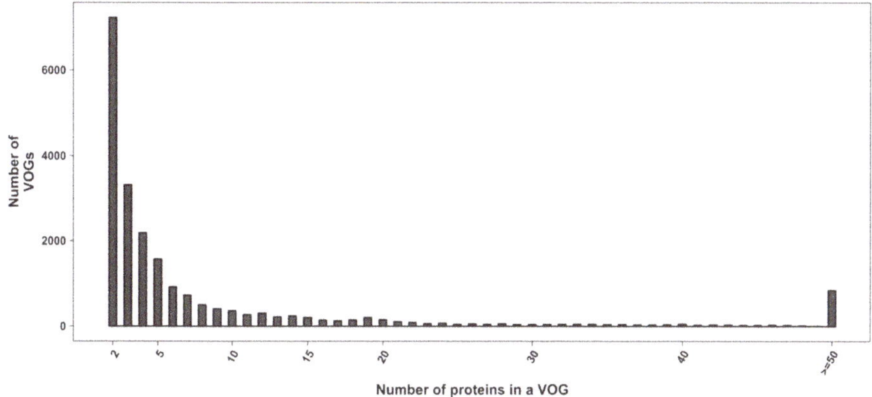

Figure 2. Distribution of VOG sizes.

2.4. Structure Prediction and subVOG Assignment

In order to assess the amount of structural RNA conservation present in subVOG candidates, multiple sequence alignments (MSAs) of proteins were calculated for each inner node of the distance trees and converted to the nucleotide alphabet. The RefSeq nucleotide and protein sequences were obtained from the VOGDB. We then employed RNALalifold from the ViennaRNA package [29], with default parameters, to determine the boundaries of locally stable structures within each MSA, and realigned these local regions using mLocARNA [43]. MLocARNA produces structure-guided multiple sequence alignments, using an adapted version of the Sankoff algorithm. The significance and conservation of the found structures was assessed with RNAz [30]. This procedure is simpler and arguably more accurate than the usual approach of applying RNAz to the entire MSA within a sliding window. RNAz classifies fragments of an MSA pre-selected by RNALalifold as containing or not containing a functional RNA secondary structural element. Realignment with mLocARNA significantly increases the precision of RNAz [30]. As no sequence of a potential subVOG can be regarded as a reference sequence, the option "no reference" was used for the subsequent RNAz analysis. The RNAz method uses the RNAfold algorithm from the ViennaRNA package to calculate secondary structures and the corresponding minimum free energy (MFE) for each individual RNA sequence in the alignment. In addition, for each aligned sequence set, RNAz calculates a consensus secondary structure and its MFE using the RNAalifold algorithm. RNAz assumes that conserved and thermodynamically stable structures are functional, in which case it outputs "RNA". Otherwise, it outputs "OTHER". For this purpose, a class probability value, combining all information on an input alignment is calculated. We used a stringent threshold of 0.9 (default 0.5) for the class probability value, which is recommended for finding high confidence structures [30]. Subsequently, the trees were scanned for subtrees containing at least one conserved structural element, that is, predicted to be functional, and the largest subtrees were designated as structurally homogenous subVOGs. We found that sequences that are only distantly related according to the neighbor-joining tree may still share conserved RNA structures. In order to account for structure-level relationships between sequences, we built covariance models for all conserved structures found within subVOGs, using the tool cmbuild from the infernal package [44], and used them to search against all sequences in the entire VOG database.

2.5. mRNA Stability

Following Tuller et al. [45] and Faure et al. [46], we employed RNAfold to calculate the folding energy of the most and the least stable 30-nucleotide segment of mRNAs ($\Delta Gmin$ and $\Delta Gmax$, respectively), as well as the average folding energy of all possible 30 nucleotide segments ($\Delta Gmean$). Faure et al. investigated the effect of mRNA stability on the translation rate and protein folding. During translation, the ribosome sequentially unfolds parts of the mRNA. These parts are typically 30 nucleotides long, which explains the choice of segment length in Faure et al. As this procedure does not take into account the actual boundaries of local structures, but rather limits all structures to the size of 30 nucleotides, we additionally calculated the three energy values for all local optimal structures found with RNALfold.

2.6. mRNA Structures and Protein Function

We investigated the relationship between protein function, described in terms of gene ontology (GO) annotation [47], and mRNA structures. Instead of using the global folding energy for classifying mRNAs as highly or lowly structured [48], we considered structural coverage—the portion of an mRNA covered by functional and conserved structures. GO terms for all VOG proteins were downloaded using QuickGO [49], where available. Based on the Evidence & Conclusion Ontology (ECO) evidence codes [50], two separate datasets were created: (i) Proteins annotated by manually or experimentally derived GO terms (ECO evidence codes: ECO:0000352, ECO:0000269), and (ii) proteins annotated by GO terms with any evidence codes. To find out whether mRNAs of proteins with certain functions tend to harbor more or fewer structures, we pooled together functionally similar GO terms with the average structural coverage of their corresponding mRNAs, using Revigo [51]. Revigo uses a semantic similarity measure to group similar GO terms together, which results in a concise list of distinct functions. To perform this analysis, we calculated the average structural coverage of all subVOG mRNAs with available GO annotation. For the experimental dataset we allowed a coverage value to be associated with a GO term if more than 50% of the sequences in a particular subVOG were annotated with this term. Within the dataset based on all evidence codes, we only allowed GO terms shared by all sequences of a subVOG. We only used mRNAs that were clustered into a subVOG. For sequences that were not part of any subVOG, we did not find conserved structures, although this does not necessarily mean that the mRNA did not contain functional structures. The distributions of standard deviations of the structural coverage values were compared within the actual and randomly generated Revigo clusters. Randomization was performed 1000 times by preserving the size of the clusters and filling them with randomly chosen GO terms.

3. Results

3.1. Overview of the Study

A graphical overview of the study is given in Figure 3. In a first step, we created distance trees for all protein sequences contained in each VOG, using the neighbor joining method, as described in Materials and Methods. All sequences of the inner nodes of each tree, representing potential subVOGs, were multiply aligned, converted to the nucleotide alphabet and processed with RNALalifold to obtain all potentially conserved local optimal structures. Each part of the alignment covering a potential structure was then realigned with the structure-guided alignment method mLocARNA and checked for functionality using RNAz. The use of structure-guided alignments as input for RNAz improves the performance, compared to pure sequence-based alignments [30]. The tree nodes containing the most sequences that yielded conserved structures were taken as final subVOGs. For all obtained subVOG structures, we computed covariance models that could be used to search for similar structures in future research.

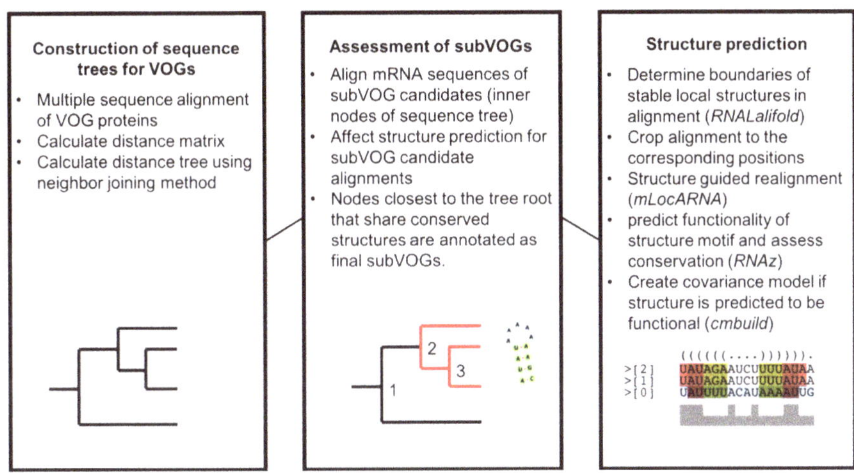

Figure 3. Overview of the analysis of conserved RNA structures in VOGs.

3.2. Structure Conservation in VOGs

The current release of the VOG database, derived from the RefSeq release 77, contains 21,200 VOGs, composed of 251,796 proteins from 6252 phages and eukaryotic viruses (Figure S1). Protein sequences in each VOG were aligned by Clustal Omega, converted to the nucleotide alphabet, and used as input for RNA structure prediction by RNALalifold. As seen in Figure 4, the number of local optimal structures conserved within entire VOGs decreases quickly with the number of aligned sequences, which may in part be the consequence of poor multiple alignment quality in large sets of sequences. Indeed, we found that proteins in smaller VOGs tend to be more closely related (Figure S2). To exclude structures found due to sequence conservation only, the potential functionality of structures was verified with RNAz. However, even those VOGs that only consist of a few sequences do not always contain conserved structures. There are 7232 VOGs with exactly two sequences, and for 1237 of these, we could not find any conserved structures. The remaining 5995 VOGs of size two had an average structural coverage of approximately 25% (Figure 5a). Out of the 13,968 VOGs with more than two sequences, 7238 VOGs were predicted to contain RNA structures conserved across the entire VOG, with an average structural coverage of approximately 18% (Figure 5b). These contain between 3 and 96 sequences, with an average of 6. On average, VOGs contain sequences from three different genera, mostly belonging to the same taxonomic family and thus also to the same order (Figure 6a–c). The 25 most diverse VOGs contain sequences from three different orders and up to 19 taxonomic families. On average, a VOG contains mRNAs from viruses that infect hosts from four different genera, belonging to three different taxonomic families and two orders. The VOG with the highest host diversity corresponds to 209 different host genera from 114 families and 64 orders.

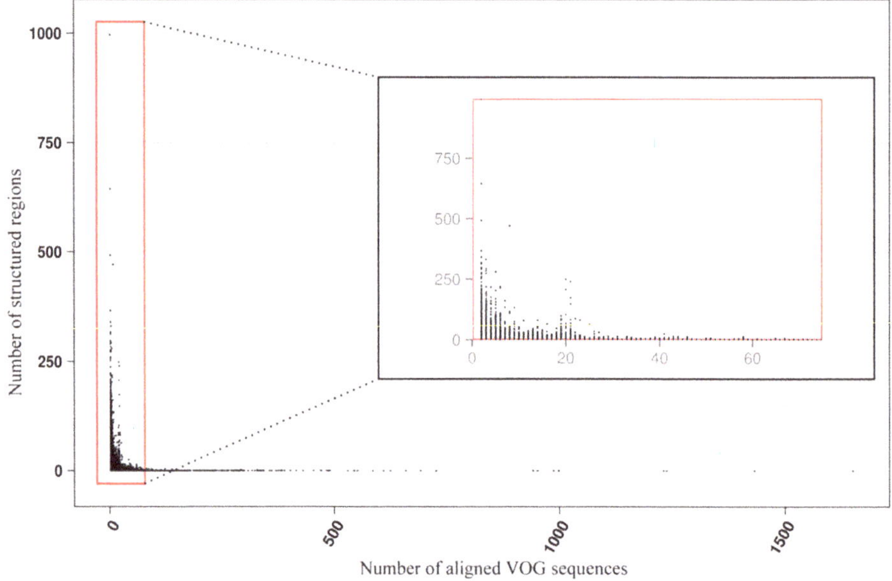

Figure 4. Number of local RNA structures as a function of VOG size.

(a)

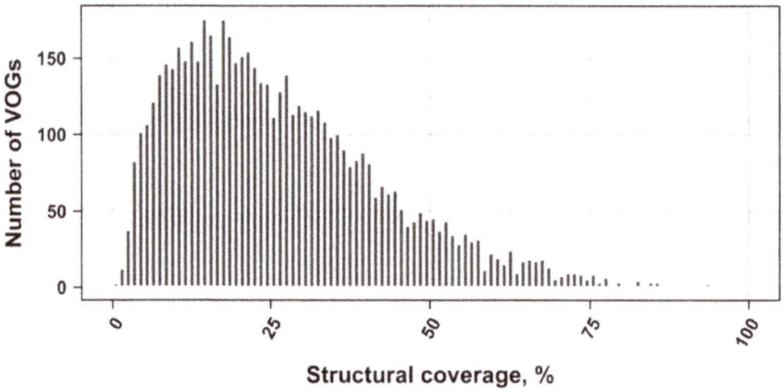

Figure 5. *Cont.*

(b)

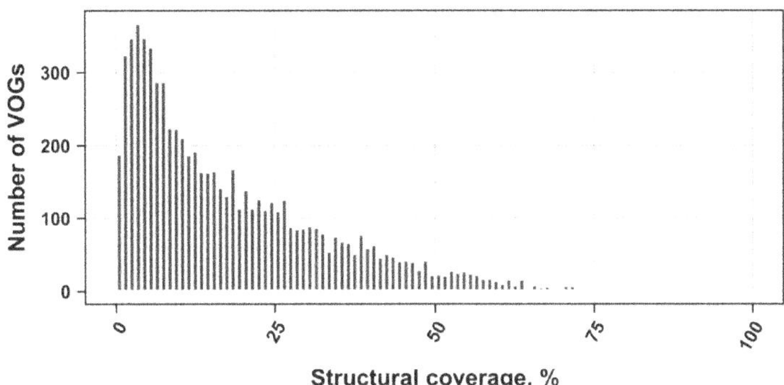

(c)

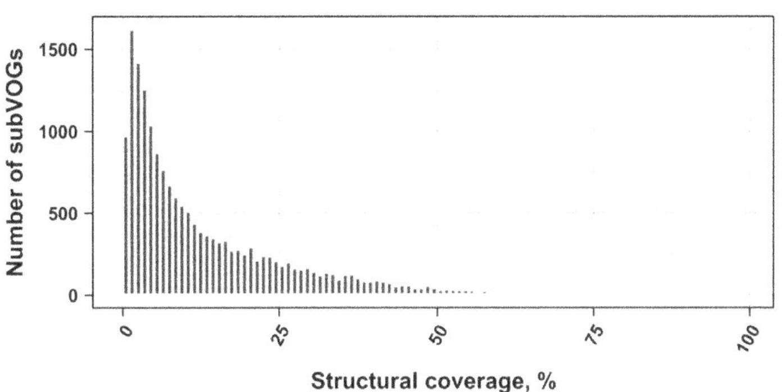

Figure 5. Coverage of VOG alignments by local optimal RNA structures. (**a**) VOGs with two sequences. (**b**) VOGs with more than two sequences, in which structures are conserved across all sequences. (**c**) subVOGs. VOGs that did not contain conserved structures, even after splitting into subVOGs, are not shown.

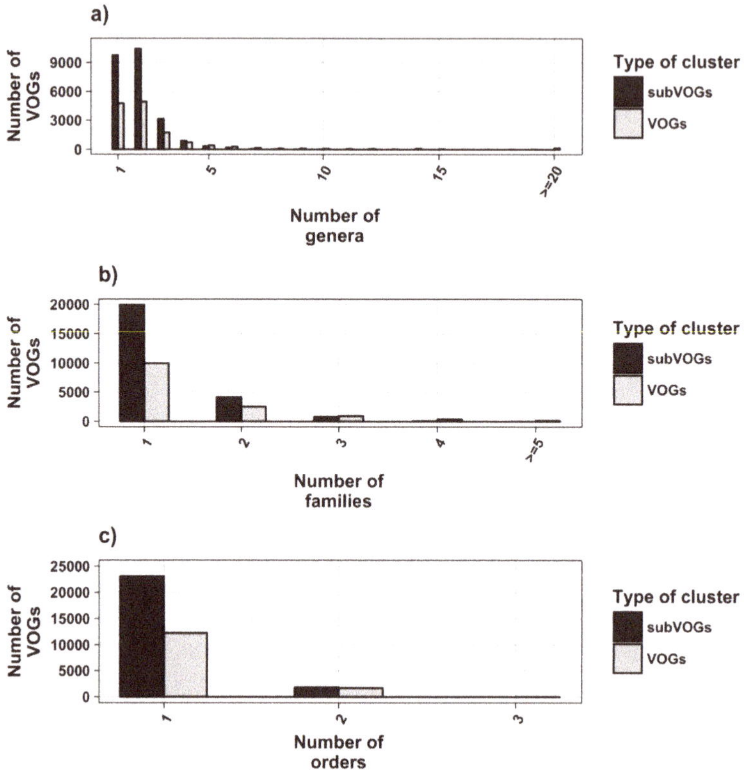

Figure 6. Taxonomic distribution of proteins in VOGs (with more than two sequences) and subVOGs.

3.3. Structure Conservation in subVOGs

We attempted to subdivide 6730 VOGs with more than two sequences and without conserved structures into structurally homogeneous subsets, which we call subVOGs, using phylogenetic trees derived by the neighbor-joining method. This procedure resulted in 17,678 subVOGs with an average structural coverage of approximately 13% (Figure 5c). The average number of genera per subVOG is 2 and the most diverse of them contains sequences from three orders and 14 families. A subVOG contains on average sequences that infect two different host genera, and the most diverse subVOG infects hosts of 42 different genera, belonging to 33 families and 20 different orders (Figure 7a–c). Thus, unsurprisingly, subVOGs, which constitute subsets of full VOGs with increased structural homogeneity, exhibit a reduced taxonomic spread, both of the viruses they contain and their hosts. A large fraction of subVOGs (63%) contains sequences from more than one genus and 21% contain sequences from more than one family. The structural coverage of subVOGs, i.e., the fraction of alignment positions that are located within conserved RNA structures, decreases with increasing taxonomic diversity of the viruses and their hosts (Figure 8). An example that demonstrates the reduction of taxonomic spread between a VOG and its corresponding subVOGs is given in Figure 9. Here, the VOG 00052, which contains 20 proteins from 12 different virus species belonging to 4 different taxonomic families, was split into four structurally homogenous subVOGs. Two of the four subVOGs consist of mRNAs belonging to the genus Avipoxvirus from the family Poxviridae, the third subVOG contains sequences from the family Mimiviridae, and the fourth subVOG consists of two mRNAs belonging to viruses from two different taxonomic families, the Ascoviridae and the Iridoviridae. For two mRNAs, we could not find structures conserved in any of the other VOG members and they are therefore not part of any subVOG.

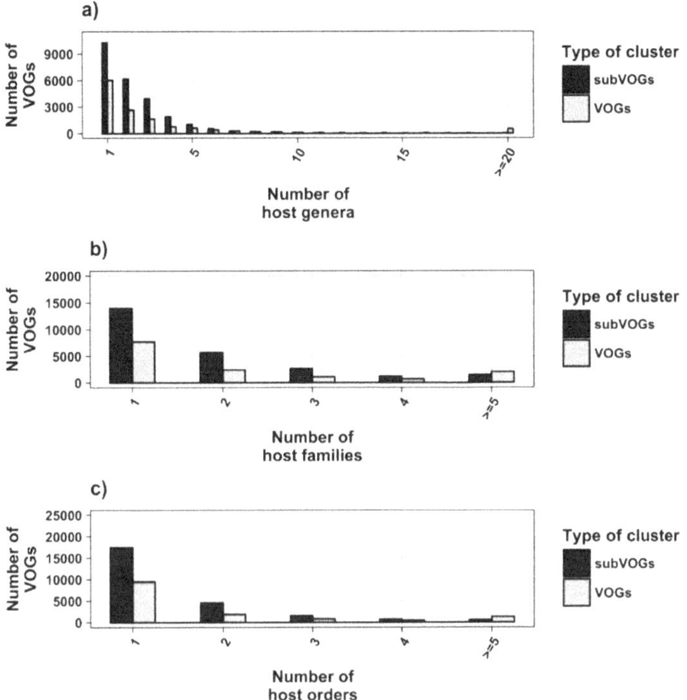

Figure 7. Taxonomic distribution of hosts in VOGs (with more than two sequences) and subVOGs.

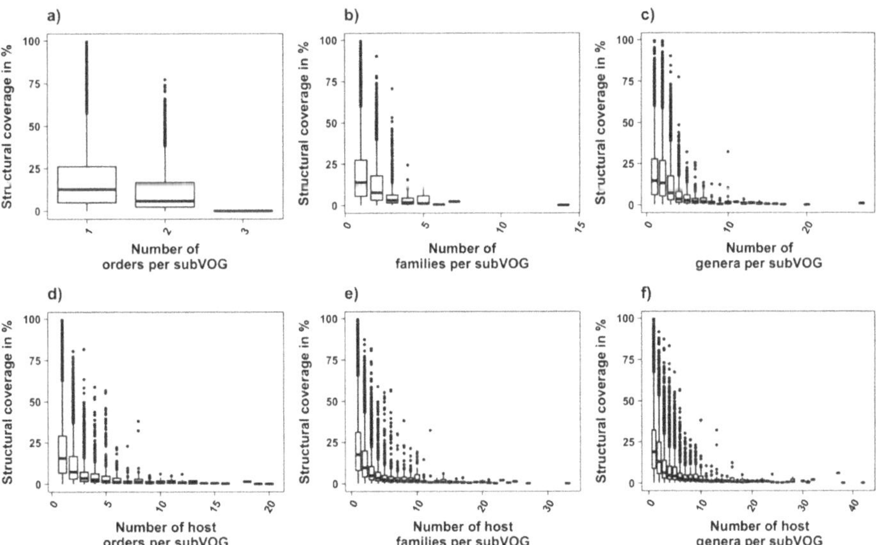

Figure 8. Structural coverage as a function of the taxonomic variety of subVOGs and their host organisms.

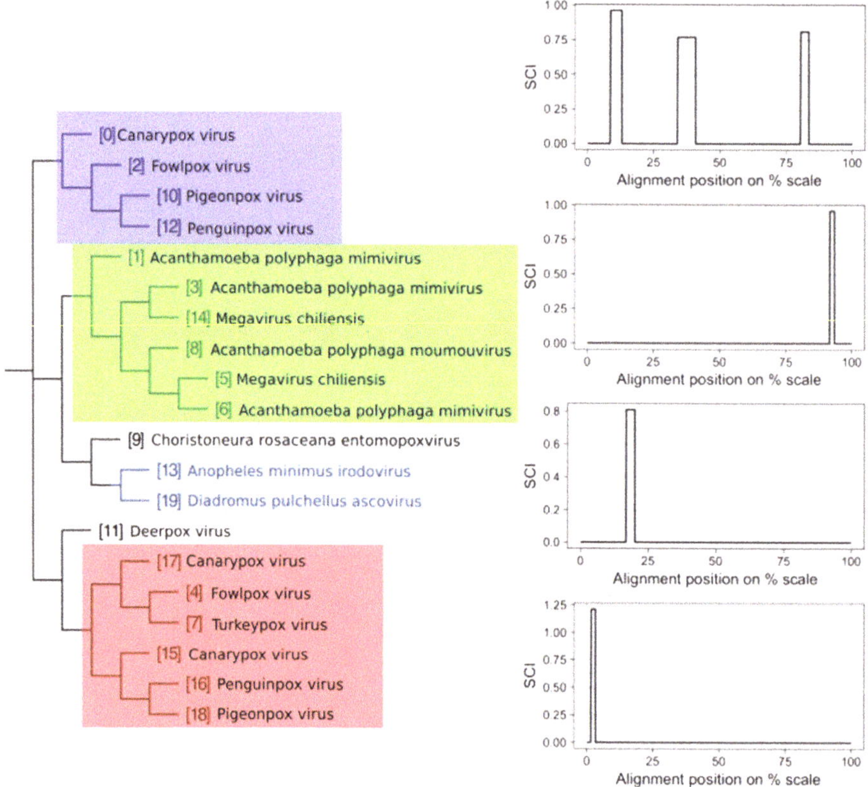

Figure 9. Example of a VOG split into structurally homogenous subVOGs. Shown is the VOG 00052 containing 20 mRNAs, encoding for Kila-N domain proteins, from 12 virus species. On the left, the neighbor-joining tree based on the pairwise sequence identity between the protein sequences is shown. Colored boxes indicate subVOGs, within which conserved structures were predicted. The tree nodes outside colored boxes did not yield any conserved structures. On the right, the structure conservation index (SCI) (black line for each subVOG alignment) is plotted against the alignment position on the percentage scale. Plots are ordered according to the subVOG position in the tree.

As an example, Figure 10 shows the subVOG 1 of VOG11160, which contains two mRNAs encoding the matrix protein 1 from the Influenza A virus (H3N2) and the Influenza B virus. There are three RNA structural motifs described in the literature for the Influenza A mRNA. Nucleotides 105 to 192 form either a multibranch structure, according to Moss et al. [23] and Jiang et al. [52], or a double hairpin structure, proposed by Jiang et al. [52]. Two consecutive stem-loop structures are formed from position 682 to 744, according to Moss et al. [23]. Despite the sequences' dissimilarity between Influenza A and B, both motifs are partly conserved, according to our RNAz analysis of the corresponding subVOG (Figure 10). Our analysis supports the second hairpin loop from the double hairpin structure, described by Jiang et al. (Figure 10a–c). From the second motif, proposed by Moss et al., we also found that the second hairpin structure was partly conserved (Figure 10d–e). The consensus structure of the first motif has a high structure conservation index (SCI) of 0.78, although the part of the alignment covering the structure has a low pairwise identity of 29%. The second motif has an SCI of 0.58 and a pairwise identity of 32%. Our analysis also revealed three further conserved stem-loop structures—position 346 to 369, 438 to 483, and 654 to 674, with SCIs and mPIDs of 0.81 and 29%, 0.66 and 48%, and 0.65 and 33%, respectively.

A recent study of secondary structures in alphaviruses by Kutchko et al. revealed that Sindbis virus mRNAs harbor many functional structures, but they are poorly conserved in the closely related Venezuelan equine encephalitis virus [53]. The corresponding subVOG containing mRNAs coding for the non-structural protein 1 includes orthologous mRNAs from 12 further alphaviruses. We identified three short structures that are conserved in all of the contained species and overlap with the functional structures described by Kutchko et al., while all other structures reported by Kutchko et al. are indeed poorly conserved in further Alphavirus species.

An example of a subVOG in which structures are conserved across mRNAs from different taxonomic families is given in Figure 11. Shown is a subVOG containing proteins from two mosaic viruses (Maracuja mosaic virus, Tobacco mosaic virus), the Bell pepper mottle virus, and the Odontoglossum ringspot virus (Figure 11a,b). The proteins are classified as replicases and RNA polymerases. The subVOG contains overall 15 locally conserved structured regions. Figure 11 shows the region covering alignment positions 4766 to 4815. The alignment covering this structure has an mPID of 72% and the structures are conserved with an SCI of 0.9.

Overall, we subdivided 21,200 VOGs containing, on average, 11 proteins (233,380 in total) into a total of 42,293 subVOGs, containing, on average, five mRNAs (201,708 in total) and three structured regions (147,087 in total). The VOGs with more than two sequences that had to be split up contain, on average, four subVOGs.

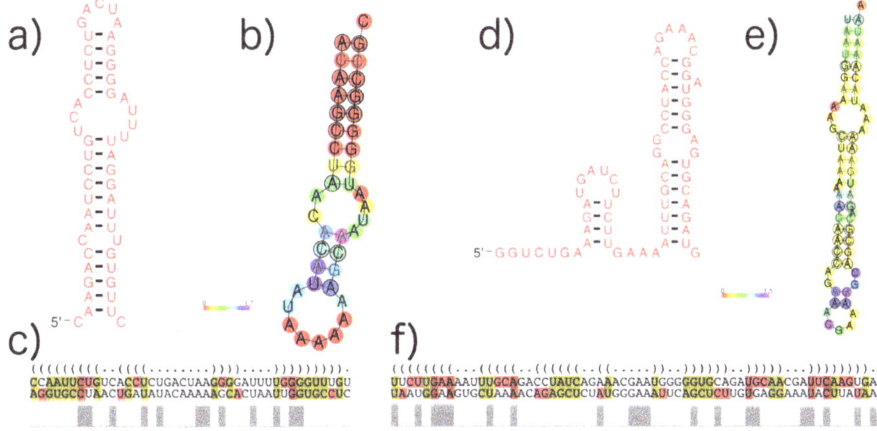

Figure 10. Structures found in Influenza A and B mRNAs encoding the matrix protein (VOG11160). Colors in MSA pictures encode compensatory mutations supporting the consensus structure. Red marks pairs with no sequence variation; ochre, green, turquoise, blue, and violet mark pairs with 2, 3, 4, 5, and 6 different types of pairs, respectively. (**a**) The second of the two consecutive stem loops of the structure proposed by Jiang et al. [52], covering positions 147–192, visualized with R2R [54]; (**b**) The predicted conserved consensus structure for nucleotides 148–188 supports the second hairpin loop of the model of Jiang et al., shown in (**a**). Colors encode the positional entropy; (**c**) Structure-guided alignment and dot bracket structure notation for the consensus structure shown in (**a**). The upper sequence corresponds to Influenza A and the lower sequence to Influenza B; (**d**) Shown are two consecutive hairpin loops for nucleotide positions 682 to 744, proposed by Moss et al. [23], visualized with R2R; (**e**) The predicted conserved structure for nucleotides 697–758 partly supports the model shown in (**e**). Colors encode the positional entropy; (**f**) Structure-guided alignment and dot bracket notation for the consensus structure shown in (**e**). The upper sequence corresponds to Influenza A and the lower sequence to Influenza B.

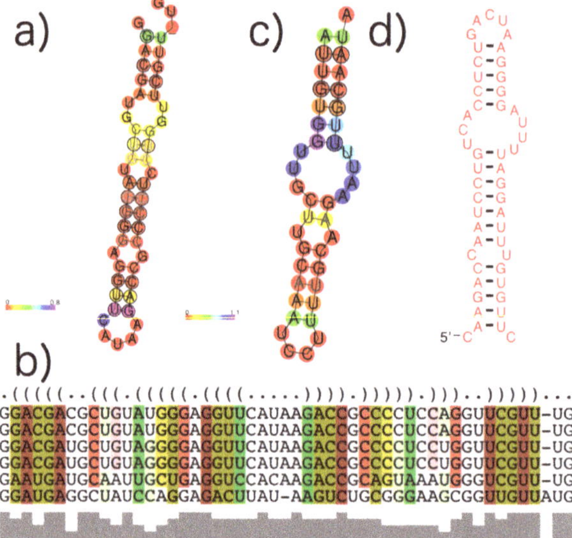

Figure 11. Example structures that were identified within subVOGs. (**a**) Structural annotation of the subVOG 30, belonging to VOG00029, which contains six mRNAs encoding a replicase protein of different Tobamovirus species. Consensus structure visualized by RNAalifold. Colors encode the positional entropy; (**b**) Structure-guided MSA and consensus structure in dot bracket notation corresponding to consensus structure shown in (**a**). Colors encode compensatory mutations supporting the consensus structure. Red marks pairs with no sequence variation; ochre, green, turquoise, blue, and violet mark pairs with 2, 3, 4, 5, and 6 different types of pairs, respectively; (**c**) Consensus structure of subVOG 64 from VOG00003, which contains four mRNAs coding for a p28-like protein of different alphabaculoviruses; (**d**) Structure found in a Heliothis virescens ascovirus 3e, by covariance model search of the structure shown in (**c**), using cmsearch in the entire sequence space of all VOGs.

3.4. subVOG Covariance Models

We built covariance models for all structures found within subVOGs and, using cmsearch, found that in many cases, structures are conserved between different subVOGs and even between different VOGs. In most cases, this was due to a shared sequence domain. For example, the subVOG 64 from VOG00003 harbors four mRNA sequences from different nucleopolyhedroviruses, belonging to the family Baculoviridae. This subVOG was predicted to contain four conserved structures. One of these structures is a highly conserved stem-loop structure (Figure 11c). This structure can also be found in an mRNA of Heliothis virescens ascovirus 3e, belonging to the family Ascoviridae, which is part of VOG01276 (Figure 11d). The two structures are highly conserved with an SCI close to 1, although they are part of different VOGs and belong to mRNAs of different virus families. The alignment of the corresponding proteins revealed that these sequences share a common domain, but the sequence similarity is below the inclusion threshold of the VOG pipeline (Figure S3).

3.5. mRNA Stability and Length

It was shown for a number of eukaryotic and prokaryotic organisms that longer mRNAs exhibit more stable RNA structures, which allows for more efficient control of co-translational protein folding [45,46]. In our dataset of viral mRNA sequences, we also found a correlation between the free energy of the most stable 30-nucleotide segment of an mRNA (ΔGmin) and mRNA length (Pearson correlation coefficient −0.27; from here on referred to as Pearson's *r*), but no correlation between the average energy of all possible 30-nucleotide windows (ΔGmean) and mRNA length (Table 1,

Figure 12a). We additionally calculated the free energy of the most and least stable local optimal segment found by RNALalifold as well as the mean energy of all found RNALalifold segments, and obtained Pearson's r values of −0.25, −0.07, and 0.29 respectively. The Pearson's r of folding energy and GC content lies between −0.5 for ΔGmax and −0.94 for ΔGmean (Table 1, Figure 12b). The number of bases that are within functional structures is positively correlated with the alignment length of subVOGS (Pearson's r 0.40, *p*-value $< 2.2^{-16}$), while this correlation becomes negative when considering the percentage of bases within structures (structural coverage) instead of the absolute value (Pearson's r −0.27, *p*-value $< 2.2^{-16}$) (Figure 13). In other words, longer mRNAs harbor more or longer structured regions, but at the same time, the percentage of positions in functional structures decreases with increasing length. The only explanation for this effect that we can think of is that there is a certain number of structured elements needed for regulatory functions, which is largely independent of the mRNA length. As expected (see Figure 8), there is a weak but significant negative correlation (Pearson's r −0.23, *p*-value $< 2.2^{-16}$) between structural coverage and the number of sequences in the MSA, with more taxonomically diverse alignments containing fewer conserved structures.

Table 1. Pearson correlation between alignment length or GC-content and the minimum (ΔGmin), maximum (ΔGmax), or mean (ΔGmean) folding energy of either all possible 30-nucleotide long-sequence windows or all local optimal structures found with RNALfold, of all mRNAs in our data set. *P*-values are given in parentheses.

Type of ΔG	Pearson Correlation Coefficient	
	ΔG vs. Sequence Length	ΔG vs. GC-Content
ΔGmin	−0.27 ($<2.2^{-16}$)	−0.73 ($<2.2^{-16}$)
ΔGmean	0.004 (0.1655)	−0.94 ($<2.2^{-16}$)
ΔGmax	0.17 ($<2.2^{-16}$)	−0.50 ($<2.2^{-16}$)
ΔGmin (RNALfold)	−0.24 ($<2.2^{-16}$)	−0.86 ($<2.2^{-16}$)
ΔGmean (RNALfold)	−0.16 ($<2.2^{-16}$)	−0.86 ($<2.2^{-16}$)
ΔGmax (RNALfold)	0.29 ($<2.2^{-16}$)	−0.07 ($<2.2^{-16}$)

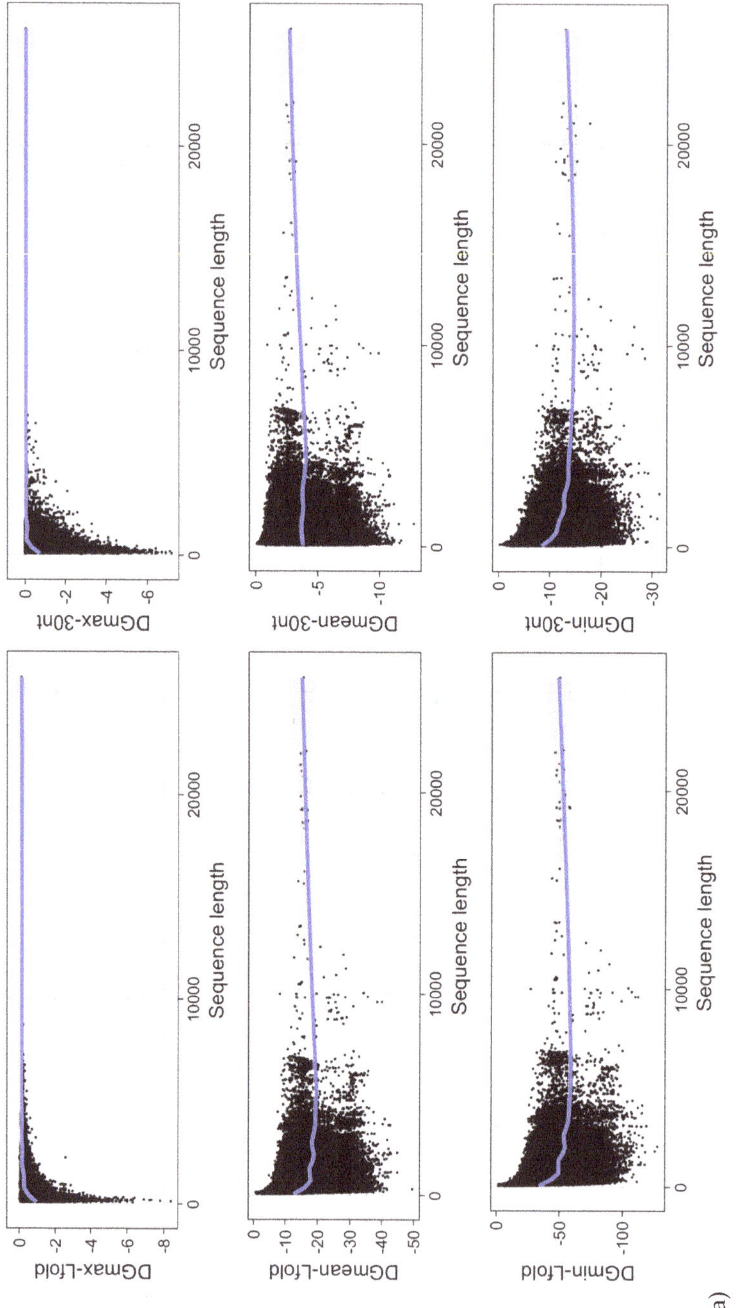

Figure 12. *Cont.*

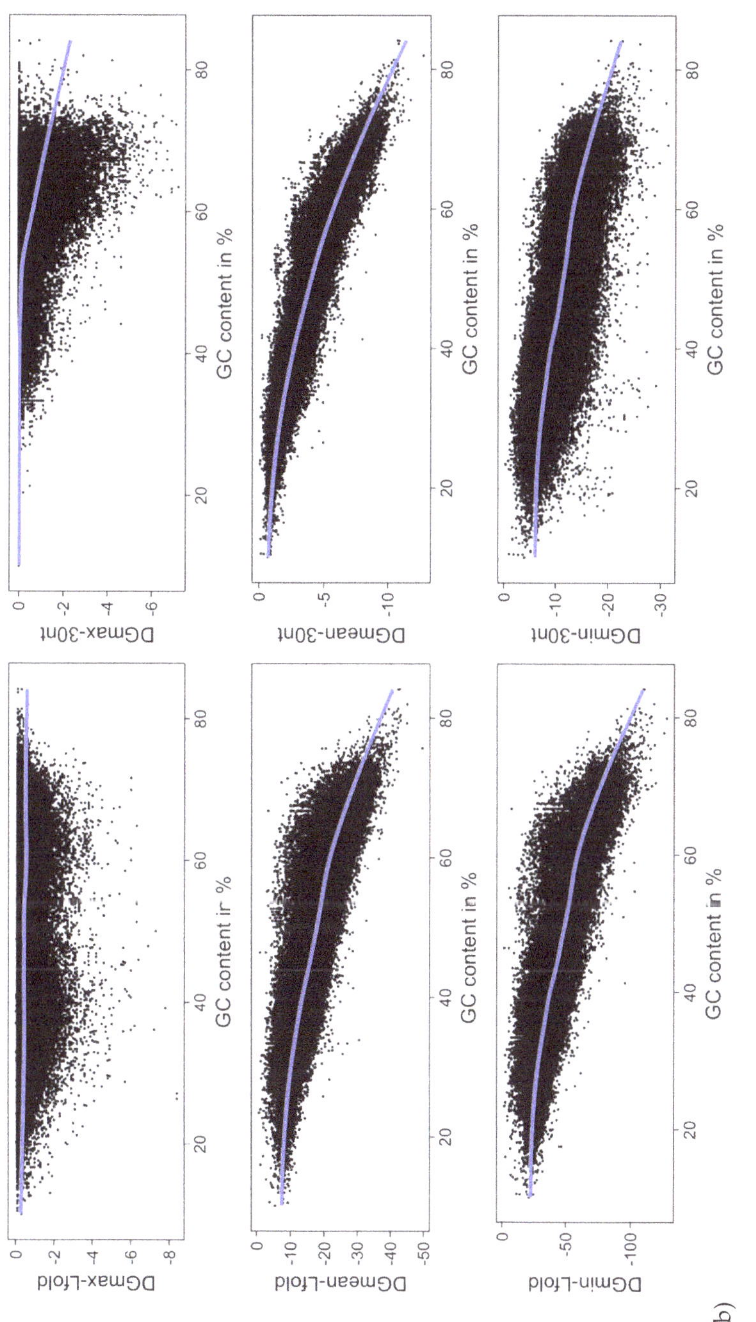

b)

Figure 12. MRNA folding energy as a function of (**a**) sequence length and (**b**) GC-content. DGmin: Minimum folding energy of either all possible 30-nucleotide windows of a sequence or all found local optimal structures using RNALfold. DGmean and DGmax: Mean and maximum of all windows, respectively.

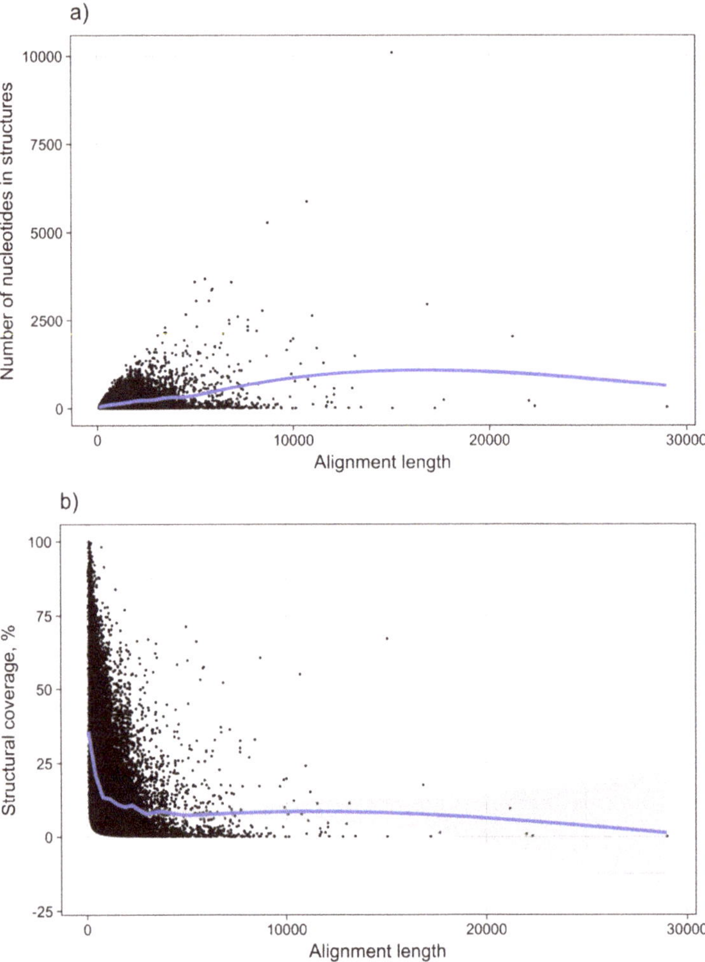

Figure 13. MRNA structure as a function of length. The graph shows the dependence of (**a**) the number of nucleotides within structures predicted to be functional, and (**b**) the structural coverage of the mRNAs in %, from the total length of mRNAs. Each point corresponds to one subVOG.

3.6. mRNA Structures and Protein Function

We analyzed the relationship between protein function and mRNA structure in viral subVOGs by comparing RNA structural coverage with gene ontology (GO) annotation. Using the QuickGO database, we identified a total of 814 VOG proteins that are manually or experimentally annotated (according to ECO evidence codes, as described in Materials and Methods) with GO terms, of which 727 are part of a subVOG, and thus harbor conserved structures according to our analysis. (For the sake of completeness, we also performed the same analysis for all GO annotated proteins, without regard for the annotation evidence codes, see Table S2). For each individual GO term, we only considered the structural coverage of mRNA sequences if that term was assigned to more than 50% of the proteins in a given subVOG. This resulted in 106 GO terms from the biological process sub-ontology and 17 terms from the molecular function sub-ontology. Note that no GO terms from the cellular component sub-ontology satisfied the criteria explained above.

Using Revigo, we derived 70 functionally similar groups of GO terms, with 57 belonging to the biological process ontology and 13 to the function sub-ontology (Table S1). The resulting GO term groups were subdivided into three categories, according to the average structural coverage of the corresponding subVOGs: Low structural coverage (up to 10%), medium structural coverage (up to 20%), and high structural coverage (more than 20%). We found that the standard deviation of the structural coverage values within the Revigo clusters was significantly smaller (Wilcoxon test p-value 1.068^{-10}), compared to randomized clusters (Figure 14). In other words, our findings suggest that mRNAs encoding the proteins with coherent functions tend to exhibit a similar structural coverage.

These findings are in line with the previous study by Vandivier et al. who found that transcripts in *Arabidopsis thaliana* with similar levels of secondary structure in their untranslated and coding regions tend to encode functionally similar proteins [48]. Likewise, Wang et al. also identified GO terms associated with highly or lowly folded mRNAs in yeast [55]. Four of the GO terms associated with highly structured mRNAs, according to Wang et al. (regulation of translation, posttranscriptional regulation of gene expression, regulation of cellular protein metabolic process, and cellular nitrogen compound biosynthetic process), correspond to highly structured viral mRNAs in our data. At the same time, none of the GO terms corresponding to lowly structured yeast mRNAs according to Wang et al. were enriched in our results. On the other hand, Fan Li et al. found that *Arabidopsis thaliana* mRNAs related to "regulation of transcription" were structurally unstable [56], while we found that mRNAs encoding the proteins related to "viral transcription" do harbor conserved RNA structures. We also found virus-specific trends not previously observed for cellular proteins, such as the high structure of viral mRNAs coding for proteins that regulate replication and transcription, suppression by viruses of host translation, or modulation by viruses of host process (Table S1). It has been reported that mRNA folding strength influences the efficiency of gene expression and that mRNAs encoding abundant proteins generally tend to be more structured [57]. In the future, once RNA-seq data for a sufficient number of viral genes becomes available, it will be interesting to investigate whether functional coherence between mRNAs with similar structural coverage is, at least in part, caused by similar expression levels.

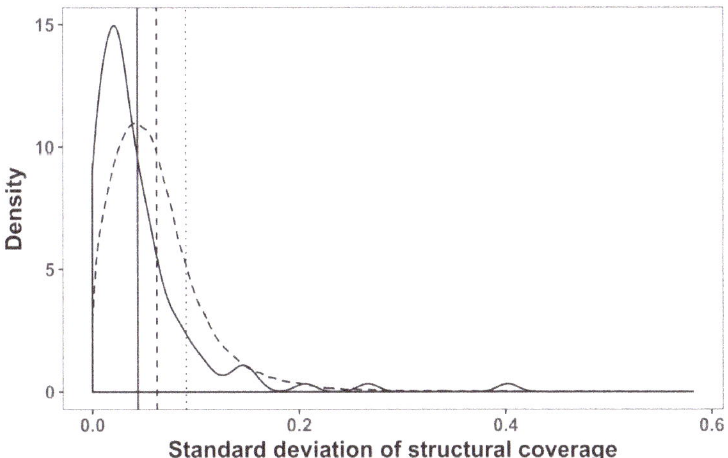

Figure 14. Distribution of standard deviations of mRNA structural coverage, mapped to GO-terms: Clustered with Revigo (solid line); randomized Revigo clusters (dashed line); not clustered (dotted line); vertical lines represent the mean of the corresponding dataset.

3.7. subVOG Online Resource

Structurally homogenous subVOGs can be accessed online (http://rnasiv.bio.wzw.tum.de) through two entry points: "Browse by VOG" and "Browse by taxonomy". The first option is a list of all VOGs, together with the consensus description of their constituent proteins. The list can be filtered with a keyword search and links to the corresponding subVOGs of each VOG are provided. The second option is an expandable taxonomic tree, based on the NCBI taxonomy [58], which allows navigation to the viral species of interest. For each species, mRNA sequences are provided, if available, interlinked to the corresponding subVOGs. Tree nodes containing only mRNAs that are not part of any subVOG are colored grey. Each subVOG contains at least two sequences that share at least one structural element predicted to be functional. If a species of interest is not contained in the subVOG database, the taxonomy tree makes it possible to find the taxonomically closest species. Web pages describing individual subVOGs contain four parts:

(i)   General information, i.e., number of mRNAs in the subVOG, the number of proteins and species in the parent VOG, as well as a consensus functional description;

(ii) Information on conserved structures among the subVOG sequences. A plot outlining the SCI for each column of the subVOG MSA gives a brief overview over the structure of the subVOG members. Also provided is a table that shows a list of all structures found, including the corresponding values of SCI, mPID, and the GC content. The consensus structure can also be visualized by Forna, and a covariance model is provided, which can be used to search for similar structures. Additionally, the RNAz results for each individual structured region can be accessed, including structure visualization, dot plots, and the local structure-guided alignments;

(iii) The global MSA for the subVOG sequences. Alignment columns colored in blue correspond to the structured regions described in the previous section. The alignment is visualized with the javascript library MSAviewer [59], which is based on Jalview [60];

(iv) The list of subVOG members, including protein names, descriptions, and taxonomy. For each protein, a link to the RefSEQ entry is provided, as well as the amino acid and nucleotide sequences. The leftmost column of the list contains a checkbox for each subVOG member, which can be used to build a subset of members and analyze the RNA structures shared by these.

4. Discussion

In this work we set out to create a possibly complete census of conserved RNA secondary structures in the coding regions of viruses and to shed light on their biological role. Using sequence comparison and structure prediction methods, we derived structurally homogenous groups of viral mRNAs from subsets of viral orthologous groups (VOGs), which we call subVOGs. We identified a total of 147,087 conserved structures in 42,293 subVOGs, which we make accessible through our database RNASIV (RNA Structures in Viruses). On average, subVOGs contain three structured regions and their structural homogeneity decreases with increasing taxonomic diversity of the viruses and their hosts. We found that 63% of all subVOGs contain mRNAs from at least two genera and 21% from more than one taxonomic family. In line with the previous studies on cellular organisms, we confirm that, in viruses, longer mRNAs tend to contain more stable structures. However, the number of structures grows only slowly with length, which implies that there is a certain minimum amount of structures required to maintain regulatory functions and control protein folding. MRNAs annotated with similar GO terms tend to have a similar structural coverage, hinting at possible commonalities in the regulatory mechanisms of functionally related proteins. It is hoped that RNASIV will be a useful resource for exploring the structure–function relationships in viral mRNAs.

Supplementary Materials: The following are available online at http://www.mdpi.com/1999-4915/11/5/401/s1: Figure S1: Virus lineages included in the VOGs; Figure S2: Mean pairwise sequence identity of VOG alignments as a function of VOG size; Figure S3: Sequence Alignment of a protein from Heliothis virescens ascovirus 3e and proteins belonging to the mRNAs of subVOG 64 of VOG00003; Table S1: Clustering of GO terms of subVOG proteins and the average structural coverage of their corresponding mRNAs; Table S2: Clustering of GO terms of subVOG proteins and the average structural coverage of their corresponding mRNAs (regardless of GO evidence codes).

Author Contributions: Conceptualization, D.F., I.H., and T.R.; methodology, D.F., R.O., and M.K.; software, M.K.; validation, D.F. and M.K.; formal analysis, D.F. and M.K.; investigation, D.F., M.K., and T.R.; resources, I.H. and T.R.; data curation, M.K. and H.-J.H.; writing—original draft preparation, D.F. and M.K.; writing—review and editing, D.F. and M.K.; visualization, D.F. and M.K.; supervision, D.F.; project administration, D.F., I.H., and T.R.; funding acquisition, D.F., I.H., and T.R.

Funding: This research was funded by the Deutsche Forschungsgemeinschaft, grant number FR1411/10-1 and the FWF-Grant I-1303.

Acknowledgments: We are grateful to Jan Zaucha for his helpful comments.

Conflicts of Interest: The authors declare no conflict of interest.

References

1. Bevilacqua, P.C.; Blose, J.M. Structures, kinetics, thermodynamics, and biological functions of RNA hairpins. *Annu. Rev. Phys. Chem.* **2008**, *59*, 79–103. [CrossRef] [PubMed]

2. Serganov, A.; Patel, D.J. Ribozymes, riboswitches and beyond: Regulation of gene expression without proteins. *Nat. Rev. Genet.* **2007**, *8*, 776–790. [CrossRef] [PubMed]

3. Gray, N.K.; Hentze, M.W. Regulation of protein synthesis by mRNA structure. *Mol. Biol. Rep.* **1994**, *19*, 195–200. [CrossRef] [PubMed]

4. Kozak, M. Regulation of translation via mRNA structure in prokaryotes and eukaryotes. *Gene* **2005**, *361*, 13–37. [CrossRef]

5. Katz, L.; Burge, C.B. Widespread selection for local RNA secondary structure in coding regions of bacterial genes. *Genome Res.* **2003**, *13*, 2042–2051. [CrossRef]

6. Mortimer, S.A.; Kidwell, M.A.; Doudna, J.A. Insights into RNA structure and function from genome-wide studies. *Nat. Rev. Genet.* **2014**, *15*, 469–479. [CrossRef]

7. Kudla, G.; Murray, A.W.; Tollervey, D.; Plotkin, J.B. Coding-sequence determinants of gene expression in Escherichia coli. *Science* **2009**, *324*, 255–258. [CrossRef]

8. Duan, J.; Wainwright, M.S.; Comeron, J.M.; Saitou, N.; Sanders, A.R.; Gelernter, J.; Gejman, P.V. Synonymous mutations in the human dopamine receptor D2 (DRD2) affect mRNA stability and synthesis of the receptor. *Hum. Mol. Genet.* **2003**, *12*, 205–216. [CrossRef]

9. Ilyinskii, P.O.; Schmidt, T.; Lukashev, D.; Meriin, A.B.; Thoidis, G.; Frishman, D.; Shneider, A.M. Importance of mRNA secondary structural elements for the expression of influenza virus genes. *OMICS* **2009**, *13*, 421–430. [CrossRef]

10. Carlini, D.B.; Chen, Y.; Stephan, W. The relationship between third-codon position nucleotide content, codon bias, mRNA secondary structure and gene expression in the drosophilid alcohol dehydrogenase genes Adh and Adhr. *Genetics* **2001**, *159*, 623–633. [PubMed]

11. Nackley, A.G.; Shabalina, S.A.; Tchivileva, I.E.; Satterfield, K.; Korchynskyi, O.; Makarov, S.S.; Maixner, W.; Diatchenko, L. Human catechol-O-methyltransferase haplotypes modulate protein expression by altering mRNA secondary structure. *Science* **2006**, *314*, 1930–1933. [CrossRef]

12. Gu, W.; Zhou, T.; Wilke, C.O. A universal trend of reduced mRNA stability near the translation-initiation site in prokaryotes and eukaryotes. *PLoS Comput. Biol.* **2010**, *6*, e1000664. [CrossRef]

13. Del Campo, C.; Bartholomäus, A.; Fedyunin, I.; Ignatova, Z. Secondary Structure across the Bacterial Transcriptome Reveals Versatile Roles in mRNA Regulation and Function. *PLoS Genet.* **2015**, *11*, e1005613. [CrossRef]

14. Shabalina, S.A.; Ogurtsov, A.Y.; Spiridonov, N.A. A periodic pattern of mRNA secondary structure created by the genetic code. *Nucleic Acids Res.* **2006**, *34*, 2428–2437. [CrossRef]

15. Simmonds, P.; Tuplin, A.; Evans, D.J. Detection of genome-scale ordered RNA structure (GORS) in genomes of positive-stranded RNA viruses: Implications for virus evolution and host persistence. *RNA* **2004**, *10*, 1337–1351. [CrossRef]

16. Meyer, I.M.; Miklós, I. Statistical evidence for conserved, local secondary structure in the coding regions of eukaryotic mRNAs and pre-mRNAs. *Nucleic Acids Res.* **2005**, *33*, 6338–6348. [CrossRef]

17. Olivier, C.; Poirier, G.; Gendron, P.; Boisgontier, A.; Major, F.; Chartrand, P. Identification of a conserved RNA motif essential for She2p recognition and mRNA localization to the yeast bud. *Mol. Cell. Biol.* **2005**, *25*, 4752–4766. [CrossRef]

18. Fricke, M.; Gerst, R.; Ibrahim, B.; Niepmann, M.; Marz, M. Global importance of RNA secondary structures in protein-coding sequences. *Bioinformatics* **2019**, *35*, 579–583. [CrossRef]

19. Thurner, C.; Witwer, C.; Hofacker, I.L.; Stadler, P.F. Conserved RNA secondary structures in Flaviviridae genomes. *J. Gen. Virol.* **2004**, *85*, 1113–1124. [CrossRef]

20. Fricke, M.; Dünnes, N.; Zayas, M.; Bartenschlager, R.; Niepmann, M.; Marz, M. Conserved RNA secondary structures and long-range interactions in hepatitis C viruses. *RNA* **2015**, *21*, 1219–1232. [CrossRef]

21. Pirakitikulr, N.; Kohlway, A.; Lindenbach, B.D.; Pyle, A.M. The Coding Region of the HCV Genome Contains a Network of Regulatory RNA Structures. *Mol. Cell* **2016**, *62*, 111–120. [CrossRef]

22. Simmonds, P.; Smith, D.B. Structural constraints on RNA virus evolution. *J. Virol.* **1999**, *73*, 5787–5794.

23. Moss, W.N.; Priore, S.F.; Turner, D.H. Identification of potential conserved RNA secondary structure throughout influenza A coding regions. *RNA* **2011**, *17*, 991–1011. [CrossRef]

24. Clyde, K.; Harris, E. RNA secondary structure in the coding region of dengue virus type 2 directs translation start codon selection and is required for viral replication. *J. Virol.* **2006**, *80*, 2170–2182. [CrossRef]

25. Goz, E.; Tuller, T. Widespread signatures of local mRNA folding structure selection in four Dengue virus serotypes. *BMC Genomics.* **2015**, *16* Suppl. 10, S4. [CrossRef]

26. Goz, E.; Tuller, T. Evidence of a Direct Evolutionary Selection for Strong Folding and Mutational Robustness Within HIV Coding Regions. *J. Comput. Biol.* **2016**, *23*, 641–650. [CrossRef]

27. Díez, J.; Jungfleisch, J. Translation control: Learning from viruses, again. *RNA Biol.* **2017**, *14*, 835–837. [CrossRef]

28. Pruitt, K.D.; Tatusova, T.; Brown, G.R.; Maglott, D.R. NCBI Reference Sequences (RefSeq): Current status, new features and genome annotation policy. *Nucleic Acids Res.* **2012**, *40*, D130–D135. [CrossRef]

29. Lorenz, R.; Bernhart, S.H.; Höner Zu Siederdissen, C.; Tafer, H.; Flamm, C.; Stadler, P.F.; Hofacker, I.L. ViennaRNA Package 2.0. *Algorithms Mol. Biol.* **2011**, *6*, 26. [CrossRef]

30. Gruber, A.R.; Findeiß, S.; Washietl, S.; Hofacker, I.L.; Stadler, P.F. RNAz 2.0: Improved noncoding RNA detection. *Pac. Symp Biocomput* **2010**, 69–79. [CrossRef]

31. O'Leary, N.A.; Wright, M.W.; Brister, J.R.; Ciufo, S.; Haddad, D.; McVeigh, R.; Rajput, B.; Robbertse, B.; Smith-White, B.; Ako-Adjei, D.; et al. Reference sequence (RefSeq) database at NCBI: Current status, taxonomic expansion, and functional annotation. *Nucleic Acids Res.* **2016**, *44*, D733–D745. [CrossRef]

32. Hulo, C.; de Castro, E.; Masson, P.; Bougueleret, L.; Bairoch, A.; Xenarios, I.; Le Mercier, P. ViralZone: A knowledge resource to understand virus diversity. *Nucleic Acids Res.* **2011**, *39*, D576–D582. [CrossRef]

33. Sievers, F.; Higgins, D.G. Clustal omega. *Curr. Protoc. Bioinformatics* **2014**, *48*, 3.13.1–3.13.16.

34. Eddy, S.R. A new generation of homology search tools based on probabilistic inference. *Genome Inform.* **2009**, *23*, 205–211.

35. Remmert, M.; Biegert, A.; Hauser, A.; Söding, J. HHblits: Lightning-fast iterative protein sequence searching by HMM-HMM alignment. *Nat. Methods* **2011**, *9*, 173–175. [CrossRef]

36. Enright, A.J.; Van Dongen, S.; Ouzounis, C.A. An efficient algorithm for large-scale detection of protein families. *Nucleic Acids Res.* **2002**, *30*, 1575–1584. [CrossRef]

37. UniProt Consortium UniProt: A hub for protein information. *Nucleic Acids Res.* **2015**, *43*, D204–D212. [CrossRef]

38. Altschul, S.F.; Gish, W.; Miller, W.; Myers, E.W.; Lipman, D.J. Basic local alignment search tool. *J. Mol. Biol.* **1990**, *215*, 403–410. [CrossRef]

39. Mihara, T.; Nishimura, Y.; Shimizu, Y.; Nishiyama, H.; Yoshikawa, G.; Uehara, H.; Hingamp, P.; Goto, S.; Ogata, H. Linking Virus Genomes with Host Taxonomy. *Viruses* **2016**, *8*, 66. [CrossRef]

40. Kazlauskas, D.; Krupovic, M.; Venclovas, Č. The logic of DNA replication in double-stranded DNA viruses: Insights from global analysis of viral genomes. *Nucleic Acids Res.* **2016**, *44*, 4551–4564. [CrossRef]

41. Saitou, N.; Nei, M. The neighbor-joining method: A new method for reconstructing phylogenetic trees. *Mol. Biol. Evol.* **1987**, *4*, 406–425.

42. Stajich, J.E.; Block, D.; Boulez, K.; Brenner, S.E.; Chervitz, S.A.; Dagdigian, C.; Fuellen, G.; Gilbert, J.G.R.; Korf, I.; Lapp, H.; et al. The Bioperl toolkit: Perl modules for the life sciences. *Genome Res.* **2002**, *12*, 1611–1618. [CrossRef]

43. Will, S.; Joshi, T.; Hofacker, I.L.; Stadler, P.F.; Backofen, R. LocARNA-P: Accurate boundary prediction and improved detection of structural RNAs. *RNA* **2012**, *18*, 900–914. [CrossRef]
44. Cui, X.; Lu, Z.; Wang, S.; Jing-Yan Wang, J.; Gao, X. CMsearch: Simultaneous exploration of protein sequence space and structure space improves not only protein homology detection but also protein structure prediction. *Bioinformatics* **2016**, *32*, i332–i340. [CrossRef]
45. Tuller, T.; Veksler-Lublinsky, I.; Gazit, N.; Kupiec, M.; Ruppin, E.; Ziv-Ukelson, M. Composite effects of gene determinants on the translation speed and density of ribosomes. *Genome Biol.* **2011**, *12*, R110. [CrossRef]
46. Faure, G.; Ogurtsov, A.Y.; Shabalina, S.A.; Koonin, E.V. Role of mRNA structure in the control of protein folding. *Nucleic Acids Res.* **2016**, *44*, 10898–10911. [CrossRef]
47. Ashburner, M.; Ball, C.A.; Blake, J.A.; Botstein, D.; Butler, H.; Cherry, J.M.; Davis, A.P.; Dolinski, K.; Dwight, S.S.; Eppig, J.T.; et al. Gene ontology: Tool for the unification of biology. The Gene Ontology Consortium. *Nat. Genet.* **2000**, *25*, 25–29. [CrossRef]
48. Vandivier, L.; Li, F.; Zheng, Q.; Willmann, M.; Chen, Y.; Gregory, B. Arabidopsis mRNA secondary structure correlates with protein function and domains. *Plant. Signal. Behav.* **2013**, *8*, e24301. [CrossRef]
49. Binns, D.; Dimmer, E.; Huntley, R.; Barrell, D.; O'Donovan, C.; Apweiler, R. QuickGO: A web-based tool for Gene Ontology searching. *Bioinformatics* **2009**, *25*, 3045–3046. [CrossRef]
50. Giglio, M.; Tauber, R.; Nadendla, S.; Munro, J.; Olley, D.; Ball, S.; Mitraka, E.; Schriml, L.M.; Gaudet, P.; Hobbs, E.T.; et al. ECO, the Evidence & Conclusion Ontology: Community standard for evidence information. *Nucleic Acids Res.* **2019**, *47*, D1186–D1194.
51. Supek, F.; Bošnjak, M.; Škunca, N.; Šmuc, T. REVIGO summarizes and visualizes long lists of gene ontology terms. *PLoS ONE* **2011**, *6*, e21800. [CrossRef] [PubMed]
52. Jiang, T.; Nogales, A.; Baker, S.F.; Martinez-Sobrido, L.; Turner, D.H. Mutations Designed by Ensemble Defect to Misfold Conserved RNA Structures of Influenza A Segments 7 and 8 Affect Splicing and Attenuate Viral Replication in Cell Culture. *PLoS ONE* **2016**, *11*, e0156906. [CrossRef]
53. Kutchko, K.M.; Madden, E.A.; Morrison, C.; Plante, K.S.; Sanders, W.; Vincent, H.A.; Cruz Cisneros, M.C.; Long, K.M.; Moorman, N.J.; Heise, M.T.; et al. Structural divergence creates new functional features in alphavirus genomes. *Nucleic Acids Res.* **2018**, *46*, 3657–3670. [CrossRef]
54. Weinberg, Z.; Breaker, R.R. R2R–software to speed the depiction of aesthetic consensus RNA secondary structures. *BMC Bioinformatics* **2011**, *12*, 3. [CrossRef] [PubMed]
55. Wang, X.; Li, P.; Gutenkunst, R.N. Systematic Effects Of mRNA Secondary Structure On Gene Expression And Molecular Function In Budding Yeast. *BioRxiv* **2017**. [CrossRef]
56. Li, F.; Zheng, Q.; Vandivier, L.E.; Willmann, M.R.; Chen, Y.; Gregory, B.D. Regulatory impact of RNA secondary structure across the Arabidopsis transcriptome. *Plant. Cell* **2012**, *24*, 4346–4359. [CrossRef]
57. Zur, H.; Tuller, T. Strong association between mRNA folding strength and protein abundance in *S. cerevisiae*. *EMBO Rep.* **2012**, *13*, 272–277. [CrossRef]
58. Federhen, S. The NCBI Taxonomy database. *Nucleic Acids Res.* **2012**, *40*, D136–D143. [CrossRef]
59. Yachdav, G.; Wilzbach, S.; Rauscher, B.; Sheridan, R.; Sillitoe, I.; Procter, J.; Lewis, S.E.; Rost, B.; Goldberg, T. MSAViewer: Interactive JavaScript visualization of multiple sequence alignments. *Bioinformatics* **2016**, *32*, 3501–3503. [CrossRef]
60. Waterhouse, A.M.; Procter, J.B.; Martin, D.M.A.; Clamp, M.; Barton, G.J. Jalview Version 2–a multiple sequence alignment editor and analysis workbench. *Bioinformatics* **2009**, *25*, 1189–1191. [CrossRef]

Review

Giant Viruses—Big Surprises

Nadav Brandes [1] and Michal Linial [2],*

[1] The Rachel and Selim Benin School of Computer Science and Engineering, The Hebrew University of Jerusalem, Jerusalem 91904, Israel; nadav.brandes@mail.huji.ac.il

[2] Department of Biological Chemistry, The Alexander Silberman Institute of Life Sciences, The Hebrew University of Jerusalem, Jerusalem 91904, Israel

* Correspondence: michall@cc.huji.ac.il; Tel.: +972-02-6585425

Received: 22 March 2019; Accepted: 23 April 2019; Published: 30 April 2019

Abstract: Viruses are the most prevalent infectious agents, populating almost every ecosystem on earth. Most viruses carry only a handful of genes supporting their replication and the production of capsids. It came as a great surprise in 2003 when the first giant virus was discovered and found to have a >1 Mbp genome encoding almost a thousand proteins. Following this first discovery, dozens of giant virus strains across several viral families have been reported. Here, we provide an updated quantitative and qualitative view on giant viruses and elaborate on their shared and variable features. We review the complexity of giant viral proteomes, which include functions traditionally associated only with cellular organisms. These unprecedented functions include components of the translation machinery, DNA maintenance, and metabolic enzymes. We discuss the possible underlying evolutionary processes and mechanisms that might have shaped the diversity of giant viruses and their genomes, highlighting their remarkable capacity to hijack genes and genomic sequences from their hosts and environments. This leads us to examine prominent theories regarding the origin of giant viruses. Finally, we present the emerging ecological view of giant viruses, found across widespread habitats and ecological systems, with respect to the environment and human health.

Keywords: Amebae viruses; viral evolution; protein domains; mimivirus; dsdna viruses; translation machinery; pandoravirus; NCLDV

---

1. Giant Viruses and the Viral World

Viruses are cell infecting agents present in almost every ecosystem. Questions the regarding viral origin and early evolution alongside all living organisms (bacteria, archaea and eukarya) are still wide open, and relevant theories remain speculative [1–5]. As viruses are exceptionally diverse and undergo rapid changes, it is impossible to construct an ancestral lineage tree for the viral world [6–10]. Instead, virus families are categorized according to the nature of their genetic material, mode of replication, pathogenicity, and structural properties [11].

At present, the viral world is represented by over 8,000 reference genomes [12]. The International Committee on Taxonomy of Viruses (ICTV) provides a universal virus taxonomical classification proposal that covers ~150 families and ~850 genera, with many viruses yet unclassified [13]. This collection provides a comprehensive, compact set of virus representatives.

Inspection of viral genomes reveals that most known viruses have genomes encoding only a few proteins. Actually, 69% of all known viruses have less than 10 proteins encoded in their genomes (Figure 1). It is a common assumption that viruses demonstrate near-optimal genome packing and information compression, presumably in order to maximize their replication rate, number of progenies, and other parameters that increase infectivity [14,15]. However, a debate is still ongoing over the generality of these phenomena [16], and there is a non-negligible percentage of larger viruses (Figure 1). On the far end of the distribution, there are viruses with hundreds of genes, most of them are considered giant viruses. While only 0.3% of the known viral proteomes contain 500 or more proteins, they encode as much as 7.5% of the total number of viral proteins (Figure 1B).

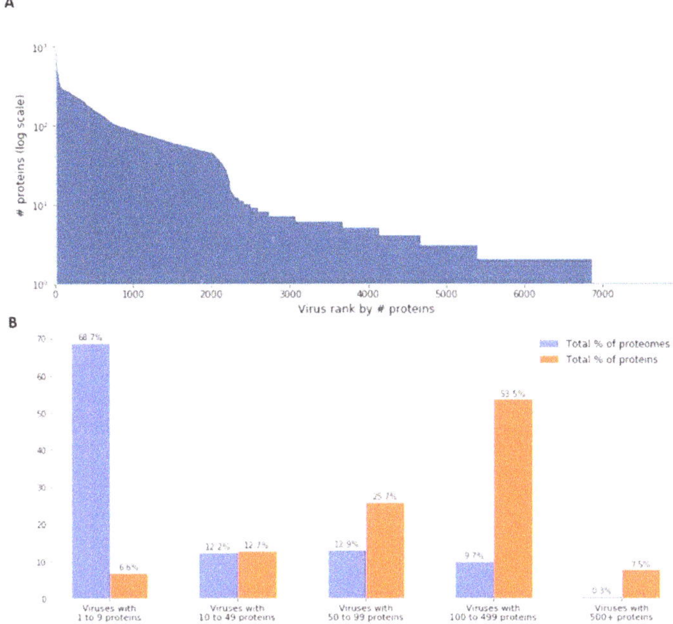

Figure 1. Number of proteins encoded by viruses. (**A**) The number of encoded proteins (y-axis) in all 7,959 viral representatives, ranked in descending order. (**B**) Partitioning of the 7959 viral proteomes by the number of encoded proteins. The 0.3% viral proteomes with the highest number of proteins (over 500) encode 7.5% of the total number of viral proteins.

2. The Discovery of Giant Viruses

The first giant virus, *Acanthamoeba polyphaga mimivirus* (APMV), was discovered in 2003 [17]. Its size was unprecedented, being on the scale of small bacteria or archaea cells [18]. Unlike any previously identified virus, APMV could be seen with a light microscope [19,20]. Initially it was mistaken for a bacterium and recognized as a virus only ten years after its isolation [21]. Up to this day, most of its proteins remain uncharacterized [22,23]. Notably, even more than a decade after the discovery of APMV, the identification of giant viruses still sometimes involves confusion, as illustrated in the discovery of the *Pandoravirus inopinatum* [24], which was initially described as an endoparasitic organism, and *Pithovirus sibericum* [25], which was also misinterpreted as an archaeal endosymbiont (see discussion in References [21,26]).

In the following years after the initial discovery of APMV, many additional giant viral species have been identified and their genomes fully sequenced. Most giant viral genomes have been obtained from large-scale metagenomic sequencing projects covering aquatic ecosystems (e.g., oceans, pools, lakes and cooling wastewater units) [27,28]; others have been sequenced from samples extracted from underexplored geographical and ecological niches (e.g., the Amazon River, deep seas and forest soils) [29–32]. Despite the accumulation of many more giant virus representatives, the fraction of uncharacterized proteins in their proteomes remains exceptionally high [33]. Many of these uncharacterized proteins were also considered orphan genes (ORFans), i.e., no significant match to any other sequence was identified. For example, 93% of the *Pandoravirus salinus* proteins, the first representative of this family [34] were reported as ORFans.

At present, there are over a hundred giant virus isolates, which reveal fascinating and unexpected characteristics. These extreme instances on the viral landscape challenge the current theories on genome size and compactness in viruses, and provide a new perspective on the very concept of a virus and viral origin [4,20,28,35–37].

3. Definition of Giant Viruses

Attempts to distinguish giant viruses from other large viruses remain somewhat fuzzy [38,39]. Any definition for giant viruses would necessarily involve some arbitrary threshold, as virus size, whether physical, genomic or proteomic, is clearly a continuum (Figure 2). Giant viruses were initially defined by their physical size as allowing visibility by a light microscope [33]. In this report, we prefer a proteomic definition, even if somewhat arbitrary. We consider giant viruses as eukaryote-infecting viruses with at least 500 protein-coding genes (Figure 2). Of the 7,959 curated viral genomes (extracted from NCBI Taxonomy complete genomes), 24 represented genomes meet this threshold. Of these, we consider the 19 eukaryote-infecting viruses to be the giant virus representatives (Table 1), excluding the five bacteria-infecting viruses.

Recall that reported proteome sizes are primarily based on automatic bioinformatics tools, which may differ from the experimental expression measurements (e.g., mimivirus APMV [40]). Moreover, physical dimensions are not in perfect correlation with the number of proteins or genome size. For example, *Pithovirus sibericum*, which was recovered from a 30,000-year-old permafrost sample [25], is one of the largest viruses by its physical dimensions (1.5 μm in length and 0.5 μm in diameter). However, it is excluded from this report, as its genome encodes only 467 proteins.

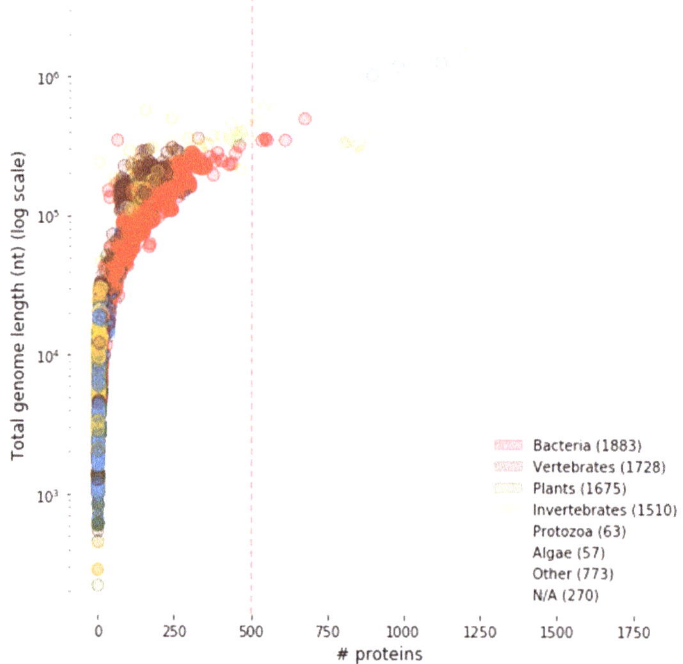

Figure 2. Distribution of viral proteome and genome sizes, colored by host taxonomy. There are 24 represented genomes that meet the threshold of ≥500 proteins (dashed red line), comprising five bacteria-infecting and 19 eukaryote-infecting viruses.

Table 1. Giant viruses.

Genome [a]	Accession	Genome Length (kb)	# of Proteins	Host [b]	Year [c]
Mi-Acanthamoeba polyphaga mimivirus	NC_014649	1181.5	979	Pz, Ver	2010
Mi-Acanthamoeba polyphaga moumouvirus	NC_020104	1021.3	894	Pz, Ver	2013
Ph-Acanthocystis turfacea chlorella virus 1	NC_008724	288.0	860	Algae	2006
Mi-Cafeteria roenbergensis virus BV-PW1	NC_014637	617.5	544	Pz	2010
Pi-Cedratvirus A11	NC_032108	589.1	574	Pz	2016
Ph-Chrysochromulina ericina virus	NC_028094	473.6	512	Algae	2015
Mi-Megavirus chiliensis	NC_016072	1259.2	1120	Pz, Ver	2011
UC-Mollivirus sibericum	NC_027867	651.5	523	Pz	2015
Ph-Orpheovirus IHUMI-LCC2	NC_036594	1473.6	1199	Algae	2017
Pa-Pandoravirus dulcis	NC_021858	1908.5	1070	Pz	2013
Pa-Pandoravirus inopinatum	NC_026440	2243.1	1839	Pz	2015
Pa-Pandoravirus macleodensis	NC_037665	1838.3	926	Pz	2018
Pa-Pandoravirus neocaledonia	NC_037666	2003.2	1081	Pz	2018
Pa-Pandoravirus quercus	NC_037667	2077.3	1185	Pz	2018
Pa-Pandoravirus salinus	NC_022098	2473.9	1430	Pz	2013
Ph-Paramecium bursaria Chlorella virus 1	NC_000852	330.6	802	Algae	1995
Ph-Paramecium bursaria Chlorella virus AR158	NC_009899	344.7	814	Algae	2007
Ph-Paramecium bursaria Chlorella virus FR483	NC_008603	321.2	849	Algae	2006
Ph-Paramecium bursaria Chlorella virus NY2A	NC_009898	368.7	886	Algae	2007

[a] Families: Mi, *Mimiviridae*; Ph, *Phycodnaviridae*; Pi, *Pithoviridae*; Pa, *Pandoraviridae*; UC, uncharacterized; [b] Pz, protozoa; Ver, vertebrates; [c] Year of genome submission to NCBI.

4. Classification of Giant Viruses and the Question of Origin

All giant viruses belong to the superfamily of nucleocytoplasmic large DNA viruses (NCLDV), which was substantially expanded following the discoveries of giant viruses [41,42]. The NCLDV superfamily had traditionally been comprised of the following families: *Phycodnaviridae, Iridoviridae, Poxviridae, Asfarviridae* and *Ascoviridae* [43,44], for which a common ancestor had been proposed [45,46]. Following the inclusion of additional giant virus taxonomy groups (*Mimiviradae, Pandoravirus* and *Marseillevirus*) into the NCLDV superfamily, there remained only a handful of genes shared by the entire superfamily. Additional disparities in virion shapes and replication modes among NCLDV has led to the conclusion that the superfamily is not necessarily a taxonomic group, and that NCLDV families are more likely to have evolved separately [47–49].

Two models have been proposed for the evolution of giant viruses. According to the reductive model, an ancestral cellular genome became reduced in size, leading to the dependence of the resulting genome on host cells. The presence of genes carrying cellular functions in almost any giant virus (e.g., translation components) [50] is consistent with this model. An alternative and more accepted theory argues for an expansion model. According to this model, current giant viruses originated from smaller ancestral viruses carrying only a few dozens of genes, and through gene duplications and horizontal gene transfer (HGT), have rapidly expanded and diversified [48,51–53]. This model agrees with metagenomic studies and the wave of giant virus discoveries in recent years, suggesting massive gene exchange between giant viruses and a variety of organisms sharing the same ecosystems (e.g., Reference [32]).

Of special interest is the degree of similarity between giant viruses and their hosts. The amebae host in particular is often described as a melting pot for DNA exchange [54] that leads to chimeric genomes. The majority of genes in giant viruses and specifically *Mimiviridae*, have originated from the cells they parasitize (mostly amoeba and bacteria). Based on phylogenetic trees, it is likely that extensive HGT events have led to their chimeric genomes. It was also suggested that the spectrum of viral hosts may be larger than anticipated, including yet unknown species [55]. Therefore, comparative genomics of giant viruses infecting the same host is unlikely to unambiguously resolve questions of gene origin, namely, whether shared genes have originated from a common viral ancestor. Thus, the degree of similarity among giant viruses infecting different hosts is of special interest. For example, the phyletic relationship between *Mimiviridae* (which infect *Acanthamoeba*) and *Phycodnaviridae* (which infect algae) was investigated, and it was found that the algae-infecting *Chrysochromulina ericina* virus (CeV, Table 1) showed moderate resemblance to the amebae-infecting mimivirus [56]. As a result, suggestions were made to reclassify CeV as a new clade of *Mimiviridae*, rather than *Phycodnaviridae*. However, a later discovery of another algae-infecting *Phycodnaviridae* virus (*Heterosigma akashiwo* virus, HaV53) has provided a coherent phyletic relationship among *Phycodnaviridae*, thereby questioning this reclassification [52].

In summary, the taxonomy of giant viruses, like all viruses, is still unstable, and rapidly updated with new discoveries [31,57]. The origin and ancestry of giant viruses have remained controversial with questions of origin also unresolved [39]. Many newly discovered giant viruses are not compatible with the notion of a single common ancestor, as some giant viruses remain taxonomically isolated [4].

5. Common Features

Despite the ongoing debate on their origin, giant viruses still share some important features. All giant viruses belong to the dsDNA group, as do all NCLDV families. The total genome size of all the giant viruses listed in Table 1 is at least 288 Kbp (Figure 2). These giant viruses are classified into several families: *Mimiviridae, Pithoviridae, Pandoraviridae, Phycodnaviridae* and the *Mollivirus* genus [21,25].

All amoebae-infecting giant viruses rely on the non-specific phagocytosis by the amoebae host [55]. Interestingly, a necessary condition for phagocytosis is a minimal particle size (~0.6 µm [58]), as amebae (and related protozoa) naturally feed on bacteria. It is likely that this minimal size for inducing phagocytosis has become an evolutionary driving force for giant viruses. This fact, together with the largely uncharacterized genomic content of giant viruses, may suggest that much of the content in the genomes of giant viruses serves only for volume filling to increase their physical size.

Giant viruses share not only the cell entry process. When they exit the host cells during lysis, as many as 1000 virions are released from each lysed host via membrane fusion and active exocytosis [59], which are relatively rare exit mechanisms in viruses.

Other than these genomic and cell-biology similarities, other features of giant viruses are mostly family-specific. For example, virion shapes and symmetries, nuclear involvement, duration of the infection cycle, and the stages of virion assembly—all substantially vary among giant viruses from different families [21,60,61].

6. Proteome Complexity and Functional Diversity

The majority of the giant virus proteomes remain with no known function (Figure 3). Actually, the fraction of uncharacterized proteins reaches 65–85% of all reported proteins in giant viral proteomes, many of them are ORFans. However, when proteomes of closely related species are considered, the fraction of ORFans obviously drops (by definition). For example, 93% of the proteins were reported as ORFans for the first representative of the *Pandoraviridae* family (*P. salinus*) [34]. But later, following the completion of five additional *Pandoravirus* proteomes (of the species *inopinatum, macleodensis, neocaledonia, dulcis,* and *quercus*), the number of *P. salinus* ORFans dropped to 29% (i.e., 71% of its genes now had a significant similarity to at least one other *Pandoravirus* protein sequence). Still, the vast majority of *Pandoravirus* proteins remain uncharacterized.

The most striking finding regarding the proteomes of giant viruses is the presence of protein functions that are among the hallmarks of cellular organisms, and are never detected in other viruses. To exemplify the complexity of proteome functions in giant viruses, we examined the proteome of the *Cafeteria roenbergensis* virus (CroV), which infects the marine plankton community in the Gulf of Mexico. *CroV* was sequenced in 2010 as the first algae-infecting virus in the *Mimiviridae* family. Unexpectedly, despite its affiliation with a recognized viral family, the majority of its proteins showed no significant similarity to any other known protein sequence. Of the remaining proteins that show significant basic local alignment search tool (BLAST) hits to other proteins from all domains of life, 45% are eukaryotic sequences, 22% are from bacteria, and the rest are mostly from other viruses, including other mimivirus strains. A similar partition of protein origin applies across other members of the *Mimiviridae* family [33].

The CroV proteome includes a rich set of genes involved in protein translation [62]. These genes include multiple translation factors, a dozen of ribosomal proteins, tRNA synthetases, and 22 sequences encoding five different tRNAs [62]. As a lack of translation potential is considered a hallmark of the virosphere, the presence of translation machinery components raised a debate on the very definition of viruses [63,64]. Similar findings were replicated in two other giant virus strains of the *Tupanvirus* genus in the same *Mimiviridae* family, which were recently isolated in Brazil [65]. The two strains have 20 open reading frames (ORFs) related to tRNA aminoacylation (aaRS), ~70 tRNA sequences decoding the majority of the codons, eight translation initiation factors, and elongation and release factors. The theory that translation optimization is an evolutionary driving force in viruses [66] may in part explain the curious presence of translation machinery in giant viruses.

In addition to translation, numerous CroV proteins are associated with the transcription machinery. Specifically, the CroV proteome contains several subunits of the DNA-dependent RNA polymerase II, initiation, elongation, and termination factors, the mRNA capping enzyme, and a poly(A) polymerase. Presumably, the virus can activate its own transcription in the viral factory foci in the cytoplasm of its host cell [47].

Another unexpected function detected in CroV is DNA repair, specifically of UV radiation damage and base-excision repair. Other DNA-maintenance functions found in CroV include helicase and topoisomerases (type I and II), suggesting regulation of DNA replication, recombination and chromatin remodeling.

Another rich set of functions related to protein maintenance include chaperons [67] and the ubiquitin-proteasome system [68]. Interestingly, some of these genes seem to be acquired from bacteria (e.g., a homolog of the Escherichia coli heat-shock chaperon). In addition, a rich collection of sugar-, lipid- and amino acid-related metabolic enzymes were also found [18,69], which occupy 13% of the CroV proteome (Figure 3).

It appears that the CroV proteome covers most functions traditionally attributed to cellular organisms, including: Protein translation, RNA maturation, DNA maintenance, proteostasis and metabolism. Although CroV exemplifies many widespread functions in giant viruses, each strain has its own unique functional composition. For example, the most abundant group of giant viruses in ocean metagenomes, the *Bodo saltans* virus (BsV), was recently identified and classified into the same microzooplankton-infecting *Mimiviridae* family [70]. Unlike the other family members, BsV does not have an elaborate translation apparatus or tRNA genes, but it carries proteins active in cell membrane trafficking and phagocytosis, yet more unprecedented functions discovered in viruses.

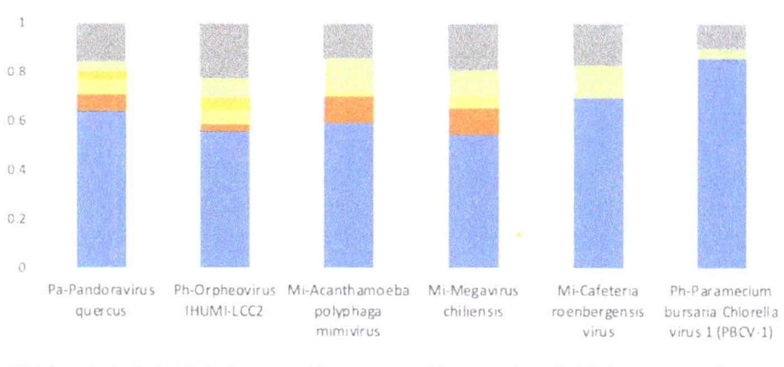

Figure 3. Protein function categories in six giant virus representatives from three families: *Mimiviridae* (Mi), *Pandoviridae* (Pa) and *Phycodnaviridae* (Ph). In all proteomes, the majority of proteins are uncharacterized. Short repeated domains such as ankyrin, F-box and MORM repeats are abundant in the proteomes of amebae-infecting giant viruses [71].

7. Virophages and Defense Mechanism in Giant Viruses

Additional important players of genome dynamics in giant viruses are the virophages [72]. These are small double-stranded DNA (dsDNA) viruses that hitchhike the replication system of giant viruses following coinfection of the host, and are considered parasites of the coinfecting giant viruses [73]. Virophages (e.g., Sputnik 1–3, Zamilion) are associated with *Mimiviridae* representatives and their specific viral strain infectivity [21,74]. Additionally, short mobile genetic agents, called transpovirions (combining features of a transposon and a virion) [73], together with other mobile elements display complex ecological interactions with their hosts. Indeed, similar to eukaryotic transposons and endogenous viruses, sequences of a virophage (*Mavirus*) of *Cafeteria roenbergensis virus* (CroV), that were integrated to the genome of the protozoan host, serves an antiviral defense mechanism, which is activated by giant virus infection [75].

An even more unexpected finding is the discovery of a nucleic acid-based immunity in mimiviruses, resembling the adaptive (clustered regularly interspaced short palindromic repeats) CRISPR-Cas system in bacteria and archaea. Despite the differences to the canonical CRISPR-Cas system, an operon-like cluster of sequences derived from the *Zamilion* virophage was identified in mimivirus and experimentally validated to govern virophage coinfection. This cluster, coined MIMIVIRE, acts as a mimivirus virophage resistance element system with an exonuclease, helicase and RNase III identified in its vicinity [76]. The homology between the MIMIVIRE-associated exonuclease to the bacterial Cas-4 exonuclease was revealed by 3D protein structure analysis [77]. This CRISPR-Cas related function in mimiviruses are assumed to degrade foreign DNA, thereby constituting an antiviral innate immune system. It is likely that the CRISPR-Cas immune system in mimiviruses contributes to its sequence diversification as well by removing unessential host sequences. Alternative mechanisms that govern viral-host infection specificity and immunity may be discovered as our knowledge on virophages and other mobile elements is expending [72,78].

Altogether, a rich network of mobile genetic elements contributes to the host-virus coevolution and interviral gene transfer [78]. Virophages and other mobile elements could facilitate gene transfer, thereby having the potential to shape the genomes of giant viruses and impact their diversity [21,79,80].

8. The Emerging Ecological View

Viruses are the most abundant entities in nature. In marine and fresh water habitats, there are millions of viruses in each milliliter of water [81]. However, the collection of virus isolates is often sporadic, especially for those without clinical or agricultural relevance. The accelerated pace in the discovery of giant viruses reflects the increasing number of sequencing projects of exotic environments, including metagenomic projects [32,82].

Giant viruses have been isolated from numerous environmental niches and distant geographic locations, revealing their global distribution and diversity. Current evidence suggests that the representation of giant viruses is underexplored, especially in soil ecosystems [31] and unique ecological niches [83,84]. In fact, ~60% of the giant viral genomes were completed after 2013 (Table 1). Many more virus–host systems, most of them reported in the last five years, still await isolation, characterization and classification [84,85].

The hosts of contemporary isolates include mainly protozoa, specifically amoeba (Table 1). However, the prevalence of amoeba as hosts may in part be attributed to sampling bias, specifically to the widespread use of amoebal coculture methods for testing ecological environments [28,85].

Despite their prevalence, the impact of giant viruses on human health deserves further investigation [22]. Some initial reports show that APMV giant virus is able to replicate in human peripheral blood cells and to induce the interferon system [86]. Sequences of numerous giant viruses were identified as part of large-scale human gut microbiome sequencing projects [87], but their abundance, compositions and ecological roles are yet to be determined [88]. Reports are accumulating on the presence of giant viral sequences in human blood, as well as antibodies against giant viral proteins. Some reports associate mimirus and marseillevirus with a broad collection of human diseases (e.g., rheumatoid arthritis, adenitis, unexplained pneumonia, lymphoma), yet causal relationship is mostly missing [89]. The presence of giant viruses in almost any environment, including extreme niches and manmade sites (e.g., sewage and wastewater plants), suggests that the ecological role of these fascinating entities and their impact on human health are yet to be fully explored.

Author Contributions: The authors contributed to writing and manuscript design.

Funding: This research received no external funding.

Acknowledgments: The authors wish to thank the support to M.L. from Yad Hanadiv. We are grateful to Elixir infrastructure tools and resources.

Conflicts of Interest: The authors declare no conflict of interest.

References

1. Holmes, E.C. What does virus evolution tell us about virus origins? *J. Virol.* **2011**, *85*, 5247–5251. [CrossRef] [PubMed]
2. Forterre, P. The origin of viruses and their possible roles in major evolutionary transitions. *Virus Res.* **2006**, *117*, 5–16. [CrossRef]
3. Yutin, N.; Koonin, E.V. Evolution of DNA ligases of nucleo-cytoplasmic large DNA viruses of eukaryotes: A case of hidden complexity. *Biol. Direct.* **2009**, *4*, 51. [CrossRef] [PubMed]
4. Forterre, P. Viruses in the 21st Century: From the Curiosity-Driven Discovery of Giant Viruses to New Concepts and Definition of Life. *Clin. Infect. Dis.* **2017**, *65*, S74–S79. [CrossRef] [PubMed]
5. Moelling, K.; Broecker, F. Viruses and Evolution - Viruses First? A Personal Perspective. *Front. Microbiol.* **2019**, *10*, 523. [CrossRef] [PubMed]
6. Moreira, D.; Lopez-Garcia, P. Ten reasons to exclude viruses from the tree of life. *Nat. Rev. Microbiol.* **2009**, *7*, 306–311. [CrossRef] [PubMed]
7. Brüssow, H. The not so universal tree of life or the place of viruses in the living world. *Philos. Trans. R. Soc. B Biol. Sci.* **2009**, *364*, 2263–2274.
8. Bamford, D.H.; Grimes, J.M.; Stuart, D.I. What does structure tell us about virus evolution? *Curr. Opin. Struct. Biol.* **2005**, *15*, 655–663. [CrossRef] [PubMed]
9. Domingo, E.; Escarmis, C.; Sevilla, N.; Moya, A.; Elena, S.; Quer, J.; Novella, I.; Holland, J. Basic concepts in RNA virus evolution. *FASEB J.* **1996**, *10*, 859–864. [CrossRef]
10. Koonin, E.V.; Yutin, N. Origin and evolution of eukaryotic large nucleo-cytoplasmic DNA viruses. *Intervirology* **2010**, *53*, 284–292. [CrossRef]
11. Murphy, F.A.; Fauquet, C.M.; Bishop, D.H.; Ghabrial, S.A.; Jarvis, A.W.; Martelli, G.P.; Mayo, M.A.; Summers, M.D. *Virus taxonomy: Classification and nomenclature of viruses*; Springer Science & Business Media: Berlin, Germany, 2012; Volume 10.
12. Stano, M.; Beke, G.; Klucar, L. viruSITE-integrated database for viral genomics. *Database (Oxford)* **2016**, *2016*. [CrossRef]
13. Lefkowitz, E.J.; Dempsey, D.M.; Hendrickson, R.C.; Orton, R.J.; Siddell, S.G.; Smith, D.B. Virus taxonomy: The database of the International Committee on Taxonomy of Viruses (ICTV). *Nucleic Acids. Res.* **2018**, *46*, D708–D717. [CrossRef] [PubMed]
14. Pybus, O.G.; Rambaut, A. Evolutionary analysis of the dynamics of viral infectious disease. *Nat. Rev. Genet.* **2009**, *10*, 540–550. [CrossRef]
15. Pérez-Losada, M.; Arenas, M.; Galán, J.C.; Palero, F.; González-Candelas, F. Recombination in viruses: Mechanisms, methods of study, and evolutionary consequences. *Infect. Genet. Evol.* **2015**, *30*, 296–307. [CrossRef] [PubMed]
16. Brandes, N.; Linial, M. Gene overlapping and size constraints in the viral world. *Biol. Direct.* **2016**, *11*, 26. [CrossRef] [PubMed]
17. La Scola, B.; Audic, S.; Robert, C.; Jungang, L.; de Lamballerie, X.; Drancourt, M.; Birtles, R.; Claverie, J.M.; Raoult, D. A giant virus in amoebae. *Science* **2003**, *299*, 2033. [CrossRef]
18. Raoult, D.; Audic, S.; Robert, C.; Abergel, C.; Renesto, P.; Ogata, H.; La Scola, B.; Suzan, M.; Claverie, J.M. The 1.2-megabase genome sequence of Mimivirus. *Science* **2004**, *306*, 1344–1350. [CrossRef]

19. Xiao, C.; Chipman, P.R.; Battisti, A.J.; Bowman, V.D.; Renesto, P.; Raoult, D.; Rossmann, M.G. Cryo-electron microscopy of the giant Mimivirus. *J. Mol. Biol.* **2005**, *353*, 493–496. [CrossRef]
20. Van Etten, J.L.; Meints, R.H. Giant viruses infecting algae. *Annu. Rev. Microbiol.* **1999**, *53*, 447–494. [CrossRef]
21. Abergel, C.; Legendre, M.; Claverie, J.M. The rapidly expanding universe of giant viruses: Mimivirus, Pandoravirus, Pithovirus and Mollivirus. *FEMS Microbiol. Rev.* **2015**, *39*, 779–796. [CrossRef]
22. Abrahão, J.S.; Dornas, F.P.; Silva, L.C.; Almeida, G.M.; Boratto, P.V.; Colson, P.; La Scola, B.; Kroon, E.G. Acanthamoeba polyphaga mimivirus and other giant viruses: An open field to outstanding discoveries. *Virol. J.* **2014**, *11*, 120. [CrossRef] [PubMed]
23. Campos, R.K.; Boratto, P.V.; Assis, F.L.; Aguiar, E.R.; Silva, L.C.; Albarnaz, J.D.; Dornas, F.P.; Trindade, G.S.; Ferreira, P.P.; Marques, J.T. Samba virus: A novel mimivirus from a giant rain forest, the Brazilian Amazon. *Virol. J.* **2014**, *11*, 95. [CrossRef] [PubMed]
24. Antwerpen, M.H.; Georgi, E.; Zoeller, L.; Woelfel, R.; Stoecker, K.; Scheid, P. Whole-genome sequencing of a pandoravirus isolated from keratitis-inducing acanthamoeba. *Genome Announc.* **2015**, *3*. [CrossRef] [PubMed]
25. Legendre, M.; Bartoli, J.; Shmakova, L.; Jeudy, S.; Labadie, K.; Adrait, A.; Lescot, M.; Poirot, O.; Bertaux, L.; Bruley, C.; et al. Thirty-thousand-year-old distant relative of giant icosahedral DNA viruses with a pandoravirus morphology. *Proc. Natl. Acad. Sci. USA* **2014**, *111*, 4274–4279. [CrossRef] [PubMed]
26. Claverie, J.M.; Abergel, C. From extraordinary endocytobionts to pandoraviruses. Comment on Scheid et al.: Some secrets are revealed: Parasitic keratitis amoebae as vectors of the scarcely described pandoraviruses to humans. *Parasitol. Res.* **2015**, *114*, 1625–1627. [CrossRef] [PubMed]
27. Andreani, J.; Verneau, J.; Raoult, D.; Levasseur, A.; La Scola, B. Deciphering viral presences: Two novel partial giant viruses detected in marine metagenome and in a mine drainage metagenome. *Virol. J.* **2018**, *15*, 66. [CrossRef]
28. Aherfi, S.; Colson, P.; La Scola, B.; Raoult, D. Giant Viruses of Amoebas: An Update. *Front. Microbiol.* **2016**, *7*, 349. [CrossRef] [PubMed]
29. Andrade, A.; Arantes, T.S.; Rodrigues, R.A.L.; Machado, T.B.; Dornas, F.P.; Landell, M.F.; Furst, C.; Borges, L.G.A.; Dutra, L.A.L.; Almeida, G.; et al. Ubiquitous giants: A plethora of giant viruses found in Brazil and Antarctica. *Virol. J.* **2018**, *15*, 22. [CrossRef]
30. Dornas, F.P.; Assis, F.L.; Aherfi, S.; Arantes, T.; Abrahao, J.S.; Colson, P.; La Scola, B. A Brazilian Marseillevirus Is the Founding Member of a Lineage in Family Marseilleviridae. *Viruses* **2016**, *8*, 76. [CrossRef] [PubMed]
31. Schulz, F.; Alteio, L.; Goudeau, D.; Ryan, E.M.; Yu, F.B.; Malmstrom, R.R.; Blanchard, J.; Woyke, T. Hidden diversity of soil giant viruses. *Nat. Commun.* **2018**, *9*, 4881. [CrossRef]
32. Backstrom, D.; Yutin, N.; Jorgensen, S.L.; Dharamshi, J.; Homa, F.; Zaremba-Niedwiedzka, K.; Spang, A.; Wolf, Y.I.; Koonin, E.V.; Ettema, T.J.G. Virus Genomes from Deep Sea Sediments Expand the Ocean Megavirome and Support Independent Origins of Viral Gigantism. *MBio* **2019**, *10*. [CrossRef] [PubMed]
33. Colson, P.; La Scola, B.; Levasseur, A.; Caetano-Anolles, G.; Raoult, D. Mimivirus: Leading the way in the discovery of giant viruses of amoebae. *Nat. Rev. Microbiol.* **2017**, *15*, 243–254. [CrossRef] [PubMed]
34. Philippe, N.; Legendre, M.; Doutre, G.; Coute, Y.; Poirot, O.; Lescot, M.; Arslan, D.; Seltzer, V.; Bertaux, L.; Bruley, C.; et al. Pandoraviruses: Amoeba viruses with genomes up to 2.5 Mb reaching that of parasitic eukaryotes. *Science* **2013**, *341*, 281–286. [CrossRef] [PubMed]
35. Sharma, V.; Colson, P.; Pontarotti, P.; Raoult, D. Mimivirus inaugurated in the 21st century the beginning of a reclassification of viruses. *Curr. Opin. Microbiol.* **2016**, *31*, 16–24. [CrossRef] [PubMed]
36. Rolland, C.; Andreani, J.; Louazani, A.C.; Aherfi, S.; Francis, R.; Rodrigues, R.; Silva, L.S.; Sahmi, D.; Mougari, S.; Chelkha, N.; et al. Discovery and Further Studies on Giant Viruses at the IHU Mediterranee Infection That Modified the Perception of the Virosphere. *Viruses* **2019**, *11*. [CrossRef] [PubMed]
37. Colson, P.; La Scola, B.; Raoult, D. Giant Viruses of Amoebae: A Journey Through Innovative Research and Paradigm Changes. *Annu. Rev. Virol.* **2017**, *4*, 61–85. [CrossRef]
38. Claverie, J.M.; Abergel, C. Mimiviridae: An Expanding Family of Highly Diverse Large dsDNA Viruses Infecting a Wide Phylogenetic Range of Aquatic Eukaryotes. *Viruses* **2018**, *10*. [CrossRef]
39. Colson, P.; Levasseur, A.; La Scola, B.; Sharma, V.; Nasir, A.; Pontarotti, P.; Caetano-Anolles, G.; Raoult, D. Ancestrality and Mosaicism of Giant Viruses Supporting the Definition of the Fourth TRUC of Microbes. *Front. Microbiol.* **2018**, *9*, 2668. [CrossRef] [PubMed]

40. Legendre, M.; Santini, S.; Rico, A.; Abergel, C.; Claverie, J.M. Breaking the 1000-gene barrier for Mimivirus using ultra-deep genome and transcriptome sequencing. *Virol. J.* **2011**, *8*, 99. [CrossRef] [PubMed]

41. Yutin, N.; Wolf, Y.I.; Raoult, D.; Koonin, E.V. Eukaryotic large nucleo-cytoplasmic DNA viruses: Clusters of orthologous genes and reconstruction of viral genome evolution. *Virol. J.* **2009**, *6*, 223. [CrossRef] [PubMed]

42. Koonin, E.V.; Yutin, N. Evolution of the Large Nucleocytoplasmic DNA Viruses of Eukaryotes and Convergent Origins of Viral Gigantism. *Adv. Virus Res.* **2019**, *103*, 167–202. [CrossRef] [PubMed]

43. Andreani, J.; Aherfi, S.; Bou Khalil, J.Y.; Di Pinto, F.; Bitam, I.; Raoult, D.; Colson, P.; La Scola, B. Cedratvirus, a Double-Cork Structured Giant Virus, is a Distant Relative of Pithoviruses. *Viruses* **2016**, *8*. [CrossRef]

44. Wilhelm, S.W.; Coy, S.R.; Gann, E.R.; Moniruzzaman, M.; Stough, J.M. Standing on the Shoulders of Giant Viruses: Five Lessons Learned about Large Viruses Infecting Small Eukaryotes and the Opportunities They Create. *PLoS Pathog.* **2016**, *12*, e1005752. [CrossRef] [PubMed]

45. Iyer, L.M.; Aravind, L.; Koonin, E.V. Common origin of four diverse families of large eukaryotic DNA viruses. *J. Virol.* **2001**, *75*, 11720–11734. [CrossRef] [PubMed]

46. Nasir, A.; Kim, K.M.; Caetano-Anolles, G. Giant viruses coexisted with the cellular ancestors and represent a distinct supergroup along with superkingdoms Archaea, Bacteria and Eukarya. *BMC Evol. Biol.* **2012**, *12*, 156. [CrossRef] [PubMed]

47. Fischer, M.G.; Allen, M.J.; Wilson, W.H.; Suttle, C.A. Giant virus with a remarkable complement of genes infects marine zooplankton. *Proc. Natl. Acad. Sci. USA* **2010**, *107*, 19508–19513. [CrossRef]

48. Yutin, N.; Wolf, Y.I.; Koonin, E.V. Origin of giant viruses from smaller DNA viruses not from a fourth domain of cellular life. *Virology* **2014**, *466–467*, 38–52. [CrossRef]

49. Koonin, E.V.; Yutin, N. Multiple evolutionary origins of giant viruses. *F1000Research* **2018**, *7*. [CrossRef]

50. Arslan, D.; Legendre, M.; Seltzer, V.; Abergel, C.; Claverie, J.M. Distant Mimivirus relative with a larger genome highlights the fundamental features of Megaviridae. *Proc. Natl. Acad. Sci. USA* **2011**, *108*, 17486–17491. [CrossRef]

51. Iyer, L.M.; Balaji, S.; Koonin, E.V.; Aravind, L. Evolutionary genomics of nucleo-cytoplasmic large DNA viruses. *Virus Res.* **2006**, *117*, 156–184. [CrossRef] [PubMed]

52. Maruyama, F.; Ueki, S. Evolution and Phylogeny of Large DNA Viruses, Mimiviridae and Phycodnaviridae Including Newly Characterized Heterosigma akashiwo Virus. *Front. Microbiol.* **2016**, *7*, 1942. [CrossRef]

53. Filee, J.; Chandler, M. Gene exchange and the origin of giant viruses. *Intervirology* **2010**, *53*, 354–361. [CrossRef]

54. Boyer, M.; Yutin, N.; Pagnier, I.; Barrassi, L.; Fournous, G.; Espinosa, L.; Robert, C.; Azza, S.; Sun, S.; Rossmann, M.G.; et al. Giant Marseillevirus highlights the role of amoebae as a melting pot in emergence of chimeric microorganisms. *Proc. Natl. Acad. Sci. USA* **2009**, *106*, 21848–21853. [CrossRef]

55. Moreira, D.; Brochier-Armanet, C. Giant viruses, giant chimeras: The multiple evolutionary histories of Mimivirus genes. *BMC Evol. Biol.* **2008**, *8*, 12. [CrossRef]

56. Gallot-Lavallee, L.; Blanc, G.; Claverie, J.M. Comparative Genomics of Chrysochromulina Ericina Virus and Other Microalga-Infecting Large DNA Viruses Highlights Their Intricate Evolutionary Relationship with the Established Mimiviridae Family. *J. Virol.* **2017**, *91*. [CrossRef]

57. Yoosuf, N.; Pagnier, I.; Fournous, G.; Robert, C.; La Scola, B.; Raoult, D.; Colson, P. Complete genome sequence of Courdo11 virus, a member of the family Mimiviridae. *Virus Genes* **2014**, *48*, 218–223. [CrossRef]

58. Rodrigues, R.A.L.; Abrahao, J.S.; Drumond, B.P.; Kroon, E.G. Giants among larges: How gigantism impacts giant virus entry into amoebae. *Curr. Opin. Microbiol.* **2016**, *31*, 88–93. [CrossRef]

59. Silva, L.; Andrade, A.; Dornas, F.P.; Rodrigues, R.A.L.; Arantes, T.; Kroon, E.G.; Bonjardim, C.A.; Abrahao, J.S. Cedratvirus getuliensis replication cycle: An in-depth morphological analysis. *Sci. Rep.* **2018**, *8*, 4000. [CrossRef]

60. Suzan-Monti, M.; La Scola, B.; Barrassi, L.; Espinosa, L.; Raoult, D. Ultrastructural characterization of the giant volcano-like virus factory of Acanthamoeba polyphaga Mimivirus. *PLoS ONE* **2007**, *2*, e328. [CrossRef]

61. Diesend, J.; Kruse, J.; Hagedorn, M.; Hammann, C. Amoebae, Giant Viruses, and Virophages Make Up a Complex, Multilayered Threesome. *Front. Cell. Infect. Microbiol.* **2017**, *7*, 527. [CrossRef]

62. Schulz, F.; Yutin, N.; Ivanova, N.N.; Ortega, D.R.; Lee, T.K.; Vierheilig, J.; Daims, H.; Horn, M.; Wagner, M.; Jensen, G.J.; et al. Giant viruses with an expanded complement of translation system components. *Science* **2017**, *356*, 82–85. [CrossRef] [PubMed]

63. Abrahao, J.S.; Araujo, R.; Colson, P.; La Scola, B. The analysis of translation-related gene set boosts debates around origin and evolution of mimiviruses. *PLoS Genet.* **2017**, *13*, e1006532. [CrossRef] [PubMed]

64. Raoult, D.; Forterre, P. Redefining viruses: Lessons from Mimivirus. *Nat. Rev. Microbiol.* **2008**, *6*, 315–319. [CrossRef] [PubMed]

65. Abrahao, J.; Silva, L.; Silva, L.S.; Khalil, J.Y.B.; Rodrigues, R.; Arantes, T.; Assis, F.; Boratto, P.; Andrade, M.; Kroon, E.G.; et al. Tailed giant Tupanvirus possesses the most complete translational apparatus of the known virosphere. *Nat. Commun.* **2018**, *9*, 749. [CrossRef] [PubMed]

66. Bahir, I.; Fromer, M.; Prat, Y.; Linial, M. Viral adaptation to host: A proteome-based analysis of codon usage and amino acid preferences. *Mol. Syst. Biol.* **2009**, *5*, 311. [CrossRef] [PubMed]

67. Barik, S. A Family of Novel Cyclophilins, Conserved in the Mimivirus Genus of the Giant DNA Viruses. *Comput. Struct. Biotechnol. J.* **2018**, *16*, 231–236. [CrossRef]

68. Yutin, N.; Colson, P.; Raoult, D.; Koonin, E.V. Mimiviridae: Clusters of orthologous genes, reconstruction of gene repertoire evolution and proposed expansion of the giant virus family. *Virol. J.* **2013**, *10*, 106. [CrossRef]

69. Van Etten, J.L. Unusual life style of giant chlorella viruses. *Annu. Rev. Genet.* **2003**, *37*, 153–195. [CrossRef] [PubMed]

70. Deeg, C.M.; Chow, C.T.; Suttle, C.A. The kinetoplastid-infecting Bodo saltans virus (BsV), a window into the most abundant giant viruses in the sea. *Elife* **2018**, *7*. [CrossRef]

71. Shukla, A.; Chatterjee, A.; Kondabagil, K. The number of genes encoding repeat domain-containing proteins positively correlates with genome size in amoebal giant viruses. *Virus Evol.* **2018**, *4*, vex039. [CrossRef]

72. Bekliz, M.; Colson, P.; La Scola, B. The Expanding Family of Virophages. *Viruses* **2016**, *8*. [CrossRef]

73. Desnues, C.; La Scola, B.; Yutin, N.; Fournous, G.; Robert, C.; Azza, S.; Jardot, P.; Monteil, S.; Campocasso, A.; Koonin, E.V.; et al. Provirophages and transpovirons as the diverse mobilome of giant viruses. *Proc. Natl. Acad. Sci. USA* **2012**, *109*, 18078–18083. [CrossRef]

74. Fischer, M.G. Sputnik and Mavirus: More than just satellite viruses. *Nat. Rev. Microbiol.* **2012**, *10*, 78. [CrossRef]

75. Fischer, M.G.; Hackl, T. Host genome integration and giant virus-induced reactivation of the virophage mavirus. *Nature* **2016**, *540*, 288–291. [CrossRef] [PubMed]

76. Levasseur, A.; Bekliz, M.; Chabriere, E.; Pontarotti, P.; La Scola, B.; Raoult, D. MIMIVIRE is a defence system in mimivirus that confers resistance to virophage. *Nature* **2016**, *531*, 249–252. [CrossRef] [PubMed]

77. Dou, C.; Yu, M.; Gu, Y.; Wang, J.; Yin, K.; Nie, C.; Zhu, X.; Qi, S.; Wei, Y.; Cheng, W. Structural and Mechanistic Analyses Reveal a Unique Cas4-like Protein in the Mimivirus Virophage Resistance Element System. *iScience* **2018**, *3*, 1–10. [CrossRef] [PubMed]

78. Koonin, E.V.; Krupovic, M. Polintons, virophages and transpovirons: A tangled web linking viruses, transposons and immunity. *Curr. Opin. Virol.* **2017**, *25*, 7–15. [CrossRef]

79. Campbell, S.; Aswad, A.; Katzourakis, A. Disentangling the origins of virophages and polintons. *Curr. Opin. Virol.* **2017**, *25*, 59–65. [CrossRef] [PubMed]

80. La Scola, B.; Desnues, C.; Pagnier, I.; Robert, C.; Barrassi, L.; Fournous, G.; Merchat, M.; Suzan-Monti, M.; Forterre, P.; Koonin, E.; et al. The virophage as a unique parasite of the giant mimivirus. *Nature* **2008**, *455*, 100–104. [CrossRef] [PubMed]

81. Suttle, C.A. Marine viruses–major players in the global ecosystem. *Nat. Rev. Microbiol.* **2007**, *5*, 801–812. [CrossRef]

82. Chatterjee, A.; Sicheritz-Ponten, T.; Yadav, R.; Kondabagil, K. Genomic and metagenomic signatures of giant viruses are ubiquitous in water samples from sewage, inland lake, waste water treatment plant, and municipal water supply in Mumbai, India. *Sci. Rep.* **2019**, *9*, 3690. [CrossRef]

83. Yoshikawa, G.; Blanc-Mathieu, R.; Song, C.; Kayama, Y.; Mochizuki, T.; Murata, K.; Ogata, H.; Takemura, M. Medusavirus, a novel large DNA virus discovered from hot spring water. *J. Virol.* **2019**. [CrossRef]

84. Wilhelm, S.W.; Bird, J.T.; Bonifer, K.S.; Calfee, B.C.; Chen, T.; Coy, S.R.; Gainer, P.J.; Gann, E.R.; Heatherly, H.T.; Lee, J.; et al. A Student's Guide to Giant Viruses Infecting Small Eukaryotes: From Acanthamoeba to Zooxanthellae. *Viruses* **2017**, *9*. [CrossRef]

85. Watanabe, T.; Yamazaki, S.; Maita, C.; Matushita, M.; Matsuo, J.; Okubo, T.; Yamaguchi, H. Lateral Gene Transfer Between Protozoa-Related Giant Viruses of Family Mimiviridae and Chlamydiae. *Evol. Bioinform. Online* **2018**, *14*. [CrossRef]

86. Silva, L.C.; Almeida, G.M.; Oliveira, D.B.; Dornas, F.P.; Campos, R.K.; La Scola, B.; Ferreira, P.C.; Kroon, E.G.; Abrahao, J.S. A resourceful giant: APMV is able to interfere with the human type I interferon system. *Microbes Infect.* **2014**, *16*, 187–195. [CrossRef]
87. Scarpellini, E.; Ianiro, G.; Attili, F.; Bassanelli, C.; De Santis, A.; Gasbarrini, A. The human gut microbiota and virome: Potential therapeutic implications. *Dig. Liver Dis.* **2015**, *47*, 1007–1012. [CrossRef]
88. Popgeorgiev, N.; Temmam, S.; Raoult, D.; Desnues, C. Describing the silent human virome with an emphasis on giant viruses. *Intervirology* **2013**, *56*, 395–412. [CrossRef]
89. Colson, P.; Fancello, L.; Gimenez, G.; Armougom, F.; Desnues, C.; Fournous, G.; Yoosuf, N.; Million, M.; La Scola, B.; Raoult, D. Evidence of the megavirome in humans. *J. Clin. Virol.* **2013**, *57*, 191–200. [CrossRef]

Article

Structure and Hierarchy of Influenza Virus Models Revealed by Reaction Network Analysis

Stephan Peter [1,2], Martin Hölzer [3,4], Kevin Lamkiewicz [3,4], Pietro Speroni di Fenizio [2], Hassan Al Hwaeer [5], Manja Marz [3,4], Stefan Schuster [6], Peter Dittrich [2,*] and Bashar Ibrahim [4,6,*]

[1] Ernst-Abbe University of Applied Sciences Jena, Carl-Zeiss-Promenade 2, 07745 Jena, Germany; stephan.peter@eah-jena.de

[2] Bio Systems Analysis Group, Department of Mathematics and Computer Science, University of Jena, Ernst-Abbe-Platz 2, 07743 Jena, Germany; pietros@gmail.com

[3] RNA Bioinformatics and High-Throughput Analysis, Friedrich Schiller University Jena, Leutragraben 1, 07743 Jena, Germany; martin.hoelzer@uni-jena.de (M.H.); Kevin.Lamkiewicz@uni-jena.de (K.L.); manja@uni-jena.de (M.M.)

[4] European Virus Bioinformatics Center, Leutragraben 1, 07743 Jena, Germany

[5] Mathematics and Computer Applications Department, Al-Nahrain University, Al-Jadriya, Baghdad 10072, Iraq; hassan1167@yahoo.com

[6] Chair of Bioinformatics, Matthias-Schleiden-Institute, University of Jena, Ernst-Abbe-Platz 2, 07743 Jena, Germany; stefan.schu@uni-jena.de

* Correspondence: peter.dittrich@uni-jena.de (P.D.); bashar.ibrahim@uni-jena.de (B.I.); Tel.: +49-3641-9-49585 (P.D. & B.I.)

Received: 26 March 2019; Accepted: 11 May 2019; Published: 16 May 2019

Abstract: Influenza A virus is recognized today as one of the most challenging viruses that threatens both human and animal health worldwide. Understanding the control mechanisms of influenza infection and dynamics is crucial and could result in effective future treatment strategies. Many kinetic models based on differential equations have been developed in recent decades to capture viral dynamics within a host. These models differ in their complexity in terms of number of species elements and number of reactions. Here, we present a new approach to understanding the overall structure of twelve influenza A virus infection models and their relationship to each other. To this end, we apply chemical organization theory to obtain a hierarchical decomposition of the models into chemical organizations. The decomposition is based on the model structure (reaction rules) but is independent of kinetic details such as rate constants. We found different types of model structures ranging from two to eight organizations. Furthermore, the model's organizations imply a partial order among models entailing a hierarchy of model, revealing a high model diversity with respect to their long-term behavior. Our methods and results can be helpful in model development and model integration, also beyond the influenza area.

Keywords: chemical organization theory; influenza A; virus dynamics modeling; complex networks analysis

---

1. Introduction

Influenza is an infectious respiratory disease, annually infecting 5–15% of the human population and causing epidemics that result in 3–5 million severe cases with 300,000–500,000 deaths each year [1]. The annual recurrence of epidemics is caused by the continuous alteration of seasonal influenza viruses, which enables them to efficiently escape the immune system even due to previous infections or vaccinations [2]. The major burden of disease in humans is caused by seasonal influenza A (IAV) and influenza B viruses, causing symptoms varying from mild respiratory disease characterized by fever, sore throat, headache and muscle pain to severe and in some cases lethal pneumonia and secondary bacterial infections [3].

The long-term spread of influenza viruses in the human population and the acute nature of influenza virus epidemics is driven by the global movement of these viruses. Differences in seasonal epidemics caused by influenza viruses are mainly driven by differences in the rates of virus evolution. The single-stranded RNA segments of influenza viruses, which are located inside the virus particle (or virion), evolve rapidly and thus can escape the host's immune response very efficiently.

Several ordinary differential equations (ODEs) models have been developed to provide insight into within-host dynamics of influenza A virus infections (for reviews, see [4–8]). These models work in a time scale of days and describe the concentration dynamics of target cells, immune system components, viral load, and sometimes co-infecting pathogens. The models differ in terms of complexity and state space dimensions, which are between 3 and 15 for the models examined here. While the low-dimensional models can be analyzed completely and in a straightforward way (e.g., by calculating their fixed points and stability analysis), the characterization of the entire behavioral spectrum of complex models is more difficult (see for example [9]).

Here, we present an approach to understand the overall structure of these models that allows them to be related to each other in a simple way. To this end, we apply chemical organization theory [10,11] to obtain a hierarchical decomposition of each model into chemical organizations. A chemical organization is a sub-set of species (i.e., dimensions or model components, like, uninfected cells or viruses) that cannot generate any other species (property of closure) and that can self-maintain its own species, i.e., any species consumed by a process within the organization can be regenerated by a process within the organization. The organizations of an ODE model are rigorously related to its long-term dynamics in the following way: Given a stationary state of the ODE model, the set of species with strictly positive concentrations must be an organization [10]. The same is true for all practically relevant periodic and chaotic attractors [12]. Note that the advantage of this approach is that decomposition into organizations is based solely on the model structure (i.e., reaction rules) and thus is independent of kinetic details, like rate constants. The relation between measured data, ODE model, and organizations is depicted in Figure 1.

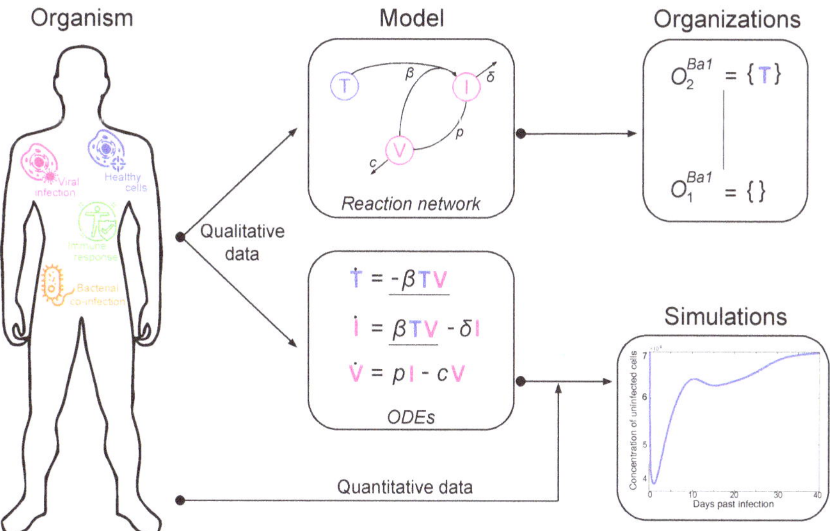

Figure 1. Relation between measured data, ordinary differential equations (ODE) model, and hierarchy of organizations.

By applying the method to twelve models of influenza A virus infection, we found different types of model structures ranging from two to eight organizations. Furthermore, the models' organizations imply a partial order among models. The resulting hierarchy of models can help to select a suitable model for certain data or serve as a framework for further model development.

We provide reaction network files for all models and a software tool for computing their organizations (https://github.com/stephanpeter/orgsflu).

2. Materials and Methods: Procedure for the Organizational Analysis

To illustrate our method, we follow a basic ODE model of influenza dynamics, namely the target cell limited model by Baccam et al. [13], called Baccam Model in the following. Note that we refer throughout this paper to a model by the first author's name of the respective publication.

The Baccam Model is based on in vivo experimental data and contains three variables: the number of susceptible and uninfected target (epithelial) cells T, the number of infected cells I, and the number of infectious-viral titer V. The dynamical behavior of the variables is given by the ODEs shown in Figure 2a. That is, target cells become infected and thus converted to infected cells at a rate βTV, infected cells die spontaneously at rate δI, virus proliferates at a rate pI and dies at a rate cV. The parameters β, δ, p and c are, as usual, positive real numbers (cf. [13] for actual values).

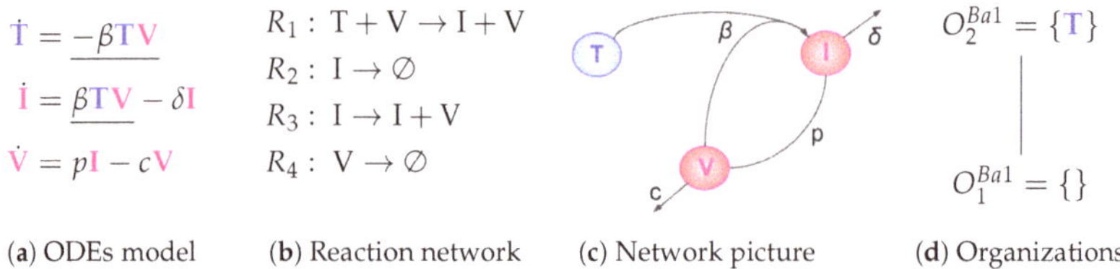

| (a) ODEs model | (b) Reaction network | (c) Network picture | (d) Organizations |

Figure 2. The **Baccam Model** [13] with three variables: uninfected (susceptible) target cells (T), infected cells (I) and infectious-viral titer (V).

Throughout this paper, the following coloring scheme for particular classes of species is used to improve readability:

- Uninfected (target) cells or those resistant/refractory to infection are marked in blue, e.g., T.
- Infected cells, partially or latently infected cells, and viruses are marked in magenta, e.g., I and V.
- Species belonging to the active immune response are marked in green. It is worth noting that the first two models analyzed in this paper [13,14] do not explicitly have immune system components.
- Bacterial co-infection species are marked in orange. These species are only occurring in Smith's model [15].
- Text referring to any other species is marked in black, e.g., transient target cell states, passive immune system, or dead cells.

For simplicity, the models' variable names also denote the related *species*. For example, V denominates not only the number of viruses in the ODE model (Figure 2a), but also the virus itself (e.g., Figure 2b).

2.1. Deriving the Reaction Network from the ODE System

In a first step, we need to obtain the reaction network underlying the ODE model. A reaction represents, for example, a cell infection by a virus, the generation of new viruses from an infected cell or the spontaneous death of a cell. The reaction rules can be derived from the ODEs in a straightforward way [16]. This step can also be performed by an online tool presented by Soliman and colleagues [16]. Note that in modeling one first creates the network and then derives the ODEs. For our analysis, we have to take the other direction.

For this purpose, we have to investigate the kinetic terms (kinetic laws) of the ODE (Figure 2a):

- The term βTV represents the reaction $R_1 : \quad T + V \rightarrow I + V$, which in turn denotes the transformation of an uninfected target cell T to an infected cell I catalysed by the virus V.
- The terms $-\delta I$ and $-cV$ represent reactions $R_2 : I \rightarrow \varnothing$ and $R_4 : V \rightarrow \varnothing$ which are the outflow of infected cells I resp. virus V.
- The term pI represents the reaction $R_3 : I \rightarrow I + V$ which is the production of viruses V catalysed by infected cells I.

The set of species $\mathcal{S} = \{T, I, V\}$ together with the set of reactions $\mathcal{R} = \{R_1, R_2, R_3, R_4\}$ constitute the so-called *reaction network* $(\mathcal{S}, \mathcal{R})$ associated with the model. The set of reactions together with their kinetic parameters are depicted in Figure 2c. Note that for clarity we use different types of underlining to highlight certain recurring kinetic terms in the ODE:

- <u>Single underline</u> for the transformation of uninfected cells into infected ones by the action of viruses.
- <u>Double underline</u> for kinetic terms involving interferon.

We call the species on the left-hand side (LHS) of a reaction R *support* of R and write $supp(R)$, e.g., $supp(R_1) = \{T, V\}$ (see Figure 2d). Analogously, we call the set of species occurring on the right-hand side (RHS) of a reaction R *products* of R and denominate this set by $prod(R)$, e.g., $prod(R_1) = \{I, V\}$.

Furthermore, let us recall the *stoichiometric matrix* $\mathbf{N} = (n_{ij})$ of a reaction network [17]. The element n_{ij} in the i-th line and j-th column of $\mathbf{N}$ denotes the net-production of the i-th species in reaction R_j. The net-production n_{ij} is the difference between the number of occurrences (stoichiometric coefficient) of species i in the RHS of reaction R_j minus the number of occurrences of species i in the LHS of reaction R_j. For example, $n_{21} = 1 - 0 = 1$ because the second species (I) appears in the first R_1 once as a reactant in the support of R_1 (LHS) but does not appear in R_1 as a product (RHS). For our example in Figure 2, the stoichiometric matrix becomes:

$$\mathbf{N} = \begin{pmatrix} -1 & 0 & 0 & 0 \\ 1 & -1 & 0 & 0 \\ 0 & 0 & 1 & -1 \end{pmatrix}. \tag{1}$$

2.2. Computing the Organizations from the Reaction Network

From the reaction network, we can compute the (chemical) *organizations* of the model. Each organization is a subset of species that is *closed* and *self-maintaining* [10,18]. In the following, let $S \subseteq \mathcal{S}$ be a subset of species and n be the total number of reactions of the reaction network ($n = 4$ in our example).

We call S **closed** if and only if all reactions $R \in \mathcal{R}$ with $supp(R) \subseteq S$ fulfill $prod(R) \subseteq S$ too [10,18]. This means that the products of a reaction R with support in S are also in S. In other words, no species outside of S can be produced by the reactions "running on" S. As an example, we assume $S = \{T, I\}$. The reactions with support in S are R_2 and R_3. However, R_3 produces species V, which is not in S. Thus, S is not closed.

We call a vector $\mathbf{v} \in \mathbb{R}^n$ *flux vector for S* if and only if it fulfills

$$\mathbf{v}_j \begin{cases} > 0, & \text{if } supp(R_j) \subseteq S, \\ = 0, & \text{if } supp(R_j) \not\subseteq S. \end{cases} \tag{2}$$

Thus, all flux vectors for S have in common that those components are strictly positive which correspond to reactions that can run on S, while all other entries are zero. As an example, consider $S_{Example} = \{T, I\}$ again. We know that the reactions R_2 and R_3 can "run on" it, i.e., they have support in $S_{Example}$. Thus, $\mathbf{v}_1 = (0, 1, 3, 0)^T$ or $\mathbf{v}_2 = (0, 5, 2, 0)^T$ are example flux vectors for $S_{Example}$.

We call S *self-maintaining* if and only if there exists (at least one) flux vector $\mathbf{v} \in \mathbb{R}^n$ for S that fulfills

$$\mathbf{N} \cdot \mathbf{v} \geq 0, \tag{3}$$

i.e., $(\mathbf{N} \cdot \mathbf{v})_i \geq 0$ for all $i = 1, \ldots, n$, where n is again the total number of reactions [10,19–21]. Roughly speaking, if S is self-maintaining, it has the potential to sustain the amount of its species above a certain level. Our example set $S_{Example}$ is not self-maintaining because $(\mathbf{N} \cdot \mathbf{v})_2 < 0$ for all flux vectors for $S_{Example}$. That is, species number 2 (the infected cells I) can not be maintained by this set.

As mentioned in the beginning of this section, we call S an *organization* if and only if it is both closed and self-maintaining. Clearly $S_{Example}$ is not an organization, as it has neither of these properties. In Figure 2d, the so-called *Hasse diagram* of organizations of this model can be seen. In it, two organizations are linked by a line if the lower one is a subset of the upper one and there is no organization in between. The Hasse diagram for the Baccam Model contains only the two organizations $O_1^B = \varnothing$ and $O_2^B = \{T\}$ (see Figure 2d). The superscript "B" stands for Baccam Model and with the subscripts we refer to the organizations within a model. The organization O_2^B represents an organism without any influenza A virus infection. Note that there is no organization with all species, i.e., representing the infected body.

2.3. The Role Organizations Play in the Dynamics

Given a fixed point x of an ODE describing a reaction system, the set of species with strictly positive concentrations in x is an organization. This is shown in [10]. This in turn means that, if a subset of species is not an organization, then the system does not have a fixed point with exactly the chosen subset of species. This is not only true for fixed points but also for other attractors [12]. Attractors are those states that a system approaches in the long-run and once reached never leaves anymore. Besides converging towards a fixed point, the long-term behavior can be also periodic oscillations or chaotic trajectories. In particular, it was shown that the long-term behavior of the system always tends at least to one organization [12]. Thus, organizations rule the long-term behavior of such dynamical reaction system. Note that the case of a system tending towards a fixed point is included in this statement as a special case. For example, the *simulation* results of the Baccam model (Figure 2a and Ref. [13]) suggest that after about two to three days the species begin to decay to finally arrive in an organization namely the empty set.

3. Results and Discussion

In the literature, there exist several mathematical models of IAV dynamics that are derived from experimental data, reviewed in Refs. [4–7]). These models differ in their complexity, e.g., the number of reactions and the number of species, depending on the available experimental data used for parameter fitting and questions to be answered. For example, models can include eclipse phases, an innate immune response, or an adaptive immune response.

After having exemplified our method in Section 2 by an analysis of the basic target cell limited model by Baccam et al. [13], we present now the full analysis for eleven additional more detailed influenza models of IAV dynamics, with up to 15 variables (species) and 45 reactions (cf. Table 1 for an overview at the end). Note that for our analysis we abstract from kinetic details, that is, the organizations are independent of particular settings of parameter values.

3.1. Target Cell Limited Model by Miao et al. (Miao Model, M, 2010)

The models by Miao et al. [14] are designed to fit experimental in vivo data from mice [6,7]. The first one ([14], Equation (1)) depends on measured time-series. The second one ([14], Equation (2)) is a simplified version of the first one, neglecting the terms depending on those time-series and still leading to a good fit within the first five days after infection [14]. Thus, we analyze this second model (Miao Model).

Compared to the basic Baccam Model, the Miao Model (Figure 3a) has the same three variables (named differently) and one additional reaction, $E_P \rightarrow 2E_P$. This reaction represents the self-replication of target cells E_P taking place at a rate $\rho_E E_P$. The full set of reactions can be found in the Appendix A (Figure A2).

In the Hasse diagram of organizations (Figure 3b), a new "full" organization O_3^M appears, which contains all three species. Thus, organization O_3^M reflects the slight difference between the two models: in the Baccam Model, uninfected target cells T are only the susceptible ones and can not increase in number, but in the Miao Model uninfected cells E_P are reproduced repeatedly by the organism. Thus, in the Baccam Model, infection is limited inherently by the limited number of uninfected target cells, while in the Miao Model the limitation of an infection in time and number of infected cells and viruses depends on other mechanisms:

$$\dot{E}_P = \rho_E E_P - \underline{\beta_\alpha E_P V}$$

$$\dot{E}_P^* = \underline{\beta_\alpha E_P V} - \delta_{E^*} E_P^*$$

$$\dot{V} = \pi_\alpha E_P^* - c_V V$$

(a) ODE model

$$O_3^M = \{all\ species\} = \{E_P, E_P^*, V\}$$
$$|$$
$$O_2^M = \{E_P\}$$
$$|$$
$$O_1^M = \{\}$$

(b) Hasse diagram of organizations

Figure 3. The **Miao Model** [14] with three variables: uninfected target cells (E_P), productively infected cells (E_P^*) and free infectious influenza viruses (V).

3.2. Target Cell Limited Model with Delayed Virus Production (Baccam II Model, Ba2, 2006)

The Baccam II Model [13,22] contains one more species than the Baccam Model presented in the methods section above. That is, there are now *two* types of infected cells: those which do not yet produce viruses I_1 and those which actively produce viruses I_2. In addition, there is only one new reaction, which transforms infected cells of type I_1 into type I_2 at rate kI_1 (Figure 4a). However, the Hasse diagram of organizations remains the same when compared with the basic Baccam Model [13].

$$\dot{T} = \underline{-\beta TV}$$

$$\dot{I}_1 = \beta TV - kI_1$$

$$\dot{I}_2 = kI_1 - \delta I_2$$

$$\dot{V} = pI_2 - cV$$

(a) ODE model

$$O_2^{Ba2} = \{T\}$$
$$|$$
$$O_1^{Ba2} = \{\}$$

(b) Hasse diagram of organizations

Figure 4. The **Baccam II Model** [13] with delayed virus production and four variables: uninfected (susceptible) target cells (T), infected cells not yet producing virus (I_1), infected cells actively producing virus (I_2) and infectious-viral titer (V).

3.3. Innate and (Simple) Adaptive Immune Response (Pawelek Model, P, 2012)

The Pawelek Model [23] contains 11 parameters and was designed to fit in vivo experimental data of horses [6,7]. The model has five variables and nine reactions. Like the basic Baccam Model, it contains uninfected target cells (T), infected cells (I), and viruses (V). Furthermore, there are two new species: interferon (F) and uninfected cells that are refractory to infections (R) because of the antiviral effect induced by interferons.

Investigating the reaction network (Figure A4 in the Appendix A) derived from the differential equations (Figure 5a), we can see that, like in the basic Baccam Model, self-replication of uninfected cells T is missing. However, due to the two new species R and F, we have five new reactions, which are neither included in the Baccam Model nor in the Miao Model. One of these five reactions is the spontaneous decay of interferon F at a rate dF. The other four new reactions describe interactions between different species:

- The rate term ϕFT represents the transformation of uninfected target cells to refractory cells catalysed by interferon.
- The reverse shift back from refractory to simple uninfected cells is represented by the term ρR.
- Furthermore, infected cells are deleted by the action of interferon at a rate κIF.
- Interferon is produced in the presence of infected cells at a rate qI.

Even though we have more species and more reactions, we get the same small pattern of organizations as in the basic Baccam Model (Figure 5b). Both models have in common that there is no self-replication of target cells. This might be one reason for the missing of other and bigger organizations which could contain species related to infection and/or immune response. This in turn means that, like the Baccam Model, this model implicitly treats infections and immune responses as phenomena that can only appear in a limited (transient) time span. The Hasse diagram of organizations (Figure 5b) tells us that the system necessarily tends towards a state of healthiness, which is represented by the organizations $O_1^P = \{\}$ and $O_2^P = \{T\}$, showing absence of infection and immune response.

$$\dot{T} = -\beta VT - \phi FT + \rho R$$

$$\dot{I} = \beta VT - \delta I - \kappa IF$$

$$\dot{R} = \phi FT - \rho R$$

$$\dot{V} = pI - cV$$

$$\dot{F} = qI - dF$$

$$O_2^P = \{T\}$$

$$O_1^P = \{\}$$

(**a**) ODE model   (**b**) Hasse diagram of organizations

Figure 5. The **Pawelek Model** [23] with five variables: (uninfected) target cells (T), productively infected cells (I), uninfected cells refractory to infections (R), free viruses (V) and interferon (F).

3.4. A Model Including Bacterial Co-Infection (Smith Model, Sm, 2016)

The Smith Model [15] contains 15 parameters and compared to experimental in vivo data from mice. It has five variables and 12 reactions. Like in the previous models, we have susceptible target cells (T) and viruses (V). Note that T is only consumed in this model but not produced. Contrarily to previous models, we have two kinds of infected cells (I_1 and I_2) here as well as bacteria (P). Bacteria P represent *bacterial co-infection* during or after virus infection. Infection is modeled as a transformation of susceptible target cells T into infected cells I_1 catalyzed by viruses V (see underlined terms in Figure 6a). Infected cells I_1 in turn spontaneously transform into I_2 at rate kI_1. Only infected cells I_2 produce viruses V at a rate pI_2. Furthermore, infected cells I_2 produce viruses V together with bacteria P. Bacteria P are self-replicating (rate term rP). Viruses V are the only species influencing bacteria P.

Figure 6b shows the Hasse diagram of organizations. The smallest one is the empty set. The biggest one is O_4^{Sm}, which contains susceptible target cells T and bacteria P. It represents an organism without viral but with bacterial infection. Between those two extreme organizations, we find $O_2^{Sm} = \{T\}$ and $O_3^{Sm} = \{P\}$. Thus, O_2^{Sm} represents the healthy organism without any infection. $O_3^{Sm} = \{P\}$ could mark the state after a viral-bacterial co-infection: after viral infection and because of the death of all target cells as well as all viruses only bacteria remain.

$$\dot{T} = -\beta TV$$

$$\dot{I}_1 = \beta TV - kI_1 - \mu PI_1$$

$$\dot{I}_2 = kI_1 - \delta I_2 - \mu PI_2$$

$$\dot{V} = pI_2(1 + P^Z) - cV$$

$$\dot{P} = rP\left(1 - \frac{P}{K_P(1 + \psi V)}\right)$$

$$- \gamma_{M_A} n^2 \frac{M_A}{P^2 + n^2 M_A} M_A^* P\left(1 - \frac{\phi V}{K_{PV} + V}\right)$$

(a) ODE model

$$O_4^{Sm} = \{T, P\}$$
$$O_2^{Sm} = \{T\} \qquad O_3^{Sm} = \{P\}$$
$$O_1^{Sm} = \{\}$$

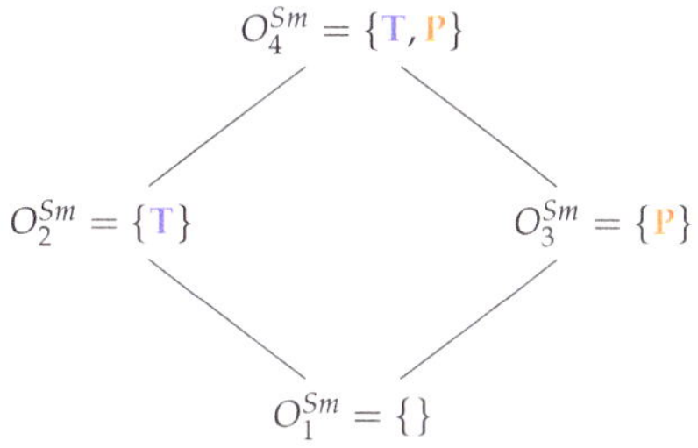

(b) Hasse diagram of organizations

Figure 6. The **Smith Model** [15] with five variables: susceptible target cells (T), two classes of infected cells (I_1 and I_2), free viruses (V), and bacteria (P).

3.5. Innate and Adaptive Immune Response (Handel Model, Ha, 2009)

The Handel Model [24] contains eight parameters and was designed to fit experimental in vivo data from mice [6,7,25]. It has seven variables (see Figure 7) and only 12 reactions (see Figure 7a):

- *Infection* is catalyzed by viruses V and transforms uninfected cells U to latently infected cells E and viruses V are consumed thereby. Latently infected cells E transform into infected cells I autonomously, which in turn transform into dead cells D autonomously too. Finally, the transformation of dead cells D into non-infected cells U closes the *circle*.
- The *remaining three species* V, F and X form an almost totally separate *subsystem* since the only interaction with the four species from the "circle" mentioned above is the catalysis of the infection by viruses V.
- The *interactions within the subsystem* $\{V, F, X\}$ consisting of viruses V and immune responses F and X are as follows:

    - Viruses V catalyze the proliferation of F and X. In the Hernandez model, proliferation of interferon F is catalyzed by infected cells instead of viruses.
    - There is no direct interaction between innate immune response F and adaptive immune response X.
    - The adaptive immune response X deletes viruses directly. Innate immune response F inhibits the self-replication of the viruses which is represented by the denominator of the fraction $\frac{pI}{1 + \kappa F}$. We ignore the inhibition because whether the rate is zero or not is independent of F.

The Hasse diagram Figure 7b shows five organizations. For the first time, it contains the empty set as well as the set of all species at the same time. Between these extremes, we find $O_2^{Ha} = \{U\}$ representing the healthy organism. The Baccam, Miao, and Pawelek models exhibit the same organization. Structurally, the Hasse diagram of the Handel Model is very similar to that of the Smith Model (Figure 6b). The first reveals an autonomy of the adaptive immune response X, whereas the latter does this same for bacteria P.

$$\dot{U} = \lambda D - \underline{bUV}$$

$$\dot{E} = \underline{bUV} - gE$$

$$\dot{I} = gE - dI$$

$$\dot{D} = dI - \lambda D$$

$$\dot{V} = \frac{pI}{1 + \kappa F} - cV - \gamma bUV - kVX$$

$$\dot{F} = wV - \delta F$$

$$\dot{X} = fV + rX$$

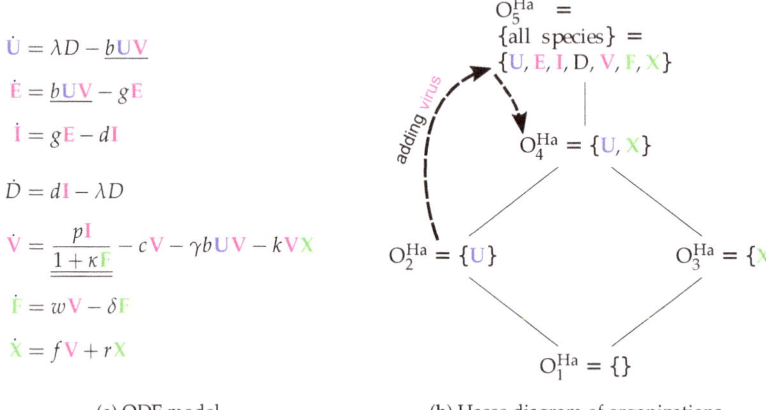

(**a**) ODE model (**b**) Hasse diagram of organizations

Figure 7. The **Handel model** [24] with seven variables: uninfected cells (U), latently infected cells (E), productively infected cells (I), dead cells (D), free viruses (V), innate immune response (F) and adaptive immune response (X). The dotted arrows denote the projection of the dynamics shown in Figure 8.

Temporal Dynamics

For the Handel Model, we perform dynamical simulation in order to show how the organizational hierarchy helps also to understand transient short-term behavior. We start at $t = -20d$ with an uninfected state, i.e., 7×10^9 uninfected cells. After 20 days, at $t = 0$, we add 10^4 virus particles. The resulting seven-dimensional trajectory in state space is shown in Figure 8. Projecting this trajectory to organizations results in a more abstract view of the dynamics, shown as a dashed curved arrow in Figure 7b. The system starts in organization O_2^{Ha} (uninfected organization), moves after adding virus particles at $t = 0$ into organization O_5^{Ha} (infected organization with immune response), and drops after 37 days into organization O_4^{Ha} (immune response active, virus absent).

The projection of a state x to an organization O follows the procedure suggested by Dittrich and Speroni d.F. [10]: First, we generate a set S of those species whose concentration is greater than a particular threshold (here: $10^0 = 1$). Then, we generate the closure of this set by adding all species that can be produced from the set. Finally, we take the largest organization O that is a subset of that closure. For example: At $t = 0$ by adding viruses to the system, we have $S = \{U, V\}$, whose closure is the set of all species, which is also an organization, here; thus, the state at $t = 0$ is projected to organization O_5^{Ha}. At $t = 60d$, we have $S = \{U, X, D\}$, whose closure is again $\{U, X, D\}$ and the largest organization contained is $O_4^{Ha} = \{U, X\}$. Thus, as can be seen in Figure 8, the system state is projected to organization O_4^{Ha}, in which it stays for $t > 37d$.

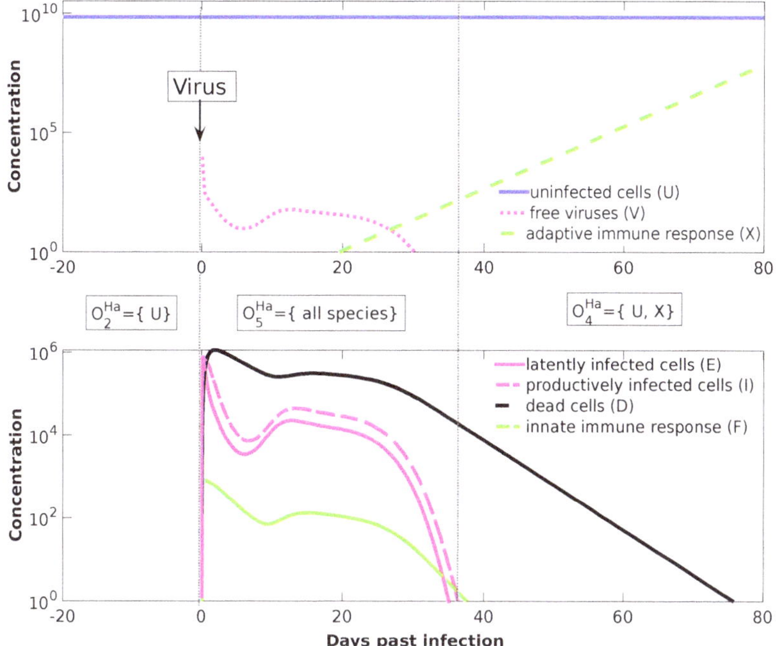

Figure 8. Temporal dynamics of the Handel model. By projecting the seven-dimensional trajectory to organizations (dotted arrows in Figure 7b) we find three phases: (Phase 1) Until day number 0, there are solely 7×10^9 uninfected cells U in the system represented by the organization $O_2^{Ha} = \{U\}$. (Phase 2) At day 0, **infection** is simulated by adding $V(0) = 10^4$ virus particles to the system. The resulting state $\{U, V\}$ is projected to organization O_5^{Ha} (all species). (Phase 3) Lastly, at day t = 37d past infection the system settles in the final organization, namely $O_4^{Ha} = \{U, X\}$, which is generated by the set $\{U, X, D\}$ (see text). The values of the model parameters are (from [24]): $\lambda = 0.25$, $b = 2.1 \times 10^{-7}$, $g = 4$, $d = 2$, $p = 5 \times 10^{-2}$, $\kappa = 1.8 \times 10^{-2}$, $c = 10$, $\gamma = 7.5 \times 10^{-4}$, $k = 1.8$, $w = 1$, $\delta = 0.4$, $f = 2.7 \times 10^{-6}$, and $r = 0.3$. Note that the number of uninfected cells U is not constant after infection as it may seem from the figure. In fact, after infection, the number of uninfected cells first decreases and than rises again [24].

3.6. Innate Immune Response and Resistance to Infection (Hernandez Model, He, 2012)

The 13 parameters of the Hernandez Model [26] were fitted to data from many different sources. The model contains seven variables and 16 reactions (see Figure 9). The species refer to viruses (V), interferon (F) and natural killer cells (K) as well as four types of respiratory tract epithelial cells: healthy/uninfected (U_H), partially infected (U_E), infected (U_I) and resistant to infection (U_R). Compared to the Pawelek Model, there are two qualitatively new species: partially infected cells U_E and killers K.

Next, we state some remarks about the reactions:

- There is an *infection* reaction catalyzed by viruses like in all previous models but with one difference: during infection, healthy cells U_H first transform to *partially infected* cells U_E and only after that they transform spontaneously to infected cells U_I at a rate $k_e U_E$.
- *Interferon* catalyzes the transformation of healthy cells to resistant cells U_R, like in the Pawelek Model. However, in the Pawelek Model, interferon removes infected cells. Here, interferon's production is catalyzed by infected cells U_I at a rate $a_I U_I$. There is no further influence of interferon on any other species.

- Infected cells are removed by natural *killers K*, which also delete partially infected cells in this model. The production of *killers K* is catalyzed by infected cells U_I at a rate $\Phi_K U_I$.
- Note that here we have an constant *inflow of healthy cells* U_H at a rate S_H (first differential equation). Thus, healthy cells cannot converge to zero.

The Hasse diagram of organizations consists only of two organizations (Figure 9b). For the first time, the empty set is not an organization because the empty set is not closed due to a constant inflow of healthy cells U_H and killers K, represented by the reaction $\varnothing \to U_H$ and $\varnothing \to K$, respectively. The smallest organization $O_1^{He} = \{U_H, K\}$ can be regarded as a state of healthiness. Contrarily, the second organization O_2^{He} contains all species (as in the Miao Model) and therefore can be interpreted as the infected organism exhibiting immune response to infection.

$$\dot{U}_H = S_H - k_I U_H V - k_R U_H F - \delta_H U_H$$

$$\dot{U}_E = k_I U_H V - k_E U_E - q_K U_E K$$

$$\dot{U}_I = k_E U_E - \delta_I U_I - q_K U_I K$$

$$\dot{U}_R = k_R U_H F - \delta_R U_R$$

$$\dot{V} = \rho_V U_I - \delta_V V$$

$$\dot{F} = a_I U_I - \delta_F F$$

$$\dot{K} = S_K + \Phi_K U_I - \delta_K K$$

$$O_2^{He} = \{all\ species\} = \{U_H, U_E, U_I, U_R, V, F, K\}$$

$$O_1^{He} = \{U_H, K\}$$

(**a**) ODE model

(**b**) Hasse diagram of organizations

Figure 9. The **Hernandez Model** [26] with seven variables: healthy cells (U_H), partially infected cell (U_E), infected cells (U_I), cells resistant to infection (U_R), virus particles (V), interferon (F) and natural killers (K).

3.7. A Model with More Detailed Immune Response (Cao Model, C, 2015)

The Cao Model [27] consists of 20 parameters and has been derived by referring to experimental data from ferrets. The model has seven variables and 26 reactions. Like in most of the previous models, we have (uninfected) target cells (T), infected cells (I), viruses (V), resistant cells (R), and interferon (F). Furthermore, there are B cells B and antibodies A.

According to the ODE shown in Figure 10a, *B cells* are only influenced by viruses: viruses support the production of B cells (rate term $m_1 V$), but the more B cells that are present, the more of them are destroyed by viruses again (term: $m_1 V B$). B cells influence only one other species, namely antibodies B, which they produce. *Antibodies A* influence only one other species, namely viruses, which are destroyed by this reaction at rate $\mu V A$. Antibodies in turn are influenced by B cells positively and by viruses negatively.

There are three organizations in this model (Figure 10b): the empty set, the healthy organism without any infection and without any immune response ($O_2^C = \{T\}$) and the organization O_3^C containing all species.

$$\dot{T} = g(T + R)(1 - \frac{T + R + I}{C_t}) - \beta'VT + \rho R - \underline{\underline{\phi FT}}$$

$$\dot{I} = \beta'VT - \delta I - \underline{\underline{\kappa I F}}$$

$$\dot{V} = \frac{pI}{\underline{\underline{1 + sF}}} - cV - \mu VA - \beta VT$$

$$\dot{R} = \underline{\underline{\phi FT}} - \rho R - \xi R$$

$$\dot{F} = qI - dF$$

$$\dot{B} = m_1 V(1 - B) - m_2 B$$

$$\dot{A} = m_3 B - rA - \mu'VA$$

$$O_3^C = \{all\ species\} = \{T, I, V, R, F, B, A\}$$

$$O_2^C = \{T\}$$

$$O_1^C = \{\}$$

(**a**) ODE model (**b**) Hasse diagram of organizations

Figure 10. The **Cao Model** [27] with seven variables: target cells (T), infected cells (I), viruses (V), resistant cells (R), interferon (F), B cells (B), and antibodies (A).

3.8. Innate Immune Response and Eclipse Phase (Saenz Model, Sa, 2010)

The Saenz Model [28] requires 12 parameters and was designed to fit experimental in vivo data from horses [6,7]. The model contains eight variables and 12 reactions (Figure 11a). There are no adaptive immune response, no dead cells, and no natural killer cells. However, the model contains viruses V and interferon F. There is an eclipse phase (E_1 and E_2) here as well as prerefractory and refractory cells. In particular, epithelial cells are represented by six species: susceptible (T), eclipse phases (E_1 *and* E_2), infectious (I), prerefractory (W), and refractory (R). Thus, the new features are the inclusion of two eclipse phases and three steps for the transformation of uninfected cells to refractory cells.

$$\dot{T} = -\underline{\beta VT} - \underline{\phi FT}$$

$$\dot{E}_1 = \underline{\beta VT} - k_1 E_1$$

$$\dot{W} = \underline{\underline{\phi FT}} - m\beta VW - aW$$

$$\dot{E}_2 = m\beta VW - k_2 E_2$$

$$\dot{R} = aW$$

$$\dot{I} = k_1 E_1 + k_2 E_2 - \delta I$$

$$\dot{V} = pI - cV$$

$$\dot{F} = nqE_2 + qI - dF$$

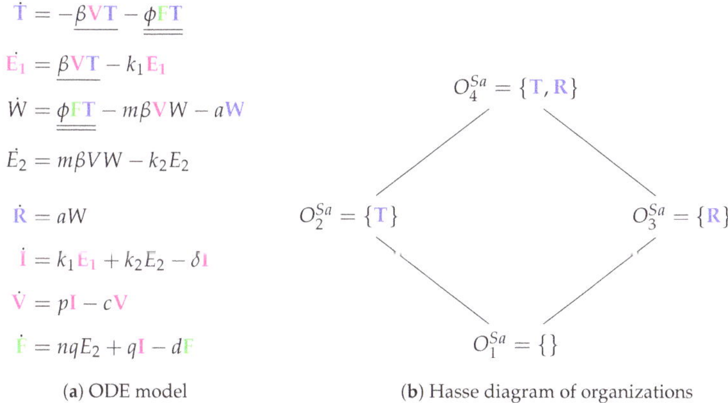

$$O_4^{Sa} = \{T, R\}$$

$$O_2^{Sa} = \{T\} \qquad O_3^{Sa} = \{R\}$$

$$O_1^{Sa} = \{\}$$

(**a**) ODE model (**b**) Hasse diagram of organizations

Figure 11. The **Saenz Model** [28] with eight variables: Epithelial cells in one of the states: susceptible (T), eclipse phase (E_1 and E_2), prerefractory (W), refractory (R) and infectious (I). The further variables are: virus cells (V) and interferon (F).

The Hasse diagram of organizations is composed by four organizations: $O_1^{Sa} = \{\}$: the empty set; $O_2^{Sa} = \{T\}$: representing a healthy organism; $O_3^{Sa} = \{R\}$: there is no consuming reaction for refractory cells R; $O_4^{Sa} = \{T, R\}$: also representing a healthy organism that contains refractory cells maybe as the remains of a previous infection.

The Hasse diagram is very similar to that from of the Handel Model. There are only two differences:

- The "full" organization is missing here. For sure, one of the reasons is that there is no reaction producing susceptible cells T. Thus, when viruses V or interferon F are present susceptible cells T can not survive and the "full" organization neither.

- Adaptive immune response is replaced by refractory cells in the organizations here.

3.9. Focusing on Innate and Adaptive Immunity (Hancioglu Model, Hcg, 2007)

The Hancioglu Model [29] contains 28 parameters, 10 variables, and 44 reactions. It has not been mathematically fitted to data but has been designed to meet specific general criteria [6,7]. The ODEs (Figure 12a) describe the dynamics of the following 10 species: viruses (V), healthy cells (H), infected cells (I), interferon (F) and resistant cells (R). The remaining species are new: antigen presenting cells (M), effector cells (E), plasma cells (P) antibodies (A) and antigenic distance (S). There are no species for an eclipse phase in this model.

Looking at the reaction network (Figure A8), we can see again a reaction for *infection*, i.e., the transformation of healthy cells H into infected cells I catalyzed by viruses V at a rate $\gamma_{HV} VH$ (single underlined in Figure 12a). Furthermore, *interferon F* is produced catalytically by antigen presenting M and infected cells I, decays spontaneously at a rate $a_F F$, and is additionally removed when converting healthy cells H into resistant cells R by the reaction $H + F \rightarrow R$ at rate $b_{HF} FH$ (double underlined in Figure 12a).

The Hancioglu Model has three organizations (Figure 12b):

- The smallest one is O_1^{Hcg}, which contains all the species responsible for the *immune response*.
- O_1^{Hcg} is a subset of O_2^{Hcg}, which additionally contains species H and R, representing the healthy organism without infection but with the immune response turned on.
- O_3^{Hcg} is the "full" organization containing all the species of the models and thus representing the organism with infection and immune response.

Thus, all the organizations represent meaningful states of the organism. However, there is no organization that only consists of healthy cells without any infection and immune response. Note that almost all the previous models except for Hernandez have such an organization.

$$\dot{H} = b_{HD}(1 - H - R - I)(H + R) + a_R R - \underline{\gamma_{HV} VH} - \underline{\underline{b_{HF} FH}}$$

$$\dot{I} = \underline{\gamma_{HV} VH} - b_{IE} EI - a_I I$$

$$\dot{V} = \gamma_V I - \gamma_{VA} SAV - \gamma_{VH} HV - \alpha_V V - \frac{a_{V1} V}{1 + a_{V2} V}$$

$$\dot{R} = \underline{\underline{b_{HF} FH}} - a_R R$$

$$\dot{M} = (b_{MD}(1 - H - R - I) + b_{MV} V)(1 - M) - a_M M$$

$$\dot{F} = b_F M + c_F I - b_{FH} HF - a_F F$$

$$\dot{E} = b_{EM} ME - b_{EI} IE + a_E(1 - E)$$

$$\dot{P} = b_{PM} MP + a_P(1 - P)$$

$$\dot{A} = b_A P - \gamma_{AV} SAV - a_A A$$

$$\dot{S} = rP(1 - S)$$

$$O_3^{Hcg} = \{all\ species\} = \{H, I, V, R, M, F, E, P, A, S\}$$

$$O_2^{Hcg} = \{H, M, F, R, E, P, A, S\}$$

$$O_1^{Hcg} = \{M, F, E, P, A, S\}$$

(a) ODE model (b) Hasse diagram of organizations

Figure 12. The **Hancioglu Model** [29] with 10 variables: viral load (V), healthy cells (H), infected cells (I), antigen presenting cells (M), interferon (F), resistant cells (R), effector cells (E), plasma cells (P), antibodies (A) and antigenic distance (S).

3.10. Model with Delay Differential Equations (Bocharov Model, Bo, 1994)

The Bocharov Model [30] contains 49 parameters and was designed to fit experimental in vivo data from humans [6,7]. It includes 10 variables and 51 reactions (Figure A9). Only here and in the Lee

Model (below) we have differential equations with *delay*, i.e., some rates depend on variable values from the past (Figure 13a). Because the delay does not matter in a steady-state, we can also neglect the delay when analyzing the chemical organizations of a delay differential equation model.

$$\dot{V}_f = \nu C + nb_{CE}\, CE - \gamma_{VF}\, V_f F - \gamma_{VM}\, V_f - \gamma_{VC}\, V_f U$$

$$\dot{C} = \sigma V_f U - b_{CE}\, CE - b_m\, C$$

$$\dot{m} = b_{CE}\, CE + b_m\, C - \alpha_m m$$

$$\dot{M}_V = \gamma_{MV} M^* V_f - \alpha_M\, M_V$$

$$\dot{H}_E = b_H^E[(1 - \frac{m}{C^*})\rho_H^E\, M_V(t - \tau_H^E)\, H_E(t - \tau_H^E) - M_V\, H_E] - b_P^{H_E}\, M_V\, H_E E + \alpha_H^E(H_E^* - H_E)$$

$$\dot{H}_B = b_H^B[(1 - \frac{m}{C^*})\rho_H^B\, M_V(t - \tau_H^B)\, H_B(t - \tau_H^B) - M_V\, H_B] - b_P^{H_B}\, M H_B B + \alpha_H^B(H_B^* - H_B)$$

$$\dot{E} = b_P^E[(1 - \frac{m}{C^*})\rho_E\, M_V((t - \tau_E))\, H_E(t - \tau_E)\, E(t - \tau_E) - M_V\, H_E E] - b_{EC}\, C_V E + \alpha_E(E^* - E)$$

$$\dot{B} = b_P^B[(1 - \frac{m}{C^*})\rho_B\, M_V(t - \tau_B)\, H_B(t - \tau_B)\, B(t - \tau_B) - M_V\, H_B B] + \alpha_E(B^* - B)$$

$$\dot{P} = b_P^P(1 - \frac{m}{C^*})\rho_P\, M_V(t - \tau^P)\, H_B(t - \tau_P)\, B(t - \tau_P) + \alpha_P(P^* - P)$$

$$\dot{F} = \rho_F P - \gamma_{FV}\, F V_f - \alpha_F F$$

$$\dot{U} = \frac{d}{dt}(C^* - C - m) = -\sigma V_f U + \alpha_m m$$

(**a**) ODE model

$$O_2^{Bo} = \{U, H_E, H_B, E, B, P, F\}$$

$$O_1^{Bo} = \{H_E, H_B, E, B, P, F\}$$

(**b**) Hasse diagram of organizations

Figure 13. The **Bocharov Model** [30] with 10 variables: infective IAV particles (V_f), IAV-infected cells (C), destroyed epithelial cells (m), stimulated macrophages (M_V), activated helper T cells providing proliferation of cytotoxic T cells (H_E), activated helper T cells providing proliferation and differentiation of B cells B (H_B), activated CTL (E), B cells (B), plasma cells (P), antibodies to IAV (F), and uninfected epithelial cells (U). Note that, for clarity, we have added U as a state variable, which is only implicitly represented as $U = C^* - C - m$ in the original model by Bocharov et al.

Note that this is by far the oldest model analyzed here. The names of the variables are a bit particular when compared to those of the previously analyzed models. As in all the other models, we have viruses V_f and infected cells C. Furthermore, there are destroyed epithelial cells m as in the Handel Model. All other species belong to the immune response. Note that only in this model there is no state variable for uninfected, healthy cells. Bocharov et al. represent these healthy cells implicitly by subtracting the amounts of infected-cells C and destroyed epithelial cells m from the initial total amount of target epithelial cells C^*. Since all the other models analyzed here have a related variable, we inserted the variable $U = C^* - C - m$ for uninfected cells together with its ODE to make the model comparable to the others.

Due to the fact that the majority of the species belongs to the immune response, this is the case for most of the reactions too. These species of the immune response form exactly the organization O_1^{Bo}, only macrophages M_V are missing.

Similarly to the Hancioglus model, the smallest organization O_1^{Bo} is an organization with immune response but without infection (C, V_f). There is only one further organization that contains only one more species than O_1^{Bo}, namely uninfected cells U. This organization we already found in three of the previous models. However, for the first time, there is no bigger organization in this model. Thus, virus infection is necessarily transient.

3.11. Complex Dual-Compartment Model (Lee Model, L, 2009)

The Lee Model [9] is the most complex model considered here. It contains 48 parameters and was designed with respect to experimental in vivo data from mice [6,7]. It has 15 variables and 37 reactions (Figure A10). Like in the Bocharov Model, Lee et al. apply delay differential equations.

Note that this model is the only one analyzed here that distinguishes between *lung compartment* and *lymphatic compartment*. There is one species representing uninfected (healthy) cells E_P and three species for modelling *infection*: E_P^*, D^* and viruses V. The remaining species belong to the *immune response*, colored black when naive to infection, while colored green when activated for infection.

Note that we write a species in the *organizations* in Figure 14b in bold text, if it is "new", that is, not contained in neither of its subset organizations. The Hasse diagram contains eight organizations. The smallest one is $O_1^L = \{E_P, D, H_N, T_N, B_N\}$ and contains exactly the uninfected cells as well as the naive part of the immune response. The biggest organization contains all species. Between these two "extreme" organizations are six further organizations containing different parts of the activated part of the immune response.

$$\dot{E}_P = \delta_E(E_0 - E_P) - \beta_E E_P V$$

$$\dot{E}_P^* = \beta_E E_P V - k_E E_P^* \gamma T_E(t - \tau_T) - \delta_{E^*} E_P^*$$

$$\dot{V} = \pi_V E_P^* - c_V V - k_V V A(t)$$

$$\dot{D} = \delta_D(D_0 - D) - \beta_D D V$$

$$\dot{D}^* = \beta_D D V - \delta_{D^*} D^*$$

$$\dot{D}_M = k_D D^*(t - \tau_D) - \delta_{D_M} D_M$$

$$\dot{H}_N = \delta_{H_N}(H_{N0} - H_N) - \pi_{H1}\frac{D_M}{D_M + \pi_{H2}}H_N$$

$$\dot{H}_E = \pi_{H1}\frac{D_M}{D_M + \pi_{H2}}H_N + \rho_{H1}\frac{D_M}{D_M + \rho_{H2}}H_E - \delta_{H1}\frac{D_M}{D_M + \delta_{H2}}H_E$$

$$\dot{T}_N = \delta_{T_N}(T_{N0} - T_N) - \pi_{T1}\frac{D_M}{D_M + \pi_{T2}}T_N$$

$$\dot{T}_E = \pi_{T1}\frac{D_M}{D_M + \pi_{T2}}H_N + \rho_{T1}\frac{D_M}{D_M + \rho_{T2}}T_E - \delta_{T1}\frac{D_M}{D_M + \delta_{T2}}T_E$$

$$\dot{B}_N = \delta_B(B_{N0} - B_N) - \pi_{B1}\frac{D_M}{D_M + \pi_{B2}}B_N$$

$$\dot{B}_A = \pi_{B1}\frac{D_M}{D_M + \pi_{B2}}B_N + \rho_{B1}\frac{D_M + hH_E}{D_M + hH_E + \rho_{B2}}B_A - \delta_{B_A}B_A - \pi_S B_A - \pi_L H_E B_A$$

$$\dot{P}_S = \pi_S B_A - \delta_S P_S$$

$$\dot{P}_L = \pi_L H_E B_A - \delta_L P_L$$

$$\dot{A} = \pi_{AS} P_S + \pi - A L P_L - \delta_A A$$

(a) ODE model

Figure 14. *Cont.*

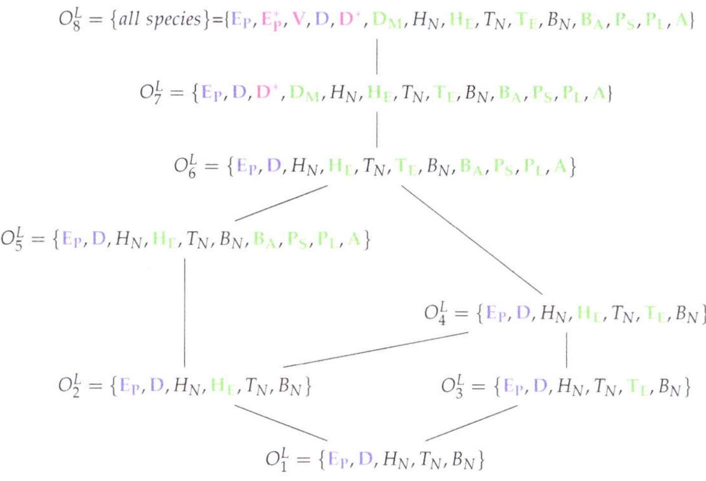

$$O_8^L = \{all\ species\} = \{E_P, E_P^+, V, D, D^*, D_M, H_N, H_E, T_N, T_E, B_N, B_A, P_S, P_L, A\}$$

$$O_7^L = \{E_P, D, D^*, D_M, H_N, H_E, T_N, T_E, B_N, B_A, P_S, P_L, A\}$$

$$O_6^L = \{E_P, D, H_N, H_E, T_N, T_E, B_N, B_A, P_S, P_L, A\}$$

$$O_5^L = \{E_P, D, H_N, H_E, T_N, B_N, B_A, P_S, P_L, A\}$$

$$O_4^L = \{E_P, D, H_N, H_E, T_N, T_E, B_N\}$$

$$O_2^L = \{E_P, D, H_N, H_E, T_N, B_N\} \qquad O_3^L = \{E_P, D, H_N, T_N, T_E, B_N\}$$

$$O_1^L = \{E_P, D, H_N, T_N, B_N\}$$

(**b**) Hasse diagram of organizations

Figure 14. The **Lee model** [9] which contains 15 variables: uninfected epithelial cells (E_P), infected epithelial cells (E_P^+), virus titer (EID_{50}/ml) (V), immature dendritic cells (D), virus-loaded dendritic cells (D^*), mature dendritic cells (D_M), naive CD4+ T cells (H_N), effector CD4+ T cells (H_E), naive CD8+ T cells (T_N), effector CD8+ T cells (T_E), naive B cells (B_N), activated B cells (B_A), short-lived plasma (antibody-secreting) B cells (P_S), long-lived plasma (antibody-secreting) B cells (P_L) and antiviral antibody titer (A). Note that here we have colored green only those species representing the immune system when activated.

3.12. Hierarchy of Influenza A Virus Models

In order to construct a hierarchical map of all investigated models, we characterize a model by a signature of organizations, which is a set of organization types. For example, the signature of the Handel Model (Figure 7b) is the set $\{\emptyset, X, X, XX, XXX\}$. An organization type like XX means that there is at least one organization that contains uninfected (target) cells (X) *and* species of the active immune response (X). The deviation of the signatures for all models is shown in Table 1. Note that we ignore species colored black. We include the empty set $\emptyset$ because this distinguishes models without any inflow from those that possess an inflow of some species.

Now, we can obtain a partial order among models by defining that a model A is smaller or equal to another model B (A ≤ B), if the signature of A is a subset of the signature of model B. For example, the Hernandez Model is smaller than the Lee Model because $\{XX, XXX\} \subseteq \{X, XX, XXX\}$. This partial order among models leads to a hierarchical map of models, which is visualized by a Hasse-diagram in Figure 15. Note that a model A that is smaller than a model B according to this partial order can possess more species and reactions than B.

In Figure 15, we can see that all models have organizations with uninfected, healthy cells (X). There are models that furthermore have infection (X) and/or immune response (X) in their organizations. There are exactly two models (Hancioglu and Hernandez Model) with immune response in all their organizations which means that these models implicitly assume immune response to be active all the time. Among the models neglecting immune response are those which have infection (Miao Model) or bacteria (X) (Smith Model) in their organizations and also those that do not (Baccam and Baccam II Model). For models involving immune response, the situation is more complex. There are those that only have healthy cells in their organizations (Pawelek and Saenz Models). This means that these models implicitly exclude infection and immune response from the long run and thus treat them as transient phenomena a priori. The Bocharov Model is the only one that exhibits only healthy cells and immune response in its organizations but no infection. The remaining five models include all kinds of species (except for bacteria of course) in their organizations.

Table 1. Overview of all models and organization types contained. An organization type like XX denotes the type of species contained in an organization, according to our coloring scheme. The set of organization types of a model is called its signature.

Model	Number of Variables	Number of Reactions	Number of Organizations	Organizations & Signature		
Baccam [13] 2006	3	4	2	$O_1^{Ba1} = \varnothing$		
				O_2^{Ba1}	X	
Miao [14] 2010	3	5	3	$O_1^{M} = \varnothing$		
				O_2^{M}	X	
				O_3^{M}	X	X
Baccam II [13] 2006	4	5	2	$O_1^{Ba2} = \varnothing$		
				O_2^{Ba2}	X	
Pawelek [23] 2012	5	9	2	$O_1^{P} = \varnothing$		
				O_2^{P}	X	
Smith [15] 2016	5	12	4	$O_1^{Sm} = \varnothing$		
				O_2^{Sm}	X	
				O_3^{Sm}	X	
				O_4^{Sm}	X	X
Handel [24] 2010	7	12	5	$O_1^{Ha} = \varnothing$		
				O_2^{Ha}	X	
				O_3^{Ha}	X	
				O_4^{Ha}	X	X
				$O_5^{Ha} = \{all\}$	X X X	
Hernandez [26] 2012	7	16	2	O_1^{He}	X X	
				$O_2^{He} = \{all\}$	X X X	
Cao [27] 2015	7	26	3	$O_1^{C} = \varnothing$		
				O_2^{C}	X	
				$O_3^{C} = \{all\}$	X X X	
Saenz [28] 2010	8	12	4	$O_1^{Sa} = \varnothing$		
				$O_2^{Sa},\, O_3^{Sa},\, O_4^{Sa}$	X	
Hancioglu [29] 2007	10	44	3	O_1^{Hcg}	X	
				O_2^{Hcg}	X X	
				$O_3^{Hcg} = \{all\}$	X X X	
Bocharov [30] 1994	10	45	2	O_1^{Bo}	X	
				O_2^{Bo}	X X	
Lee [9] 2009	15	37	8	O_1^{L}	X	
				$O_2^{L},\, O_3^{L},\, O_4^{L},\, O_5^{L},\, O_6^{L}$	X X	
				$O_7^{L},\, O_8^{L} = \{all\}$	X X X	

By looking at the hierarchy of models, it becomes evident that there is space for more models. Above the Smith and Handel Model, there could be one in which virus infection as well as bacterial coinfection can be simultaneously persistent ("fully persistent models" denote such hypothetical models in Figure 15). Another extreme case would be a "fully-transient model" in which we have only transient dynamics and all species would finally tend to zero. Such a model would be the smallest one in our partial order of models (Figure 15).

The derived hierarchical map of models might be used to choose the most appropriate model for a particular domain and data set: The model should contain at least one organization for each set of species that were experimentally observed to survive in the long run. If there are several models with such organizations, the one with the smallest organizations might be chosen to provide maximum efficiency in modeling. Table 1 as well as Figure 15 might guide the selection process, complementing established quantitative selection methods, such as those using the area under the viral load curve [31].

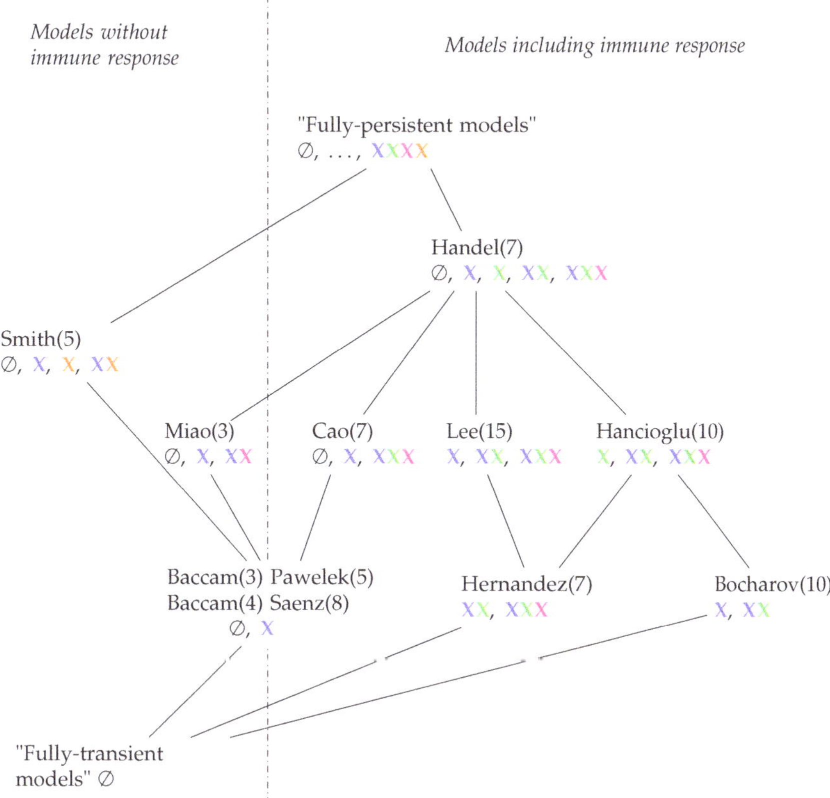

Figure 15. Hasse-diagram of the hierarchy of IAV models with respect to their long-term behaviour. In brackets (), we added the number of species of each model. Underneath (marked by colors) the kinds of species contained in the organizations belonging to each model. The meaning of the four colors is as follows: Species belonging to the healthy state of the organism are colored blue, those belonging to the immune response are colored green, those belonging to infection like infected cells and viruses are colored magenta, and bacteria from bacterial co-infection are colored orange. Horizontally, the diagram consists of four lines. The models in the lowest line contain organizations with exactly two different kinds of species (colors) (including the empty set). In the second line above, there are three different combinations of species (colors) to be found in each model. There is only one model in each of the highest two lines: The Smith model [4] is the only one with bacteria and contains four different combinations of colors. In the Handel Model, there are even five different combinations of colors out of $2^4 = 16$ possible combinations.

4. Conclusions

By analyzing twelve published IAV immune system models, we have shown that we can compute, independently of quantitative kinetic data like rate constants or kinetic laws, all chemical organizations of a typical IAV model, which provides a hierarchical decomposition of the model and an overview of its potential long-term behavior.

It turned out that the derived organizations are meaningful with respect to the model's domain. That is, the composition of species inside an organization can be related to a particular state of the organism, like "healthy", "infected", or "virus controlled by active immune response", and it is possible to annotate organizations accordingly.

By deriving an organizational signature from a model's organizations, we obtained a novel classification scheme and a hierarchy of models with respect to their qualitative long-term behavior. Although the classification via organizational signature is quite coarse-grained, the analysis revealed still a high diversity of models. That is, the models have different potentiality with respect to which variables persist in the long turn and which vanish. Furthermore, the hierarchy map of models contains various empty territories, suggesting space for potential future models.

We envision as a practical use that our method and results can help to select the right model for a particular situation, to relate other models to the present ones, to obtain an overview of the potential long-term dynamics of complex models, and to support model development, for example, by providing a rapid consistency check. Note that measured long-term as well as transient data can be explained with respect to organizations, by defining a projection from a system state to an organization, as demonstrated for the Handel model (Figures 7 and 8).

In addition, our approach is not limited to IAV models, but can be directly applied to other viruses in the same way since their dynamics are similarly modeled by ODEs [32]. Furthermore, our approach is open to include other dynamically relevant components like treatment and vaccination strategies. Another task for future work would be to study the transient virus immune system in more detail, for example, by mapping the basin of attraction of each organization or by systematically analyzing the transition dynamics in-between organizations [33,34].

Author Contributions: S.P. performed the implementation and computations. S.P., P.D. and B.I. analyzed the results with the help of M.H. and K.L. S.P., P.D., and B.I. completed the analysis and final conclusions. S.P., P.D. and B.I. wrote the paper with critical input from M.H., K.L., P.S.d.F., H.A.H., M.M. and S.S. P.D. and B.I. supervised the project and conceived the study. All authors reviewed and approved the final version of the paper.

Funding: S.P. and P.D. acknowledge funding from the Zeiss Foundation, project "Eine virtuelle Werkstatt für die Digitalisierung in den Wissenschaften" (Durchbrüche 2018). B.I. and StS acknowledge funding by the German Research Foundation (DFG) within the Collaborative Research Center (CRC) 1127 ChemBioSys (SFB 1127, Project C07) and M.H. within the CRC 1076 AquaDiva (Project A06). K.L. acknowledges the funding by the Bundesministerium für Bildung und Forschung (BMBF) (Project 03ZZ0820A).

Conflicts of Interest: The authors declare no conflict of interest

Appendix A

Here you find the sets of reactions for all example models presented in this paper. Written in brackets behind each reaction is its corresponding term from the differential equations:

$$
\begin{aligned}
R_1: \quad & T + V \rightarrow I + V \quad (\beta TV) \\
R_2: \quad & I \rightarrow \varnothing \quad (\delta I) \\
R_3: \quad & I \rightarrow I + V \quad (pI) \\
R_4: \quad & V \rightarrow \varnothing \quad (cV)
\end{aligned}
$$

Figure A1. Reactions of Baccam model [13].

$$R_1: \quad E_P \to 2E_P \quad (\rho_E E_P)$$
$$R_2: \quad E_P + V \to E_P^* + V \quad (\beta_\alpha E_P V)$$
$$R_3: \quad E_P^* \to \varnothing \quad (\delta_{E^*} E_P^*)$$
$$R_4: \quad E_P^* \to E_P^* + V \quad (\pi_\alpha E_P^*)$$
$$R_5: \quad V \to \varnothing \quad (c_V V)$$

Figure A2. Reactions of Miao model [14].

$$R_1: \quad T + V \to I_1 + V \quad (\beta T V)$$
$$R_2: \quad I_1 \to I_2 \quad (k I_1)$$
$$R_3: \quad I_2 \to \varnothing \quad (\delta I_2)$$
$$R_4: \quad I_2 \to I_2 + V \quad (p I_2)$$
$$R_5: \quad V \to \varnothing \quad (c V)$$

Figure A3. Reactions of Baccam model [14] with delayed virus production.

$$R_1: \quad T + V \to I + V \quad (\beta V T)$$
$$R_2: \quad F + T \to F + R \quad (\phi F T)$$
$$R_3: \quad R \to T \quad (\rho R)$$
$$R_4: \quad I \to \varnothing \quad (\delta I)$$
$$R_5: \quad I + F \to F \quad (\kappa I F)$$
$$R_6: \quad I \to I + V \quad (p I)$$
$$R_7: \quad V \to \varnothing \quad (c V)$$
$$R_8: \quad I \to I + F \quad (q I)$$
$$R_9: \quad F \to \varnothing \quad (d F)$$

Figure A4. Reactions of Pawelek model [23].

$$R_1: \quad D \to U \quad (\lambda D)$$
$$R_2: \quad U + V \to E + V \quad (b U V)$$
$$R_3: \quad E \to I \quad (g E)$$
$$R_4: \quad I \to D \quad (d I)$$
$$R_5: \quad I \to I + V \quad \left(\frac{p I}{1 + \kappa F}\right)$$
$$R_6: \quad V \to \varnothing \quad (c V)$$
$$R_7: \quad U + V \to U \quad (\gamma b U V)$$
$$R_8: \quad V + X \to X \quad (k V X)$$
$$R_9: \quad V \to V + F \quad (w V)$$
$$R_{10}: \quad F \to \varnothing \quad (\delta F)$$
$$R_{11}: \quad V \to V + X \quad (f V)$$
$$R_{12}: \quad X \to 2X \quad (r X)$$

Figure A5. Reactions of Handel model [24].

$$R_1: \quad \emptyset \rightarrow U_H \qquad (S_H)$$
$$R_2: \quad U_H + V \rightarrow U_E + V \qquad (k_I U_H V)$$
$$R_3: \quad U_H + F \rightarrow U_R + F \qquad (k_R U_H F)$$
$$R_4: \quad U_H \rightarrow \emptyset \qquad (\delta_H U_H)$$
$$R_5: \quad U_E \rightarrow U_I \qquad (k_E U_E)$$
$$R_6: \quad U_E + K \rightarrow K \qquad (q_K U_E K)$$
$$R_7: \quad U_I \rightarrow \emptyset \qquad (\delta_I U_I)$$
$$R_8: \quad U_I + K \rightarrow K \qquad (q_K U_I K)$$
$$R_9: \quad U_R \rightarrow \emptyset \qquad (\delta_R U_R)$$
$$R_10: \quad U_I \rightarrow U_I + V \qquad (\rho_V U_I)$$
$$R_11: \quad V \rightarrow \emptyset \qquad (\delta_V V)$$
$$R_12: \quad U_I \rightarrow U_I + F \qquad (a_I U_I)$$
$$R_13: \quad F \rightarrow \emptyset \qquad (\delta_F F)$$
$$R_14: \quad \emptyset \rightarrow K \qquad (S_K)$$
$$R_15: \quad U_I \rightarrow U_I + K \qquad (\Phi_K U_I)$$
$$R_16: \quad K \rightarrow \emptyset \qquad (\delta_K K)$$

Figure A6. Reactions of Hernandez model [26].

$$R_1: \quad V + T \rightarrow V + E_1 \qquad (\beta VT)$$
$$R_2: \quad F + T \rightarrow F + W \qquad (\phi FT)$$
$$R_3: \quad E_1 \rightarrow I \qquad (k_1 E_1)$$
$$R_4: \quad V + W \rightarrow V + E_2 \qquad (m\beta VW)$$
$$R_5: \quad W \rightarrow R \qquad (aW)$$
$$R_6: \quad E_2 \rightarrow I \qquad (k_2 E_2)$$
$$R_7: \quad I \rightarrow \emptyset \qquad (\delta I)$$
$$R_8: \quad I \rightarrow I + V \qquad (pI)$$
$$R_9: \quad V \rightarrow \emptyset \qquad (cV)$$
$$R_10: \quad E_2 \rightarrow E_2 + F \qquad (nq E_2)$$
$$R_11: \quad I \rightarrow I + F \qquad (qI)$$
$$R_12: \quad F \rightarrow \emptyset \qquad (dF)$$

Figure A7. Reactions of Saenz model [28].

$$R_1: \quad I \to I + V \qquad (\gamma_V I)$$
$$R_2: \quad S + A + V \to S + A \qquad (\gamma_{VA} SAV)$$
$$R_3: \quad H + V \to H \qquad (\gamma_{VH} HV)$$
$$R_4: \quad V \to \varnothing \qquad (\alpha_V V)$$
$$R_5: \quad V \to \varnothing \qquad \left(\frac{a_{V1} V}{1 + a_{V2} V}\right)$$
$$R_6: \quad H \to 2H \qquad (b_{HD} H)$$
$$R_7: \quad R \to R + H \qquad (b_{HD} R)$$
$$R_8: \quad H \to \varnothing \qquad (-b_{HD} H^2)$$
$$R_9: \quad R + H \to R \qquad (-2b_{HD} HR)$$
$$R_{10}: \quad R + H \to R \qquad (-b_{HD} R^2)$$
$$R_{11}: \quad I + H \to I \qquad (-b_{HD} IH)$$
$$R_{12}: \quad I + R + H \to I + R \qquad (-b_{HD} IR)$$
$$R_{13}: \quad R \to H \qquad (a_R R)$$
$$R_{14}: \quad V + H \to V + I \qquad (\gamma_{HV} VH)$$
$$R_{15}: \quad F + H \to F + R \qquad (b_{HF} FH)$$
$$R_{16}: \quad E + I \to E \qquad (b_{IE} EI)$$
$$R_{17}: \quad I \to \varnothing \qquad (a_I I)$$
$$R_{18}: \quad \varnothing \to M \qquad (b_{MD})$$
$$R_{19}: \quad M \to \varnothing \qquad (-b_{MD} M)$$
$$R_{20}: \quad H + M \to H \qquad (-b_{MD} H)$$
$$R_{21}: \quad H + M \to H + 2M \qquad (b_{MD} HM)$$
$$R_{22}: \quad R + M \to R \qquad (-b_{MD} R)$$
$$R_{23}: \quad R + M \to R + 2M \qquad (b_{MD} RM)$$
$$R_{24}: \quad I + M \to I \qquad (-b_{MD} I)$$
$$R_{25}: \quad I + M \to I + 2M \qquad (b_{MD} IM)$$
$$R_{26}: \quad V \to V + M \qquad (b_{MV} V)$$
$$R_{27}: \quad V + M \to V \qquad (-b_{MV} VM)$$
$$R_{28}: \quad M \to \varnothing \qquad (a_M M)$$
$$R_{29}: \quad M \to M + F \qquad (b_F M)$$
$$R_{30}: \quad I \to I + F \qquad (c_F I)$$
$$R_{31}: \quad H + F \to H \qquad (b_{FH} HF)$$
$$R_{32}: \quad F \to \varnothing \qquad (a_F F)$$
$$R_{33}: \quad M + E \to M + 2E \qquad (b_{EM} ME)$$
$$R_{34}: \quad I + E \to I \qquad (b_{EI} IE)$$
$$R_{35}: \quad \varnothing \to E \qquad (a_E)$$
$$R_{36}: \quad E \to \varnothing \qquad (-a_E E)$$
$$R_{37}: \quad M + P \to M + 2P \qquad (b_{PM} MP)$$
$$R_{38}: \quad \varnothing \to P \qquad (a_P)$$
$$R_{39}: \quad P \to \varnothing \qquad (-a_P P)$$
$$R_{40}: \quad P \to P + A \qquad (b_A P)$$
$$R_{41}: \quad S + A + V \to S + V \qquad (\gamma_{AV} SAV)$$
$$R_{42}: \quad A \to \varnothing \qquad (a_A A)$$
$$R_{43}: \quad P \to P + S \qquad (rP)$$
$$R_{44}: \quad P + S \to P \qquad (-rPS)$$

Figure A8. Reactions of Hancioglu model [29].

$$R_1: \quad C \to C + V_f \quad ()$$
$$R_2: \quad C + E \to C + E + V_f \quad ()$$
$$R_3: \quad V_f + F \to F \quad ()$$
$$R_4: \quad V_f \to \varnothing \quad ()$$
$$R_5: \quad V_f \to \varnothing \quad ()$$
$$R_6: \quad V_f + C \to 2V_f + C \quad ()$$
$$R_7: \quad m + V_f \to m + 2V_f \quad ()$$
$$R_8: \quad V_f + U \to V_f + C \quad ()$$
$$R_9: \quad V_f + C \to V_f + U \quad ()$$
$$R_{10}: \quad m + V_f + C \to m + V_f + U \quad ()$$
$$R_{11}: \quad C + E \to E + m \quad ()$$
$$R_{12}: \quad C \to M_V \quad ()$$
$$R_{13}: \quad m \to U\varnothing \quad ()$$
$$R_{14}: \quad V_f \to V_f + M_V \quad ()$$
$$R_{15}: \quad M_V \to \varnothing \quad ()$$
$$R_{16}: \quad M_V + H_E \to M_V + 2H_E \quad ()$$
$$R_{17}: \quad m + M_V + H_E \to m + M_V \quad ()$$
$$R_{18}: \quad M_V + H_E \to M_V \quad ()$$
$$R_{19}: \quad M_V + H_E + E \to M_V + E \quad ()$$
$$R_{20}: \quad \varnothing \to H_E \quad ()$$
$$R_{21}: \quad H_E \to \varnothing \quad ()$$
$$R_{22}: \quad M_V + H_B \to M_V + 2H_B \quad ()$$
$$R_{23}: \quad m + M_V + H_B \to m + M_V \quad ()$$
$$R_{24}: \quad M_V + H_B \to M_V \quad ()$$
$$R_{25}: \quad M_V + H_B + B \to M_V + B \quad ()$$
$$R_{26}: \quad \varnothing \to H_B \quad ()$$
$$R_{27}: \quad H_B \to \varnothing \quad ()$$
$$R_{28}: \quad M_V + H_E + E \to M_V + H_E + 2E \quad ()$$
$$R_{29}: \quad m + M_V + H_E + E \to m + M_V + H_E \quad ()$$
$$R_{30}: \quad M_V + H_E + E \to M_V + H_E \quad ()$$
$$R_{31}: \quad C + E \to C \quad ()$$
$$R_{32}: \quad \varnothing \to E \quad ()$$
$$R_{33}: \quad E \to \varnothing \quad ()$$
$$R_{34}: \quad M_V + H_B + B \to M_V + H_B + 2B \quad ()$$
$$R_{35}: \quad m + M_V + H_B + B \to m + M_V + H_B \quad ()$$
$$R_{36}: \quad M_V + H_B + B \to M_V + H_B \quad ()$$
$$R_{37}: \quad \varnothing \to B \quad ()$$
$$R_{38}: \quad B \to \varnothing \quad ()$$
$$R_{39}: \quad M_V + H_B + B \to M_V + H_B + B + P \quad ()$$
$$R_{40}: \quad m + M_V + H_B + B + P \to m + M_V + H_B + B \quad ()$$
$$R_{41}: \quad \varnothing \to P \quad ()$$
$$R_{42}: \quad P \to \varnothing \quad ()$$
$$R_{43}: \quad P \to P + F \quad ()$$
$$R_{44}: \quad V_f + F \to V_f \quad ()$$
$$R_{45}: \quad F \to \varnothing \quad ()$$

Figure A9. Reactions of Bocharov model [30].

$$R_1: \quad \emptyset \to E \qquad ()$$
$$R_2: \quad E \to \emptyset \qquad ()$$
$$R_3: \quad E + V \to V + E_p^* \qquad ()$$
$$R_4: \quad E_p^* + T_E \to T_E \qquad ()$$
$$R_5: \quad E_p^* \to \emptyset \qquad ()$$
$$R_6: \quad E_p^* \to E_p^* + V \qquad ()$$
$$R_7: \quad V \to \emptyset \qquad ()$$
$$R_8: \quad V + A \to A \qquad ()$$
$$R_9: \quad \emptyset \to D \qquad ()$$
$$R_{10}: \quad D \to \emptyset \qquad ()$$
$$R_{11}: \quad D + V \to V + D^* \qquad ()$$
$$R_{12}: \quad D^* \to D^* + D_M \qquad ()$$
$$R_{13}: \quad D^* + D_M \to D^* \qquad ()$$
$$R_{14}: \quad D_M \to \emptyset \qquad ()$$
$$R_{15}: \quad \emptyset \to H_N \qquad ()$$
$$R_{16}: \quad H_N \to \emptyset \qquad ()$$
$$R_{17}: \quad D_M + H_N \to D_M + H_E \qquad ()$$
$$R_{18}: \quad D_M + H_E \to D_M + 2H_E \qquad ()$$
$$R_{19}: \quad D_M + H_E \to D_M \qquad ()$$
$$R_{20}: \quad \emptyset \to T_N \qquad ()$$
$$R_{21}: \quad T_N \to \emptyset \qquad ()$$
$$R_{22}: \quad D_M + T_N \to D_M + T_E \qquad ()$$
$$R_{23}: \quad D_M + T_E \to D_M + 2T_E \qquad ()$$
$$R_{24}: \quad D_M + T_E \to D_M \qquad ()$$
$$R_{25}: \quad \emptyset \to B_N \qquad ()$$
$$R_{26}: \quad B_N \to \emptyset \qquad ()$$
$$R_{27}: \quad D_M + B_N \to D_M + B_A \qquad ()$$
$$R_{28}: \quad D_M + B_A \to D_M + 2B_A \qquad ()$$
$$R_{29}: \quad H_E + B_A \to H_E + 2B_A \qquad ()$$
$$R_{30}: \quad B_A \to \emptyset \qquad ()$$
$$R_{31}: \quad B_A \to P_S \qquad ()$$
$$R_{32}: \quad H_E + B_A \to H_E + P_L \qquad ()$$
$$R_{33}: \quad P_S \to \emptyset \qquad ()$$
$$R_{34}: \quad P_L \to \emptyset \qquad ()$$
$$R_{35}: \quad P_S \to P_S + A \qquad ()$$
$$R_{36}: \quad P_L \to P_L + A \qquad ()$$
$$R_{37}: \quad A \to \emptyset \qquad ()$$

Figure A10. Reactions of Lee model [9].

References

1. Stöhr, K. Influenza—WHO cares. *Lancet Infect. Dis.* **2002**, *2*, 517. [CrossRef]
2. Petrova, V.N.; Russell, C.A. The evolution of seasonal influenza viruses. *Nat. Rev. Microbiol.* **2018**, *16*, 47–60. [CrossRef] [PubMed]

3.   Krammer, F.; Smith, G.J.; Fouchier, R.A.; Peiris, M.; Kedzierska, K.; Doherty, P.C.; Palese, P.; Shaw, M.L.; Treanor, J.; Webster, R.G.; et al. Influenza. *Nat. Rev. Dis. Prim.* **2018**, *4*. [CrossRef] [PubMed]

4.   Smith, A.M.; Perelson, A.S. Influenza A virus infection kinetics: Quantitative data and models. *Wiley Interdiscip. Rev. Syst. Biol. Med.* **2011**, *3*, 429–445. [CrossRef] [PubMed]

5.   Beauchemin, C.A.; Handel, A. A review of mathematical models of influenza A infections within a host or cell culture: Lessons learned and challenges ahead. *BMC Public Health* **2011**, *11* (Suppl. 1), S7. [CrossRef] [PubMed]

6.   Dobrovolny, H.M.; Reddy, M.B.; Kamal, M.A.; Rayner, C.R.; Beauchemin, C.A. Assessing mathematical models of influenza infections using features of the immune response. *PLoS ONE* **2013**, *8*, e57088. [CrossRef] [PubMed]

7.   Boianelli, A.; Nguyen, V.K.; Ebensen, T.; Schulze, K.; Wilk, E.; Sharma, N.; Stegemann-Koniszewski, S.; Bruder, D.; Toapanta, F.R.; Guzman, C.A.; et al. Modeling Influenza Virus Infection: A Roadmap for Influenza Research. *Viruses* **2015**, *7*, 5274–5304. [CrossRef] [PubMed]

8.   Handel, A.; Liao, L.E.; Beauchemin, C.A. Progress and trends in mathematical modelling of influenza A virus infections. *Curr. Opin. Syst. Biol.* **2018**, *12*, 30–36. [CrossRef]

9.   Lee, H.Y.; Topham, D.J.; Park, S.Y.; Hollenbaugh, J.; Treanor, J.; Mosmann, T.R.; Jin, X.; Ward, B.M.; Miao, H.; Holden-Wiltse, J.; et al. Simulation and prediction of the adaptive immune response to influenza A virus infection. *J. Virol.* **2009**, *83*, 7151–7165. [CrossRef]

10. Dittrich, P.; Speroni di Fenizio, P. Chemical Organization Theory. *Bull. Math. Biol.* **2007**, *69*, 1199–1231. [CrossRef]

11. Matsumaru, N.; Centler, F.; di Fenizio, P.S.; Dittrich, P. Chemical organization theory applied to virus dynamics. *IT-Inf. Technol.* **2006**, *48*, 154–160.

12. Peter, S.; Dittrich, P. On the Relation between Organizations and Limit Sets in Chemical Reaction Systems. *Adv. Complex Syst.* **2011**, *14*, 77–96. [CrossRef]

13. Baccam, P.; Beauchemin, C.; Macken, C.A.; Hayden, F.G.; Perelson, A.S. Kinetics of influenza A virus infection in humans. *J. Virol.* **2006**, *80*, 7590–7599. [CrossRef]

14. Miao, H.; Hollenbaugh, J.A.; Zand, M.S.; Holden-Wiltse, J.; Mosmann, T.R.; Perelson, A.S.; Wu, H.; Topham, D.J. Quantifying the early immune response and adaptive immune response kinetics in mice infected with influenza A virus. *J. Virol.* **2010**, *84*, 6687–6698. [CrossRef]

15. Smith, A.M.; Smith, A.P. A Critical, Nonlinear Threshold Dictates Bacterial Invasion and Initial Kinetics During Influenza. *Sci. Rep.* **2016**, *6*, 38703. [CrossRef]

16. Soliman, S.; Heiner, M. A unique transformation from ordinary differential equations to reaction networks. *PLoS ONE* **2010**, *5*, e14284. [CrossRef] [PubMed]

17. Heinrich, R.; Schuster, S. *The Regulation of Cellular Systems*; Springer Science & Business Media: Berlin, Germany, 2012.

18. Fontana, W.; Buss, L.W. "The arrival of the fittest": Toward a theory of biological organization. *Bull. Math. Biol.* **1994**, *56*, 1–64.

19. Kreyssig, P.; Wozar, C.; Peter, S.; Veloz, T.; Ibrahim, B.; Dittrich, P. Effects of small particle numbers on long-term behaviour in discrete biochemical systems. *Bioinformatics* **2014**, *30*, 475–481. [CrossRef] [PubMed]

20. Ibrahim, B. Toward a systems-level view of mitotic checkpoints. *Prog. Biophys. Mol. Biol.* **2015**, *117*, 217–224. [CrossRef]

21. Kreyssig, P.; Escuela, G.; Reynaert, B.; Veloz, T.; Ibrahim, B.; Dittrich, P. Cycles and the qualitative evolution of chemical systems. *PLoS ONE* **2012**, *7*, e45772. [CrossRef]

22. Smith, A.P.; Moquin, D.J.; Bernhauerova, V.; Smith, A.M. Influenza virus infection model with density dependence supports biphasic viral decay. *Front. Microbiol.* **2018**, *9*, 1554. [CrossRef]

23. Pawelek, K.A.; Huynh, G.T.; Quinlivan, M.; Cullinane, A.; Rong, L.; Perelson, A.S. Modeling within-host dynamics of influenza virus infection including immune responses. *PLoS Comput. Biol.* **2012**, *8*, e1002588. [CrossRef] [PubMed]

24. Handel, A.; Longini, I.M., Jr.; Antia, R. Towards a quantitative understanding of the within-host dynamics of influenza A infections. *J. R. Soc. Interface* **2009**, *7*, 35–47. [CrossRef] [PubMed]

25. Handel, A.; Longini, I.M., Jr; Antia, R. Neuraminidase inhibitor resistance in influenza: Assessing the danger of its generation and spread. *PLoS Comput. Biol.* **2007**, *3*, e240. [CrossRef]

26. Hernandez-Vargas, A.E.; Meyer-Hermann, M. Innate immune system dynamics to influenza virus. *IFAC Proc. Vol.* **2012**, *45*, 260–265. [CrossRef]

27. Cao, P.; Yan, A.W.; Heffernan, J.M.; Petrie, S.; Moss, R.G.; Carolan, L.A.; Guarnaccia, T.A.; Kelso, A.; Barr, I.G.; McVernon, J.; et al. Innate Immunity and the Inter-exposure Interval Determine the Dynamics of Secondary Influenza Virus Infection and Explain Observed Viral Hierarchies. *PLoS Comput. Biol.* **2015**, *11*, e1004334. [CrossRef] [PubMed]

28. Saenz, R.A.; Quinlivan, M.; Elton, D.; Macrae, S.; Blunden, A.S.; Mumford, J.A.; Daly, J.M.; Digard, P.; Cullinane, A.; Grenfell, B.T.; et al. Dynamics of influenza virus infection and pathology. *J. Virol.* **2010**, *84*, 3974–3983. [CrossRef] [PubMed]

29. Hancioglu, B.; Swigon, D.; Clermont, G. A dynamical model of human immune response to influenza A virus infection. *J. Theor. Biol.* **2007**, *246*, 70–86. [CrossRef] [PubMed]

30. Bocharov, G.A.; Romanyukha, A.A. Mathematical model of antiviral immune response. III. Influenza A virus infection. *J. Theor. Biol.* **1994**, *167*, 323–360. [CrossRef]

31. Cao, P.; McCaw, J. The mechanisms for within-host influenza virus control affect model-based assessment and prediction of antiviral treatment. *Viruses* **2017**, *9*, 197. [CrossRef]

32. Zitzmann, C.; Kaderali, L. Mathematical Analysis of Viral Replication Dynamics and Antiviral Treatment Strategies: From Basic Models to Age-Based Multi-Scale Modeling. *Front. Microbiol.* **2018**, *9*, 1546. [CrossRef] [PubMed]

33. Mu, C.; Dittrich, P.; Parker, D.; Rowe, J.E. Organisation-Oriented Coarse Graining and Refinement of Stochastic Reaction Networks. *IEEE/ACM Trans. Comput. Biol. Bioinform.* **2018**, *15*, 1152–1166. [CrossRef] [PubMed]

34. Henze, R.; Mu, C.; Puljiz, M.; Kamaleson, N.; Huwald, J.; Haslegrave, J.; di Fenizio, P.S.; Parker, D.; Good, C.; Rowe, J.E.; et al. Multi-scale stochastic organization-oriented coarse-graining exemplified on the human mitotic checkpoint. *Sci. Rep.* **2019**, *9*, 3902. [CrossRef] [PubMed]

Article

Evaluation of Sequencing Library Preparation Protocols for Viral Metagenomic Analysis from Pristine Aquifer Groundwaters

René Kallies [1,*], Martin Hölzer [2,3], Rodolfo Brizola Toscan [1], Ulisses Nunes da Rocha [1], John Anders [1,4], Manja Marz [2,3,5] and Antonis Chatzinotas [1,5]

[1]   Helmholtz Centre for Environmental Research - UFZ, Department of Environmental Microbiology, 04318 Leipzig, Germany; rodolfo.toscan@ufz.de (R.B.T.); ulisses.rocha@ufz.de (U.N.d.R.); johnanders@posteo.de (J.A.); antonis.chatzinotas@ufz.de (A.C.)
[2]   Friedrich Schiller University Jena, RNA Bioinformatics and High-Throughput Analysis, 07743 Jena, Germany; martin.hoelzer@uni-jena.de (M.H.); manja@uni-jena.de (M.M.)
[3]   European Virus Bioinformatics Center, 07743 Jena, Germany
[4]   Bioinformatics Group, Department of Computer Science and Interdisciplinary Center for Bioinformatics, University Leipzig, 04081 Leipzig, Germany
[5]   German Centre for Integrative Biodiversity Research (iDiv) Halle-Jena-Leipzig, 04103 Leipzig, Germany
*   Correspondence: rene.kallies@ufz.de; Tel.: +49-(0341)-235-1375

Received: 31 March 2019; Accepted: 27 May 2019; Published: 28 May 2019

Abstract: Viral ecology of terrestrial habitats is yet-to be extensively explored, in particular the terrestrial subsurface. One problem in obtaining viral sequences from groundwater aquifer samples is the relatively low amount of virus particles. As a result, the amount of extracted DNA may not be sufficient for direct sequencing of such samples. Here we compared three DNA amplification methods to enrich viral DNA from three pristine limestone aquifer assemblages of the Hainich Critical Zone Exploratory to evaluate potential bias created by the different amplification methods as determined by viral metagenomics. Linker amplification shotgun libraries resulted in lowest redundancy among the sequencing reads and showed the highest diversity, while multiple displacement amplification produced the highest number of contigs with the longest average contig size, suggesting a combination of these two methods is suitable for the successful enrichment of viral DNA from pristine groundwater samples. In total, we identified 27,173, 5,886 and 32,613 viral contigs from the three samples from which 11.92 to 18.65% could be assigned to taxonomy using blast. Among these, members of the *Caudovirales* order were the most abundant group (52.20 to 69.12%) dominated by *Myoviridae* and *Siphoviridae*. Those, and the high number of unknown viral sequences, substantially expand the known virosphere.

Keywords: viral metagenome; groundwater; aquifer; AquaDiva; sequencing library preparation

---

1. Introduction

Groundwater systems are important compartments of the global hydrological cycle. They donate about 30% of all freshwater sources [1] and provide important ecosystem services. For example, purification and storage of water, active biodegradation of anthropogenic contaminants and nutrient recycling [2]. Many of these services are directly linked to the presence of microorganisms [2,3]. Studies in particular in marine systems have significantly contributed to a better understanding of viruses and their impacts on the mortality, diversity and genetic landscape of their microbial hosts [4–6]. However, only recently, and only in a limited number of surveys, has the potential role of viruses been explored in terrestrial subsurface systems [7–11].

In theory, metagenomics enables the identification and genomic characterisation of all (micro)organisms present in a sample, including viruses [12]. However, the proportion of viral sequences within a metagenome is usually far lower than for other organisms, leading to limitations in their detection. Especially, pristine aquifers are characterised by low microbial biomass and low abundances of virus particles [9,13,14], which might make their detection even more difficult. Size filtration or density-based enrichment methods are therefore widely used to concentrate virus particles from environmental samples [15,16]. However, a significant obstacle in applying metagenomics for pristine aquifers is the still too low amount of DNA required for the direct sequencing of such samples, making amplification techniques mandatory to further enrich viral nucleic acids. It is however widely known that DNA amplification is a source of bias that may lead to inaccurate conclusions after sequence analysis [17]. Three amplification techniques are commonly used to enrich low amounts of DNA [17,18], i.e., (i) linker amplification shotgun libraries (LASL) [19,20]; (ii) sequence-independent, single-primer amplification (SISPA) [21,22]; and (iii) multiple displacement amplification (MDA) [23,24]. Each method has its own potential source of bias. LASL relies on DNA fragmentation and subsequent linker ligation to blunt-end repaired DNA molecules prior to amplification, using primer oligos that bind to the linker sequences [19,25]. Linker ligation efficiency might be one source of bias, especially for very low amounts of DNA [26]. However, previous studies demonstrated that as little as a few pg to ng of DNA is sufficient for low amplification biases [26,27]. LASL may, in addition, be inefficient in recovering ssDNA viruses due to the double-stranded nature of linker DNA molecules [28] though this has recently been overcome with an adapted LASL protocol [29]. SISPA is built upon the use of pseudo-degenerated primer oligonucleotides, containing a stretch of random nucleotides at their 3′-end and a defined sequence at their 5′-end [21], and has successfully been applied to recover both RNA and DNA virus sequences [22,30]. It has, however, been reported that SISPA has a strong amplification bias resulting in an uneven sequencing read distribution and hence overrepresentation of some genome parts while other parts were completely uncovered. In addition, SISPA negatively affects the detection of low abundant genomes [31]. MDA works under isothermal conditions [32] with very low amounts of input DNA, random hexamer primer oligonucleotides and high fidelity as well as strand displacement functions of the phi29 polymerase [23]. Several sources of bias have been identified for phi29 amplification, including chimera formation [33], discontinuous amplification of linear DNA molecules [34] and preferential amplification of circular ssDNA molecules [35]. Recent studies evaluated different library preparation protocols using low input-DNA to assess the reconstruction of microbial communities from metagenomes [36,37]. Similar studies have been performed for the identification of virus sequences from, for example, seawater and human samples [17,35]. Despite these advances, to our knowledge no study has to date assessed and benchmarked sequence library preparation protocols for the identification of viral sequences from pristine aquifer groundwaters.

The Hainich Critical Zone Exploratory (Hainich CZE) in central Germany is an infrastructure designed to, among others, investigate the diversity, identity and abundance of microorganisms in the Hainich aquifers. In addition, analysis of metabolic potential and activities of microorganisms will be linked to physico-chemical parameters in spatial and temporal scales [38]. Here we sampled three carbonate-rock aquifer assemblages of the Hainich CZE, which represent a pristine and uncontaminated aquifer [38]. One problem in obtaining viral sequences from groundwater samples is the low amount of DNA (usually a pico- to few nanograms) that was extracted from isolated virus particles. The aim of this project is therefore two-fold. The first aim is to evaluate different DNA amplification techniques that may offer a sufficient amount of DNA for high-throughput sequencing. In addition, these methods should have a low amplification bias to reflect the natural diversity of the analysed samples. The second aim consisted of evaluating different viral sequence recovery tools to provide first insights on which viruses are present in the Hainich CZE groundwater aquifers.

2. Materials and Methods

2.1. Sample Collection

Groundwater samples were taken from three Hainich CZE aquifer wells in Thuringia, Germany, within the framework of the Collaborative Research Centre AquaDiva (http://www.aquadiva.uni-jena.de) (CRC 1076) [38]. The sampling site was located in the agriculturally used midslope and footslope regions of the Hainich low-mountain range. The three wells were drilled to depths of 50 m (H53), 65 m (H52) and 88 m (H51). H53 and H52 reflect anoxic conditions while oxic conditions prevailed for H51. A detailed description of hydrochemical and geostructural parameters can be found elsewhere [39,40].

Ten liters of groundwater (with approximately 2.3×10^5 (SD: 1.2×10^4) viral particles per milliliter) were collected from each well during a sampling campaign in May 2015. Water was filtered through 200 nm pore filters using a cross-flow system (Sartorius, Göttingen, Germany). Samples were then enriched for viral particles by filtration through 35 kDa filters using the same system. Approximately 60 mL were retained and further concentrated by ultracentrifugation at $22,000 \times g$ for 2 h and 4 °C. The viral particle containing pellet was resuspended in 500 µL TM buffer (50 mM Tris HCl, 10 mM Magnesium sulfate at pH 7.5). One volume of chloroform was given to the samples to remove microsized prokaryotes. The upper phase, intended for DNA extraction, was treated with DNase I to remove free DNA.

2.2. DNA Extraction, Library Construction and Sequencing

Viral DNA was extracted as described previously [20]. Viral DNA concentration was determined using the Qubit® dsDNA HS Assay Kit (Thermo Fisher Scientific, Waltham, MA, USA) resulting in total DNA amounts of 31.8 ng (H51), 5.4 ng (H52) and 25.9 ng (H53). DNA was divided into four parts to prepare four libraries for each sample. Non-amplified shotgun libraries (NASL): using a Covaris ultrasonicator, DNA was sheared to approx. 350 bp fragments and libraries were prepared with a TruSeq DNA PCR-Free Library Prep Kit (Illumina, San Diego, CA, USA) according to the manufacturer's instructions. Linker amplification shotgun libraries (LASL): DNA was sheared to approx. 350 bp fragments as mentioned above and LASL was performed with a NEBNext Ultra DNA Lib Prep Kit (New England Biolabs, Ipswich, MA, USA) as recommended by the manufacturer including 12 PCR cycles to enrich adaptor-ligated DNA. Single-primer amplification (SISPA): PCR was performed by ten cycles using random octamer primers that were linked to a specific primer sequence followed by amplification using a 1:9 mixture of random octamers and a primer targeting the specific primer sequence as described previously [41]. Multiple displacement amplification (MDA): DNA was subjected to phi29 amplification at 25 °C for 8 h using the illustra GenomiPhi V2 DNA Amplification Kit (Thermo Fisher Scientific) as described in the manual. PCR amplicons for the latter two libraries were purified using the Sureclean reagent (Bioline, Luckenwalde, Germany), fragmented as described above and libraries were prepared as described for NASL. Sequencing was performed on one lane of an Illumina HiSeq 2500 system to generate 100-bp paired-end reads.

2.3. Sequencing Read Processing and Assembly

PhiX contaminants were removed, SISPA primer sequences were clipped and raw sequencing reads were quality checked using Trimmomatic [42] and low-quality bases were trimmed from both ends. Reads were screened with a 4-base wide sliding window until the remaining sequences had a Phred-score of at least 15 and a minimum length of 36 nt. Sequencing read redundancy was identified by clustering at 90% sequencing read identity using CD-hit v.4.6 [43,44].

Sequencing reads were independently assembled for each sampling site and library preparation using metaSPAdes [45,46] and SOAPdenovo-Trans [47]. In addition, cross-assemblies were performed for each sampling site including all reads from LASL, SISPA and MDA libraries. We used the transcriptome assembler SOAPdenovo-Trans in addition to SPAdes because recent analyses revealed this assembly tool as very efficient for the assembly of RNA virus genomes [48]. Further analyses suggested this might be also true for the assembly of DNA virus genomes.

2.4. Viral Contig Recovery

Three viral sequence identification tools were used to recover viral contigs, i.e., VirSorter [49], VirFinder [50] and VrAP (https://www.rna.uni-jena.de/research/software/vrap-viral-assembly-pipeline/). VirSorter is based on the identification of viral hallmark genes, enrichment in hypothetical proteins and other viral signatures [49]. Only contigs identified as VirSorter categories 1 and 2 (higher confidence predictions) were retained for further analysis. VirFinder is a kmer based tool for the identification of viral contigs from metagenomes with improvements especially for the detection of short viral contigs [50]. Contigs with a *p*-value < 0.01 were used for further analysis. These two detection tools were completed by using VrAP, a novel de novo genome assembly pipeline especially designed for viruses. The pipeline is able to assemble complete genomes of viruses representing new strains and species, as well as prototypes of new genera and families. VrAP is based on the genome assembler SPAdes [45] combined with an additional read correction [51,52] and several filter steps. The pipeline classifies the contigs to distinguish host from viral sequences by annotation and open reading frame (ORF) density scores. By applying the ORF density method we were able to identify potential novel viruses without any sequence homology to known references (manuscript in preparation).

2.5. Virome Diversity Measures and Comparison of Library Preparation Methods

Nonpareil [53–55] was used with default settings to estimate diversity and coverage of virome datasets. Viral reads present in one or more datasets reflecting LASL, SISPA and MDA per sampling site were identified as follows. Redundancy was removed for each dataset by CD-hit-est clustering at 95% identity. A database was created containing all viral contigs and, using Bowtie2 [56], read cluster per library preparation method and sampling site were mapped to the database. Mapped clusters were extracted, counted and overlapping information were generated using SAMtools [57]. Viral contigs were compared between sites by an all-versus-all clustering approach (95% identity) with CD-hit-est-2D [44].

Venn diagrams were computed in R [58] using the package "venneuler" (https://cran.r-project. org/web/packages/venneuler/index.html).

2.6. Viral Taxonomic Assignment

All viral contigs per sampling site, i.e., contigs identified from all virus identification tools and library preparation methods, were combined (resulting in three datasets) and redundancies were removed by clustering with CD-hit-est at 95% nt identity. Open reading frames (ORFs) were translated from these contigs using prodigal [59] and aligned to a viral RefSeq protein database (February 2019) using DELTA BLAST [60] with an e-value cut off of 10^{-3}. Hits were sorted by *e*-value and bit score and ORFs with most significant hits were aligned to the respective contigs using an in-house python script (Supplementary Information), resulting in one hit per contig. Gene sharing networks based on shared protein clusters (PCs) between viral genomes were calculated with vConTACT2 [61,62] on the iVirus platform [63] and were displayed with Cytoscape [64]. DNA contamination from cellular organisms was determined using EMIRGE [65].

2.7. Data Availability

Sequence read raw data have been made available at Sequence Read Archive accession: PRJNA530103.

3. Results

3.1. Raw Sequencing Output Statistics

The first aim of this study was to evaluate different DNA amplification techniques that may result in a sufficient amount of viral DNA for high throughput sequencing. We therefore compared three DNA amplification methods, i.e., LASL, SISPA and MDA. NASL was used as control.

MDA produced highest (quality trimmed) sequencing read numbers followed by SISPA and LASL as compared to NASL that exhibited lowest read numbers (Table S1). Significant differences (ANOVA) in quality-trimmed sequencing output were observed between NASL-MDA, NASL-SISPA and SISPA-MDA (Table 1).

Table 1. *P*-values of analysis of variance (ANOVA) of raw sequencing read and read cluster numbers between the different library preparation methods.

	Number of Raw Reads			
Library Preparation	NASL	LASL	SISPA	MDA
NASL	n/a	>0.05	0.008	0.002
LASL		n/a	>0.05	0.023
SISPA			n/a	>0.05
MDA				n/a

	Clusters at 90% Read Identity							
	Relative proportion				Number of clusters			
Library Preparation	NASL	LASL	SISPA	MDA	NASL	LASL	SISPA	MDA
NASL	n/a	<0.001	>0.05	>0.05	n/a	0.018	>0.05	0.008
LASL		n/a	<0.001	<0.001		n/a	>0.05	>0.05
SISPA			n/a	>0.05			n/a	>0.05
MDA				n/a				n/a

NASL: non-amplified shotgun library; LASL: linker amplification shotgun libraries; SISPA: single-primer amplification; MDA: multiple displacement amplification.

Read quality of all libraries was >97% except for libraries H51 LASL and H52 LASL for which 33.78% and 28.13% of the reads were discarded after quality trimming. However, no significant differences (ANOVA) in quality between any of the library preparation methods was observed.

PCR amplification bias may influence the evenness among sequencing reads. For example, GC-rich primers and primers with GC-stretches at their 3′-end, both present in a random primer mix, may anneal more efficiently to a target sequence than AT-rich primer oligos do. As a result, amplicons amplified from such target sequences may be favored during the amplification process what in turn leads to high numbers of identical or related DNA molecules. We therefore clustered all quality-trimmed sequencing reads with a 90% cut-off to remove this redundancy. LASL libraries produced the lowest redundancy (41.7 to 60.7% relative proportion of clusters to sequencing read numbers), with significant differences not only to non-amplified libraries (10.8 to 21.3 relative proportion of clusters to sequencing read numbers) but MDA libraries (9.1 to 17.0% relative proportion of clusters to sequencing read numbers) and SISPA libraries (5.9 to 7.4 relative proportion of clusters to sequencing read numbers) (Table 1). These data suggest an amplification bias during PCR with random primer oligomers. However, MDA libraries (together with LASL libraries) still resulted highest average numbers of read clusters (Table S1). The presence of many repetitive and homopolymeric sequencing reads (possibly sequencing artefacts) may explain the low proportion of clustered reads in NASL libraries.

3.2. Data Set Comparison

We used Nonpareil, i.e., a kmer based approach that examines the degree of overlap among individual sequence reads, to determine redundancy [53–55] among the individual reads to further assess the average coverage created from the different library data sets. NASL, SISPA and MDA libraries seem to reach a nearly full coverage while LASL libraries vary between ~20 to 80% coverage (Figure 1). However, diversity among libraries increased from NASL to SISPA, MDA and LASL, the latter being the most diverse libraries (Figure 1). These results strongly indicate the target discrimination of SISPA and MDA during PCR that results in uneven coverage of the viral metagenomes and in addition, may fail to target low abundant sequences. NASL sequencing reads dominantly consisted of repetitive and homopolymeric sequences (see also below), with most likely too low an input of DNA explaining the observed Nonpareil curve for these libraries.

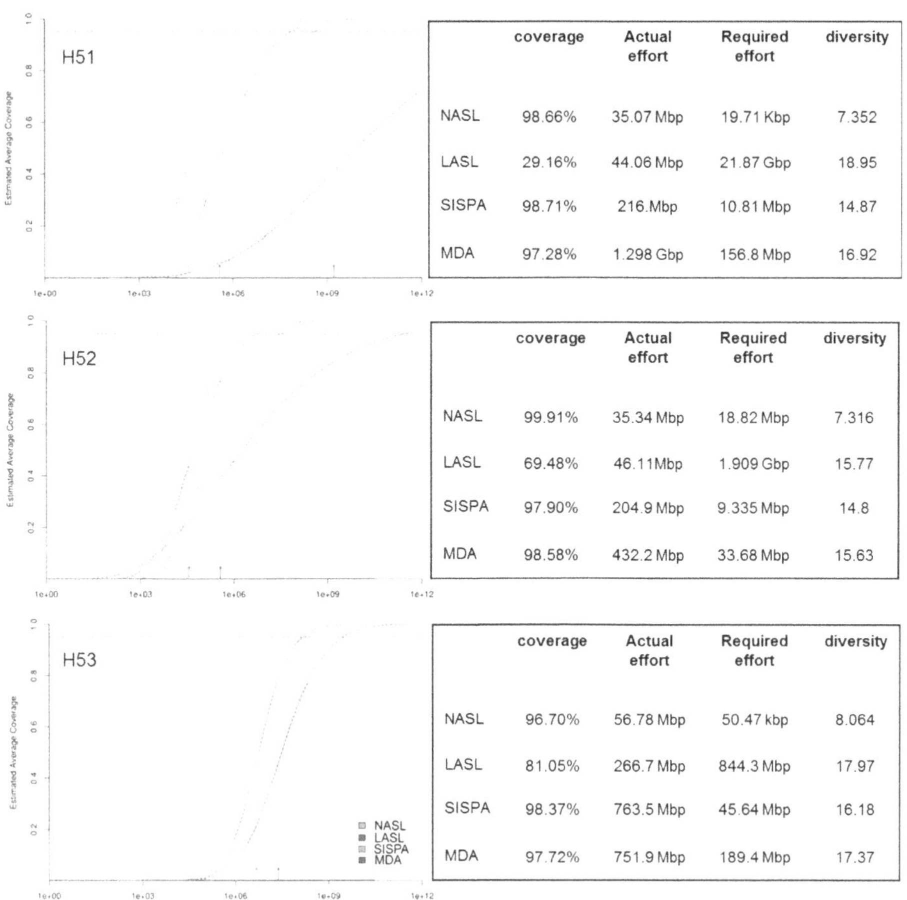

Figure 1. Comparison of Hainich groundwater viromes diversity and coverage as function of sequencing effort using Nonpareil curves [53–55]. Estimated coverage is shown as dotted lines, true coverage as solid lines. Estimated diversity is shown with arrows on the *x*-axis. Horizontal dotted line shows 95% coverage. Viral metagenome coverage, actual sequencing effort, required sequencing effort and kmer-based diversity for each library are shown in the right panel.

We were further interested in both, the number of viral reads that were exclusively detected by one of the library preparation methods and those reads that were identified from more than one library preparation method. For this, redundancy removed reads (i.e., reads that clustered at 90% identity) of LASL, SISPA and MDA libraries were independently mapped to viral contigs per individual sampling site (i.e., all viral contigs that were identified by the three virus identification tools and cross-assemblies) and counted. MDA libraries produced most reads (average: 350 k) followed by LASL (average: 143 k). Least reads were identified from SISPA libraries (average: 64 k). Overlapping information (reads found in more than one library) was rather low with 0.27 to 4.64% of reads present in all three libraries while 0.07 to 14.77% of reads were identified by two libraries (Figure 2). These data indicate target sequence discrimination between each of the library preparation methods.

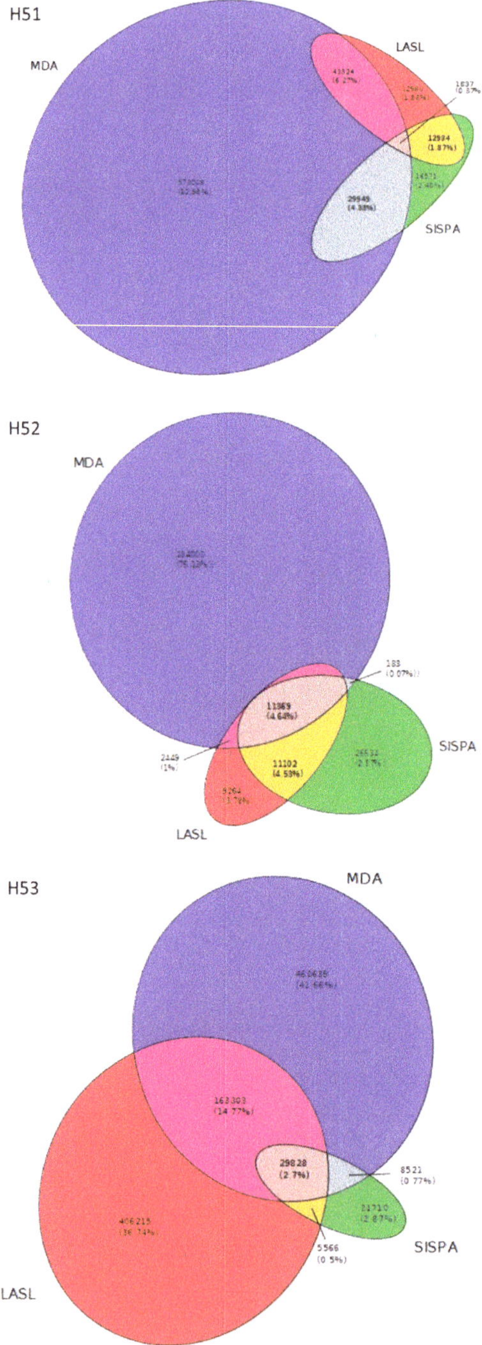

Figure 2. Overlap of sequencing read cluster (90% identity) information identified by library preparation methods, independently shown for each sampling site. Non amplified sequencing libraries (NASL) were not included in the analysis due to the homopolymeric and repetitive nature of sequences obtained from these libraries.

3.3. Assembly Statistics and Evaluation of Viral Identification Tools

Using both, SPAdes and SOAPdenovo-Trans, assemblies from non-amplified libraries completely failed due to repetitive and homopolymeric sequences. We therefore excluded these datasets from further analysis. Contig numbers tend to be higher for LASL and MDA libraries than for SISPA libraries (LASL-SISPA: $p = 0.062$, SISPA-MDA: $p = 0.087$; statistical test: one-way ANOVA) when assembled with SPAdes. Similar results were observed for SOAPdenovo-Trans assemblies (LASL-SISPA: $p = 0.052$, SISPA-MDA: $p = 0.028$; statistical test: one-way ANOVA). In addition, MDA library assemblies produced longer contigs (N50) when compared to LASL and SISPA ($p = 0.001$ (SPAdes), $p < 0.001$ (SOAPdenovo-Trans); statistical test: one-way ANOVA) (Figure 3, Table S2). A comparison (student's t-test) of the two assembly tools showed that SOAPdenovo-Trans may tend to produce longer contigs ($p = 0.084$), while there is no significant difference in the average contig size (N50) ($p = 0.2972$).

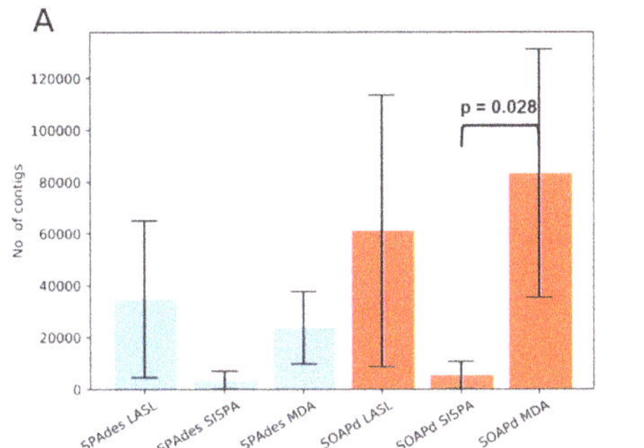
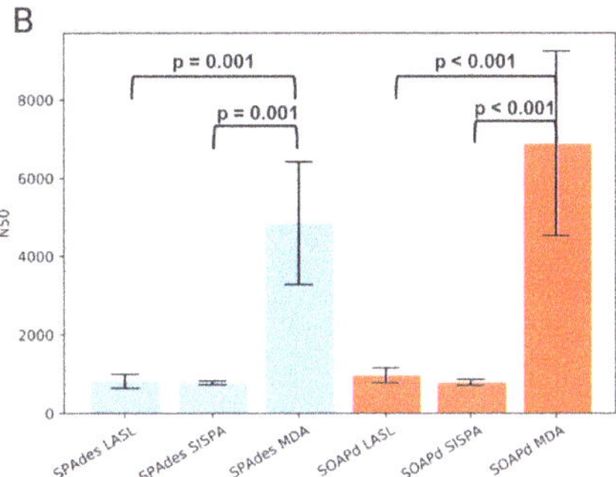

Figure 3. Number of contigs (**A**) and N50 (**B**) produced by sequence library preparation methods and assembly tools. Differences between library preparation methods were tested using analysis of variance (ANOVA). SOAPd: SOAPdenovo-Trans, NASL: non-amplified shotgun library, LASL: linker amplification shotgun libraries, SISPA: single-primer amplification, MDA: multiple displacement amplification.

Viral contigs (as identified by VirSorter, VirFinder and VrAP obtained from cross-assemblies) were clustered at 95% identity to determine a core set of sequences among the sampling sites. Only 37 contigs (0.5%) were shared by the three viromes indicating there is at least a minor common core set in the groundwater aquifers. The amount of shared contigs increased from 0.85% (HS1 and HS2) and 1.04% (HS2 and HS3) to 2.85% (HS1 and HS3) when two viromes were compared. However, the majority of viral contigs is exclusive for the respective virome (Figure 4). The overall viral contig number from HS2 is rather low compared to HS1 and HS3, most likely due to the lower amount of DNA extracted from this sample. This might explain the lower contig overlap of HS2 with HS1 and HS3, respectively, than the overlap of HS1 and HS3.

We used three different viral sequence identification tools that are based on the detection of viral hallmark genes (VirSorter), kmer distribution (VirFinder) and orf density (VrAP) (see more detailed description in the Materials and Methods section). VirFinder and VrAP significantly identified a higher number of viral contigs than VirSorter (One-way ANOVA $p < 0.001$). The size of viral contigs obtained by VirSorter and VirFinder were in contrast significantly higher than for VrAP (one-way ANOVA $p < 0.05$). However, each tool identified viral contigs that were not recognized by the other two revealing an advantage in the use of several identification tools for the recovery of viral sequences.

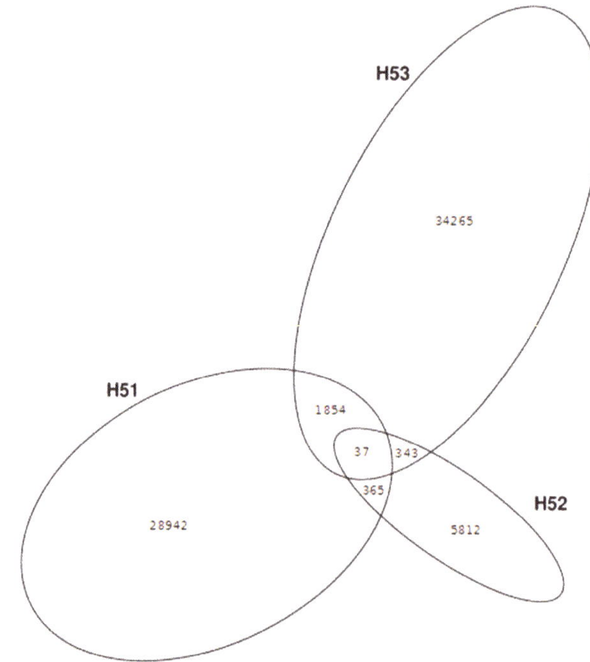

Figure 4. The venn diagram presents numbers of unique and shared viral contigs among the different viromes. Cross-assembled viral contigs (as identified by VirSorter, VirFinder and VrAP) were compared between sites by an all-versus-all clustering approach (95% identity) with CD-hit-est-2D [44].

3.4. First Insights into Viral Taxonomic Composition of Hainich Groundwater

Using cross-assembled contigs (assemblies including sequencing reads from LASL, SISPA and MDA per sampling site) and a set of three viral sequence recovery tools, we identified 27,173 (H51), 5,886 (H52) and 32,613 (H53) viral contigs from the Hainich groundwater samples (Figure 3; Table S3). These contigs were assembled from 31.19% (H51), 52.08% (H52) and 28.41% (H53) of the quality trimmed sequence reads. Among these reads we identified 19 Small subunit ribosomal ribonucleic acid sequences (8 bacterial 16S, 11 unclassified) demonstrating a low contamination with DNA from cellular organisms. Only 14.81% (H51), 18,65% (H52) and 11.92% (H53) of the viral populations could be assigned to taxonomy using delta-blast (Figure 5A). Most of them were assigned to dsDNA viruses dominated by the order *Caudovirales* (H51: 69.12%, H52: 58.20%, H53: 62.40%). Within the *Caudovirales*, members of the *Myoviridae* (40 to 46.5%) and *Siphoviridae* (40.7 to 41.9%) families were most abundant (Figure 5B, Figure S1). These findings are not surprising since *Caudovirales* have previously been presented as the most abundant group of viruses in environmental ecosystems [8,66,67]. Other identified dsDNA virus sequences belonged, for example, to the amoeba infecting giant virus families *Marseilleviridae* and *Mimiviridae*, to the algae infecting *Phycodnaviridae* family whose hosts has been shown to be present in groundwater [68], and invertebrate-infecting viruses such as *Iridoviridae* and *Poxviridae* (Figure 5B). Surprisingly, we identified only a small number of circular ssDNA viruses (Figure 5B). These viruses have been revealed as an abundant group in other environments [69,70]. We used Phi29 polymerase in MDA that preferentially amplifies circular ssDNA [35] and one could expect a bias towards overrepresentation of circular ssDNA genomes. Although this study is only a first snapshot into the Hainich groundwater virome we speculate that circular ssDNA viruses are rare in this environment. A small fraction of these DNA viromes was assigned to RNA viruses, most likely due to PCR errors and incomplete/erroneous virus reference databases.

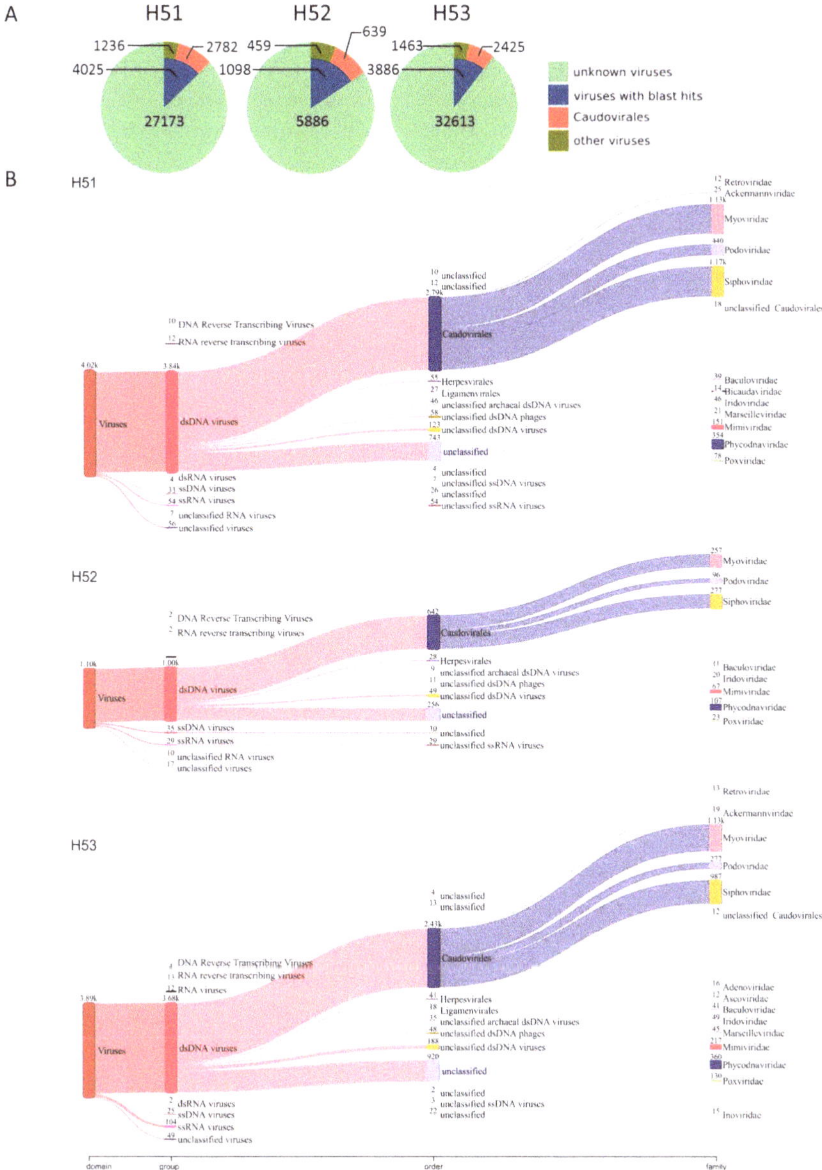

Figure 5. Taxonomic assignment of viral contigs identified from cross-assemblies. (**A**) pie charts present relative and absolute abundance of viral contigs after blastp analysis. (**B**) Taxonomic profile of viral contigs as classified by blastp (viral contigs with blast hits in figure **A**). Data were visualized with Pavian [71].

However, a high number of blast-based taxonomy assigned contigs could not be affiliated to deeper taxonomic levels but have similarity to unclassified viruses present in the viral RefSeq database (Figure 5B). These findings, together with the huge number of unknown viral contigs (without any blast hit) reveal substantial genomic and taxonomic diversity in Hainich groundwater viromes, as observed also in other environments [66,72]. To further investigate the similarity of Hainich groundwater viromes to viral RefSeq database, we used a genome-based network analysis of their shared protein content (Figure 6) [61,62]. This analysis groups viral contigs at the approximately genus level into viral clusters [61,62,73]. In total, 539 viral clusters were identified. Of those, Hainich viral contigs were found in 191 clusters, 183 of them were exclusive to Hainich viromes, among those 95 clusters exclusive to H51, 8 clusters to H52 and 23 clusters to H53. In addition, approximately 34% (H51), 64% (H52) and 63% (H53) of viral protein clusters were present in at least one other Hainich groundwater sample, suggesting some sequence conservation across these samples.

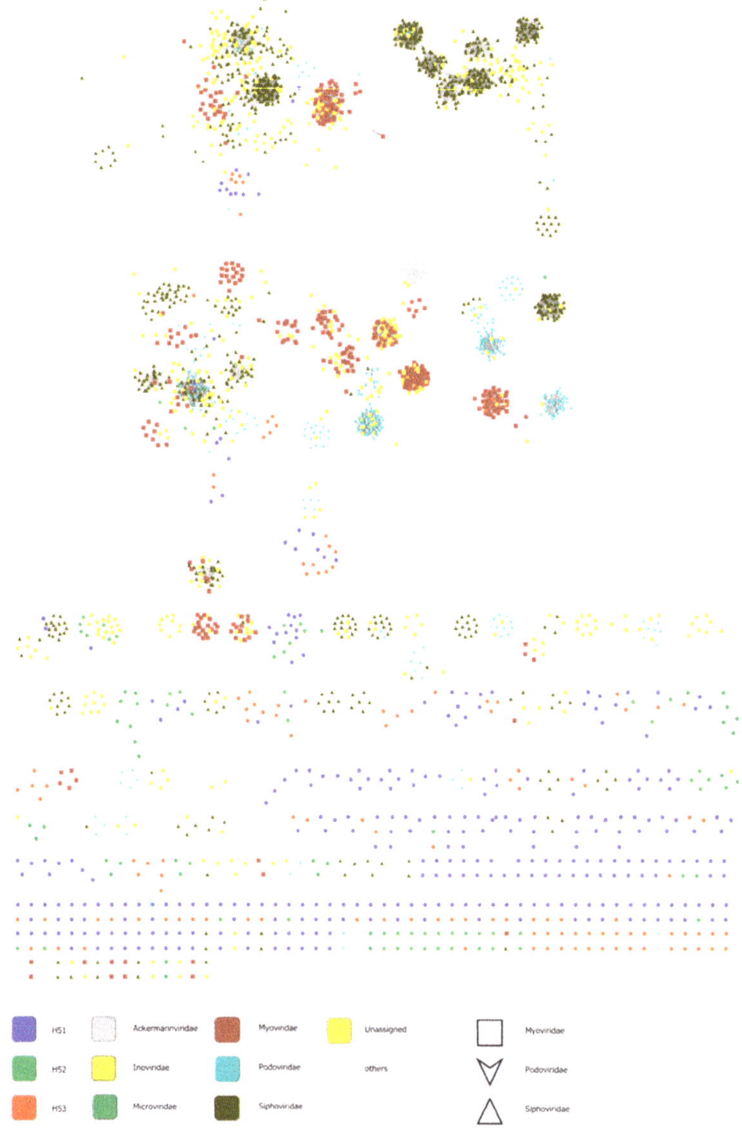

Figure 6. A network analysis of shared predicted protein content between viral RefSeq database and Hainich viral populations. Nodes (circles) indicate contigs and shared edges (lines) indicate shared protein content. Data were analysed using vConTACT2 [61,62] and displayed with cytoscape [64].

4. Discussion

Viruses play a key role in ecosystems, with most of them infecting microbes. They directly affect their hosts by lysis and horizontal gene transfer, and hence are responsible for changes in microbial community structure and composition what in turn has consequences on biogeochemical cycles and food web structures [4–6,74]. Viral metagenomics has been increasingly used to unravel viral community composition and interactions with their hosts from different ecosystems, such as marine environments and soil [66,67,72]. The terrestrial subsurface including groundwater ecosystems is at present yet underexplored [7,8,10,11,75]. A common problem is the relatively low biomass present in these difficult to obtain samples, which in return, results in only low amounts of DNA not sufficient for standard preparation of metagenome sequencing libraries [9,13,14]. Efforts have been undertaken to overcome this problem, including DNA enrichment using different DNA amplification techniques [17–19,22]. Each of these methods has its own advantages and limitations making it difficult to provide a standard protocol. Benchmark tests should therefore be performed when investigating new sample types.

Sampling procedure, virus particle isolation and nucleic acid extraction protocols are potential sources of bias [17] that have to be considered prior to sampling. Here, we focused on non-enveloped DNA viruses that passed a pore size of 200 nm after filtration and performed a benchmark study to find a method of choice to enrich viral DNA that is sufficient for sequencing. We furthermore intended to get a first snapshot of the viruses present in Hainich groundwater aquifers.

We used three DNA amplification methods, i.e., LASL, SISPA and MDA to compare one another and with NASL, using three groundwater samples. Although NASL resulted in some sequencing output none of the reads could be used for further analysis (assembly, virus sequence identification) due to their repetitive and homopolymeric nature; demonstrating that direct sequencing of NASL is not feasible with low DNA amounts. According to the Nonpareil curves, LASL was the method with the lowest amplification bias since the curves were located rightward in the plots indicating a higher diversity than for SISPA and MDA (Figure 1). Nonpareil curves for SISPA and MDA simulate a nearly full sequence coverage that emerge from redundant sequence information (Table 1, Table S1). False sequence coverage interpretation could be a result when data analysis exclusively rely on these library preparation methods. In addition, LASL resulted in the highest number of unique sequencing reads as compared to SISPA and MDA. MDA on the other hand outperformed LASL and SISPA in terms of viral contig numbers and their average contig size. In addition, MDA performed (at least in two samples) much better for taxonomic assignments in the case of *Caudovirales* families, which were dominant among the viral contigs with taxonomic affiliation (Figure S1). Considering the amount of unique viral reads per method and their low overlap (Figure 2), together with the results from cross-assemblies, it became apparent that none of the here tested DNA enrichments methods could completely detect viral sequences from pristine groundwater. However, SISPA even underperformed in terms of sequencing output, diversity and assembly statistics. Metagenomic benchmark studies using both, microbial mock communities and marine samples demonstrated the use of Mondrian and Illumina Nextera XT technologies produced high quality metagenomes from even femtogram-input DNA libraries [36,37]. These library preparation methods are comparable with the LASL protocol used in this study because all these methods use linker ligation on fragmented or tagmented DNA prior to amplification for generation of sequencing libraries. The low bias introduced by LASL on virus enriched groundwater samples from our work is consistent with these previous studies on prokaryotic metagenomes. In addition, other studies on viromes from marine and human samples showed substantial differences with respect to diversity, assembly output, types and ratio of viral sequences between LASL and MDA [18] and an outperformance of MDA over SISPA [17]. However, these studies observed an overrepresentation of circular sequences in MDA libraries as compared to LASL and SISPA. In contrast, our data identified only a few contigs that belong to circular ssDNA viruses (see also discussion below). We therefore suggest the combined use of LASL and MDA for future analysis of viral communities from pristine groundwater aquifers.

SOAPdenovo-Trans produced more contigs than SPAdes. However, average contig size was similar (Figure 3, Table S2). A combination of the assembly output seems to produce most comprehensive results but might also introduce unnecessary redundancy. Assembly for metagenomic data is already difficult, but appear to be more complex for viruses with their possibly more uneven genome coverage. Specialized tools are needed for the (de novo) assembly of viral sequences from metagenomic data [76]. The lower number of contigs for H52 could be a result of the lower amount of DNA extracted from this sample as compared to H51 and H53. Future studies will reveal whether there is a correlation between input DNA amount and contig numbers, including replicates and different yields of DNA input.

There is a high number of virus identification tools available, with all of them having their limitations [77]. We decided to use VirSorter [49], VirFinder [50] and VrAP. The latter two do not rely on database matches, increasing the chance to detect novel viruses not related to those present in public databases. Using our dataset, each tool exclusively identified some viral contigs demonstrating a combination of different virus identification tools increases the number of recovered viral contigs as also suggested previously [59,78]. However, the number of viral contigs was lower than the total number of contigs (compare Table S2 and Table S3). The experimental procedure included several steps to enrich virus particles, i.e., size filtration, chloroform treatment to remove most small-sized bacteria and digestion of free DNA that is not protected by a protein shell. Although some non-viral sequences might still be present after such methodology, one could assume the majority of the dataset consists of viral sequences and consequently includes a high number of viral contigs not recovered by one of the detection tools. Efforts should be undertaken, e.g., using machine learning, to overcome these likely limitations [78,79].

Like in many environmental studies, the taxonomy of most viral contigs remained unknown as demonstrated by blast and network analysis (Figure 5, Figure 6) [8,10,66]. Members of the order *Caudovirales* were dominating among viral contigs with taxonomic assignment. This group of tailed viruses infects a wide variety of bacteria and has been shown as one major group present in environmental ecosystems [8,66,72,80,81]. Another group of commonly highly abundant viruses, i.e., circular ssDNA viruses of the families *Microviridae* and *Circoviridae* [69,70,82], were almost entirely absent in our dataset. This is in contrast to previous results from groundwater aquifers where these viruses even dominated over dsDNA viruses among the classified sequences [10]. A technical bias seems to be unlikely since MDA is known for preferential amplification of these target sequences [35]. Future analyses including spatial and temporal variation will reveal whether these viruses are rare in pristine groundwater. We further identified viruses infecting algae, invertebrates and microeukaryotes, among the latter, contigs similar to giant viruses from the *Mimiviridae* family. These viruses should, by default, not be detected after 200 nm pore size filtration. A possible explanation could be sequence similarity of conserved mimivirus orfs, such as polymerases, to yet unknown viruses [83,84].

We show viral metagenome libraries can be produced from pristine aquifer groundwaters and suggest a combination of LASL and MDA to enrich viral DNA from these samples and to diminish an amplification bias that may occur during enrichment. We further identified new viral sequences that will help to understand the role of viruses in pristine groundwaters.

Supplementary Materials: The following are available online at http://www.mdpi.com/1999-4915/11/6/484/s1, Figure S1: Bubble plot shows families of the Caudovirales order present in each virome., Table S1: Overview of raw sequencing read pair numbers, sequence quality and sequencing read clusters for each prepared sequencing library., Table S2: Assembly statistics for LASL, SISPA and MDA libraries per sampling site., Table S3: Overview of identified viral contigs as per virus identification tool, assembly software, sequencing library and sampling site., Supplementary information: Python script to assign orfs to contigs.

Author Contributions: Designed the experiments: R.K., M.H., M.M. and A.C.; performed the experiments: R.K. and M.H.; data analysis: R.K., M.H., R.B.T., U.N.d.R. and J.A.; prepared the manuscript: R.K., M.H., U.N.d.R., M.M. and A.C. All authors read and approved the final manuscript.

Funding: This research was funded by Deutsche Forschungsgemeinschaft, grant number CRC 1076 "AquaDiva" and grant number SPP-1596 "Ecology and species barriers in emerging viral diseases".

Acknowledgments: We are very grateful to Nicole Steinbach Anett Heidtmann for excellent technical assistance, to Robert Lehmann, Friedrich-Schiller-Universität Jena, who helped with sampling and to Ben Bolduc, Ohio State University, for his help in using vConTACT2. We further thank Ivonne Görlich and Marco Groth from the Core Facility DNA sequencing of the Leibniz Institute on Aging - Fritz Lipmann Institute in Jena for their help with DNA sequencing. MH appreciates the support of the Joachim Herz Foundation by the add-on fellowship for interdisciplinary life science.

Conflicts of Interest: The authors declare no conflict of interest.

References

1. Danielopol, D.L.; Pospisil, P.; Rouch, R. Biodiversity in groundwater: A large-scale view. *Trends Ecol. Evol.* **2000**, *15*, 223–224. [CrossRef]

2. Griebler, C.; Avramov, M. Groundwater ecosystem services: A review. *Freshw. Sci.* **2015**, *34*, 355–367. [CrossRef]

3. Griebler, C.; Lueders, T. Microbial biodiversity in groundwater ecosystems. *Freshw. Biol.* **2009**, *54*, 649–677. [CrossRef]

4. Suttle, C.A. Viruses in the sea. *Nature* **2005**, *437*, 356. [CrossRef] [PubMed]

5. Suttle, C.A. Marine viruses — major players in the global ecosystem. *Nat. Rev. Microbiol.* **2007**, *5*, 801. [CrossRef] [PubMed]

6. Breitbart, M. Marine Viruses: Truth or Dare. *Annu. Rev. Mar. Sci.* **2011**, *4*, 425–448. [CrossRef]

7. Daly, R.A.; Borton, M.A.; Wilkins, M.J.; Hoyt, D.W.; Kountz, D.J.; Wolfe, R.A.; Welch, S.A.; Marcus, D.N.; Trexler, R.V.; MacRae, J.D.; et al. Microbial metabolisms in a 2.5-km-deep ecosystem created by hydraulic fracturing in shales. *Nat. Microbiol.* **2016**, *1*, 16146. [CrossRef]

8. Daly, R.A.; Roux, S.; Borton, M.A.; Morgan, D.M.; Johnston, M.D.; Booker, A.E.; Hoyt, D.W.; Meulia, T.; Wolfe, R.A.; Hanson, A.J.; et al. Viruses control dominant bacteria colonizing the terrestrial deep biosphere after hydraulic fracturing. *Nat. Microbiol.* **2019**, *4*, 352–361. [CrossRef]

9. Kyle, J.E.; Eydal, H.S.C.; Ferris, F.G.; Pedersen, K. Viruses in granitic groundwater from 69 to 450 m depth of the Äspö hard rock laboratory, Sweden. *ISME J.* **2008**, *2*, 571. [CrossRef]

10. Smith, R.J.; Jeffries, T.C.; Roudnew, B.; Seymour, J.R.; Fitch, A.J.; Simons, K.L.; Speck, P.G.; Newton, K.; Brown, M.H.; Mitchell, J.G. Confined aquifers as viral reservoirs. *Environ. Microbiol. Rep.* **2013**, *5*, 725–730. [CrossRef]

11. Pan, D.; Watson, R.; Wang, D.; Tan, Z.H.; Snow, D.D.; Weber, K.A. Correlation between viral production and carbon mineralization under nitrate-reducing conditions in aquifer sediment. *ISME J.* **2014**, *8*, 1691–1703. [CrossRef]

12. Wooley, J.C.; Ye, Y. Metagenomics: Facts and Artifacts, and Computational Challenges. *J. Comput. Sci. Technol.* **2009**, *25*, 71–81. [CrossRef] [PubMed]

13. Wilhartitz, I.C.; Kirschner, A.K.T.; Brussaard, C.P.D.; Fischer, U.R.; Wieltschnig, C.; Stadler, H.; Farnleitner, A.H. Dynamics of natural prokaryotes, viruses, and heterotrophic nanoflagellates in alpine karstic groundwater. *Microbiol. Open* **2013**, *2*, 633–643. [CrossRef] [PubMed]

14. Roudnew, B.; Lavery, T.J.; Seymour, J.R.; Smith, R.J.; Mitchell, J.G. Spatially varying complexity of bacterial and virus-like particle communities within an aquifer system. *Aquat. Microb. Ecol.* **2013**, *68*, 259–266. [CrossRef]

15. Ruby, J.G.; Bellare, P.; Derisi, J.L. PRICE: Software for the targeted assembly of components of (Meta) genomic sequence data. *G3 Bethesda Md* **2013**, *3*, 865–880. [CrossRef] [PubMed]

16. Rose, R.; Constantinides, B.; Tapinos, A.; Robertson, D.L.; Prosperi, M. Challenges in the analysis of viral metagenomes. *Virus Evol.* **2016**, *2*, vew022. [CrossRef] [PubMed]

17. Parras-Moltó, M.; Rodríguez-Galet, A.; Suárez-Rodríguez, P.; López-Bueno, A. Evaluation of bias induced by viral enrichment and random amplification protocols in metagenomic surveys of saliva DNA viruses. *Microbiome* **2018**, *6*, 119. [CrossRef] [PubMed]

18. Kim, K.-H.; Bae, J.-W. Amplification methods bias metagenomic libraries of uncultured single-stranded and double-stranded DNA viruses. *Appl. Environ. Microbiol.* **2011**, *77*, 7663–7668. [CrossRef] [PubMed]

19. Breitbart, M.; Salamon, P.; Andresen, B.; Mahaffy, J.M.; Segall, A.M.; Mead, D.; Azam, F.; Rohwer, F. Genomic analysis of uncultured marine viral communities. *Proc. Natl. Acad. Sci. USA* **2002**, *99*, 14250–14255. [CrossRef] [PubMed]

20. Thurber, R.V.; Haynes, M.; Breitbart, M.; Wegley, L.; Rohwer, F. Laboratory procedures to generate viral metagenomes. *Nat. Protoc.* **2009**, *4*, 470–483. [CrossRef]

21. Froussard, P. A random-PCR method (rPCR) to construct whole cDNA library from low amounts of RNA. *Nucleic Acids Res.* **1992**, *20*, 2900. [CrossRef] [PubMed]

22. Djikeng, A.; Halpin, R.; Kuzmickas, R.; Depasse, J.; Feldblyum, J.; Sengamalay, N.; Afonso, C.; Zhang, X.; Anderson, N.G.; Ghedin, E.; et al. Viral genome sequencing by random priming methods. *BMC Genom.* **2008**, *9*, 5. [CrossRef] [PubMed]

23. Dean, F.B.; Nelson, J.R.; Giesler, T.L.; Lasken, R.S. Rapid amplification of plasmid and phage DNA using Phi 29 DNA polymerase and multiply-primed rolling circle amplification. *Genome Res.* **2001**, *11*, 1095–1099. [CrossRef] [PubMed]

24. Angly, F.E.; Felts, B.; Breitbart, M.; Salamon, P.; Edwards, R.A.; Carlson, C.; Chan, A.M.; Haynes, M.; Kelley, S.; Liu, H.; et al. The marine viromes of four oceanic regions. *PLoS Biol.* **2006**, *4*, e368. [CrossRef] [PubMed]

25. Henn, M.R.; Sullivan, M.B.; Stange-Thomann, N.; Osburne, M.S.; Berlin, A.M.; Kelly, L.; Yandava, C.; Kodira, C.; Zeng, Q.; Weiand, M.; et al. Analysis of high-throughput sequencing and annotation strategies for phage genomes. *PloS ONE* **2010**, *5*, e9083. [CrossRef] [PubMed]

26. Solonenko, S.A.; Ignacio-Espinoza, J.C.; Alberti, A.; Cruaud, C.; Hallam, S.; Konstantinidis, K.; Tyson, G.; Wincker, P.; Sullivan, M.B. Sequencing platform and library preparation choices impact viral metagenomes. *BMC Genom.* **2013**, *14*, 320. [CrossRef] [PubMed]

27. Duhaime, M.B.; Deng, L.; Poulos, B.T.; Sullivan, M.B. Towards quantitative metagenomics of wild viruses and other ultra-low concentration DNA samples: A rigorous assessment and optimization of the linker amplification method. *Environ. Microbiol.* **2012**, *14*, 2526–2537. [CrossRef] [PubMed]

28. Székely, A.J.; Breitbart, M. Single-stranded DNA phages: From early molecular biology tools to recent revolutions in environmental microbiology. *FEMS Microbiol. Lett.* **2016**, *363*. [CrossRef]

29. Roux, S.; Solonenko, N.E.; Dang, V.T.; Poulos, B.T.; Schwenck, S.M.; Goldsmith, D.B.; Coleman, M.L.; Breitbart, M.; Sullivan, M.B. Towards quantitative viromics for both double-stranded and single-stranded DNA viruses. *PeerJ* **2016**, *4*, e2777. [CrossRef]

30. Drexler, J.F.; Corman, V.M.; Müller, M.A.; Maganga, G.D.; Vallo, P.; Binger, T.; Gloza-Rausch, F.; Cottontail, V.M.; Rasche, A.; Yordanov, S.; et al. Bats host major mammalian paramyxoviruses. *Nat. Commun.* **2012**, *3*, 796. [CrossRef]

31. Karlsson, O.E.; Belák, S.; Granberg, F. The Effect of Preprocessing by Sequence-Independent, Single-Primer Amplification (SISPA) on Metagenomic Detection of Viruses. *Biosecurity Bioterrorism Biodefense Strategy Pract. Sci.* **2013**, *11*, S227–S234. [CrossRef] [PubMed]

32. Blanco, L.; Bernad, A.; Lázaro, J.M.; Martín, G.; Garmendia, C.; Salas, M. Highly efficient DNA synthesis by the phage phi 29 DNA polymerase. Symmetrical mode of DNA replication. *J. Biol. Chem.* **1989**, *264*, 8935–8940. [PubMed]

33. Lasken, R.S.; Stockwell, T.B. Mechanism of chimera formation during the Multiple Displacement Amplification reaction. *BMC Biotechnol.* **2007**, *7*, 19. [CrossRef] [PubMed]

34. Zhang, K.; Martiny, A.C.; Reppas, N.B.; Barry, K.W.; Malek, J.; Chisholm, S.W.; Church, G.M. Sequencing genomes from single cells by polymerase cloning. *Nat. Biotechnol.* **2006**, *24*, 680. [CrossRef] [PubMed]

35. Kim, K.-H.; Chang, H.-W.; Nam, Y.-D.; Roh, S.W.; Kim, M.-S.; Sung, Y.; Jeon, C.O.; Oh, H.-M.; Bae, J.-W. Amplification of uncultured single-stranded DNA viruses from rice paddy soil. *Appl. Environ. Microbiol.* **2008**, *74*, 5975–5985. [CrossRef] [PubMed]

36. Rinke, C.; Low, S.; Woodcroft, B.J.; Raina, J.-B.; Skarshewski, A.; Le, X.H.; Butler, M.K.; Stocker, R.; Seymour, J.; Tyson, G.W.; et al. Validation of picogram- and femtogram-input DNA libraries for microscale metagenomics. *PeerJ* **2016**, *4*, e2486. [CrossRef]

37. Bowers, R.M.; Clum, A.; Tice, H.; Lim, J.; Singh, K.; Ciobanu, D.; Ngan, C.Y.; Cheng, J.-F.; Tringe, S.G.; Woyke, T. Impact of library preparation protocols and template quantity on the metagenomic reconstruction of a mock microbial community. *BMC Genom.* **2015**, *16*, 856. [CrossRef]

38. Küsel, K.; Totsche, K.U.; Trumbore, S.E.; Lehmann, R.; Steinhäuser, C.; Herrmann, M. How Deep Can Surface Signals Be Traced in the Critical Zone? Merging Biodiversity with Biogeochemistry Research in a Central German Muschelkalk Landscape. *Front. Earth Sci.* **2016**, *4*, 32. [CrossRef]

39. Kohlhepp, B.; Lehmann, R.; Seeber, P.; Küsel, K.; Trumbore, S.E.; Totsche, K.U. Aquifer configuration and geostructural links control the groundwater quality in thin-bedded carbonate–siliciclastic alternations of the Hainich CZE, central Germany. *Hydrol. Earth Syst. Sci.* **2017**, *21*, 6091–6116. [CrossRef]

40. Kumar, S.; Herrmann, M.; Thamdrup, B.; Schwab, V.F.; Geesink, P.; Trumbore, S.E.; Totsche, K.-U.; Küsel, K. Nitrogen Loss from Pristine Carbonate-Rock Aquifers of the Hainich Critical Zone Exploratory (Germany) Is Primarily Driven by Chemolithoautotrophic Anammox Processes. *Front. Microbiol.* **2017**, *8*, 1951. [CrossRef]

41. Palacios, G.; Quan, P.-L.; Jabado, O.J.; Conlan, S.; Hirschberg, D.L.; Liu, Y.; Zhai, J.; Renwick, N.; Hui, J.; Hegyi, H. Panmicrobial oligonucleotide array for diagnosis of infectious diseases. *Emerg. Infect. Dis.* **2007**, *13*, 73. [CrossRef] [PubMed]

42. Bolger, A.M.; Lohse, M.; Usadel, B. Trimmomatic: A flexible trimmer for Illumina sequence data. *Bioinformatics* **2014**, *30*, 2114–2120. [CrossRef] [PubMed]

43. Li, W.; Godzik, A. Cd-hit: A fast program for clustering and comparing large sets of protein or nucleotide sequences. *Bioinformatics* **2006**, *22*, 1658–1659. [CrossRef] [PubMed]

44. Fu, L.; Niu, B.; Zhu, Z.; Wu, S.; Li, W. CD-HIT: Accelerated for clustering the next-generation sequencing data. *Bioinformatics* **2012**, *28*, 3150–3152. [CrossRef] [PubMed]

45. Bankevich, A.; Nurk, S.; Antipov, D.; Gurevich, A.A.; Dvorkin, M.; Kulikov, A.S.; Lesin, V.M.; Nikolenko, S.I.; Pham, S.; Prjibelski, A.D. SPAdes: A new genome assembly algorithm and its applications to single-cell sequencing. *J. Comput. Biol.* **2012**, *19*, 455–477. [CrossRef]

46. Nurk, S.; Meleshko, D.; Korobeynikov, A.; Pevzner, P.A. metaSPAdes: A new versatile metagenomic assembler. *Genome Res.* **2017**, *27*, 824–834. [CrossRef]

47. Xie, Y.; Wu, G.; Tang, J.; Luo, R.; Patterson, J.; Liu, S.; Huang, W.; He, G.; Gu, S.; Li, S. SOAPdenovo-Trans: De novo transcriptome assembly with short RNA-Seq reads. *Bioinformatics* **2014**, *30*, 1660–1666. [CrossRef]

48. Hölzer, M.; Marz, M. De novo transcriptome assembly: A comprehensive cross-species comparison of short-read RNA-Seq assemblers. *GigaScience* **2019**, *8*, giz039. [CrossRef]

49. Roux, S.; Enault, F.; Hurwitz, B.L.; Sullivan, M.B. VirSorter: Mining viral signal from microbial genomic data. *PeerJ* **2015**, *3*, e985. [CrossRef]

50. Ren, J.; Ahlgren, N.A.; Lu, Y.Y.; Fuhrman, J.A.; Sun, F. VirFinder: A novel k-mer based tool for identifying viral sequences from assembled metagenomic data. *Microbiome* **2017**, *5*, 69. [CrossRef]

51. Song, L.; Florea, L.; Langmead, B. Lighter: Fast and memory-efficient sequencing error correction without counting. *Genome Biol.* **2014**, *15*, 509. [CrossRef] [PubMed]

52. Magoč, T.; Salzberg, S.L. FLASH: Fast length adjustment of short reads to improve genome assemblies. *Bioinformatics* **2011**, *27*, 2957–2963. [CrossRef] [PubMed]

53. Rodriguez-r, L.M.; Konstantinidis, K.T. Nonpareil: A redundancy-based approach to assess the level of coverage in metagenomic datasets. *Bioinformatics* **2013**, *30*, 629–635. [CrossRef] [PubMed]

54. Rodriguez, L.M.; Konstantinidis, K.T. Estimating coverage in metagenomic data sets and why it matters. *ISME J.* **2014**, *8*, 2349. [CrossRef] [PubMed]

55. Rodriguez-R, L.M.; Gunturu, S.; Tiedje, J.M.; Cole, J.R.; Konstantinidis, K.T. Nonpareil 3: Fast Estimation of Metagenomic Coverage and Sequence Diversity. *MSystems* **2018**, *3*, e00039-18.

56. Langmead, B.; Salzberg, S.L. Fast gapped-read alignment with Bowtie 2. *Nat. Methods* **2012**, *9*, 357–359. [CrossRef]

57. Li, H.; Handsaker, B.; Wysoker, A.; Fennell, T.; Ruan, J.; Homer, N.; Marth, G.; Abecasis, G.; Durbin, R. The sequence alignment/map format and SAMtools. *Bioinformatics* **2009**, *25*, 2078–2079. [CrossRef] [PubMed]

58. RStudio Team. RStudio: Integrated Development for R. Available online: http://www.rstudio.com (accessed on 27 May 2019).

59. Hyatt, D.; Chen, G.-L.; LoCascio, P.F.; Land, M.L.; Larimer, F.W.; Hauser, L.J. Prodigal: Prokaryotic gene recognition and translation initiation site identification. *BMC Bioinformatics* **2010**, *11*, 119. [CrossRef]

60. Boratyn, G.M.; Schäffer, A.A.; Agarwala, R.; Altschul, S.F.; Lipman, D.J.; Madden, T.L. Domain enhanced lookup time accelerated BLAST. *Biol. Direct* **2012**, *7*, 12. [CrossRef]

61. Bolduc, B.; Jang, H.B.; Doulcier, G.; You, Z.-Q.; Roux, S.; Sullivan, M.B. vConTACT: An iVirus tool to classify double-stranded DNA viruses that infect Archaea and Bacteria. *PeerJ* **2017**, *5*, e3243. [CrossRef]

62. Bin Jang, H.; Bolduc, B.; Zablocki, O.; Kuhn, J.H.; Roux, S.; Adriaenssens, E.M.; Brister, J.R.; Kropinski, A.M.; Krupovic, M.; Lavigne, R.; et al. Taxonomic assignment of uncultivated prokaryotic virus genomes is enabled by gene-sharing networks. *Nat. Biotechnol.* **2019**. [CrossRef] [PubMed]

63. Bolduc, B.; Youens-Clark, K.; Roux, S.; Hurwitz, B.L.; Sullivan, M.B. iVirus: Facilitating new insights in viral ecology with software and community data sets imbedded in a cyberinfrastructure. *Isme J.* **2016**, *11*, 7. [CrossRef] [PubMed]

64. Shannon, P.; Markiel, A.; Ozier, O.; Baliga, N.S.; Wang, J.T.; Ramage, D.; Amin, N.; Schwikowski, B.; Ideker, T. Cytoscape: A software environment for integrated models of biomolecular interaction networks. *Genome Res.* **2003**, *13*, 2498–2504. [CrossRef] [PubMed]

65. Miller, C.S.; Baker, B.J.; Thomas, B.C.; Singer, S.W.; Banfield, J.F. EMIRGE: Reconstruction of full-length ribosomal genes from microbial community short read sequencing data. *Genome Biol.* **2011**, *12*, R44. [CrossRef]

66. Emerson, J.B.; Roux, S.; Brum, J.R.; Bolduc, B.; Woodcroft, B.J.; Jang, H.B.; Singleton, C.M.; Solden, L.M.; Naas, A.E.; Boyd, J.A.; et al. Host-linked soil viral ecology along a permafrost thaw gradient. *Nat. Microbiol.* **2018**, *3*, 870–880. [CrossRef] [PubMed]

67. Roux, S.; Brum, J.R.; Dutilh, B.E.; Sunagawa, S.; Duhaime, M.B.; Loy, A.; Poulos, B.T.; Solonenko, N.; Lara, E.; Poulain, J.; et al. Ecogenomics and potential biogeochemical impacts of globally abundant ocean viruses. *Nature* **2016**, *537*, 689. [CrossRef] [PubMed]

68. Reisser, W. The Hidden Life of Algae Underground. In *Algae and Cyanobacteria in Extreme Environments*; Seckbach, J., Ed.; Springer Netherlands: Dordrecht, The Netherlands, 2007; pp. 47–58, ISBN 978-1-4020-6112-7.

69. Rosario, K.; Duffy, S.; Breitbart, M. Diverse circovirus-like genome architectures revealed by environmental metagenomics. *J. Gen. Virol.* **2009**, *90*, 2418–2424. [CrossRef]

70. Tucker, K.P.; Parsons, R.; Symonds, E.M.; Breitbart, M. Diversity and distribution of single-stranded DNA phages in the North Atlantic Ocean. *Isme J.* **2010**, *5*, 822. [CrossRef]

71. Breitwieser, F.P.; Salzberg, S.L. Pavian: Interactive analysis of metagenomics data for microbiomics and pathogen identification. *bioRxiv* **2016**. [CrossRef]

72. Paez-Espino, D.; Eloe-Fadrosh, E.A.; Pavlopoulos, G.A.; Thomas, A.D.; Huntemann, M.; Mikhailova, N.; Rubin, E.; Ivanova, N.N.; Kyrpides, N.C. Uncovering Earth's virome. *Nature* **2016**, *536*, 425. [CrossRef]

73. Roux, S.; Hallam, S.J.; Woyke, T.; Sullivan, M.B. Viral dark matter and virus–host interactions resolved from publicly available microbial genomes. *eLife* **2015**, *4*, e08490. [CrossRef] [PubMed]

74. Wegner, C.-E.; Gaspar, M.; Geesink, P.; Herrmann, M.; Marz, M.; Küsel, K. Biogeochemical Regimes in Shallow Aquifers Reflect the Metabolic Coupling of the Elements Nitrogen, Sulfur, and Carbon. *Appl. Environ. Microbiol.* **2019**, *85*, e02346-18. [CrossRef] [PubMed]

75. Anderson, R.E.; Brazelton, W.J.; Baross, J.A. Is the genetic landscape of the deep subsurface biosphere affected by viruses? *Front. Microbiol.* **2011**, *2*, 219. [CrossRef] [PubMed]

76. Hölzer, M.; Marz, M. Chapter Nine—Software Dedicated to Virus Sequence Analysis "Bioinformatics Goes Viral." In *Advances in Virus Research*; Beer, M., Höper, D., Eds.; Academic Press: Cambridge, MA, USA, 2017; Volume 99, pp. 233–257, ISBN 0065-3527.

77. Nooij, S.; Schmitz, D.; Vennema, H.; Kroneman, A.; Koopmans, M.P.G. Overview of Virus Metagenomic Classification Methods and Their Biological Applications. *Front. Microbiol.* **2018**, *9*, 749. [CrossRef] [PubMed]

78. Hurwitz, B.L.; Ponsero, A.; Thornton, J.; U'Ren, J.M. Phage hunters: Computational strategies for finding phages in large-scale 'omics datasets. *Virus Res.* **2018**, *244*, 110–115. [CrossRef] [PubMed]

79. Bzhalava, Z.; Tampuu, A.; Bała, P.; Vicente, R.; Dillner, J. Machine Learning for detection of viral sequences in human metagenomic datasets. *BMC Bioinform.* **2018**, *19*, 336. [CrossRef] [PubMed]

80. Wommack, K.E.; Colwell, R.R. Virioplankton: Viruses in Aquatic Ecosystems. *Microbiol. Mol. Biol. Rev.* **2000**, *64*, 69. [CrossRef] [PubMed]

81. Hurwitz, B.L.; Sullivan, M.B. The Pacific Ocean Virome (POV): A Marine Viral Metagenomic Dataset and Associated Protein Clusters for Quantitative Viral Ecology. *PLoS ONE* **2013**, *8*, e57355. [CrossRef]

82. Roux, S.; Krupovic, M.; Poulet, A.; Debroas, D.; Enault, F. Evolution and diversity of the Microviridae viral family through a collection of 81 new complete genomes assembled from virome reads. *PLoS ONE* **2012**, *7*, e40418. [CrossRef]

83. Earl, P.L.; Jones, E.V.; Moss, B. Homology between DNA polymerases of poxviruses, herpesviruses, and adenoviruses: Nucleotide sequence of the vaccinia virus DNA polymerase gene. *Proc. Natl. Acad. Sci. USA* **1986**, *83*, 3659–3663. [CrossRef]
84. Villarreal, L.P.; DeFilippis, V.R. A Hypothesis for DNA Viruses as the Origin of Eukaryotic Replication Proteins. *J. Virol.* **2000**, *74*, 7079. [CrossRef] [PubMed]

Article

Purple: A Computational Workflow for Strategic Selection of Peptides for Viral Diagnostics Using MS-Based Targeted Proteomics

Johanna Lechner [1,†], Felix Hartkopf [1,†], Pauline Hiort [1], Andreas Nitsche [2], Marica Grossegesse [2], Joerg Doellinger [3], Bernhard Y. Renard [1,*] and Thilo Muth [1]

[1] Bioinformatics Unit (MF 1), Department for Methods Development and Research Infrastructure, Robert Koch Institute, 13353 Berlin, Germany; LechnerJ@rki.de (J.L.); HartkopfF@rki.de (F.H.); p.hiort@web.de (P.H.); MuthT@rki.de (T.M.)
[2] Centre for Biological Threats and Special Pathogens, Highly Pathogenic Viruses (ZBS1), Robert Koch Institute, 13353 Berlin, Germany; NitscheA@rki.de (A.N.); GrossegesseM@rki.de (M.G.)
[3] Centre for Biological Threats and Special Pathogens, Proteomics and Spectroscopy (ZBS 6), Robert Koch Institute, 13353 Berlin, Germany; DoellingerJ@rki.de (J.D.)
* Correspondence: RenardB@rki.de; Tel.: +49-(0)30-18754-2561
† These authors contributed equally to this work.

Received: 19 March 2019; Accepted: 4 June 2019; Published: 8 June 2019

Abstract: Emerging virus diseases present a global threat to public health. To detect viral pathogens in time-critical scenarios, accurate and fast diagnostic assays are required. Such assays can now be established using mass spectrometry-based targeted proteomics, by which viral proteins can be rapidly detected from complex samples down to the strain-level with high sensitivity and reproducibility. Developing such targeted assays involves tedious steps of peptide candidate selection, peptide synthesis, and assay optimization. Peptide selection requires extensive preprocessing by comparing candidate peptides against a large search space of background proteins. Here we present Purple (Picking unique relevant peptides for viral experiments), a software tool for selecting target-specific peptide candidates directly from given proteome sequence data. It comes with an intuitive graphical user interface, various parameter options and a threshold-based filtering strategy for homologous sequences. Purple enables peptide candidate selection across various taxonomic levels and filtering against backgrounds of varying complexity. Its functionality is demonstrated using data from different virus species and strains. Our software enables to build taxon-specific targeted assays and paves the way to time-efficient and robust viral diagnostics using targeted proteomics.

Keywords: virus proteomics; mass spectrometry; virus diagnostics; data analysis; targeted proteomics; peptide selection; parallel reaction monitoring

---

1. Introduction

Virus infections present serious health threats to millions of individuals worldwide. For public health, the accurate detection of pathogenic viruses is time-critical because reducing the time for diagnosis and treatment lowers the risk of disease transmission and patient mortality. Fast and robust diagnostic assays are therefore required to rapidly detect re-emerging and newly emerging viruses (e.g., Influenza, Ebola, Zika, or Hepatitis C virus). These diagnostic methods need to cover a broad spectrum of potentially disease-causing viral agents.

Classical diagnostic strategies for detecting viral infection can be divided into two different categories: on the one hand, virus detection can be established by targeted methods, such as agent-specific polymerase chain reaction (PCR) or immunological techniques. On the other hand, detection approaches exist that provide an open view, such as electron microscopy or next-generation sequencing (NGS). Besides their unbiased view, the latter methods have the advantage of identifying multiple pathogens in a single experimental run. Due to its specificity (hybridization and sequencing) and sensitivity (qPCR), the detection of nucleic acids is the gold standard in diagnostics. Conversely, the detection of viral proteins is used less frequently in diagnostic settings and is usually based on interaction with affine binding molecules such as antibodies or aptamers. However, producing these binding molecules is generally time-consuming and laborious, as is the validation of their specificity.

While in clinical microbiology the analysis of subproteomes (<12 kDa) using matrix assisted laser desorption/ionization-time of flight (MALDI-TOF) mass spectrometry (MS) has become a standard method for the identification of bacteria and fungi, no comparable proteomic approach exists in virology for technical reasons [1]. In recent years, MS-based targeted proteomics has evolved into a technique for detecting proteins in complex samples with high sensitivity, quantitative accuracy, and reproducibility [2,3]. Targeted proteomics is commonly used to test hypotheses on a subset of proteins of interest, in contrast to discovery shotgun proteomics. The latter provides global proteome profiling of thousands of proteins in a sample, however, at the expense of sensitivity and reproducibility. Unlike discovery methods, targeted methods of selected/multiple reaction monitoring (SRM/MRM) [4] and parallel reaction monitoring (PRM) [5] nowadays allow for detecting and analyzing preselected proteins and peptides in sensitive, specific, and time-efficient manner. Furthermore, the development of targeted proteomics assays has become easier in the past few years, owing to the advances of analytical methods, instrumental capabilities, and computational workflows [6].

Targeted MS-based proteomics assay development typically involves (i) peptide candidate selection, (ii) peptide synthesis, and (iii) assay optimization. This procedure now enables the transfer of a process highly similar to the design of multiplex PCRs to the proteome level for detecting pathogens. While MS-based targeted assays have not been used for detecting viruses in any diagnostic setting yet, promising findings could already be achieved for identifying and quantifying pathogenic bacterial species. For example, targeted proteomics methods were successfully used in previous studies on *Streptococcus pyogenes* [7] and *Mycobacterium tuberculosis* [8].

Although targeted proteomics has gained much popularity with many use cases in experimental research by now, relatively few research-oriented algorithms and software tools have been developed that support the user-defined selection of peptides for designing targeted SRM or PRM assays. In this context, Skyline [9] is a powerful and widely used software for designing targeted proteomics assays. Besides its wide applicability to different targeted methods and its intuitive use, it also has some internal limitations: first, Skyline is dependent on the operating system Windows, and can therefore not be used under a Linux cluster server environment, and second, it does perform only exact string matching during the peptide selection process without considering any homologies between related organisms. PeptidePicker [10] is a web-based workflow to select peptides by providing, amongst further options, the protein accession number and was designed for human and mouse proteomes. PeptideManager [11] is a tool developed to select peptide candidates as protein surrogates from a defined proteome. It was optimized for the use case of xenografts, i.e., human tumors orthotopically implanted into a different species. While this software allows for constructing a peptide database from any species-specific proteome, sequence homologies, and multiple taxonomic levels are disregarded. Picky [12]—a web-based method designer for targeted assays—only provides support for human and mouse sequences, while it relies on synthetic peptide data from the human-focused ProteomeTools project [13,14]. In the context of targeted metaproteomics, the Unique Peptide Finder of the UniPept web application [15] was developed to select unique peptides for user-defined taxa. Furthermore, various computational tools have been developed to predict proteotypic peptides for targeted proteomics experiments [16–18]. These methods often make use of machine learning training setups that incorporate the probability of observing a peptide in a standard proteomics analysis, referred to as peptide detectability [19] or observability [20], and commonly involve physicochemical properties of the proteins to select high-responding peptides [21]. To our best knowledge, however, no software tool is currently available to select taxon-specific peptides for targeted proteomics assays that also accounts for sequence homologies between different species or strain proteomes. Effectively considering homologies is crucial for accurate taxon-specific diagnostics, because proteins measured in virus samples frequently have a high sequence similarity either in closely related strains or due to highly conserved functional domains.

Here we present Purple (Picking unique relevant peptides for viral experiments), a platform-independent software that returns a set of taxon-specific peptides, after the user has specified the viral target (i.e., a particular virus species or genus), as candidates for targeted proteomics experiments. Equipped with a user-friendly graphical user interface and a threshold-based filtering strategy for homologous sequences, it simplifies the design of MS-based targeted proteomics assays for the end user. Purple enables peptide candidate selection and considers background sequence information, i.e., proteins that are not related to a specific virus target, at various taxonomic levels. Thus, all peptide candidates are validated against a user-defined database of virus proteomes. While the design of MS-based targeted assays requires further steps, our software greatly facilitates the cumbersome, yet important task of peptide selection and thereby paves the way to time-efficient and robust pathogen screening and viral diagnostics. Purple is open source software available at https://gitlab.com/rki_bioinformatics/Purple.

2. Materials and Methods

2.1. Purple Workflow

Purple is implemented in Python (version 3.6) and makes use of additional Python libraries such as tqdm (https://github.com/tqdm/tqdm) for process bar calculation and Biopython [22] to calculate the molecular weight of peptides. Purple is available as portable standalone version that already includes all required libraries or Purple can be installed using pip or conda, which are managing dependencies. The workflow of Purple is depicted in an overview diagram (Figure 1). Purple requires the input of protein sequence databases and a configuration file. The databases are automatically rearranged into a target and a background database. The "exact matching" step is used to remove exact sequence matches with the background from the target peptide set. The remaining target peptides are used to detect and remove homologous peptides. A result file containing the final unique peptides is created together with various intermediate result files. These are outputs of all Purple processing tasks, namely (i) digested peptides, (ii) exact matching peptides, (iii) non-homologous matching peptides and (iv) background shared peptides.

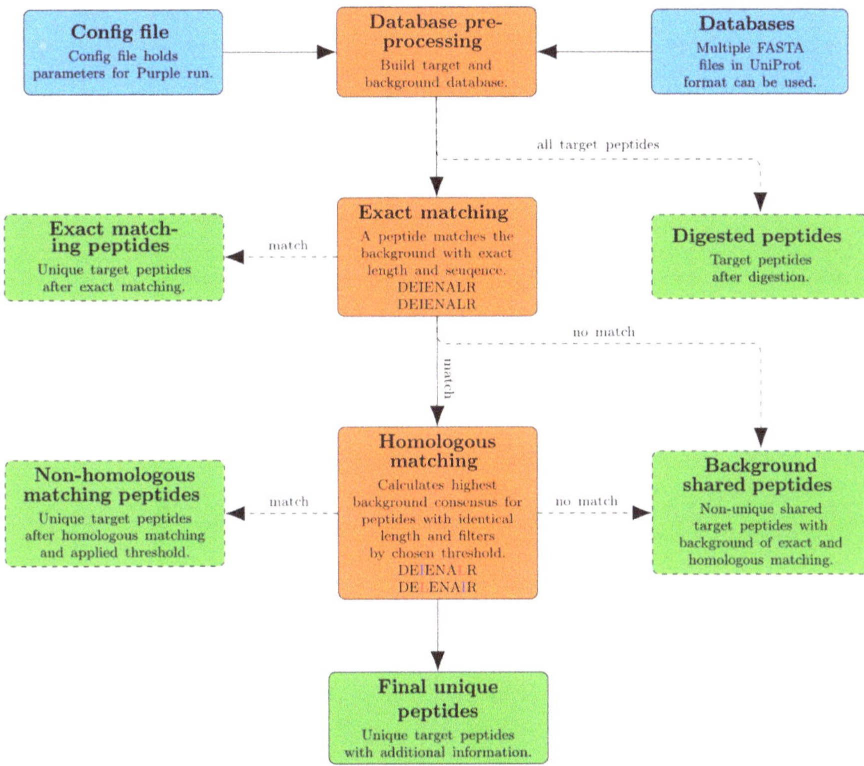

Figure 1. Overview of the Purple workflow. A configuration file and a directory path to the location of FASTA databases serve as input (blue). In the database preprocessing step, the databases are separated into target and background (orange). Any target peptides exactly matching to the background database are removed. In the homologous matching step, any target peptides that have similar sequences are filtered out (orange). All intermediate and final results are exported automatically to a user-defined output folder (green).

2.1.1. Preprocessing (Target Selection)

The selection of a target virus proteome is handled by input and preprocessing routines in Purple. For target selection, protein sequence databases in FASTA format serve as main input and are required to be provided in UniProt format. To select the database input, a directory needs to be specified by the user and multiple FASTA files can be considered for the processing. Two options of database specification are available in Purple: the first option is to explicitly define target species names as a list, which leads to the merging of all provided input databases. Each protein entry that contains one of the defined target species names in the protein header is considered as a target protein. The protein sequences not matching the defined target species are used as background database. The second option is to select a specific FASTA file in the database directory as target database. All remaining databases in the directory are then automatically merged to a single background database. As the background database may still consist of proteins originating from one of the target species, each protein in the background database is checked once more: if a protein header matches any species in the specified target database file, the protein entry is removed from further processing accordingly.

Both options result in two types of databases, namely a target and a background database. In the following, each protein sequence in these databases is *in silico*-digested using the enzymatic rule of trypsin with optional proline digestion. The in silico digest step results in multiple peptides for each protein entry, and peptide sequences beyond the user-defined length boundaries are filtered out. In addition, preprocessing includes the option of removing protein fragments and also allows replacing each isoleucine by leucine: this option was implemented because these amino acids share identical molecular masses and are therefore commonly not distinguishable in mass spectrometry. When the preprocessing is completed, both a target and a background database are provided for further analysis, which in this stage consist of peptides instead of proteins.

2.1.2. Exact Matching

Exact matching presents the first actual processing step in Purple: here, each of the previously *in silico*-digested target peptides is compared against the provided background database (see previous paragraph). In this procedure, target and background peptides of identical length are compared and only those target peptides that are not contained in the background are considered further; thus, peptide sequences with one or more exact sequence matches in the background database are filtered out at this stage, because they are not unique to the user-defined taxa of the target space. This procedure is performed iteratively until all *in silico*-digested peptides have been evaluated. The remaining peptides that have not been filtered out are stored as unique peptide candidates for further processing and are exported as intermediate result of the exact matching step.

2.1.3. Homologous Matching

Homologous matching is performed subsequently to the exact matching step. The goal is to evaluate each of the unique peptide candidates concerning its potential sequence consensus to homologous proteins in the background. The rationale behind this approach is that the more similar a target peptide is to the background, the less appropriate it is as candidate for a taxon-specific targeted assay. To assess the similarity of each peptide to the background proteomes, a sequence background consensus metric is introduced (see next paragraph). The target peptides that are discarded either during the exact or the homologous matching step are exported as so-called "shared" peptides. Shared peptides have either an exact sequence match with the background or have background consensus value above a user-defined threshold. To keep track of all processed data, target peptides with a background consensus below the threshold are exported as well.

2.1.4. Background Consensus Metric and Threshold Generation

Owing to mutational effects on conserved viral proteins, peptides can often be shared within a virus genus or family with minor sequence variations between them. This is problematic for targeted assays because such peptide candidates are not specific for species- or strain-level identification. To remove such taxon-unspecific peptides from the final sequence set, the background consensus metric $f(A, B)$ is used in Purple as the essential part of the homologous matching. Basically, the background consensus presents the Hamming distance of a target peptide A and background peptide B of the same length in relation to the length of the peptide n (Equation (2)). An amino acid is shared if the same amino acid ($d(x, y)$) is at the same position in A and B (Equation (1)).

$$d(x,y) = \begin{cases} 1, & if \ x = y \\ 0, & if \ x \neq y \end{cases} \tag{1}$$

In other words, the background consensus is the sum of shared amino acids at a specific position i divided by the number of amino acids in both (target and background) peptides. Even though the Hamming distance is a simple metric, it provides a proof-of-concept and validation of Purple, as adding more sophisticated methods should only slightly improve the homologous matching while increasing the computational effort and complexity.

$$For\ A = \{a_1, a_2, \ldots, a_n\}\ and\ B = A = \{b_1, b_2, \ldots, b_n\}\ and\ n = |A| = |B|:$$
$$f(A, B) = \frac{\sum_{i=1}^{n} d(a_i, b_i)}{n}, for\ a_i \in A\ and\ b_i \in B \tag{2}$$

This metric is applied to each of the target peptides that are compared to all background peptides of the same length. For each target peptide, the maximum consensus is stored when being below a user-defined background consensus threshold. A target peptide with a high background consensus is likely to originate from a homologous protein or common protein domain. Therefore, the consensus metric evaluates the conservation of peptides in the target and background database. A low background consensus marks target peptides that are unique in sequence in the target species. All peptides with a high background consensus below the previously chosen threshold are exported into the final results file and the remaining shared peptides are exported as part of the intermediate output. The results are supplemented with the peptide weight, the number of occurrences in the target database, as well as species and proteins names. This enables the user to conduct further analysis with the previously retrieved unique peptides. The Purple documentation is available for a complete description of all output files and more details about the data interpretation.

2.2. Graphical User Interface

A graphical user interface (GUI) was developed for using Purple (Figure 2). This interface allows researchers with less expertise in handling bioinformatics methods on the command line to use Purple in a efficient and user-friendly manner. The Purple GUI makes software configuration and execution straightforward and complex tasks can be rapidly accomplished. Any parameter can be adjusted in the GUI, and the background consensus threshold can be set by the user. Furthermore, the processing status can be inspected in a logging panel and a file menu provides options for saving and loading configuration files. Note that configuration files are optional in Purple and a default configuration is provided; thus, only system-specific parameters must be set in the GUI. Using configuration files makes each task reproducible and the GUI-integrated configuration file choice allows for switching between multiple settings easily. Figure 3 shows the final output in the tab separated values (TSV) format that can be further processed and visualized using common spreadsheet software.

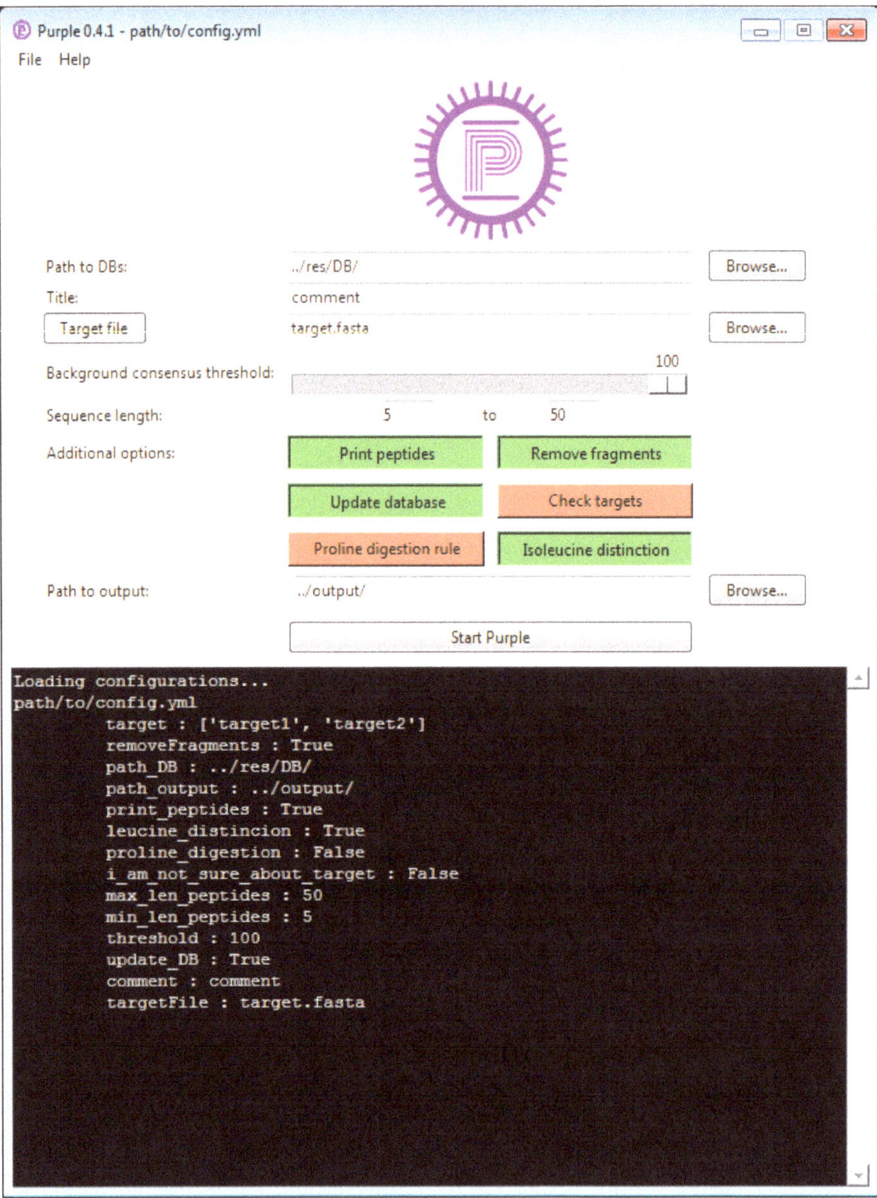

Figure 2. The graphical user interface of Purple. In the top file menu, configurations files can be loaded and saved. The top menu also includes a link to the documentation and manual. The listed GitLab page provides direct user support from the developers via an issue tracking system. The upper panel shows default parameters and allows modifying the configuration settings and processing start. The lower panel displays the current processing status with logging information on the current run, configuration, and progress of the analysis.

	A	B	C	D	E	F	G	H	I
	peptide	peptide weight	highest background consensus	occurrences	species	proteins	protein names	fasta entries	descriptions
1									
2	PSFLA	533.6171	80.00%	1	['Cowpox viru	1	['RP132_CWPXI	1	['>sp|P17474
3	LYVGGGISNDQTTT	2149.2249	70.00%	1	['Cowpox viru	1	['KBTB1_CWPX	1	['>sp|Q8QM(
4	NISNLLDDDILCDVI	2158.4717	36.84%	1	['Cowpox viru	1	['KBTB2_CWPX	1	['>sp|Q8QMƝ
5	QLCLVCHDTK	1159.3794	50.00%	1	['Cowpox viru	1	['KBTB2_CWPX	1	['>sp|Q8QMƝ
6	YNVCNPCILVYNIN1	6135.9066	14.55%	1	['Cowpox viru	1	['KBTB2_CWPX	1	['>sp|Q8QMƝ
7	LLPDMPIALSSYGM	5877.5567	16.98%	1	['Cowpox viru	1	['KBTB2_CWPX	1	['>sp|Q8QMƝ
8	YDTVNNIWETLPNF	2412.6495	35.00%	1	['Cowpox viru	1	['KBTB2_CWPX	1	['>sp|Q8QMƝ
9	PGVVSHEDDIYVVC	1888.1022	41.18%	1	['Cowpox viru	1	['KBTB2_CWPX	1	['>sp|Q8QMƝ
10	YIENK	665.735	80.00%	1	['Cowpox viru	1	['SPI1_CWPXB !	1	['>sp|P42927
11	NIVTSVDMMVSTK	1424.6826	53.85%	1	['Cowpox viru	1	['SPI1_CWPXB !	1	['>sp|P42927
12	NDLQYVHINELFGG	4999.556	25.00%	1	['Cowpox viru	1	['SPI1_CWPXB !	1	['>sp|P42927
13	ESFGNFSIIELPYVGI	4776.24	23.26%	1	['Cowpox viru	1	['SPI2_CWPXB !	1	['>sp|P07385
14	HQESEPASVPTSSR	1511.5499	42.86%	1	['Cowpox viru	1	['A18_CWPXB T	1	['>sp|Q8QM1
15	IIPIDNGSNMLILNP	3034.5026	28.57%	1	['Cowpox viru	1	['IL1BP_CWPXB	1	['>sp|Q0452:
16	NETYCDMMSLNLT	3896.3331	23.53%	1	['Cowpox viru	1	['IL1BP_CWPXB	1	['>sp|Q0452:
17	NDAGYYTCVLK	1246.3886	54.55%	1	['Cowpox viru	1	['IL1BP_CWPXB	1	['>sp|Q0452:
18	YTYGDK	745.7766	66.67%	1	['Cowpox viru	1	['IL1BP_CWPXB	1	['>sp|Q0452:
19	INPVK	569.694	80.00%	1	['Cowpox viru	1	['IL1BP_CWPXB	1	['>sp|Q0452:

Figure 3. Graphical representation of the Purple output. The tabular TSV output of Purple can be imported into various spreadsheet software tools. This exemplary table shows the peptide sequence, the calculated theoretical mass weight (Da), the highest background consensus, and the number of peptide occurrences in the target proteome. The species, protein name and full description of the associated protein are stored in a list for further analysis. In addition, the number of proteins and FASTA entries are listed separately, because they can diverge, e.g., when a protein has multiple sequence variants.

2.3. Data

2.3.1. Target Virus Databases

To evaluate the performance of Purple, selected target virus species from sequence databases were used. This section provides an overview on the virus species used with respect to database composition and further background information on the virus type. The virus species were selected based on their relevance for current or upcoming diagnostic settings.

Arenaviruses

Arenaviruses are enveloped RNA viruses with an average diameter of 120 nanometers that have a bisegmented negative-strand RNA genome. The Latin term "arena" refers to the grainy ribosomal particles acquired from the virus-host cells that can be viewed in cross-section with electron microscopy imaging. Arenaviridae is a virus family whose members are generally associated with causing chronic infections in rodents and zoonotically acquired severe diseases, such as lymphocytic choriomeningitis or hemorrhagic fever, in humans. In this work, nine disease-causing Old and New World arenavirus species are taken as targets for evaluating the performance of Purple (Table 1). Besides Lymphocytic choriomeningitis virus, strain members of which cause aseptic meningitis, encephalitis, or meningoencephalitis, all listed arenaviruses are causative agents for viral hemorrhagic fever (VHF).

Table 1. Alphabetically ordered list of arenavirus species used for the performance benchmarking. The reader is referred to [23] for further details on these arenaviruses.

Virus Species	Abbreviation	NW/OW [2]	NW - Clade [3]	No. Proteins	No. Peptides [1]
Chapare mammarenavirus	CHAV	NW	B	4	252
Guanarito mammarenavirus	GTOV	NW	B	4	244
Junin mammarenavirus	JUNV	NW	B	4	246
Lassa virus	LASV	OW	-	4	242
Lujo mammarenavirus	LUJV	OW [4]	-	4	250
Lymphocytic choriomeningitis virus	LCMV	OW	-	4	245
Machupo virus	MACV	NW	B	4	237
Sabia mammarenavirus	SABV	NW	B	4	248
Whitewater Arroyo mammarenavirus	WWAV	NW	A/rec	4	240

[1] Number of in silico-digested peptide sequences, [2] New World (NW)/ Old World (OW), [3] New World clade [4] Based on genome sequence clustering, Lujo mammaarenavirus shows its own cluster [23].

Cowpox virus

Cowpox virus (CPXV) is a large double-stranded DNA virus with a proteome of over 200 proteins [24] that belongs to the genus Orthopoxvirus (OPV) of the Poxviridae family. CPXV has been described as the source of the first vaccine used by Edward Jenner, who was the first to scientifically describe the vaccination process against the smallpox-causing variola virus. Recent findings based on a conducted analysis on the smallpox vaccine gave evidence of the suspected role of horsepox (instead of cowpox) in the origin of the vaccine [25,26]. Since the pathogenicity and zoonotic potential of CPXV are investigated at the Robert Koch Institute, detailed data acquired from MS measurements were available (see Section 2.3.3). For performance evaluations, CPXV is further beneficial because this virus species has several close relatives. In addition to the cowpox strains Brighton Red and Grishak-90, four very close relatives with high sequence similarity are given: a genome comparison performed with BLAST [27] showed that variola virus, monkeypox virus, horsepox virus, and vaccinia virus share sequence identities of up to 98% (Supplementary Table S1).

Vaccinia virus (VACV Copenhagen and VACV Western Reserve)

Vaccinia virus is a member of the Orthopoxvirus (OPV) genus [28] and has been used for vaccination against smallpox since the 19th century. Due to the high sequence similarity of members of the OPV genus, it is possible to provide cross-protection vaccination by one member of the OPV genus. Hence, the classification can be an issue, because it can be challenging to find peptides to reliably classify a species or a strain. In this work, we investigate whether it is possible to distinguish between the two strains VACV Copenhagen and VACV Western Reserve by finding strain-specific peptides using Purple. Similar to CPXV, experimental data was publically available (see Section 2.3.3).

2.3.2. Background Virus Databases

The target databases mentioned above are species-specific and therefore cannot represent all available virus proteomes. From the target databases, Purple only yields to species-specific unique peptides. To extend this space to all virus proteomes and subsequently be able to find unique peptides in that relation, we added a database that consists of all reviewed virus proteins available on UniProt/Swiss-Prot [29]. In contrast to the target databases, this database is used exclusively as a background database. At the time of writing, UniProt/Swiss-Prot contains 16,846 reviewed viral proteins, which results in 301,387 in silico-digested tryptic peptides. In this work, we evaluate Purple with and without the use of the larger background database.

2.3.3. Background Human Databases

To account for samples mixed with human proteins we added a human database to the background. This database originates from UniProt/Swiss-Prot [29] and enables Purple to discard human peptides. Subsequently, this reduces false positives in experiments using virus-infected human samples. The database consists of 20,428 proteins and was used exclusively for the CPXV analysis in this work.

2.3.4. Experimental Data

The MS/MS datasets used for the benchmarking of Purple originate from a previous study published by Doellinger et al. in 2015 [24] (PRIDE project accession: PXD003013). In this work, a subset of the data available was used including three CPXV Brighton Red, three VACV Copenhagen, and three VACV Western Reserve MS/MS raw files. These raw files were acquired by an LTQ Orbitrap in data-dependent manner. Further experimental details are listed and described in the above-mentioned publication. Subsequently, three CPXV Brighton Red raw files were converted into MGF files using the MSConvert function of ProteoWizard [30] with the peak picking parameter of MS-level two and with zero sampling removal activated. Table 2 shows the number of MS/MS spectra for each virus strain (CPXV Brighton Red, VACV Copenhagen and VACV Western Reserve). For peptide and protein identification, these spectra were searched against proteome databases using the MS-GF+ [31] (version v20181015) database search engine. The database search was performed with eight threads, an activated decoy search, a chosen precursor with mass tolerance of five ppm, optimized for Orbitrap instruments, and trypsin was selected as digestion enzyme. The sequence databases used for protein identification are described in detail in Section 2.3.1. The database searches produced mzid output files that were converted into TSV files using the build-in MS-GF+ conversion tool. Afterwards, the results were filtered by applying a 1% false discovery rate (FDR) threshold at the PSM-level.

Table 2. This table shows the number of spectra from each sample replicate for CPXV Brighton Red, VACV Copenhagen, and VACV Western Reserve virus species/strains.

Species/Strain	No. Spectra in Replicate 1	No. Spectra in Replicate 2	No. Spectra in Replicate 3	No. Total Spectra
CPXV Brighton Red	19,396	19,352	18,920	57,668
VACV Copenhagen	19,740	19,265	19,170	58,175
VACV Western Reserve	19,421	19,453	19,076	57,950

3. Results

We here present three different use cases to illustrate the possibilities of targeted proteomics using Purple in viral diagnostic settings. The first analysis focuses on the species-level resolution for arenaviruses, the second evaluates the taxonomic classification using cowpox data from shotgun proteomics measurements, and the third tests the capabilities of strain-level differentiation using experimental data from two closely related vaccinia virus strains.

3.1. Analysis of Species-Level Resolution using Nine Arenavirus Species

In the first analysis, we aimed to evaluate the species-level resolution of our diagnostic approach using sequence data from the Arenaviridae family. For this purpose, we investigated the resolution of Purple by evaluating different viral species as target organisms against a proteome background of similar species and viruses in general. We used nine arenavirus species (MACV, JUNV, SABV, CHAV, GTOV, LASV, LCMV, WWAV, and LUJV; see Table 1) with proteomes containing four proteins, namely (1) RNA-directed RNA polymerase L, (2) nucleoprotein N, (3) pre-glycoprotein polyprotein GP complex and (4) RING finger protein Z. As background proteomes, we added all reviewed virus proteins available on UniProt/Swiss-Prot to remove frequently occurring peptides (e.g., from conserved sequences of functional domains). The removal of target peptides from similar virus proteomes intends to eliminate false positive detections (i.e., to increase the specificity). Since the protein sequences differ strongly between the arenavirus species, we expected to retrieve sufficient unique peptides for each species that serve as candidates for designing a targeted assay. For a benchmarking, we examined the relative amount of taxon-specific target peptides for each of the arenavirus species using both exact and homologous matching mode (Tables 3 and 4). The homologous matching was performed to evaluate the impact of sequence homologies for the arenaviruses and between these and all other virus species.

Table 3. This table shows the number of taxon-specific peptides from nine arenavirus species after (i) *in silico* digest, (ii) exact matching, and (iii) homologous matching (80% background consensus threshold). Each target species was compared against the background of eight remaining arenavirus species proteomes. The second column provides the number of nonspecific peptides, i.e., the ones being shared with the background.

Species	No. Digested Peptides	No. Background Shared	No. Exact Matching	No. Homologous Matching
MACV	237	119	178	**118**
SABV	248	127	191	**121**
LUJV	250	24	241	**226**
CHAV	252	121	197	**131**
GTOV	244	75	205	**169**
JUNV	246	123	187	**123**
LASV	242	35	227	**207**
LCMV	245	31	232	**214**
WWAV	240	31	226	**209**

Table 4. This table shows the number of taxon-specific peptides from nine arenavirus species after (i) *in silico* digest, (ii) exact matching, and (iii) homologous matching (80% background consensus threshold). Each target species was compared against the background of eight remaining arenavirus species proteomes and additionally against all reviewed virus proteomes (from UniProt/Swiss-Prot). The second column provides the number of nonspecific peptides, i.e., the ones being shared with the background.

Species	No. Digested Peptides	No. Background Shared	No. Exact Matching	No. Homologous Matching
MACV	237	143	162	**94**
SABV	248	144	183	**104**
LUJV	250	52	229	**198**
CHAV	252	137	190	**115**
GTOV	244	118	189	**126**
JUNV	246	139	171	**107**
LASV	242	126	171	**116**
LCMV	245	110	187	**135**
WWAV	240	130	181	**110**

First, we investigated the ratios of taxon-specific unique peptides and in silico-digested peptides with a background database consisting of the four arenavirus proteins, as mentioned above. The exact matching yielded to taxon-specific peptide ratios between 75.1% (MACV) and 96.4% (LUJV) (Figure 4). This can be explained by the high sequence diversity between the nine arenavirus species: when generating multiple sequence alignments (MSA) of these species for their four proteins, overall, a low consensus of the sequences was found (Supplementary Data S1–S4). When applying a background consensus threshold of 80%, significantly fewer taxon-specific peptides were obtained with relative numbers between 48.8% and 90.4% for SABV and LUJV, respectively (Figure 4). Overall, the mean decrease in the ratio of all species is 16.6% and the strongest ratio decrease can be found for MACV (25.3%), SABV (28.2%), CHAV (26.2%), and JUNV (26.0%). These four species are all New World arenaviruses and part of the clade B (see Table 2). The close relationship of these four virus species (as shown in the phylogenetic tree in Figure 5) causes high numbers of shared peptides which explains the decline in taxon-specific peptides. The Old World arenavirus LUJV shows the highest taxon-specific peptide ratio after homologous matching (90.4%) and even after homologous analysis against all virus proteomes (79.2%). This illustrates that LUJV has the lowest sequence similarity with the other arenaviruses. The low similarity can be explained by the isolated geographical distribution of LUJV in Southern Africa [32]. In 2008, an outbreak of LUJV led to a high case fatality rate of 80% (4/5 cases), and a follow-up analysis of its genome confirmed that LUJV is a novel virus species being only distantly related to known arenaviruses and groups genetically closer to Old World viruses not associated with VHF [33].

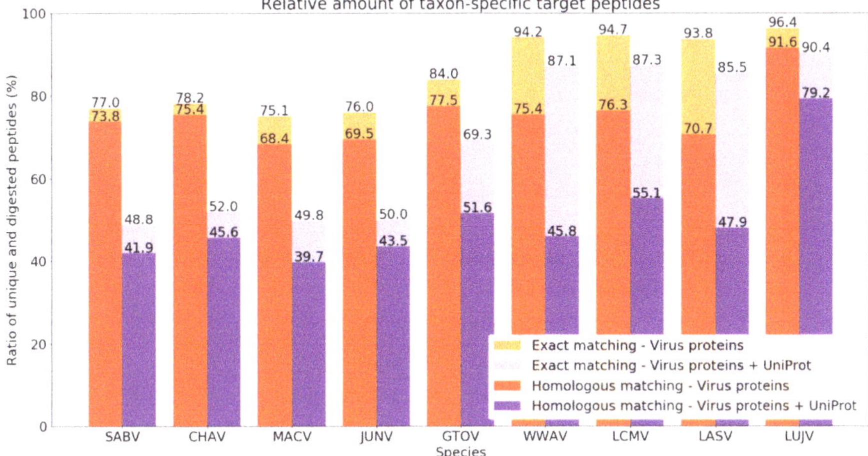

Figure 4. Relative amount of taxon-specific target peptides from nine arenavirus species proteomes. The ratio of unique to in silico-digested peptides is shown for exact (lighter colors) and homologous (darker colors) matching mode with a background consensus threshold of 80%. Orange bars show the results for the database consisting of four virus proteins for each arenavirus species. Purple bars indicate results that were generated when adding protein sequences from all reviewed virus proteomes (from UniProt/Swiss-Prot) as additional background.

Next, we assessed the protein sequence coverage on the basis of Purple-selected unique peptides for all four arenavirus proteins (RNA-directed RNA polymerase L; Nucleoprotein N; Pre-glycoprotein polyprotein GP complex GLYC; RING finger protein Z). We evaluated two different backgrounds here: (i) a small background with the arenavirus proteomes (containing the four proteins) of the remaining eight non-target species and (ii) a large background containing all arenavirus proteomes combined with all reviewed virus proteomes from UniProt/Swiss-Prot (see Section 2.3.2).

The analysis of the protein sequence coverage shows that L, GLYC and Z are relatively well covered by the taxon-specific peptides across all nine species for the small background (Figure 6). Nucleoprotein NCAP has the highest variability in protein coverage with an interquartile range (IQR) of 35.22% on the small background, suggesting that NCAP is the best-conserved protein among the considered arenavirus species. When taking a closer look at the results of the larger background analysis with all reviewed virus proteins, it can be found that the coverage decreases for all four proteins. The NCAP protein shows the lowest median in protein coverage (20.18%). This shows that NCAP has the lowest sequence consensus of taxon-specific peptides with other virus proteomes, indicating that it is the best-conserved of the four proteins. Indeed, the other three proteins (L, GLYC, and Z) have above 40% sequence coverage, thus more taxon-specific peptides can be obtained from these proteins. This analysis shows that, depending on the use case, it may make sense to investigate individual proteins instead of whole proteomes. For example, proteins with low sequence coverage based on taxon-specific peptides may be excluded.

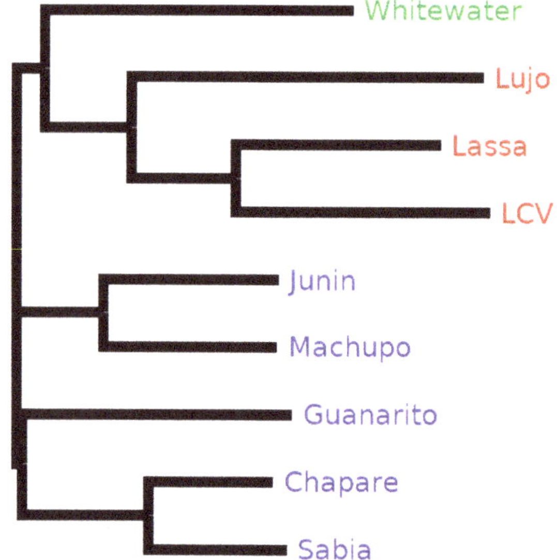

Figure 5. Phylogenetic tree of the pre-glycoprotein polyprotein GP complex (GLYC) of nine arenaviruses. The Whitewater strain is the only New World clade A/rec arenavirus (green). Lujo (LUJV), Lassa (LASV), and Lymphocytic choriomeningitis (LCV) are geographical Old World arenaviruses (red). Junin (JUNV), Machupo (MACV), Guanarito (GTOV), Chapare (CHAV), and Sabia (SABV) are members of the New World arenaviruses clade B (blue). The neighbor-joining tree without distance corrections was created using CLUSTAL Omega [34] for the multiple sequence alignment and the tree visualization software FigTree (http://tree.bio.ed.ac.uk/software/figtree/).

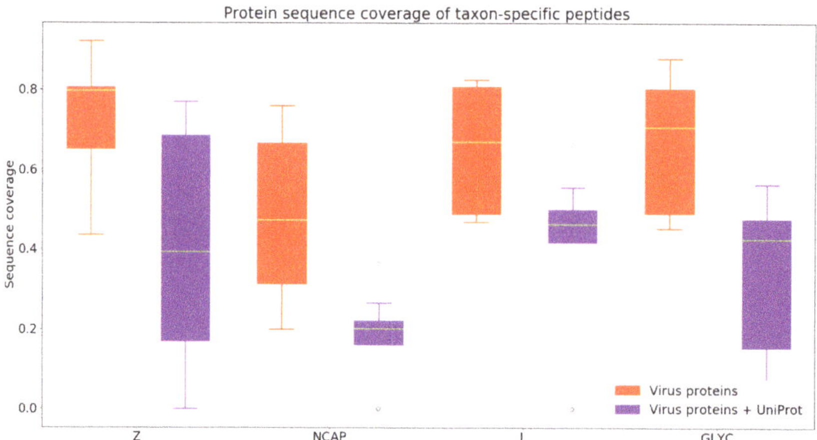

Figure 6. Protein sequence coverage of taxon-specific peptides selected by Purple on proteins for nine arenavirus species proteomes. The four proteins of the arenavirus proteomes are RNA-directed RNA polymerase L (L), nucleoprotein N (NCAP), pre-glycoprotein polyprotein GP complex (GLYC), and RING finger protein Z (Z). The coverage of selected peptides is displayed for homologous matching when applying a background consensus threshold of 80%.

3.2. Evaluating Species-Level Classification Based on Detected Peptides from Viral Shotgun Proteomics Measurements

To evaluate the peptide selection method in Purple on experimental data, we used representative MS/MS datasets derived from human cowpox virus (CPXV) samples. The main goal was to test whether peptides identified in a typical shotgun proteomics experiment can be used for differentiating viruses at the species level. We also aimed for estimating the expected accuracy gain for taxonomic classification when using a targeted proteomics assay on the basis of peptides suggested by Purple.

In a pre-analysis, we performed a Purple run using CPXV as target proteome to select species-specific peptides. For the peptide selection process, 18 reviewed (from UniProt/Swiss-Prot) and 208 unreviewed (from UniProt/TrEMBL) CPXV-specific protein sequences were used as target database, which is part of the PRIDE project (see Section 2.3.4). We used this combined database consisting of reviewed and unreviewed protein sequences because the available reviewed protein sequences for the Brighton Red strain yielded to a very limited number of peptide identifications during the database search (Supplementary Table S2). All available virus proteomes (a total of 16,846 sequences) and all reviewed human proteins were taken as background. These proteomes were obtained from UniProt/Swiss-Prot (see Section 2.3 for database details).

The Purple run resulted in 1509 in silico-digested peptides after exact matching and 885 peptides after homologous matching (using a background consensus threshold of 80%). The distribution of the homologous background consensus shows a normal distribution below 50% (Supplementary Figure S2). 3986 peptides were discarded, because they were shared with other (i.e., non-CPXV) viral proteomes or the human proteome. The remaining 885 CPXV-specific peptides have a mean background consensus of 53.9%, which means that on average around half of the amino acids of each peptide are equal to residues of peptides in the background.

Next, we searched experimental MS/MS spectra from CPXV samples using the search algorithm MS-GF+ [31] against a CPXV and human sequence database for peptide identification (see Section 2.3). In this analysis, CPXV datasets from MS measurements of three technical replicates, each with ~19,000 MS/MS spectra, were evaluated. The database search resulted in 4028, 4125, and 3967 identified peptides per sample replicate with sequence duplicates removed. More than twice the amount of CPXV peptides were identified as human peptides in this sample before applying a FDR filtering. After applying an FDR threshold of 1%, 1067, 1028, and 1004 CPXV peptides were identified (Table 5). Subsequently, the identified peptides (below 1% FDR threshold) were compared against the set of taxon-specific CPXV peptides suggested by Purple using both exact and homologous matching mode. Between 83 and 94 peptides selected by Purple were detected in the MS/MS experiments (without applying any FDR threshold). When filtered by 1% FDR, the peptides decreased to numbers between 78 and 84. Consequently, this analysis demonstrates that it would be possible to reliably identify CPXV for these three sample replicates.

Table 5. This table shows the number of peptides from the cowpox virus (CPXV) Brighton Red strain after (i) database search with duplicates removed (CPXV); (ii) database search with duplicates removed (human); (iii) intersection of peptides obtained from Purple and peptide identifications from database search; (iv) database search, duplicates removed and filtered by 1% FDR threshold; and (iiv) intersection of peptides suggested by Purple and peptide identifications from FDR-filtered database search. The CPXV Brighton Red strain was compared against the background of all reviewed virus proteomes and the reviewed human proteome. In addition, the second column specifies the sample replicate data that was used for the database search.

Strain	Replicate	No. Database Search (CPXV)	No. Database Search (HUMAN)	No. Intersection	No. Database Search Filtered	No. Intersection Filtered
Brighton Red	1	4028	10319	94	1067	84
Brighton Red	2	4125	10286	83	1028	78
Brighton Red	3	3967	10068	92	1004	84

When considering the results of all three replicates, it can be observed that 61 CPXV-specific peptides were detected without any applied FDR threshold (Figure 7A). Filtered by 1% FDR, 56 peptides across all replicates can be used to specifically identify the species within the sample as a member of CPXV (Figure 7B).

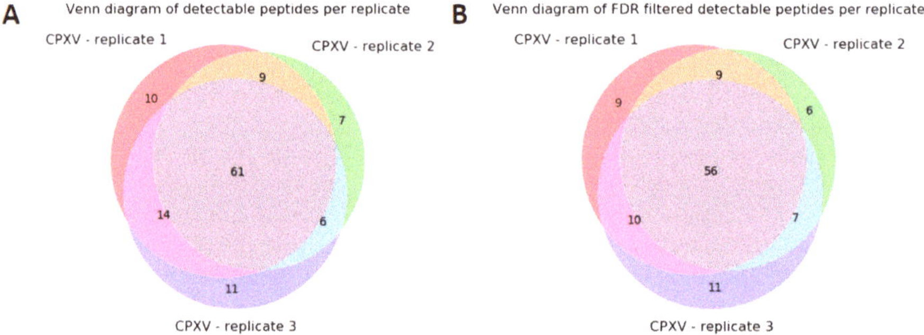

Figure 7. Intersection of detectable peptides of CPXV sample replicates. These Venn diagrams show the intersection of the detectable peptides in replicates 1–3. The subfigures depict the number of peptides without applying any false discovery rate (FDR) threshold (**A**) and filtered by 1% FDR (**B**).

When examining the peptides shared by the target and background proteomes, it can be found that the Cowpox virus shares ~3000 peptide sequences per strain with the Vaccinia virus strains and Variola virus strains (Figure 8). Other Orthopoxviruses were found as well, although the number of peptides is low, due to fewer proteins of these strains in the background database. The CPXV Brighton Red strain-specific peptides are small in number because most matches originate from the Cowpox virus species proteome without giving any details about a particular strain. Around 500 peptides were shared with the human proteome and were consequently discarded.

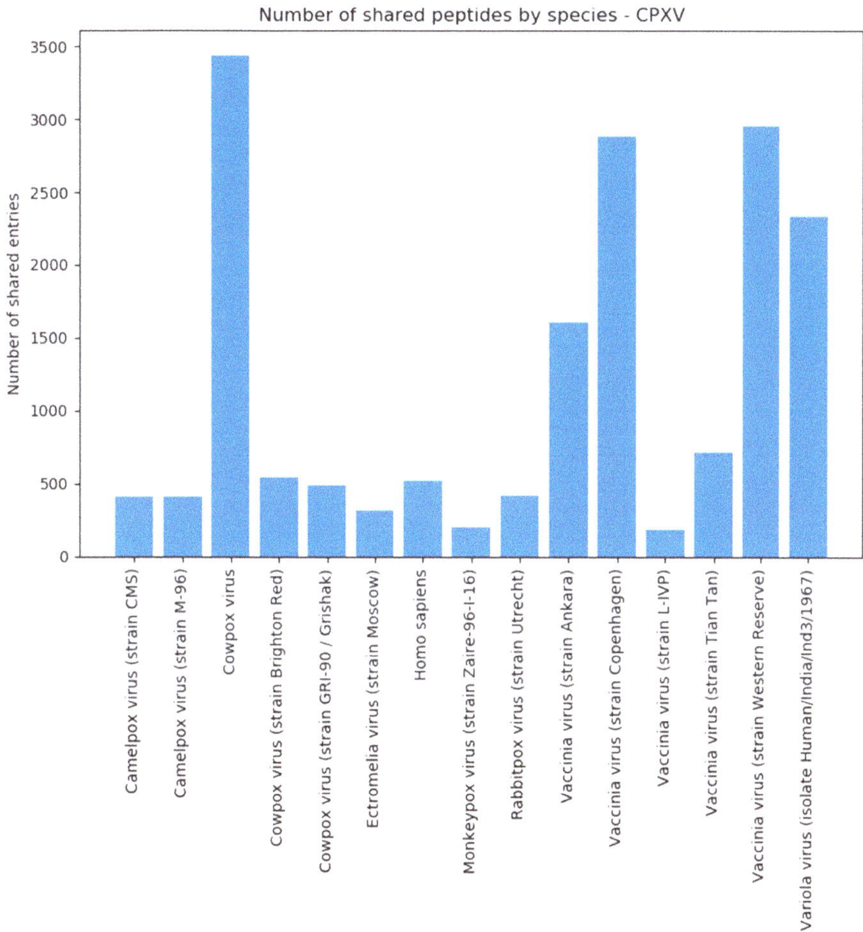

Figure 8. Number of shared CPXV peptides by species/strain assigned. This plot shows the number of shared peptides that Purple detected in the background for a species/species after the CPXV Brighton Red target analysis. All species that contribute less than 0.5% to the total amount of shared peptides were removed.

3.3. Comparison of Strain vs. Strain and Strain vs. All Virus Level Resolution

Next, we conducted a performance evaluation using two different, yet highly similar Vaccinia virus strains, namely VACV Copenhagen and VACV Western Reserve. The objective was to test whether Purple can retrieve strain-specific peptides that are then used in the targeted proteomics assay for accurate taxonomic classification. In this analysis, the target database contained sequences from one of the two VACV virus strains (either Copenhagen or Western reserve). Consequently, the background database contained the remaining VACV strain and all reviewed virus proteins available on UniProt. This procedure was repeated with the remaining VACV strains as target. The goal was to find strain-specific peptides to accurately detect the virus strain. We used a background consensus threshold of 80% to filter out homologous peptides. Afterwards, experimental data (see Section 2.3.3) was used to validate the results and to show if the selected strain-specific peptides are found in the acquired tandem mass spectrometry (MS/MS) data. For peptide identification, we used the software MS-GF+ [31] with an 1% FDR threshold (see Section 2.3.3).

In the case of VACV Copenhagen, Purple discarded 3848 peptides because a perfect sequence match was present in the background with a peptide of another strain or virus (Table 6). Equally, 3971 VACV Western Reserve peptides are marked as shared with the background and discarded. After exact matching, 498 and 341 strain-specific peptides could be obtained for VACV Copenhagen and VACV Western Reserve, respectively. The homologous matching removed additional 157 (VACV Copenhagen) and 172 (VACV Western Reserve) peptides from the set of unique peptides. The remaining 352 (VACV Copenhagen) and 169 (VACV Western Reserve) peptides can be used to uniquely identify the strain in a mixture of all reviewed virus proteins available on UniProt/Swiss-Prot.

Table 6. This table shows the number of taxon-specific peptides from the VACV Copenhagen and VACV Western Reserve strain after (i) *in silico* digest, (ii) exact matching, and (iii) homologous matching (80% background consensus threshold). Each target strain was compared against the background of the other strain and all reviewed virus proteomes. The second column provides the number of nonspecific peptides, i.e., the ones being shared with the background.

Species	No. Digested Peptides	No. Background Shared	No. Exact Matching	No. Homologous Matching
Copenhagen	4200	3848	498	**352**
Western Reserve	4140	3971	341	**169**

In addition, we categorized the shared peptides by virus species to check for close relationships in the background. For VACV Copenhagen, it can be observed that most peptide matches are found in the Vaccinia species (Figure 9), owing to a high protein sequence similarity of involved Vaccinia strains. Other contributing species are Camelpox virus, Cowpox virus, Monkeypox virus, Rabbitpox virus, and Ectromelia virus. All these viruses are, as expected, members of the orthopoxvirus genus. Similar findings could be observed for the results of the VACV Western Reserve strain (Supplementary Figure S1). Note here that Figure 9 shows the number of peptides and if a species is underrepresented in the databases, it will affect the outcome concerning the number of peptides that contribute to the shared peptides.

To evaluate the detectability of taxon-specific peptides for the given DDA experiments, we performed database searches for peptide identification using three different technical replicates of VACV Copenhagen. Without any FDR cut-off, we could identify between 60 and 66 strain-specific peptides selected by Purple (Table 7). However, when filtered by an FDR of 1% the number of peptides decreased drastically and only one or two taxon-specific peptides were confirmed in the shotgun proteomics data. It was possible to identify Replicate 1 and 2 as VACV Copenhagen by using the peptide sequence ILFWPYIEDELR. The number of peptides can be increased by switching to a targeted proteomics approach and by considering PTMs or by an improved homologous matching. The three technical replicates of the VACV Western Reserve strain resulted in fewer peptides in the intersection with the database search results (between 32 and 42), but when filtered by 1% FDR, the number of peptides was increased up to 11-fold (with nine to 11 peptides) in comparison to the VACV Copenhagen replicates. Six peptides were detected, and their sequences were identical among all three replicates.

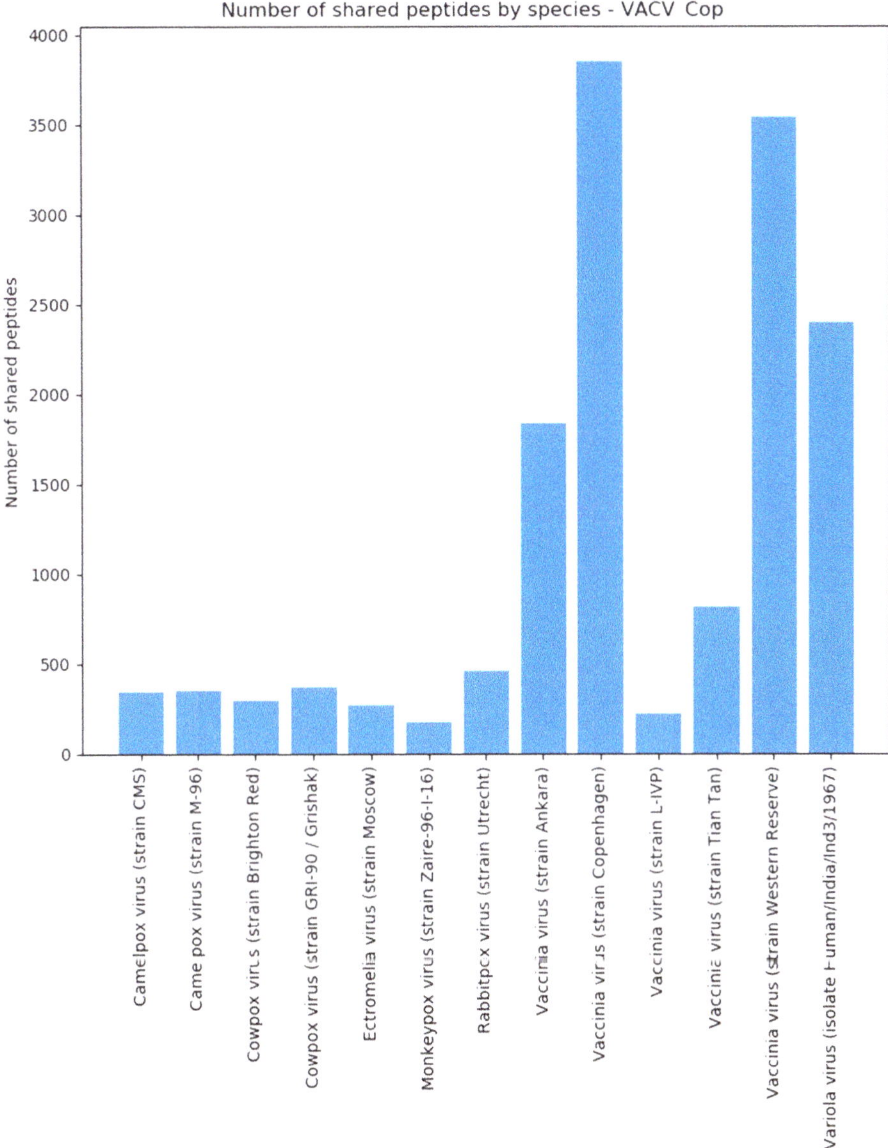

Figure 9. Number of shared peptides by species. This plot shows the number of shared peptides that Purple detected in the background for a species after the VACV Copenhagen analysis. All species that contribute less than 0.5% to the total amount of shared peptides were removed here.

Table 7. This table shows the number of peptides from VACV Copenhagen and VACV Western Reserve strain after (i) database search with duplicates removed; (ii) intersection of peptides obtained by Purple and database search; (iii) database search, duplicates removed and filtering by FDR; and (iv) intersection of peptides obtained by Purple and filtered database search. Each target strain was compared against the background of the other strain and all reviewed virus proteomes. The second column specifies the replicate data that was used for the database search.

Strain	Replicate	No. Database Search	No. Intersection	No. Database Search Filtered	No. Intersection Filtered
Copenhagen	1	3585	66	825	2
Copenhagen	2	3507	62	800	1
Copenhagen	3	3525	60	828	1
Western Reserve	1	3636	35	841	9
Western Reserve	2	3736	42	800	11
Western Reserve	3	3507	32	809	9

In conclusion, we were able to identify every strain in each sample with an applied FDR of 1%. For VACV Western Reserve, the number of peptides was higher than for the VACV Copenhagen strain. The number of detectable peptides could be increased by improving scoring and filtering or by switching from shotgun to targeted proteomics methods or by considering PTMs.

Figure 10 reveals a normal distributed homologous consensus in the interval from 10% to 50%. This is caused by random matches with background peptides and these peptides should be unique for the strain. We could not observe a distinct distribution above 50%. This could be improved by moving from identity to a similarity-based matching, as this would differentiate peptides with the same amount of matching consensus residuals.

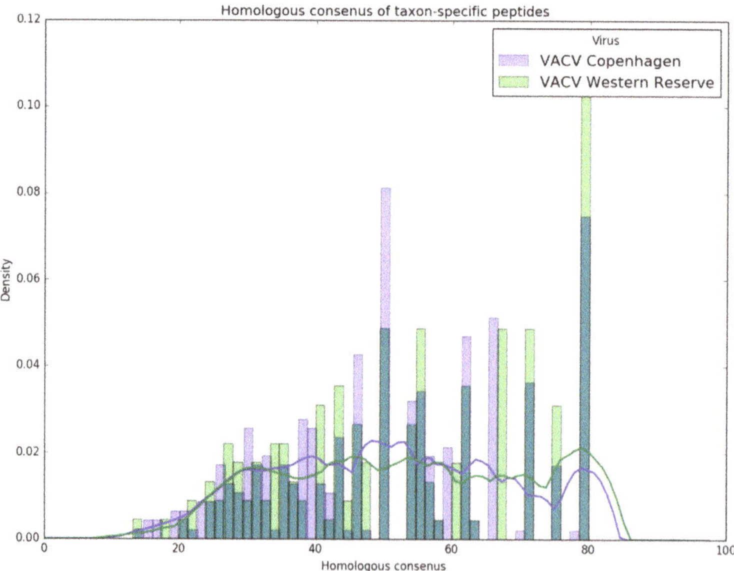

Figure 10. Histogram and density plot of homologous consensus. This histogram shows the distribution of the homologous consensus for the VACV Copenhagen (blue) and Western Reserve (green) strains. Additionally, the kernel density was calculated utilizing the Epanechnikov kernel and a Silverman bandwidth estimation.

4. Discussion

The main goal of our developed Purple software is to provide taxon-specific peptides for a targeted proteomics assay. These targeted assays can be used in a diagnostic setting to identify a virus species/strain or even a whole virus family in a sample in sensitive and time-efficient manner. In this work, we validated the software in three different benchmarking experiments.

Purple enabled us to retrieve taxon-specific peptides to distinguish between arenavirus species proteomes that are very similar in their sequences (see Section 3.1). Accordingly, we observed a comparable decrease in the ratio of unique to *in silico*-digested peptides for New and Old World arenaviruses based on differences between their proteomes (Figure 4). This effect could also be recognized also on the clade level for the New World viruses.

The data analysis of CPXV (see Section 3.2) resulted in 56 taxon-specific peptides (Figure 7). These peptides were present in each MS/MS sample replicate and can be used to uniquely identify CPXV in a mixed biological sample, although its proteome is very similar to other Orthopoxvirus species and strains (Figure 8). By changing to a Brighton Red strain-specific target database, a reliable determination of the strain would be possible as well. This underlines that Purple relies on a correct and complete database to yield to the best possible results. Missing or incorrectly assigned protein sequences could result in incorrect selected unique peptides or discarded ones. Furthermore, although many spectra in the shotgun proteomics experiment were assigned to human peptides, this does not present a limitation for the targeted proteomics approach, because unique virus peptides selected by Purple can be detected using a targeted (e.g., PRM-based) assay in specific and sensitive manner; for example, in a recently published study [35], a PRM-based assay was used to identify dengue virus species directly from clinical serum samples. Nevertheless, to validate the resulting set of peptides, it would be recommended to test them on other CPXV samples and to check if the peptides are detectable in these samples likewise. In addition, the selected background database might be incomplete, e.g., when proteome references were missed to be included for the Purple analysis. In this case, it is useful to validate Purple-selected peptides using secondary tools such as Unipept [36] for resolving the taxonomic origin of any tryptic peptide based on the complete UniProt database. Furthermore, false negatives may result from issues during sample preparation or poor instrument performance. Therefore, these parameters need to be controlled in diagnostic PRM assays, e.g., by using internal standards and running further quality control samples.

It can be crucial in virus infection scenarios to accurately distinguish between specific strains. To cover these cases, we examined the strain-level resolution of our tool using data of VACV Copenhagen and VACV Western Reserve strains (see Section 3.3). Purple was able to find a reliable amount of strain-specific peptides (Table 7). The intersection between the Purple-selected peptides and the peptide identification from the database search showed that it is possible to detect these peptides. In general, strain-level identification was possible even for an applied FDR threshold of 1%, however, it became apparent that the shotgun proteomics approach becomes limited due to the spurious numbers of identified peptides. The number of peptides could be increased by adjusting the FDR filtering or by using a targeted proteomics approach with higher sensitivity.

In comparison to other tools, Purple offers several advantages, such as cross-platform compatibility on multiple operating systems. Purple allows a homology-based analysis of multiple proteome databases at once and produces an aggregated and summarized export on various levels. In addition, Purple is not limited to specific organisms, but can be used with general UniProt databases, also including eukaryotic and bacterial databases. High sequence similarity between strains and horizontal gene transfer may complicate taxon-specific classification for bacterial samples. However, Purple could help to overcome complications and can be helpful for creating targeted assays for bacterial detection as well. The graphical user interface and compatibility with all UniProt databases enables researchers without bioinformatics background to find taxon-specific peptides in an easy and straightforward manner.

A potential improvement to the software would be to move from a sequence identity-based metric based on the Hamming distance to similarity-based matching for the homologous matching mode. In this case, amino acid substitutions are not weighted equally, for example by using a PAM or BLOSUM matrix [37]. This similarity-based metric might allow a more accurate homologous matching in Purple. For example, an approach based on a structural alignment as introduced by Ogata et al. [38] might be useful. Further potential improvements with useful features in Purple include adding plots for better data exploration and a tabular view for inspecting the results (that are currently exportable as text files to spreadsheet software).

In summary, the most promising application of Purple is to select taxon-specific peptides for creating tailored SRM or PRM assays with high sensitivity and specificity. This application will allow for new time- and cost-efficient diagnostic methods in healthcare and further biological applications. It could even be used to identify multiple organisms in a single sample in the context of targeted metaproteomics [39].

Purple is available for download on our GitLab website (https://gitlab.com/rki_bioinformatics), by using the Python package manager pip (https://pypi.org/project/purple-bio/) or via the Bioconda channel (https://anaconda.org/bioconda/purple-bio) [40]. The software is available as graphical user interface version, Python package and command line version for Windows, Linux, and MacOS. In addition, user support, tutorials, and the documentation manual can be found on the GitLab webpages.

Supplementary Materials: The following are available online at http://www.mdpi.com/1999-4915/11/6/536/s1, Table S1: Genome sequence similarities of cowpox virus; Table S2: Number of peptides from CPXV Brighton Red strain processing; Figure S1: Number of shared peptides by species for VACV Western Reserve; Figure S2: Histogram and density plot of homologous consensus—CPXV; Data S1: MSA of the pre-glycoprotein polyprotein GP complex (GPC gene); Data S2: MSA of nucleocapsid protein (N gene); Data S3: MSA of RNA-directed RNA polymerase L (L gene); Data S4: MSA of RING finger protein Z (Z gene).

Author Contributions: Conceptualization: J.D., B.Y.R., and T.M.; methodology: J.D., J.L., P.H., and T.M.; software: F.H., J.L., and P.H.; validation: F.H. and J.L.; formal analysis: F.H., J.L., and T.M..; investigation: F.H. and J.L..; resources: A.N., B.Y.R., J.D., and M.G.; data curation: F.H.; writing—original draft preparation: F.H. and J.L.; writing—review and editing: A.N., B.Y.R., J.D., M.G., P.H., and T.M.; visualization: F.H.; supervision: B.Y.R. and T.M.; project administration: T.M.

Funding: This research was funded by the German Research Foundation (DFG), grand number RE3474/2-2, and intramural funds (#8321566) of the Robert Koch Institute.

Conflicts of Interest: The authors declare no conflict of interest.

References

1. Singhal, N.; Kumar, M.; Kanaujia, P.K.; Virdi, J.S. MALDI-TOF mass spectrometry: An emerging technology for microbial identification and diagnosis. *Front. Microbiol.* **2015**, *6*, 791. [CrossRef] [PubMed]

2. Ebhardt, H.A.; Root, A.; Sander, C.; Aebersold, R. Applications of targeted proteomics in systems biology and translational medicine. *Proteomics* **2015**, *15*, 3193–3208. [CrossRef] [PubMed]

3. Deutsch, E.W. The PeptideAtlas project. *Methods Mol. Biol.* **2010**, *604*, 285–296. [PubMed]

4. Picotti, P.; Aebersold, R. Selected reaction monitoring-based proteomics: Workflows, potential, pitfalls and future directions. *Nat. Methods* **2012**, *9*, 555–566. [CrossRef] [PubMed]

5. Peterson, A.C.; Russell, J.D.; Bailey, D.J.; Westphall, M.S.; Coon, J.J. Parallel reaction monitoring for high resolution and high mass accuracy quantitative, targeted proteomics. *Mol. Cell. Proteom.* **2012**, *11*, 1475–1488. [CrossRef] [PubMed]

6. Borràs, E.; Sabidó, E. What is targeted proteomics? A concise revision of targeted acquisition and targeted data analysis in mass spectrometry. *Proteomics* **2017**, *17*, 17–18.

7. Karlsson, C.; Malmström, L.; Aebersold, R.; Malmström, J. Proteome-wide selected reaction monitoring assays for the human pathogen *Streptococcus pyogenes*. *Nat. Commun.* **2012**, *3*, 1301. [CrossRef]

8. Peters, J.S.; Calder, B.; Gonnelli, G.; Degroeve, S.; Rajaonarifara, E.; Mulder, N.; Blackburn, J.M. Identification of quantitative proteomic differences between *Mycobacterium tuberculosis* lineages with altered virulence. *Front. Microbiol.* **2016**, *7*, 813. [CrossRef]

9.   MacLean, B.; Tomazela, D.M.; Shulman, N.; Chambers, M.; Finney, G.L.; Frewen, B.; MacCoss, M.J. Skyline: An. open source document editor for creating and analyzing targeted proteomics experiments. *Bioinformatics* **2010**, *26*, 966–968. [CrossRef]

10. Mohammed, Y.; Domański, D.; Jackson, A.M.; Smith, D.S.; Deelder, A.M.; Palmblad, M.; Borchers, C.H. PeptidePicker: A scientific workflow with web interface for selecting appropriate peptides for targeted proteomics experiments. *J. Proteom.* **2014**, *106*, 151–161. [CrossRef]

11. Demeure, K.; Duriez, E.; Domon, B.; Niclou, S.P. PeptideManager: A peptide selection tool for targeted proteomic studies involving mixed samples from different species. *Front. Genet.* **2014**, *5*, 305. [CrossRef] [PubMed]

12. Zauber, H.; Kirchner, M.; Selbach, M. Picky: A simple online PRM and SRM method designer for targeted proteomics. *Nat. Methods* **2018**, *15*, 156–157. [CrossRef] [PubMed]

13. Zolg, D.P.; Wilhelm, M.; Schnatbaum, K.; Zerweck, J.; Knaute, T.; Delanghe, B.; Yu, P. Building ProteomeTools based on a complete synthetic human proteome. *Nat. Methods* **2017**, *14*, 259–262. [CrossRef] [PubMed]

14. Zolg, D.P.; Wilhelm, M.; Schmidt, T.; Medard, G.; Zerweck, J.; Knaute, T.; Kuster, B. ProteomeTools: Systematic characterization of 21 post-translational protein modifications by liquid chromatography tandem mass spectrometry (LC-MS/MS) using synthetic peptides. *Mol. Cell. Proteom.* **2018**, *17*, 1850–1863. [CrossRef] [PubMed]

15. Mesuere, B.; Van der Jeugt, F.; Devreese, B.; Vandamme, P.; Dawyndt, P. The unique peptidome: Taxon-specific tryptic peptides as biomarkers for targeted metaproteomics. *Proteomics* **2016**, *16*, 2313–2318. [CrossRef] [PubMed]

16. Mallick, P.; Schirle, M.; Chen, S.S.; Flory, M.R.; Lee, H.; Martin, D.; Kuster, B. Computational prediction of proteotypic peptides for quantitative proteomics. *Nat. Biotechnol.* **2007**, *25*, 125–131. [CrossRef] [PubMed]

17. Eyers, C.E.; Lawless, C.; Wedge, D.C.; Lau, K.W.; Gaskell, S.J.; Hubbard, S.J. CONSeQuence: Prediction of reference peptides for absolute quantitative proteomics using consensus machine learning approaches. *Mol. Cell. Proteom.* **2011**, *10*, M110-003384. [CrossRef]

18. Qeli, E.; Omasits, U.; Goetze, S.; Stekhoven, D.J.; Frey, J.E.; Basler, K.; Ahrens, C.H. Improved prediction of peptide detectability for targeted proteomics using a rank-based algorithm and organism-specific data. *J. Proteom.* **2014**, *108*, 269–283. [CrossRef]

19. Tang, H.; Arnold, R.J.; Alves, P.; Xun, Z.; Clemmer, D.E.; Novotny, M.V.; Radivojac, P. A computational approach toward label-free protein quantification using predicted peptide detectability. *Bioinformatics* **2006**, *22*, e481–e488. [CrossRef]

20. Sanders, W.S.; Bridges, S.M.; McCarthy, F.M.; Nanduri, B.; Burgess, S.C. Prediction of peptides observable by mass spectrometry applied at the experimental set level. *BMC Bioinform.* **2007**, *8*, S23. [CrossRef]

21. Fusaro, V.A.; Mani, D.R.; Mesirov, J.P.; Carr, S.A. Prediction of high-responding peptides for targeted protein assays by mass spectrometry. *Nat. Biotechnol.* **2009**, *27*, 190–198. [CrossRef] [PubMed]

22. Cock, P.J.; Antao, T.; Chang, J.T.; Chapman, B.A.; Cox, C.J.; Dalke, A.; De Hoon, M.J. Biopython: Freely available python tools for computational molecular biology and bioinformatics. *Bioinformatics* **2009**, *25*, 1422–1423. [CrossRef] [PubMed]

23. Olayiwola, J.O.; Bakarey, A.S. Epidemiological trends of Lassa fever outbreaks and insights for future control in Nigeria. *Int. J. Trop. Dis. Heal.* **2017**, *24*, 1–14. [CrossRef]

24. Doellinger, J.; Schaade, L.; Nitsche, A. Comparison of the cowpox virus and Vaccinia virus mature virion proteome: Analysis of the Species- and strain-specific proteome. *PLoS ONE* **2015**, *10*, e0141527. [CrossRef] [PubMed]

25. Schrick, L.; Tausch, S.H.; Dabrowski, P.W.; Damaso, C.R.; Esparza, J.; Nitsche, A. An early American smallpox vaccine based on horsepox. *N. Engl. J. Med.* **2017**, *377*, 1491–1492. [CrossRef] [PubMed]

26. Esparza, J.; Schrick, L.; Damaso, C.R.; Nitsche, A. Equination (inoculation of horsepox): An. early alternative to vaccination (inoculation of cowpox) and the potential role of horsepox virus in the origin of the smallpox vaccine. *Vaccine* **2017**, *35*, 7222–7230. [CrossRef] [PubMed]

27. Altschul, S.F.; Madden, T.L.; Schäffer, A.A.; Zhang, J.; Zhang, Z.; Miller, W.; Lipman, D.J. Gapped BLAST and PSI-BLAST: A new generation of protein database search programs. *Nucleic Acids Res.* **1997**, *25*, 3389–3402. [CrossRef]

28. Jacobs, B.L.; Langland, J.O.; Kibler, K.V.; Denzler, K.L.; White, S.D.; Holechek, S.A.; Baskin, C.R. Vaccinia virus vaccines: Past, present and future. *Antivir. Res.* **2009**, *84*, 1–13. [CrossRef]

29. Apweiler, R.; Bairoch, A.; Wu, C.H.; Barker, W.C.; Boeckmann, B.; Ferro, S.; Martin, M.J. UniProt: The Universal protein knowledgebase. *Nucleic Acids Res.* **2004**, *32*, D115–D119. [CrossRef]

30. Kessner, D.; Chambers, M.; Burke, R.; Agus, D.; Mallick, P. ProteoWizard: Open source software for rapid proteomics tools development. *Bioinformatics* **2008**, *24*, 2534–2536. [CrossRef]

31. Kim, S.; Pevzner, P.A. MS-GF+ makes progress towards a universal database search tool for proteomics. *Nat. Commun.* **2014**, *5*, 5277. [CrossRef] [PubMed]

32. Fehling, S.; Lennartz, F.; Strecker, T. Multifunctional nature of the arenavirus RING finger protein Z. *Viruses* **2012**, *4*, 2973–3011. [CrossRef] [PubMed]

33. Briese, T.; Paweska, J.T.; McMullan, L.K.; Hutchison, S.K.; Street, C.; Palacios, G.; Nichol, S.T. Genetic detection and characterization of Lujo virus, a new hemorrhagic fever-associated arenavirus from southern Africa. *PLoS Pathog.* **2009**, *5*, e1000455. [CrossRef] [PubMed]

34. Chojnacki, S.; Cowley, A.; Lee, J.; Foix, A.; Lopez, R. Programmatic access to bioinformatics tools from EMBL-EBI update: 2017. *Nucleic Acids Res.* **2017**, *45*, W550–W553. [CrossRef] [PubMed]

35. Wee, S.; Alli-Shaik, A.; Kek, R.; Swa, H.L.; Tien, W.P.; Lim, V.W.; Gunaratne, J. Multiplex targeted mass spectrometry assay for one-shot flavivirus diagnosis. *Proc. Natl. Acad. Sci. USA* **2019**, *116*, 6754–6759. [CrossRef]

36. Mesuere, B.; Devreese, B.; Debyser, G.; Aerts, M.; Vandamme, P.; Dawyndt, P. Unipept: Tryptic peptide-based biodiversity analysis of metaproteome samples. *J. Proteome Res.* **2012**, *11*, 5773–5780. [CrossRef]

37. Henikoff, S.; Henikoff, J.G. Amino acid substitution matrices from protein blocks. *Proc. Natl. Acad. Sci. USA* **1992**, *89*, 10915–10919. [CrossRef] [PubMed]

38. Ogata, K.; Ohya, M.; Umeyama, H. Amino acid similarity matrix for homology modeling derived from structural alignment and optimized by the Monte Carlo method. *J. Mol. Gr. Model.* **1998**, *16*, 178–189. [CrossRef]

39. Saito, M.A.; Dorsk, A.; Post, A.F.; McIlvin, M.R.; Rappé, M.S.; DiTullio, G.R.; Moran, D.M. Needles in the blue sea: Sub-species specificity in targeted protein biomarker analyses within the vast oceanic microbial metaproteome. *Proteomics* **2015**, *15*, 3521–3531. [CrossRef]

40. Dale, R.; Grüning, B.; Sjödin, A.; Rowe, J.; Chapman, B.A.; Tomkins-Tinch, C.H.; Köster, J. Bioconda: A sustainable and comprehensive software distribution for the life sciences. bioRxiv 2017, bioRxiv:207092. *bioRxiv* **2017**. bioRxiv:207092.

MDPI
St. Alban-Anlage 66
4052 Basel
Switzerland
Tel. +41 61 683 77 34
Fax +41 61 302 89 18
www.mdpi.com

Viruses Editorial Office
E-mail: viruses@mdpi.com
www.mdpi.com/journal/viruses